Shark Biology and Conservation

Shark
Biology and Conservation

Essentials for Educators, Students, and Enthusiasts

SECOND EDITION

Daniel C. Abel, R. Dean Grubbs, Tristan Guttridge, *and* **Joshua K. Moyer**

Illustrated by

Elise Pullen *and* **Marc Dando**

With contributions from

Callie Crawford, Richard Lund, Eileen D. Grogan, Kara E. Yopak, Igbal Elhassan, Vera Schluessel, Alifa Bintha Haque, James Gelsleichter, Philip Matich, and Matthew Smukall

Johns Hopkins University Press
Baltimore

First printed in the United States of America on acid-free paper
9 8 7 6 5 4 3 2 1

Johns Hopkins University Press
2715 North Charles Street
Baltimore, Maryland 21218
www.press.jhu.edu

Library of Congress Cataloging-in-Publication Data

Names: Abel, Daniel C. author | Grubbs, R. Dean author | Guttridge, Tristan L. author | Moyer, Joshua K., 1987- author | Pullen, Elise illustrator | Dando, Marc
Title: Shark biology and conservation : essentials for educators, students, and enthusiasts / Daniel C. Abel, PhD, R. Dean Grubbs, PhD, Tristan Guttridge, PhD, Joshua K. Moyer, PhD ; illustrated by Elise Pullen and Marc Dando ; with contributions from Callie Crawford, Igbal Elhassan, James Gelsleichter, Eileen Grogan, Alifa Bintha Haque, Dick Lund, Philip Matich, Vera Schlüssel, Matthew Smukall, and Kara Yopak.
Description: [Second edition]. | Baltimore, Maryland : Johns Hopkins University Press, [2026] | Previous edition: 2020. | Includes bibliographical references and index. | Summary: "Shark Biology and Conservation is a comprehensive and accessible guide to the diversity, evolution, ecology, behavior, physiology, anatomy, and conservation of sharks, combines scientific depth with clear explanations, making it valuable to both specialists and general readers. Drawing on their extensive field experience, the authors share engaging stories while presenting traditional and modern research techniques, from molecular methods to advanced tracking technologies. They explore often-overlooked aspects of shark physiology, including heart function, sensory biology, digestion, metabolism, and reproduction, as well as big-picture ecological topics like habitat preferences and migration patterns. The book also examines sharks' extraordinary adaptations, evaluates conservation strategies, and warns against the misuse of science. Illustrated with hundreds of original photographs and detailed drawings, this work offers a rich and authoritative resource for marine biologists, educators, students, and anyone captivated by these remarkable and threatened animals"—Provided by publisher.
Identifiers: LCCN 2025038459 | ISBN 9781421453880 hardcover | ISBN 9781421453897 ebook
Subjects: LCSH: Sharks | Sharks—Conservation
Classification: LCC QL638.9 .A245 2026
LC record available at https://lccn.loc.gov/2025038459

A catalog record for this book is available from the British Library.

Special discounts are available for bulk purchases of this book. For more information, please contact Special Sales at specialsales@jh.edu.

EU GPSR Authorized Representative
LOGOS EUROPE, 9 rue Nicolas Poussin,
17000, La Rochelle, France
E-mail: Contact@logoseurope.eu

CONTENTS

PREFACE

This is not a coffee table book, but if you leave it in plain view, we suspect people will thumb through it to take a closer look at our exceptional photographs and artwork. Nor is it a field guide, but in a pinch it can help you identify more common species or narrow your choices. And this book is not a scholarly volume, though you would be excused for concluding the opposite due to the breadth of coverage, scientific citations, and terminology not typically found in coffee table books and field guides.

So exactly *what kind of book is this?* As the title denotes, we wrote *Shark Biology and Conservation: Essentials for Educators, Students, and Enthusiasts* for those in need of or curious about accurate information on sharks presented in what we hope is an accessible but not watered-down format. In doing so, we do not disparage coffee table books, field guides, or scholarly tomes—a glimpse of our bookcases would reveal shelves of these. But no books in these categories have it both ways; that is, be popular and scholarly at the same time. So, that is in part our goal and challenge.

One of the reasons we wrote this book is an attempt to supplant fear of sharks (known as *galeophobia* and *selachophobia*) with respect, and myth with knowledge. We hope to do so by providing a comprehensive overview of sharks, including recent scientific advances that will feed your fascination, calm any fears you might harbor, and explain why sharks are critical to ocean health. We will also broaden the discussion to include the rays, or *batoids*, an overlooked group closely related to sharks, which are equally fascinating and, in a few cases, are far more endangered than the vast majority of sharks.

The idea for this book arose during Coastal Carolina University's *Biology of Sharks* course, a three-week course held in part on wondrous Winyah Bay and at the extraordinary Bimini Biological Field Station (BBFS). The class was offered annually from 1996 to 2025, making it perhaps the longest running shark biology course globally.

The main drawback to the course was the lack of a suitable textbook. To be sure, there are several exceptional books on sharks and their relatives. However, most shark-related nonfiction books for nonspecialists are field guides, coffee table books, personal narratives, natural histories, stories of shark attacks, and so on, written mainly for a general audience.

Highly specialized books on sharks are more technical and their content and writing are more accessible to graduate students and specialists than to others. These provide an exhaustive survey and synthesis of facts and concepts and contain complex graphs and diagrams. Both of us, along with others in the shark research community, use these texts regularly, but generally not as textbooks for our undergraduate courses. Indeed, without this assemblage of more technical books, this book would not have been possible.

The books closest to meeting our needs were *Sharks, Skates, and Rays*[1] by William Hamlett; *Biology of Sharks and Their Relatives*[2], edited by Jeffrey Carrier et al.; and Peter Klimley's *The Biology of Sharks and Rays.*[3] We highly recommend these outstanding books but, as informative and thorough as they are, they did not hit the sweet spot we were seeking.

Moreover, as we contemplated writing the book for students in our shark biology course, we realized that the shark booklist had largely overlooked a surprisingly expansive market: educators, students, advanced enthusiasts, field biologists, naturalists, and marine biologists who might not have a background in fish biology or sharks. What the market is missing for this group is a comprehensive, systematic overview of the diversity, evolution, ecology, behavior, physiology, anatomy, and conservation of sharks and their relatives written in a style that is sufficiently detailed but not too technical or intimidating.

The field of shark biology has blossomed in the last several decades. The American Elasmobranch Society, the professional organization for biologists and fishery scientists specializing in sharks, skates, and rays, has an impressive membership and holds well-attended annual national meetings. Similar societies, for example, the European Elasmobranch Association, exist both regionally and globally.

Courses in shark biology (we count at least 20) have appeared on numerous college campuses. There is even a shark MOOC (massive open online course) called *Sharks! Global Biodiversity, Biology, and Conservation*, which has enrolled over 30,000 students over the past six years! These courses all have coincided with renewed public interest in this group of organisms, whose populations in some cases are threatened by a number of human impacts and with whom there is an enormous depth of fascination.

This book draws on our vast combined experience as shark biologists and educators, as well as our extensive connections to the close-knit shark biology community (and the generous and enthusiastic contributions from many of them). We hope you have fun reading this book. Even more, we will deem our time writing it worthwhile if you use this book to expand your knowledge of these wondrous beasts and make it a springboard to educate others, stoke your enthusiasm and passion, and work so that there is always a place for sharks, and all of their relatives (including humans) to safely swim.

This second edition of *Shark Biology and Conservation* is a revised and expanded effort, with corrections of embarrassing mistakes (did we really write that?), updates throughout, new content, and an infusion of fresh blood from new co-authors and contributors. Noteworthy revisions include our treatment of evolution, ecology, and functional anatomy, and the updated art program. As in the first edition, to keep the book affordable (have you seen the prices of similar books?), we made tough choices about which topics to include and how detailed to make them. That said, with some of the extra space the publisher allowed for the second edition, we added new boxes that we call *Research, Conservation, and*

Education Focuses in several chapters, in which voices that expand the authors' perspectives discuss aspects of their work. We are particularly excited about these fresh insights into shark biology.

NOTES

1. Hamlett, W.C. (ed.). 1999. *Sharks, skates, and rays: the biology of elasmobranch fishes.* Johns Hopkins U. Press.
2. Carrier, J.C. et al. (eds.). 2022. *Biology of sharks and their relatives.* CRC Press.
3. Klimley, A.P. 2013. *The biology of sharks and rays.* U. of Chicago Press.

Dedication

In April 2019, our friend and mentor Samuel "Doc" Gruber died. No topic in this book was outside of his expertise, and shark science would not be as advanced as it is if there had not been a Doc. While our sadness at his loss is profound, we hope that dedicating this book to Doc honors his outsized legacy.

ACKNOWLEDGMENTS

The acknowledgment section is the last part of this book that we wrote. After expending every remaining ort of creativity, knowledge, energy, humor, and so on, in our arsenal writing *Shark Biology and Conservation*, we find ourselves struggling to recognize and offer our gratitude to our family, friends, colleagues, students, editors, and staff at Johns Hopkins University Press, and numerous others, in proportion to their exponential contributions to this book. Some clichés exist because they are truths, and this one is not *fake news*: we could not have written a grocery list, much less a book, without their presence in our lives and their support.

Let us begin by thanking those colleagues and friends who stepped up, sometimes with unexpected zeal for the project, when we asked, or even volunteered before we could hit them up for favors. All these talented professional and amateur photographers willingly offered their works; we thank, in no particular order: Katie Grudecki, Grant Johnson, Jeff Carrier, Robert Johnson, George Boneillo, Emily Marcus, Marcus Drymon, Lesley Rochat, Annie Guttridge, Chelle Blais, Matt Smukall, Steven Kessel, Craig O'Connell, Nick Wegner, Lai Chin, Andrew Raak, Bryan Keller, D. Ross Robertson, Caroline Collatos, Matt Larsen, James St. John, Gavin Naylor, Charles Cotton, Steven Kajiura, Dave Itano, Jason Romine, Joshua Bruni, Ken Goldman, Shmulik Blum, Matt Potenski, Erin Burge, Bob Crimian, Eugene Kitsios, David Gandy, Callie Crawford, Kathyrn Dickson, Sandra Brooke, Chugey Sepulveda, Lance Jordan, Brendan Talwar, Charles Messing, Jim Luken, Mackellar Violich, Joel Blessing, Nick Whitney, Amanda Brown, Michael Scholl, Laura Stone, Tanya Houppermans, and Trey Spearman. In some cases, we were saved by Creative Commons photos, and the photographers are acknowledged in figures in which these have been used. We thank Greenpeace and the Florida Fish and Wildlife Conservation Commission for photographs, as well as Shutterstock, Kelvin Aitken of Marine Themes, and Barbel Knieper of Biosphoto.

For permission to use graphics and photographs, we also thank Oxford University Press, NOAA and NOAA Fisheries, and the Bimini Biological Field Station.

Skillful and talented scientific illustrators are rare and expensive to engage. We were very fortunate to have not one, but two principal artists for this book, neither of whom will earn near what they are worth for their contributions to *Shark Biology and Conservation*. Almost all the illustrations except those in chapter 3 were the result of extremely fortuitous serendipity. In the fall of 2017, Elise Pullen became one of Dan's graduate students. To his surprise, she was already a professional artist. Her work is so detailed and accurate that it may surprise readers that her inaugural entry into scientific illustration was in the book's first edition, but we think you will agree that it will not be her last! And her professionalism, enthusiasm, flexibility, and attention to detail ensure a great career for her.

All the magnificent shark watercolors for chapter 3 were done by artist Marc Dando, whose reputation as an exceptional wildlife illustrator is long-established. We think you will agree that *exceptional* is an understatement in Marc's case. Marc gleefully met our demanding schedule and was among the easiest people to work with, in spite of his numerous other impending deadlines.

We also thank other illustrators and colleagues who provided illustrations. These include Mark Grace, Matthew Kolmann, the late Kurt Smolen, James Gelsleichter, Roi Gurka, and Read Frost.

Even after friends and colleagues provided photos and line art for the book, we had numerous gaps to fill. We discovered that we were naïve to think that we could write a book with all of the world-class photographs and artwork we would need without a budget. Thanks to the previously acknowledged friends and colleagues listed above, we *almost* did. A generous grant from the Save Our Seas Foundation, and the active involvement and enthusiasm of then-CEO Michael Scholl, allowed us to engage our two main illustrators, plus obtain some additional photographs. *Shark Biology and Conservation* would likely be destined to be a bargain bin book, wasting the gorgeous photos and illustrations from friends and colleagues, without Michael and the Save Our Seas Foundation's integral role.

We are also grateful to the anonymous reviewers of both the prospectus for this book and the submission manuscript. We hope that the finished book makes them proud.

This book would never have materialized were it not for the opportunities provided by the Bimini Biological Field Station. We thank the staff, past and present, for accommodating our courses and providing access to the wondrous natural world surrounding Bimini to us and more than a thousand of our students.

And a whale-shark-sized thanks and endless espressos on demand to our editor for the first edition, the unflappable, understated, erudite, and witty Tiffany Gasbarrini. When we submitted a preliminary manuscript for the first edition with nearly 500 illustrations, 150 more than our contractual agreement, instead of the venomous response we deserved, she wrote that she was *somewhat discomfited* with the overage. She gave us mostly free rein throughout the process to write the book we wanted, and even expanded our word and illustration quotas in this second edition. A true editorial hall-of-famer!

Except for Tiffany Gasbarrini, we have crossed paths with almost all of the Johns Hopkins University Press production and marketing team only by email, and the only reason we know that they are not Artificial Intelligence is because of how exceptional a book they produced, one that AI will never be able to match. We owe this team our most profound gratitude for the pride they have obviously taken in publishing our book. When cloning of people becomes a thing, we would urge our publisher to replicate Assistant Acquisitions Editor Ezra Rodriguez first, whose general excellence, professionalism, and responsiveness throughout allowed us to focus on our principal job of writing. Some of the Johns Hopkins University Press team that worked on this book have remained anonymous to us. We thank you, as well as Diem Do, Anne-Marie Bono, Adeline Galich, Jennifer D'Urso, Teena Lawrence, Nishtha Singhal, Molly Seamans, Katie Feild, Jenn Paulson, Kait Howard, and Kris Lykke.

Daniel C. Abel

I thank Coastal Carolina University (CCU), especially my College of Science colleagues. I am also very appreciative of boat captains Sam Gary, Edwin Jayroe, Jamie Phillips, and Kris Dobinson for their expertise, sound advice, good humor, fervor about safety and, of disproportionate worth, for always ensuring the presence of cupholders for my oat milk lattes wherever I found myself on our boats. My graduate students and the legions of undergraduate volunteers willing to awake at ungodly hours just to be a part of the CCU Shark Project, validated my decision to plant my roots at CCU. Their enthusiasm for sharks as well as for my writing fueled this effort. Bryan Keller reviewed the manuscript for the first edition and Vivian Turner reviewed the section on shark locomotion. Numerous friends through their interest and words of encouragement, provided inspiration beyond which they could know. I will miss the wisdom and gusto of two friends no longer with us, Terry Munson and William Holliday. Other friends won't know the value of their lending a word of support, but I do. So, I add, with gratitude, Hartmut and MaryJo Fege, Robert and Susan Sturgis, James Borton and Nena Rice, Pat Munson, Al Stephenson, and newest friends Ron and Signe Cann.

I also thank my co-author Dean Grubbs. Dean and I have been friends and teaching colleagues for about 30 years. Dean is many things: consummate educator, passionate but scientifically informed conservationist, farm animal lover, and perhaps the most knowledgeable and insightful shark biologist on the planet. I would not have ever envisioned writing a book on sharks if Dean had not been a part of it.

If this book is better than the first edition, it is primarily because of the expanded roles of co-author Tristan Guttridge and new co-author Josh Moyer, and the updated illustrations of Elise Pullen. I am pleased beyond limit that they lent their expertise to this book!

More than gratitude is also owed to the professorial giants who modeled for me what a teacher-scholar-citizen should be. There are too many to name; I frequently wonder where I would be without the following educators: Reid Wiseman, Norman Chamberlain, Bill Anderson, Jim Smiley, Marion Doig, Chip Biernbaum, Felicia Coleman, and Chris Koenig of the College of Charleston; and Jeff Graham and Ralph Shabetai, my PhD co-advisors from Scripps Institute of Oceanography and the School of Medicine at the University of California at San Diego; respectively. They taught me how to think and comport myself like a scientist, and how to write effectively. Chris, who died in December of 2024, continues to inspire me, both from the seeds of wisdom he planted nearly four decades ago about science and life and, more recently, for recommending that I use honey in my caipirinhas. If my undergraduate and graduate students profited in meaningful ways from their association with me, they should thank all of the above, but none more so than Chris, to whom I am dedicating my contributions to this book.

Then there are my late parents, Harris and Ruth; my brothers, Billy and Alan; and my sister, Sara, all of whom made sacrifices and contributions that nurtured my love of knowledge and marine life in my childhood and beyond. And finally, my wife Mary, daughter Juliana and her husband Tim, and son Louis. For them, words are inadequate, but I am confident they know how they have sustained me to work tirelessly toward the goal of a peaceful, sustainable planet.

R. Dean Grubbs

I am eternally grateful to my academic mentors that encouraged but also challenged me throughout my career. This includes my undergraduate advisor Dan Diresta, who encouraged me to lead a scientific reading group on sharks. I thank my friend and confidant, John Morrissey, a masterful lecturer who strongly encouraged me to pursue graduate school, gave me my first opportunities to teach, and helped me push forward when my confidence in this career path wavered. None of my career would have been possible without the opportunities for research and education provided by Sonny Gruber, with whom I conducted my first shark research and developed my first shark biology courses. Dr. Gruber was my academic father, and I was privileged to maintain a relationship with him for 30 years until his passing in 2019. I am also indebted to Arthur Myrberg, who, in addition to being Sonny Gruber's doctoral advisor, was my favorite undergraduate professor. Whatever skill I have as an educator comes from being a very poor mimic of the two best I have known, Art Myrberg and John Morrissey. I am also incredibly grateful to my doctoral advisor, Jack Musick, a renowned scientist and naturalist cut in the mold of David Starr Jordan. In a time when naturalists have become scarce in academia, Jack encouraged those tendencies as well as my propensity to engage in disparate lines of research. He showed me that there are still places for our ilk in academic careers. I thank my post-doctoral advisor Kim Holland, who, among many things, taught me the value of balancing one's professional and personal life, a skill I continue to struggle to hone. Finally, I also thank Captains Tony Pinello and Durand Ward, friends and confidants that taught me much about life and the sea.

I thank the Florida State University Coastal and Marine Lab (FSUCML) administration that has given me the freedom to pursue my research and educational interests unfettered, and the FSUCML staff, without whom none of my work would be possible. I am also thankful to the many colleagues with whom I have collaborated on many projects, some that are highlighted in this book. This list is too long to include as I have been extremely fortunate to have a large number of incredible collaborators. As the academic generations progress, I am also extremely grateful to the many undergraduate and graduate students that have chosen to conduct research in my lab. They teach me as much as I them and continually make me proud and make me look better than I am.

I also thank my co-author, collaborator, co-instructor, and friend Dan Abel. Dan is an incredibly talented and humorous writer and educator, and over the many years that he and I taught our shark courses, we mused over the possibility of writing this book. Dan (and I) knew it would only happen if he took the initiative to push it forward and I am grateful to him for all of the hard work he put in to seeing this project through and to produce a book of which we are very proud.

Finally, I am especially grateful to my parents Audrey and Ralph Grubbs, who exposed me to marine life through fishing and snorkeling on nearly every family vacation, encouraged my crazy childhood notion to become a marine biologist who studies sharks, and sacrificed to help make it possible. I would not be where I am if not for them.

Tristan Guttridge

I would like to start by thanking Dan and Dean for inviting me to contribute to the first edition of this book and for expanding my role alongside Josh in the second version. It seems only yesterday that I met you both in Bimini during a Coastal Carolina University shark biology course, but incredibly it has been 20 years! I am truly honored to be able to share authorship on this special book with you both, especially since it was inspired by my PhD mentor "the legendary Doc Gruber" and the many shark biology courses that you led in Bimini. I was thrilled to be part of those at every career stage and will forever remember the huge deepwater Tiger Sharks, baby Lemon Sharks in Aya's (my almost getting Doc bit as I dropped the dipnet, *eeeek*), the wall of students at the reef shark dive, and the many fascinating lectures and dissections! Thank you for believing in me across those 20 years and for trusting me to contribute to this *Shark Bible*.

My career with sharks started in Bimini as a student under Doc Gruber. He remains an inspiration to me, and I will cherish our many discussions/debates on shark behavior and working alongside him as Director at BBFS from 2012 to 2018.

To my parents Peter and Yvonne, sisters Leonie and Justine, you always believed in me and motivated me to follow my dreams. I am forever indebted to you all for your guidance, love and support. And finally, to my wife Annie, children Isla and Leonard, our journey is just beginning, but what an adventure it has been so far. Annie, your energy and passion for life and sharks are infectious and luckily for me you are gifted behind a camera too! Thank you for contributing many amazing photos that are displayed throughout this book, and for instilling in me that *you can always find time* and to believe in myself.

Joshua K. Moyer

First and foremost, I thank my co-authors whose first edition of this book was a boon for students and educators alike. Knowing the utility and quality of the first edition enhanced the delight and honor that it was to work with Dan, Dean, Tristan, and our illustrator extraordinaire Elise on the second edition.

Many individuals influenced my contributions to this volume, but three merit special mention for their effects on my growth as a scientist and educator. The first is Queen of the Ratfish, Dominique Didier. Dominique showed me the rejuvenating powers of self-directed research when academics weighed heavily on me as an undergraduate. She also introduced me to the second individual I would like to recognize, Willy Bemis. Every study I've undertaken, manuscript I've written, and lecture I've delivered since getting to know Willy has been crafted with his scientific and educational style in mind. The third person who merits acknowledgment is Duncan Irschick. Duncan's practical approaches to teaching and research were invaluable reference points as I charted my own professional course.

I would also like to thank Willy Bemis, Al Richmond, and Glenn Northcutt for many of the volumes, especially rare and classic texts in vertebrate anatomy and ichthyology, that I referenced while crafting portions of this book. These were invaluable resources for which I am very grateful.

Greg Watkins-Colwell, Dan Brinkman, and Charmaine Wong each provided access to specimens or artifacts at Yale University's Peabody Museum. John Galbraith, Lisa Natanson, Michelle Passerotti, and Christian Jones all contributed specimens I have used in the development of lab and field courses. Many exercises from those courses formed the basis for content I contributed to this edition. I also thank the leadership and volunteers of the Atlantic Shark Institute who facilitated field work, thereby enabling me to capture photographs, make behavioral observations, and collect data, some of which appear in this book.

I am indebted to Emily Moyer, who is a reliable sounding board, ready to provide an honest answer when I read a freshly written block of text and ask, "Does that make sense?" She is also one of the best Manta Ray dive buddies I could hope for.

My parents, Sally and Brad Moyer, deserve acknowledgment for subsidizing my first series of shark lectures, which I began at my sixth birthday party. Although the content was largely derivative and my grasp of chondrichthyan systematics was tenuous at best, they allowed me to finish and then brought me cake. Several decades later, I no longer finish my lectures with cake, but I am still just as grateful for their support.

Finally, I would like to dedicate my part in this edition to Eugenie Clark. I wrote to Dr. Clark as a ten year old, asking, "What should I do to become a marine biologist?" She made the time to write back and outlined considerations, but she never suggested that my success hinged on anything more than interest and work ethic. If I followed through, there would be a place in the field of chondrichthyan biology waiting for me, just as there is a place today for anyone ready to learn, work, and sink their teeth into shark science!

Authors' Note

Readers of this book know exactly which fish we mean when we say *Great White Shark*. There is only one such species, and it is easily distinguished from even closely related species like the Shortfin Mako and Porbeagle. Scientifically, the Great White Shark, which more accurately is called the White Shark, is known as *Carcharodon carcharias* everywhere in the world, in accordance with the rules of scientific nomenclature that assign a unique two-part name to every species.

But do you instinctively picture one specific animal called a *Sand Shark*? In the Southeast United States, the name *Sand Shark* is used by local residents for numerous species that include Blacktip, Sandbar, Dusky, Silky, Atlantic Sharpnose, and Blacknose Sharks, and even guitarfish, which are not sharks but rather are rays, a closely related group. Or perhaps even what is pictured in the figure below. Basically, any shark that is not a hammerhead (and even a ray, the Atlantic Guitarfish) and is found nearshore in the Southeast United States, especially near sandy beaches, has been called a Sand Shark.

And that is more than merely a trivial problem, since assigning one name to numerous different species can easily lead to cases of mistaken identity with serious consequences. In chapter 11, for example, we discuss a case where grouping the Dusky Smoothhound and Spiny Dogfish both as *dogfish* in managing a fishery nearly led to potentially catastrophic consequences for the latter.

The problem is this: in science, a species is a precisely defined, unique entity, with its own distinct genetic makeup, distribution, life history, behavior, environmental requirements, and so on. Calling two different species by the same common name denies at least one of them its unique biological heritage.

A *species* is a real concept in nature, not a human construct. For sexually reproducing organisms, a species consists of individuals that will interbreed with each other, are reproductively isolated from (i.e., do not interbreed with) other species, and which will produce offspring that are viable (capable of surviving until adulthood) and fertile (capable of having offspring).

And every species is assigned its own unique binomial (two-part) name, by a rigorous process specified in the *International Code of Zoological Nomenclature* (ICZN). Each and every reader of this book, excluding AI (!), is a *Homo sapiens* and nothing else, no matter your race or nationality. *Homo sapiens* is the species name. *Homo* is our genus, and *sapiens* is our specific epithet.

Among sharks, we have studied *Carcharhinus plumbeus,* which is variously known as a Sandbar Shark, Brown Shark, and yes, Sand Shark. There are over 550 named species of sharks, and a total of over 1300 species of sharks and shark relatives—the skates, rays, and chimaeras (or ghost sharks).

The result of using a standard system of nomenclature is an organized classification that scientists and policymakers all over the world can reliably use when studying an organism, determining its habitat requirements, assessing the health of its populations, making policy decisions, and so on.

This takes us to our dilemma: whether to use common or scientific names, or both, in this book. Whether, on the one hand, to risk alienating readers happy with the use of a common name and perhaps intimidated or otherwise put-off by difficult-to-pronounce names derived from Latin or Greek, or on the other hand to betray our fealty to scientific accuracy.

We lied. It really is not a dilemma to us. We will use scientific names at least the first time a species is mentioned in every chapter (except chapter 1), but not in every instance. When we do use them, we will also include widely accepted common names.

Speaking of common names, the latest convention, in accordance with the American Fisheries Society's authoritative *Common and Scientific Names of Fishes from the United States, Canada, and Mexico,* is to capitalize the first letters of all common names. Although not all publications use this convention, we do in this book, except when we refer to a group of sharks in the same genus or family. For instance, it would be hammerheads or threshers, but Scalloped Hammerhead and Bigeye Thresher.

There is a second major issue as well that we need to address. This is a book about sharks, but much of what is true about sharks also applies to their close relatives, the skates and rays, with whom sharks are classified as *elasmobranchs.* Rays and skates are so closely related to sharks that the former are sometimes called *flat sharks.* Sharks, skates, and rays together are closely related to a more obscure group called *chimaeras* or *ghost sharks,* and collectively sharks, skates, rays, and chimaeras are known as *chondrichthyans.*

So, how to proceed? Chances are, you bought this book because you wanted to read about sharks (*Sharks,* after all, is in the title) so we will focus on sharks. Occasionally we will broaden the discussion by referring to elasmobranchs or sharks and their relatives. And there are times when we discuss only skates and rays, given their close relationship to sharks, and when we do this, we usually call them *rays* or *batoids.*

There, now that *that* is settled, let us focus on this magnificent assemblage of cartilaginous fishes.

PART I

Introduction, Evolution, Diversity, and Classification

1 / Introduction

CHAPTER OUTLINE

Introduction: What Is a Shark?

Most of the eight billion or so of us inhabiting Earth know a shark when we see one. Not so fast! Everyone knows the *iconic* species, e.g., White Sharks, Great Hammerheads, Bull Sharks, and Grey Reef Sharks. After all, these and a few other sharks are ubiquitous across the media landscape, so much so that they are nearly impossible to dislodge from our mind's eye, about as difficult as extricating ourselves from continuously replaying that earworm of a song *Baby Shark* (apologies if we just instigated the latter). But these iconic species no more represent sharks as a group than Chihuahuas and Great Danes represent dogdom, Model T Fords and stretch limos epitomize cars, or *Baby Shark* represents music. In fact, if you were creating a t-shirt with creatures truly emblematic of shark biodiversity, you'd abandon the Great White for a Hawaiian Shortspine Shortspur, the Great Hammerhead for a Gulper, a Bull Shark for the Ninja Lantern Shark, and the Grey Reef Shark for a Demon Catshark. Good luck searching online for such an article of clothing featuring our alternate list or, more relevantly, information on their biology and ecology. Nearly two-thirds of the more than 550 species of living sharks are no larger than 1 m (3.3 ft) and over half of the species live deeper than 200 m (660 ft). A typical shark is not large, gray, fast-swimming, and found along coastline, but rather is small, brown, more sluggish, and lives in the deep ocean. So much for stereotypes!

Not only do many of us have the wrong impression of shark biodiversity, but we also virtually completely ignore their closest relatives, the skates and rays, even though their biodiversity—over 650 species—exceeds that of sharks. Were it not for the sad, unfortunate, and unlikely accident in which a stingray barb killed wildlife media personality Steve Irwin, who was more commonly known as the Crocodile Hunter, most people would not give the rays a second thought. Batoids, the collective name for skates and rays, diverged from the shark evolutionary tree about 230 million years ago, but they exhibit so many of the characteristics that define sharks that shark biologists refer to them as pancake or flat sharks. (Some batoid enthusiasts who feel slighted by this reference call skates and rays "magnificent shark pancakes of wonder," and they lightheartedly disparage sharks as "sausage rays."). Together with sharks, batoids are classified as *elasmobranchs* (see box 1.1) and, like their shark relatives, about half of these species also live in the deep sea.

Whether Great White or Hawaiian Shortspine Shortspur, or shark or batoid, these animals share a suite of characteristics that define who they are and how they differ from other groups. We will start our journey into shark biology here.

External Anatomy and Distinguishing Features

We begin with naming of parts,[1] so examine Figure 1.1, the basic external anatomy of a shark. Many of these vary among species, thus keep these in mind as we discuss the evolution and diversity of the chondrichthyan fishes in subsequent chapters.

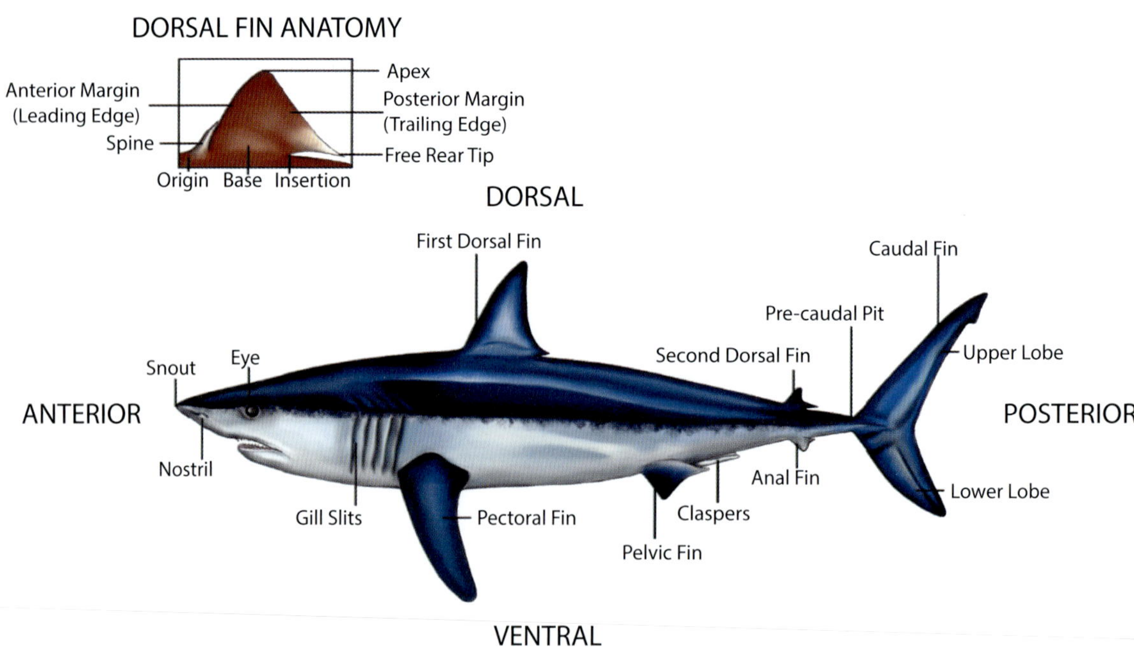

Figure 1.1. External anatomy of a shark (in this case a male Shortfin Mako), with a close-up of a dorsal fin (from a Gulper Shark) showing terminology useful in describing sharks. The Mako has no spiracles.

Sharks have both median fins (along the centerline) and bilateral fins (on both sides). Most sharks have three median fins: two dorsal fins and a single anal fin. Some groups have lost a dorsal fin through evolution and possess only one, for example the cow sharks, which include the sixgill and sevengill sharks, in the order Hexanchiformes.[2] The entire ~135 species of dogfish sharks, a diverse group in the order Squaliformes, have lost their anal fin. No batoids possess an anal fin, and many possess no defined dorsal fins.

There are two sets of paired bilateral fins: the more anterior pectorals and the posterior pelvics. These fins can also technically be called *limbs* (although no shark biologist would do so commonly), and all vertebrates have them except where they are lost or vestigial, as in snakes. In terrestrial vertebrates, we call the pectorals *arms* or *wings*, and the pelvics, *legs*.

Although sharks and rays are each other's closest relatives, there are anatomical differences, the most recognizable being the placement of the gill slits (five in most species), which are lateral in the sharks and ventral in the rays. Of the approximately 1200 different species of sharks and rays, only nine—eight sharks and a single ray—have more than five pairs of gill slits.

Many sharks have *spiracles*, paired, bilateral openings behind the eyes that connect the mouth to the water environment. In the requiem sharks (members of the family Carcharhinidae) like the Blue, Bull, and Lemon Sharks, the spiracles are vestigial, meaning that the structure is rudimentary and has lost its core function. Spiracles are prominent among dogfish and benthic sharks, as well as in most batoids.

In most rays, the primary source of incurrent water for the gills is through the spiracles (which makes sense, since inhaling using their mouths, which are close to the substrate in benthic species, would result in inspiring sediment[3]). Exceptions are pelagic batoids; that is, those that actively swim in open water. Mantas, for example, swim with their mouths open, as do most pelagic sharks, a phenomenon appropriately called *ram ventilation*, and in doing so irrigate their gills with oxygenated water. No need for spiracles if you are already swimming, and merely opening your mouth serves the same purpose, a sort of anatomical multitasking that eliminates the need for additional respiratory muscles, yet another example of evolution at its finest.

Other external features in some shark species include a precaudal pit (which allows increased tail flexibility), one or more ridges and keels (rigid lateral extensions) along the body, and spines on the leading edge of one or more of the dorsal fins. All of these have hydrodynamic functions, and the spines are also defensive.

Now that you know some of the terminology of the external features, here are the distinguishing features that together define a shark. Some of these we necessarily mentioned above, and some are not unique to sharks, but together these delineate sharks (and rays) and no other organisms.

First, a shark is a *fish*; that is, a vertebrate with fins for locomotion and gills for breathing. Biologists distinguish between two major groups of fishes, *bony* and *cartilaginous* (box 1.1). The former group, including tuna, mullet, carp, etc., numbers approximately 35,000 or more species and is the most *speciose* (species-rich) and

BOX 1.1

Who's Who in the World of Sharks

In the Linnaean system of nomenclature, every species is given a unique two-part name that is recognized globally and unambiguously as that, and only that, species; for example, *Carcharhinus plumbeus* is most commonly known as the Sandbar or Brown Shark.

Recall the classic taxonomic hierarchy of life: Domain—Kingdom—Phylum—Class—Order—Family—Genus—Species.* Thus *C. plumbeus*** is also, starting with the most inclusive category, a eukaryan, animal, chordate, vertebrate (or craniate), gnathostome, chondrichthyan, elasmobranch, galeomorph, carcharhiniform, and carcharhinid in the genus *Carcharhinus*.

Sandbar Shark on a longline near Kaneohe, Hawai'i.

Domain: **Eukarya**

Refers to organisms that are eukaryotic (i.e., possess membrane-bound organelles and a nucleus surrounded by a nuclear envelope with two membranes).

Kingdom: **Animalia**

Organisms that are multicellular, with mode of nutrition known as heterotrophic (i.e., not photosynthetic) by ingestion (as opposed to absorption, like Fungi).

Phylum: **Chordata**

Organisms with a notochord (skeletal supporting rod) and pharyngeal gill slits at some point during development, as well as a single, hollow, dorsal nerve chord.

Subphylum: **Vertebrata** (or Craniata)

The most prominent characteristic is a skull and vertebral column. Vertebrates also exhibit bilateral symmetry, a brain, and sensory organs.

Superclass: **Gnathostomata**

"Super" in this case is not a superlative. Gnathostomes are the jawed vertebrates, and include all vertebrate classes except the jawless fishes (hagfish and lampreys).

Class: **Chondrichthyes** (or Elasmobranchiomorpha)

Chondrichthyan fish (sharks, rays, and chimaeras) possess an endoskeleton made of cartilage, employ internal fertilization using claspers, lack a swim bladder, and have a heterocercal tail and ampullae of Lorenzini.

Subclass: **Elasmobranchii** (also called Euselachii)

Elasmobranchs (*ee-las'-mo-branks*), the sharks and rays, are the sister-group (i.e., the closest related group) to the subclass Holocephali, or the chimaeras, and together these groups form the class Chondrichthyes.

Superorder: **Galeomorphii**

Galeomorphii is one of three superorders and the sister-group to the other two, Squalomorphii and Batoidea. Diverse group of small, demersal (living on the bottom) and large, pelagic predators.

Order: **Carcharhiniformes**

The Carcharhiniformes (*kar-kar-eye'-ni-form-eez*) is the largest order of sharks, with about 296 species of sharks in nine families, commonly called the *ground sharks*.

Family: **Carcharhinidae**

The Carcharhinidae (*kar-kar-eye'-ni-dee*) is the second largest family of sharks, with about 60 species. The family is commonly known as the *requiem sharks*.

Genus: **Carcharhinus**

Carcharhinus (*kar-kar-eye'-nis*) is the genus of 36 species of carcharhinid sharks, including the Sandbar, Blue, Oceanic Whitetip Sharks, and so on.

Species: ***Carcharhinus plumbeus***

Carcharhinus plumbeus (kar-kar-ey'e-nis and *plum'-bee-us*), written in italics or underlined, and all in lower case except the first letter, is the unique name assigned to the shark we commonly refer to as the Sandbar Shark. This two-word appellation is the *species*. Technically, *Carcharhinus* is the genus, and *plumbeus* is the specific epithet; the term *species* is reserved for, and only for, both words together: *Carcharhinus plumbeus*. Note that the term genus-species, though commonly used, is wrong, and would be the meaningless *Carcharhinus plumbeus*.

Although there are exceptions, there are standardized endings for most of the above taxonomic categories. These are:

- -morph = superclass
- -morphii = superorder
- -iformes = order
- -oidei = suborder
- -idae = family
- -inae = subfamily
- -ini = tribe
- -i or -ii = patronym (named for a male)
- -ae = matronym (named for a female)
- -ensis = location (named for a location)

---

* If needed, or for entertainment, consult the Internet for mnemonic methods of recalling the hierarchy of major taxonomic categories. On mnemonic-device.com, we found the following: *Drunken Kangaroos Punch Children On Family Game Shows* and *Do Koalas Prefer Chocolate Or Fruit, Generally Speaking?* We selected these as examples so we can sell more books in Australia.

** Note that we abbreviated *Carcharhinus* here using only the letter C, followed by a period, which is permissible once you have typed out the entire scientific name previously.

diverse assemblage of all vertebrates, whereas there are only about 1300 different kinds of cartilaginous fish; that is, sharks, rays, plus a relatively obscure but ancient group known as *holocephalans, chimaeras,* or *ghost sharks.*

Thirteen hundred species of cartilaginous fishes may seem like a lot of variety until you realize that there are more birds (> 10,000 species), amphibians (> 7000), mammals (> 5000), and even catfish (3900 species), and minnows and carp (3000 species). We explain this disparity in chapter 3. Also, don't conclude that the comparatively low biodiversity of sharks and other chondrichthyans equates to a lack of ecological value, which we discuss in chapters 9 and 10.

The essence of the difference between bony and cartilaginous fishes is an anatomical feature you cannot discern externally: the material comprising the internal skeleton. Bony fishes, which are in the class Osteichthyes (*oste* = bone; *ichthyes* = fish), have a skeleton composed of dense, rigid, *ossified* tissue, that is, hardened by calcium and other mineral salts, and containing nerves and blood vessels.

Sharks, members of the class Chondrichthyes[4] (*chondr* = cartilage), possess a skeleton that is tough yet lighter and more flexible than bone, but which lacks nerves and blood vessels (fig. 1.2). Another way of depicting the skeleton, in this case a Lemon Shark, is through the use of the versatile imaging technique *Computed Tomography* (CT) (See *Research, Conservation, and Education Focus* Box 1.2). Chemically, cartilage is composed principally of protein and sugar molecules, specifically *collagen,* an abundant fibrous protein, and *proteoglycans,* protein-sugar combinations.

Mammals, including humans, also have cartilage, three different kinds in fact, in places like the nose, ears, bronchial tubes, trachea, ribs, ends of long bones, and the discs between vertebrae. Cartilage comprises about 0.6% of a typical mammal's body weight but from 6% to 8% in sharks.

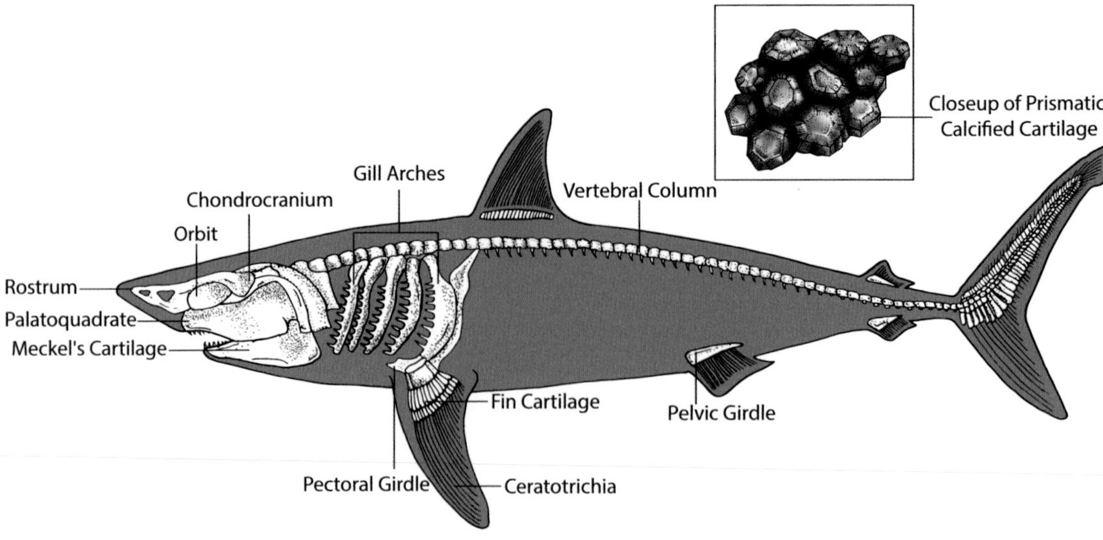

Figure 1.2. Cartilaginous skeleton of a shark (In part adapted from Whitely 2015. http://www.fossilsofnj.com/shark/cartilage.htm. Accessed 9/16/24).

BOX 1.2

Research, Conservation, and Education Focus

A Look Inside: Computed Tomography in Shark Science

Callie Crawford

Dept. of Biology
Coastal Carolina University
Conway, SC, US

Computed tomography (CT) is an imaging tool that uses x-rays to create 3D volumes of objects by computationally combining sequential 2D x-ray images to reconstruct the 3D image or model. Although most commonly connected with medical imaging, CT is also used across a variety of fields of study and industry, including zoology, engineering, paleontology, botany, and of course, as we are going to explore here, shark research. Using similar technology, sometimes even the same machines, as human medical scans or other instruments designed for research that can give researchers a closer look at much higher resolution, we can explore internal structures without the need, or at least prior to dissecting, breaking, or opening objects. CT scanning can be completed at low resolution such as that used within the medical fields, microscale for fine scale studies, and all the way to nanoscale to look at structures at the cellular level. Within shark research, CT scanning has been used across subfields including, but not limited to, exploring morphological diversity, soft tissue studies, paleontology, and in the development of new educational resources. Here we will discuss some examples of CT scanning in recent works.

Morphological Diversity

Recent large-scale projects that have used CT scanning technology to expand our understanding of vertebrate diversity, including shark and ray diversity, include the Chondrichthyan Tree of Life,[1] the Open Vertebrate (Overt) Project,[2] and the Scan All Fishes initiative.[3] Through some of these endeavors, whole scans have been made available to educators and researchers worldwide, opening access to the diversity of sharks and their relatives. The use of CT scanning at the whole animal scale allows researchers to non-destructively sample a wide variety of species to reveal internal anatomy, relatively quickly make counts of structures such as vertebrae and teeth (both functional and non-erupted), place landmarks for 3D morphometric studies, and even make digital models for printing. In some scans, researchers have come across prey items (see figure), parasites, and developing offspring which can be referred to as "natural history bycatch." This "natural history bycatch" could be used to understand more about prey items and reproductive strategies in less common or elusive species. With all the data available through these scanning projects, there are several questions ready to be explored.

Soft Tissue Studies

CT scanning is used not only to look at skeletal tissue (although those structures are generally easier to visualize), it can also be used to measure and model soft tissues. With the use of different staining methods, such as diceCT,[4] which increase the radiopacity (how opaque a material is to x-rays), of tissues, researchers are able to use CT scanning to study soft tissues including muscle, nervous tissue, and organs not easily visible without stains. Some recent works using staining methods to analyze soft tissues in sharks include sensory studies on the olfactory rosettes in sharks and chimaeras,[5] visualization optimization of the brain in sharks and bony fishes,[6] and cranial anatomy including skeletal and soft tissues such as musculature and sensory tissues.[7]

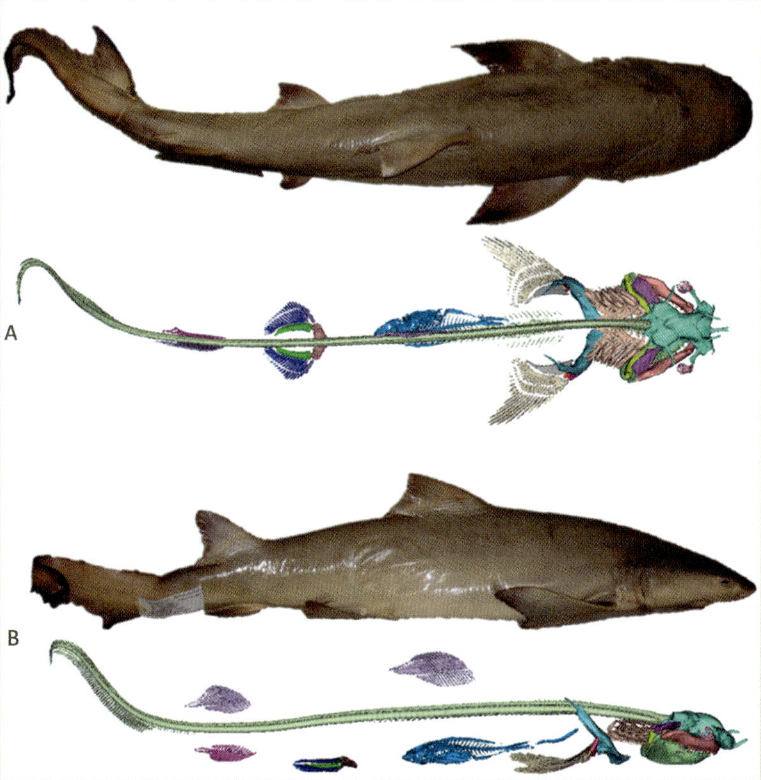

Dorsal (A) and lateral (B) views of a Lemon Shark, *Negaprion brevirsotris*, with completed segmentation and specimen photo in the same views. This Lemon Shark consumed a small fish (*blue*) prior to being collected, illustrating "natural history bycatch" which is not discernible from the specimen photographs. This specimen is from the Grice Marine Lab ichthyology collection at the College of Charleston. (Photos and scan courtesy of Callie Crawford, Gavin Naylor, and John Maisey as part of the CToL project [NSF DEB Award #1132229]).

Paleontology

Chapter 2 informs us that the fossil record for cartilaginous fishes is sparser than that of bony fishes. Even with this relative paucity of fossils, paleontologists have uncovered a great deal about the diversity of extinct sharks, their relatives, and how the chondrichthyan lineage evolved. In the previous sections we talked about how CT scanning can be used to study extant sharks; this technology has also been used by paleontologists to further study fossil sharks and their relatives. Within paleontology, CT scanning has been used in conjunction with more traditional methods for studying fossils to increase evidence for distinguishing separate species and tracing evolutionary changes. Adding CT to the myriad of tools and instruments used in paleontology has made it possible to more readily explore the internal structures, such as endocasts and ears of early sharks,[8] and create 3D reconstructions[9,10,11] of specimens without having to completely remove the fossil from the matrix surrounding it. These 3D reconstructions can increase the characters visible to researchers, allowing for more detailed analyses of shared and disparate traits which together can help solidify deep relationships that are sometimes uncertain, or even debated amongst researchers. By expanding the information paleontologists can get from fossils, CT scanning is opening more opportunities to study evolution through comparison, both among fossils and modern sharks.

Educational Resources

So far we have focused on how CT scanning is supporting advances in research, but these tools are also expanding how we teach the diversity, evolution, and anatomy of sharks. Open access tools such as Morphosource[12] and Sketchfab[13] have been used in lesson plans for K–12 courses, using these web-based data repositories and by downloading accessible models for 3D printing.[14,15] At the college level, CT scanning has been used in the development of active learning kits designed to be hands-on opportunities for students learning shark cranial morphology, neural anatomy, and biomechanics.[16] As technology access in the classroom continues to increase, more and more classrooms are likely to have opportunities to use CT scans and 3D printed models in teaching students of all ages about the wonderful diversity of sharks and their relatives.

---

1. Chondrichthyan Tree of Life. https://sharksrays.org (Accessed 9/17/25).

2. Blackburn, D.C. *et al.* 2024. BioSci. 74(3): 169–186.

3. #ScanAllFish. https://www.adamsummers.org/scanallfish (Accessed 9/17/25).

4. Gignac, P.M. *et al.* 2016. J. Anat. 228 (6): 889–909.

5. Simonitis, L.E. *et al.* 2024. Integr. Comp. Biol. icae103.

6. Camilieri-Asch, V. *et al.* 2020. Eneuro, 7(4).

7. Staggl, M.A., *et al.* 2023. J. Anat. 242(5): 872–890.

8. Bronson, A.W. *et al.* 2018. Pap. Palaeon. 4(3): 349–362.

9. Bronson, A.W. *et al.* 2024. Geodiver. 46(4): 101–117.

10. Coates, M.I., *et al.* 2018. Proc. Roy. Soc. B: Biol. Sci. 285(187): 20172418.

11. Maisey, J.G. *et al.* 2019. Doliodus and Pucapampellids: Contrasting Perspectives on Stem Chondrichthyan Morphology. pp. 87–109 in Johnson, Z. *et al.* (eds) *Evolution and development of fishes*, Cambridge University Press. 280 pp.

12. MorphoSource. https://www.morphosource.org/ (Accessed 9/17/25).

13. Sketchfab. https://sketchfab.com/ (Accessed 9/17/25).

14. Abel, A. and Ziman, R. 2023. Edu. Sci. 14(1): 34.

15. Grant, C.A., *et al.* 2016. Paleon. Soc. Pap. 22: 197–209.

16. 3d Anatomy Studios. https://3danatomystudios.com/(Accessed 9/24/24). Login-Required.

The cartilage of both chondrichthyans and humans is not as hard as bone, because it lacks bone's high levels of calcium and phosphorous salts. However, in sharks, at certain stress points the cartilage is strengthened by incorporating calcium in the form of the mineral apatite. Because these crystals are arranged in a pattern resembling a prism, this type of cartilage is known as *prismatic calcified cartilage* (fig. 1.2).

In addition to being less rigid, cartilage is also lighter than bone. The density[5] of bone in humans is about 1.85 g/ml, and in bony fishes, like mackerel and pike, ranges from 1.3 to 1.5. The density of shark cartilage is about 1.1. For reference, the density of pure water is 1.000, seawater is about 1.025, honey is 1.36, and concrete is 2.4. So, honey is denser than cartilage!

One major benefit of cartilage to sharks is that its reduced density in part compensates for their absence of a swim bladder, a gas-filled internal structure found in most bony fishes that allows them to adjust their buoyancy, and in the process save precious energy and expand their range of living spaces, which we discuss more fully in chapter 3.

Sharks are relatively muscle-bound (fig. 1.3) and are heavier than seawater, with most having whole body densities between 1.03 and 1.09, even with an appreciably less-dense skeleton and other adaptations that we discuss later.

Figure 1.3. Preserved Horn Shark (head at left) with skin removed to show underlying dense musculature. Note the white septa separating the skeletal muscle blocks (*myomeres*) (Physiological Research Lab, Scripps Institution of Oceanography).

Does the lightness of the shark skeleton make it weaker, as one might reasonably conclude? Surprisingly, the answer is *probably not*. Because cartilage is light and flexible, it is often inferred to be inferior to bone as a structural material and incapable of tolerating the same high forces that impact bone during vigorous swimming. This seems like an unwarranted assertion, given that many sharks are powerful swimmers.

Now we know that the inference is indeed false, at least in some sharks. A 2006 study[6] found that the vertebrae in several pelagic sharks were as *stiff* and *strong* as bone, which the authors attributed to increased mineralization of the vertebrae as well as how the minerals were arranged. The hardness of bone thus does not equal mechanical superiority, just a different pathway to evolutionary success.

Finally, the widely held assumption that cartilage is an evolutionarily older structural material than bone has been challenged by the finding in fossils that early shark cartilage may have been built on a matrix that included bone. That sharks as a group are no older than bony fish would also seem to counter that assumption.

Let us examine other distinguishing characteristics of sharks, and in most cases rays as well. Many of these differences may be obvious to you: a shark just *looks like a shark* and a barracuda, sea robin, etc., does not. Refer to Figure 1.4 as you read about these shark features that as a group distinguish them from bony fishes:

- Five to seven bilateral gill slits
- Heterocercal caudal fin
- Internal fertilization and claspers
- Serial tooth replacement
- Placoid scales
- Simple chondrocranium (skull, braincase)
- Tribasic cartilage support of pectoral fins
- Ceratotrichia (soft, unsegmented fin rays)
- Ampullae of Lorenzini
- Absence of swim bladder
- Nictitating membrane in some taxa
- Retention of urea
- Rectal gland
- Oil-filled liver
- Intestinal valve
- Pericardio-peritoneal canal

Five to Seven Bilateral Gill Slits

One of the quickest ways to identify whether a fish is a shark is to count gill slits. Most sharks have five gill slits on each side, and eight species have six or seven,

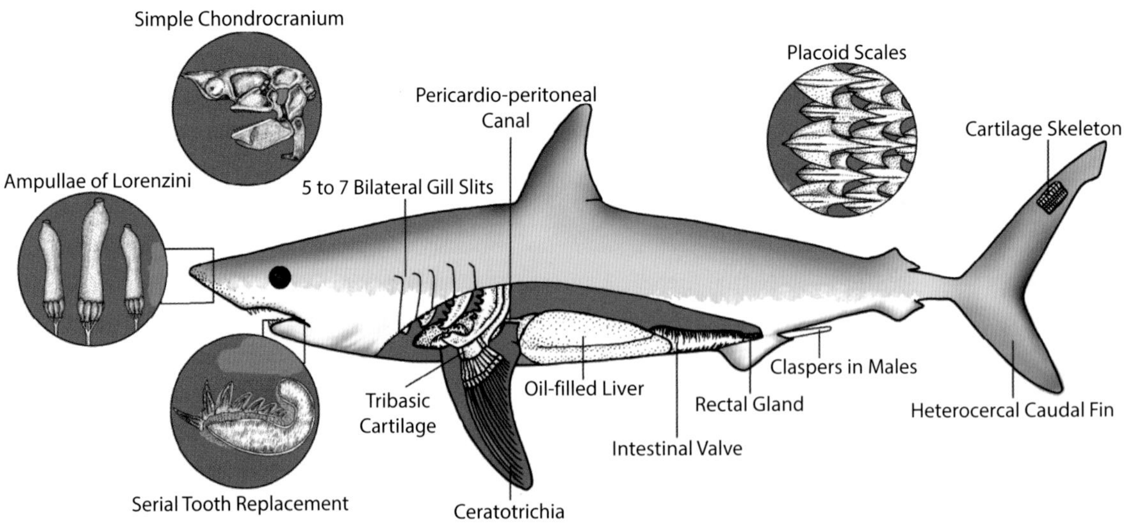

Figure 1.4. Features that distinguish a typical shark from a typical bony fish.

whereas all bony fishes have only a single bilateral flap, the operculum, covering the gills, which makes the inch-long fish in pet stores called a Red-tailed *Shark* in fact a freshwater bony fish in the minnow family (Cyprinidae). Clever marketing that fooled us as kids. All but one species of batoid possess five bilateral gill slits, but they are located ventrally rather than laterally.

Heterocercal Caudal Fin

Sharks have an asymmetrical caudal fin in which the upper lobe is typically longer than the lower lobe, whereas the lobes of the tail fin in most bony fishes are symmetrical and equal in length. The *heterocercal* (meaning *different tail*) caudal fin of sharks, which varies in size and shape among species, is characterized by the vertebral column extending almost to the tip of the fin's upper lobe, which adds stiffening support for generating thrust. Some primitive bony fishes (e.g., sturgeon), also have a heterocercal tail, whereas in others, like the tarpon, the fin is externally symmetrical, but the vertebral column extends slightly up into the upper lobe. In more advanced bony fishes, the caudal vertebrae have fused over evolution at the base of the fin and do not extend into their *homocercal* (meaning *same tail*) caudal fin.

Internal Fertilization and Claspers

All sharks use internal fertilization, an adaptation that sets them apart reproductively from the bony fishes, which we discuss in chapter 6. To transfer sperm to the female, male sharks have special organs known as *claspers* that they insert into the female cloaca.[7] Among bony fishes, only a handful (e.g., mosquitofish, mollies, and four-eyed fishes) fertilize internally; the overwhelming majority fertilize externally. When you consider that internal fertilization is largely a terrestrial phenomenon (except for marine reptiles and mammals, which evolved from terrestrial ancestors), that sharks evolved that ability independently early in their evolutionary history makes claspers evolutionary marvels!

The term *claspers* dates back to Aristotle, who incorrectly inferred that males used these modifications of the pelvic fin exclusively to clasp onto females during external fertilization. Although Aristotle's interpretation of the use of the claspers was incorrect, he accurately concluded that some coupling mechanism was required for males to transfer semen to females while both were swimming. We think readers could use their imagination to acknowledge that difficulty.

Figure 1.5 shows claspers from two sharks. Note the hook and spine at the tip of the splayed claspers from the Atlantic Sharpnose Shark. *Cloacal grapplers* has been proposed as a more accurate term for the structures.

Claspers were in the news (that's a phrase you don't see much) several years ago when the Underwater Times[8] reported that a Malaysian fisher caught a shark with what looked to the world like webbed feet. Had an unsuspecting fisher serendipitously unearthed one of the proverbial missing links? Well, no. In fact, the *webbed feet* were the claspers from a male shark splayed out as they would be when attached inside a female shark's cloaca, a condition that sometimes also occurs after death in male sharks.

Serial Tooth Replacement

Sharks also exhibit serial tooth replacement (fig. 1.6), in which their teeth are not firmly embedded in bone like those of humans and most other vertebrates, but rather are loosely rooted in collagen, one of the most abundant fibrous proteins used to connect and support parts of the body. These teeth are renewed periodically using a conveyor belt system that may replace entire rows of teeth or individual teeth. Serial tooth replacement is a form of *polyphyodonty*, a phenomenon among animals in which teeth are continually replaced. Most fish groups have polyphyodont dentition, along with amphibians and many reptiles (mammals, with the exception of manatees, elephants, and kangaroos, all of which have polyphyodont dentition, have only two generations of teeth, which is known as *diphyodonty*). In the other polyphyodont groups, teeth are replaced individually, as opposed to the unique conveyor-belt seriality of tooth replacement in sharks.

Sharks have from one to as many as 10 rows of functional teeth (e.g., White Shark and Dusky Smoothhound, respectively), with several rows of teeth in varying stages of maturity lying in wait on the very slow-moving collagen conveyor belt. In rays, there may be as many as 13–15 rows of functional teeth, which form plates for prey-handling.

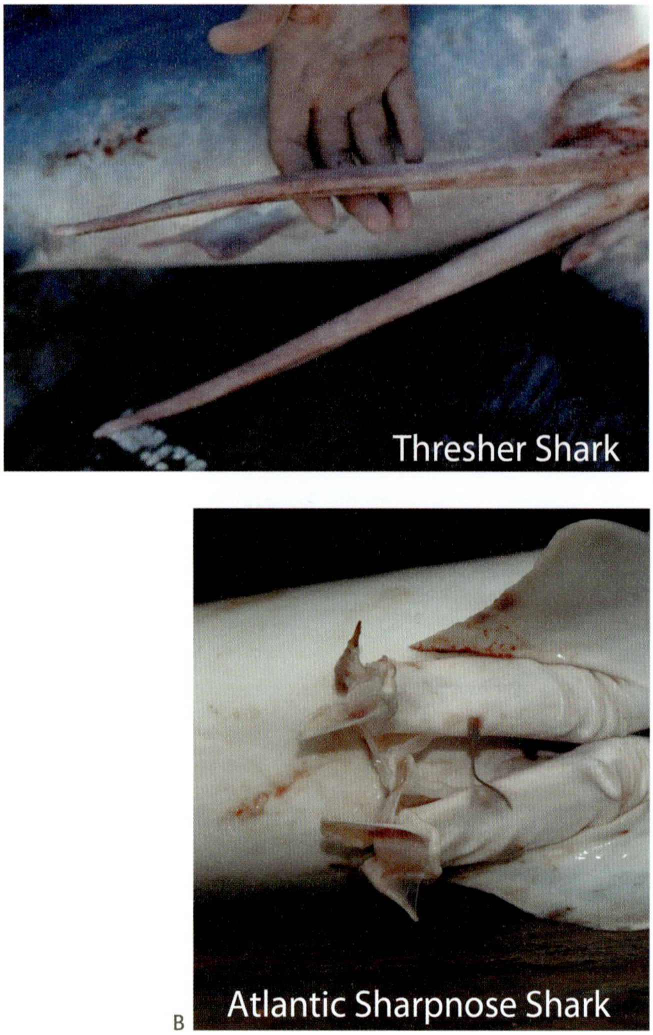

A

B

Figure 1.5. Claspers from thresher shark (A) and Atlantic Sharpnose Shark (B). Note the splayed claspers from the latter, displaying specialized structures for securing the clasper in the female's cloaca.

A row of teeth of suitable cutting, penetrating, grasping, or crushing ability is thus always available, and if one or more teeth on the front line are lost, replacements will soon be in transit. A shark may produce and lose thousands of teeth over its lifetime, and the replacement time is typically measured in weeks.

Surely it is wasteful to continually make and discard teeth instead of producing one permanent set, especially since one of the central tenets of evolution is that the supply of energy is often limiting and thus conserving energy is paramount to success. Evolution, however, often entails compromises, and natural selection can be thought of as a continuous experiment that tests what works and what does not. The individual with the suite of characteristics that works best, that is, that enables the bearer to eat sufficiently, avoid being eaten, and reach

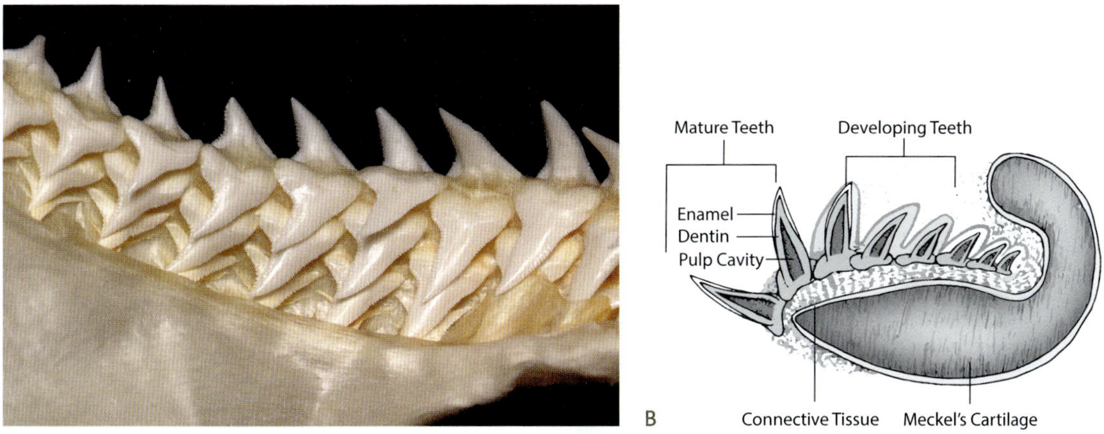

Figure 1.6. Examples of serial tooth replacement. (A) Jaw of Blue Shark showing both functional and replacement teeth. The functional teeth can be replaced individually. (B) Drawing of a sagittal section through a shark jaw showing teeth in various developmental stages (Redrawn from fig. 3, Moyer, J.K. *et al.* 2015. J. Morphol. 276: 797–817). (C) Adult Lemon Shark temporarily missing several teeth after biting the wooden rub rail of a research vessel.

maturity and successfully mate and produce viable offspring, is the one that passes the evolutionary test. The individual with characteristics less successful in these areas likely does not live to pass its genes to the next generation. You may know this concept as *survival of the fittest.*

Then *yes,* shedding teeth may waste the mineral and energy resources that were used to produce them, but the advantages of serial tooth replacement as part of the complete package of shark adaptations, outweighed these costs. Starvation is one of the biggest threats to predators, particularly terrestrial ones but also some marine forms. Turtles and birds may do well without teeth, but if a lion or tiger loses one of its long canine teeth, or these teeth become worn down, its chances of surviving are significantly diminished. If a Tiger Shark loses several teeth, no problem, polyphyodonty to the rescue!

Placoid Scales

The body of sharks is covered with *placoid* scales, or *dermal denticles* (fig. 1.7). Structurally, these are like miniature teeth, with an inner pulp cavity, surrounding layer of dentin (hard, calcified tissue), and a hardened outer layer of enamel.

Quick deviation: anyone who regularly handles sharks occasionally contracts *shark burn,* a rash that begins to sting shortly after rubbing against a shark, especially along the shark's body from back-to-front. Think *rug burn* or *artificial turf*

A 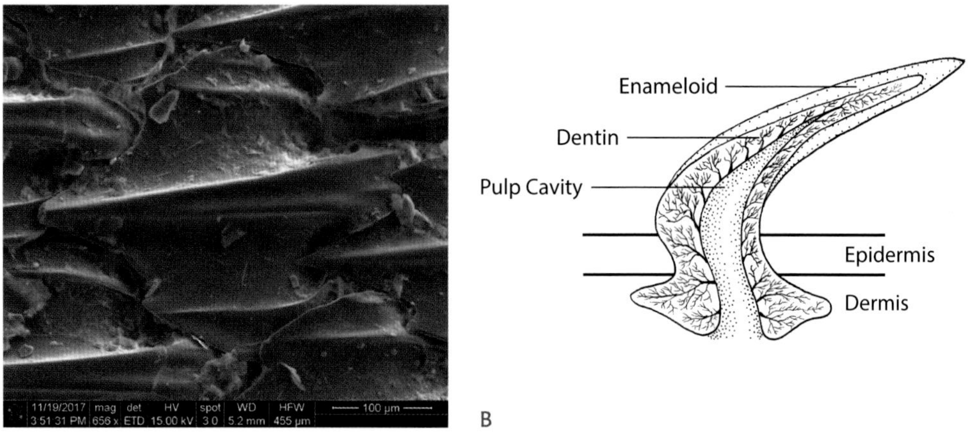 B

Figure 1.7. (A) Scanning electron photomicrograph of scales of the recently described shark named Genie's Dogfish. The ridges, known as *riblets*, play a role in reducing drag (Courtesy of Charles Cotton). (B) Labeled cross-section of a generalized shark scale (dermal denticle) (Redrawn from fig. 5, Moyer, J.K. *et al.* 2015. J. Morphol. 276: 797–817).

burn. If you are the recipient of shark burn, because scales are miniature teeth, you can legitimately (well, sort of) claim that you survived being bitten by a shark.

Like their teeth, scales are shed and replaced in sharks, although more slowly. Bony fishes' scales are composed of the fibrous protein collagen and mineralized bone, and are permanent. The scales have varying functions among sharks, including reducing drag while swimming, protecting against ectoparasites, and safeguarding females from male bites during mating season. We discuss some of these in chapter 4.

Simple Chondrocranium

The skeletal components of the head of tetrapods comprise a neurocranium (or braincase) that protects the brain, jaws, and other facial elements. Because the neurocranium of a shark is a single, sutureless piece of cartilage, we call it the *chondro*cranium. In contrast, the neurocrania of bony fishes and humans consist of multiple pieces of dermal bone connected by sutures. Similarly, the left and right cartilaginous elements of a shark's upper jaw, or palatoquadrate, differ from the upper jaws of many bony fishes, which have premaxilla and maxilla making up each side of their upper jaws (fig. 1.8). The mosaic of bony elements observed across fish heads show a remarkable evolutionary plasticity. Since sharks do not have the same suite of cranial structures, the degree to which evolution can tinker with the underlying anatomical architecture is more limited compared to bony fishes.

Tribasic Cartilage Support of Pectoral Fins

In the evolution of sharks, an important advancement occurred when the inflexible, solid skeletal support of each pectoral fin, as well as pelvic fins, was replaced by three basal cartilages (named *propterygium*, *mesopterygium*, and *metapterygium*;

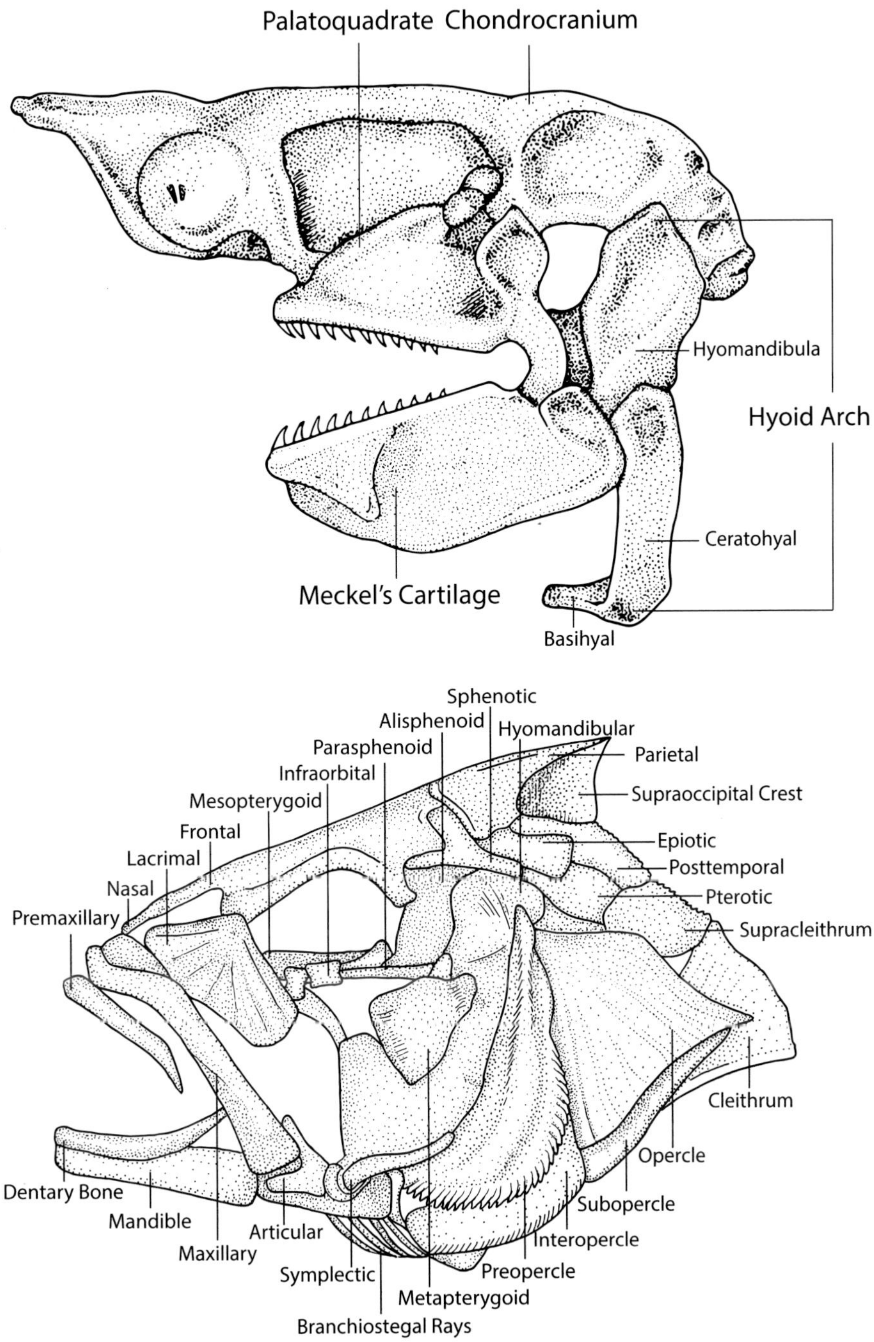

Figure 1.8. Generalized heads of a bony fish (bottom) and shark. The bony fish head is complex. Numerous bones provide the raw material for the evolution of diverse mouth and head shapes, and hence different feeding styles. In contrast, the reduced number of cartilage components in the shark head has limited their feeding styles (Redrawn from fig. 8.3, Hyman, L. and Wake, M. 1979. *Hyman's Comparative Vertebrate Anatomy.* [3rd ed.]. University of Chicago Press and Day, F. 1889. *The Fauna of British India, including Ceylon and Burma. Fishes*. Taylor and Francis).

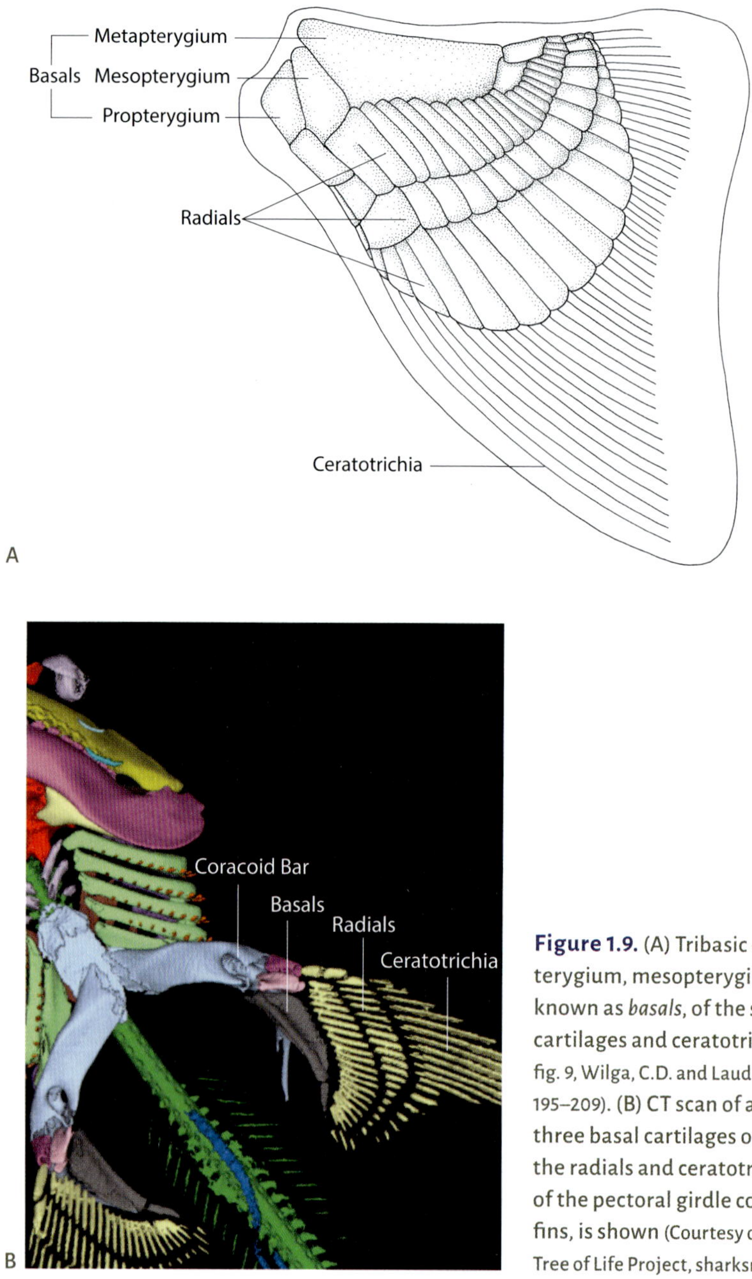

Figure 1.9. (A) Tribasic cartilage support (metapterygium, mesopterygium, and propterygium), also known as *basals*, of the shark pectoral fin. Radial cartilages and ceratotrichia are shown (Redrawn from fig. 9, Wilga, C.D. and Lauder, G.V. 2001. J. Morphol. 249: 195–209). (B) CT scan of a Lemon Shark showing these three basal cartilages of the pectoral fin along with the radials and ceratotrichia. The coracoid bar, part of the pectoral girdle connecting the two pectoral fins, is shown (Courtesy of Gavin Naylor, Chondrichthyan Tree of Life Project, sharksrays.org).

fig. 1.9). These cartilages supported and allowed slightly higher mobility of the pectoral fins and thus increased maneuverability over the more fixed pectorals of ancestors, an evolutionary event contributing to the success of modern sharks.

Ceratotrichia

Ceratotrichia (figs. 1.9 and 1.10) provide the support for the distal (outer) part of the fins of sharks. These elongate rays are soft, flexible, unsegmented, and are

Figure 1.10. CT scan showing ceratotrichia (light yellow) comprising the wings (i.e., pectoral fins) of a ray (Courtesy of Gavin Naylor, Chondrichthyan Tree of Life Project, sharksrays.org).

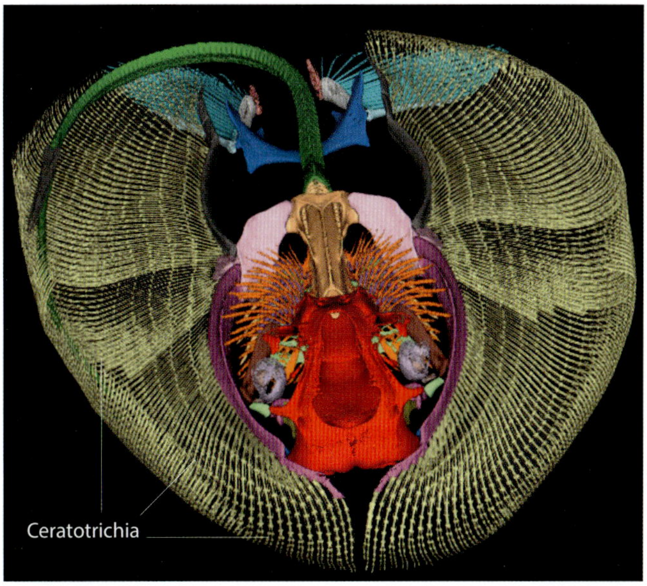

constructed of an elastic protein similar to the keratin of hair, nails, and claws. Unfortunately for sharks, the essential ingredient of the popular Asian delicacy, shark fin soup, is the ceratotrichia.

Ampullae of Lorenzini

Sharks also possess a means of detecting extremely minute electrical fields using a series of receptors located on the head known as *ampullae of Lorenzini* (see fig. 1.11). A few primitive bony fishes (e.g., paddlefish) possess very similar organs. These structures, which we describe more fully in chapter 5, allow sharks to detect and locate prey from distances of about 0.5 m (1.6 ft), where the sensory environment might be otherwise confusing because of murky water, widely diffused odors, etc.

Absence of Swim Bladder

Sharks lack a swim bladder, and thus must either swim to maintain their position in the water column, or rest on the bottom. Refer to chapter 3 to see the significance of a swim bladder to bony fishes, and of its absence to sharks.

Nictitating Membrane

Most members of the largest order of sharks, the Carcharhiniformes, possess a nictitating eyelid, or membrane, in the lower portion of each eye (fig. 1.12). It is prominent in the carcharhiniform families Carcharhinidae, (requiem sharks), Sphyrnidae (hammerheads), and the monotypic Galeocerdidae (Tiger Shark). It is rudimentary or absent in most of the other carcharhiniforms. Some birds, lizards, frogs, and even a few mammals (e.g., seals, Polar Bears, camels, and Aardvarks)

Figure 1.11. Ampullae of Lorenzini (pores appearing as darkened dots) on the underside of the head of a Bonnethead (Courtesy of George Boneillo).

Figure 1.12. Nictitating membrane covering the eye of a Tiger Shark.

have functional nictitating membranes as well. The structure, which is covered in scales in sharks, rises to cover and protect the eye as the shark is about to eat, or when it is threatened. White and cow sharks have different methods of protecting their eyes, which we discuss later.

Retention of Urea

If you have ever prepared a shark steak to eat (which we discourage except for a few sustainably fished species), an essential step is soaking or marinating the meat before cooking to remove some of the chemicals. Otherwise, the flesh may be unpalatable because the liver of sharks converts nitrogen-containing wastes (e.g., those remaining after proteins and other compounds are digested), into the waste-products ammonia and urea, whose taste elicits, in scientifically accurate verbiage, the following reaction: *Blech!*

Sharks are not distasteful to discourage predators but more so as part of an adaptive strategy to maintain the overall concentration of solutes (dissolved substances) in their blood and other extracellular fluids (the liquid outside of the cells) at about the same level as that of the seawater in which they live (see chapter 8).

Rectal Gland

Sharks also have a small finger-like or bean-shaped structure called a rectal, or digitiform, gland (fig. 1.13). This structure also aids in eliminating the excess sodium and chloride which diffuse into the shark's body as a result of the higher concentrations of these two ions in seawater compared to the internal fluid environment of sharks. We explain this phenomenon in chapter 8.

Oil-filled Liver

Earlier we mentioned that sharks are heavier than water, and they compensate in part by *dynamic* lift created by the caudal fins and in some sharks, the head, while swimming. The presence of an oil-filled liver (see fig. 7.5, which helps

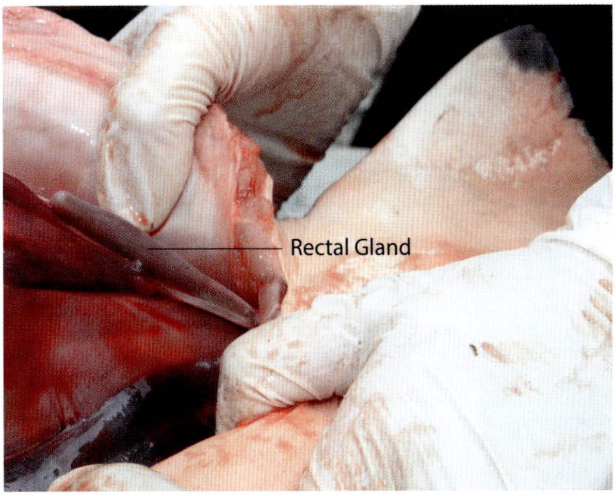

Figure 1.13. Rectal gland of a Blacktip Shark.

decrease the overall density, or heaviness, of a shark's body, provides buoyancy, or *static* lift; that is, lift that does not require movement. The density of oil in the shark liver is between 0.86 and 0.95 g/ml (recall the density of seawater is about 1.025).

Intestinal Valve

Sharks also have an unusual adaptation of the intestine that increases its surface area for digestion and absorption of nutrients. In most vertebrates, this same result is accomplished by increasing the length of the intestine (your small intestine is 3–5 m, or 9.8–16.4 ft, long), then convoluting, or folding, it onto itself to fit into the abdominal space. Sharks instead increase the surface area entirely within the compact length of the intestine with an intestinal valve (fig. 1.14), that has spirals, rings, or is tightly wound to resemble a scroll (depending on the species), which in each case increases the surface area for absorption in the intestine and saves space in the process.

Pericardio-peritoneal Canal

Finally, sharks have a pericardio-peritoneal canal, a tongue-twisting but aptly named duct that connects the pericardial space surrounding the heart with the peritoneal, or abdominal, space surrounding the visceral organs like the stomach, pancreas, liver, spleen, etc. (fig. 1.15).

A pericardio-peritoneal canal is present in the embryological development of all vertebrates, but persists only in larval lampreys, adult elasmobranchs, hagfish, and sturgeon. This canal plays a major and often unrecognized role in heart function by protecting the heart from being dangerously compressed (e.g., by increased forces when the shark is eating) and by enhancing heart function in other

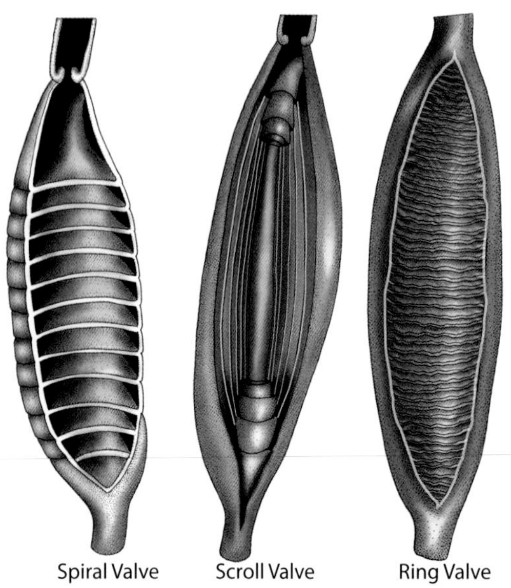

Figure 1.14. Spiral, scroll, and ring intestinal valves, characteristic of Spiny Dogfish, Blacktip Sharks, and Shortfin Makos, respectively. In all three cases, the surface area for digestion and absorption is increased.

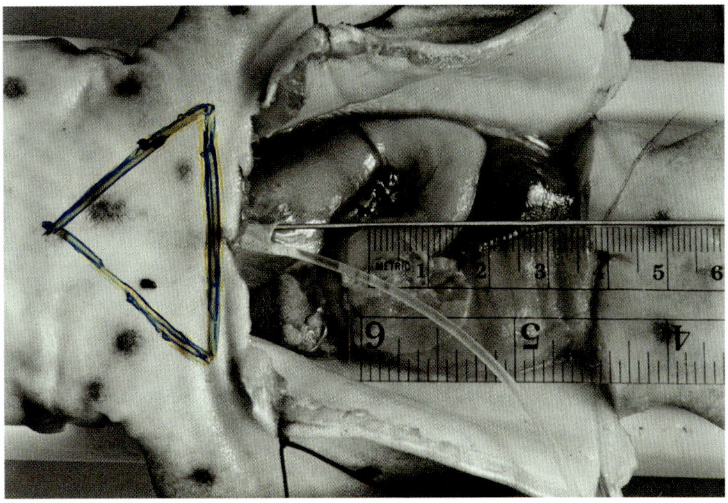

igure 1.15. Ventral view of the open
odominal space in a Horn Shark
iowing the pericardio-peritoneal
inal (with tube and wire inserted).
ne liver has been removed. The
iangle drawn on the photo at left
enotes the approximate ventral
itline of the pericardial space, in
hich the heart resides and where the
ericardio-peritoneal canal originates.

ways (see chapter 7). You likely will not see the pericardio-peritoneal canal listed in other references as a distinguishing feature of sharks (not all distinguishing features are unique), but it indeed is and reasons for its absence, other than its challenging pronunciation, are unclear.

Life-History Characteristics of Sharks

For a shark (or any organism), the name of the game in evolution is surviving long enough to reproduce, thereby passing your genes, which have proven effective in getting you through life's exigencies to this point, to your offspring. If *they* survive to reproduce, then the offspring pass on their successful characteristics, creating a multi-generational blueprint for evolutionary success.

Organisms employ one of two main life-history strategies or their numerous intermediate stages, in this *quest* for evolutionary success. On the one hand are organisms like Pollack or fruit flies, which grow quickly, become mature early, and lay lots of quickly hatching eggs.

Close to the other extreme are the sharks, whose suite of more conservative life-history characteristics has suited them well in their ~450-million-year evolutionary history. At the same time, for many sharks these same characteristics make them particularly vulnerable to pressures, e.g., overfishing, habitat destruction, and climate change impacts, that can reduce their populations and make recovery difficult.

Sharks as a group

- are very slow growing,
- are long-lived,
- reach sexual maturity late in life,
- have a long gestation period (the time developing embryos are retained in the female or in the egg),

- typically use specific mating and nursery areas, and
- have relatively low fecundity (the number of offspring produced by a female of a species; see chapter 9).

Where Is the Sharkiest Place on the Planet?

A mockumentary about sharks from the 1980s states "Sharks are found in only two places—the northern and southern hemispheres." It is not hyperbole to assert that, where there is enough water, there is shark habitat, although there are some qualifications. Sharks are found in marine environments from tropical coral reefs and mangrove ecosystems to cold polar waters, from estuaries to the deep sea and open ocean. Perhaps surprisingly, a small number of species inhabit lakes and rivers. Big sharks, for example Lemon and Bull Sharks, can occupy small creeks, and small sharks, like the Lantern Sharks, inhabit large spaces. Sharks like the Blue Shark, Oceanic Whitetip Shark, and Silky Shark live in the open ocean's surface layer, whereas the Portuguese Dogfish, which holds the depth record for sharks, has been caught at 12,000 ft (3700 m).

Among these habitats, which is the sharkiest? The answer depends on what *sharkiest* means. If the term refers to the greatest biomass or number of sharks, then the winner of the title might be the NW Atlantic Ocean, specifically, the highly productive Grand Banks and adjacent waters, where one of the most numerically abundant shark species, the Spiny Dogfish, or Spurdog, resides. The species was heavily overfished, but it is recovering as a result of effective management policies (see chapter 11).

If we interpret *sharkiest* as *most biodiverse,* then we have numerous candidates. In general, the richest areas are along the fringes of continents; that is, over the continental shelves, as opposed to the open ocean, and particularly in the mid-latitudes (between 25° and 40°) in both hemispheres. Other hotspots include regions where cold, nutrient-rich deep water upwells to the surface (e.g., NW Africa) and seamounts (undersea mountains that do not breech the surface), which are like underwater island oases. Figure 1.16 shows biodiversity hotspots.

A 2011 paper[9] listed the following as global hotspots for sharks based on species-richness (the number of species): Japan (with a surprising 124 species of sharks), Taiwan, the east and west coasts of Australia (> 170 species), SE Africa (about 100 species), SE Brazil, and SE USA.

Sharks in Our Lives

Sharks in Human History and Culture

Sharks have permeated human cultures for millennia. Early coastal populations encountered sharks and used them as natural resources like many other fishes. Shark meat was eaten and traded. We know this because archeologists have found the remains of carcharhinid sharks and dasyatid stingrays in ancient trade locations in what is now modern-day Israel, Lebanon, and Turkey.

Figure 1.16. Simplified representation of shark biodiversity. Note that the highest diversities occur along continental shelves in the mid-latitudes, whereas open oceans are far less species-rich (Courtesy of Gavin Naylor, Chondrichthyan Tree of Life Project, sharksrays.org).

These sites date back to the Bronze Age of the Middle East,[10] which was more than 3,500 years ago!

As fishing brought coastal sharks and humans closer, advances in maritime technology and navigation enabled humans to venture further out to sea where they would encounter and catch more sharks and observe different behaviors. Both fishing and exploration increased human awareness of sharks, culminating in observations about their anatomy and behavior that made it clear that sharks were no ordinary fish (We mean no offense to the many ichthyologists who study "ordinary" fishes.). Sharks soon took on special roles in human mythologies[11] and fascinated early naturalists.

Anthropologists studying the mythological and cultural significance of sharks often reference ancient Polynesian[12] cultures, and with good reason. By 1300 BCE, the ancestors of Polynesian peoples had reached the Solomon Islands and were sailing eastward into the Pacific. By 900 CE, a group of Polynesians had reached Hawaii,[13] where sharks flourished as mythological figures. Ancient Hawaiians revered some sharks as *aumākua*, which were deified ancestors that acted as guardians. Other sharks, the *niuhi*, were large and dangerous.[14] Teeth of these sharks, such as the White Shark (*Carcharodon carcharias*) and the Tiger Shark (*Galeocerdo cuvier*), were fastened to wooden war clubs, swords, and knives to enhance their efficacy as cutting implements (fig. 1.17). Many Hawaiian gods and goddesses were associated with sharks. For example, Pele, the volcano goddess, had an older brother, Kamohoali'i, a shark god. Ka'ahupahau, a redheaded maiden who turned into a shark, lived in what is now called Pearl Harbor, making peace between humans and their elasmobranch neighbors.

Although Polynesian mythology is for many westerners a go-to example of spiritually significant sharks, it is far from the only one. Another notable example

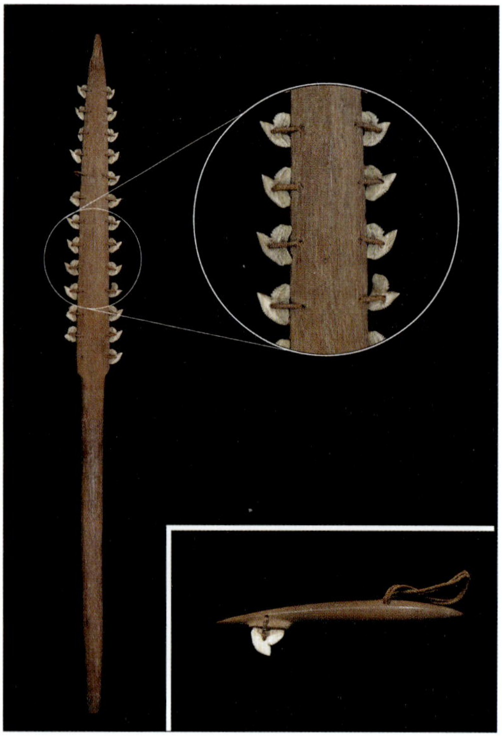

Figure 1.17. A Micronesian sword lined with Tiger Shark (*Galeocerdo cuvier*) teeth and a single-handed cutting implement from Hawaii, also with a large Tiger Shark tooth (Specimens from the anthropology collection of Yale University's Peabody Museum).

is the mythology of the Haida people of the Pacific Northwest, which tells of the "dogfish woman." As the moniker suggests, this figure could transform herself into a dogfish, specifically the Pacific Spiny Dogfish (*Squalus suckleyi*). How do we know this? The beautiful artwork of the Haida people depict a shark with two dorsal fins, each with prominent spines,[15] and the presence of dogfish in the totem poles of Haida people is a testament to the shark's significance in Haida culture.

Of course, not all of humans' historic attraction to sharks was of a spiritual nature. Early naturalists studied sharks as biological entities, a practice that goes back to the fourth century BCE. The Greek polymath Aristotle (384–322 BCE) was fascinated by sharks.[16] He distinguished sharks, *selachia*, from other fishes by their cartilaginous skeletons and uncovered gills (meaning they lack an operculum). He noted that many sharks were viviparous, giving birth to live young. More than 2,000 years before the concept of *r* vs. *K* reproductive strategies was introduced in the 1960s (see chapter 9), Aristotle noted the connection between sharks' smaller litter sizes relative to other fishes and the higher survival rate of sharks' larger, more developed offspring.

Following Aristotle's contribution to an early understanding of sharks was no easy task, but several Romans made noteworthy contributions in their own rights. For example, Pliny the Elder (23–79 CE) provided some of the earliest descriptions of interactions between sharks and divers. Soon thereafter, Claudius Aelianus (circa 175–235 CE), in his "On the Nature of Animals", wrote of three kinds of sharks: the large, dangerous sharks; small sharks that "live in the mud;" and dogfish. Formal taxonomic classification of species was still more than 1,700 years into the future.

As the West entered the Middle Ages, academic attention toward sharks in that part of the world waned. It is worth noting though that collectively, historians of shark science owe a debt to Muslim and Jewish scholars who preserved the works of Greek and Roman naturalists. By the end of the Middle Ages, Latin translations of Aristotle's seminal works were made from Arabic.

Like many areas of human endeavor, the study of sharks benefited from the Renaissance. Education was revitalized in Europe, and students of medicine and anatomy appreciated sharks as comparative specimens. This set the stage for some truly remarkable insights from sixteenth and seventeenth century individuals like Nicolas Steno (1638–1686 CE). Steno was a Danish anatomist and naturalist whose contributions to our scientific understanding of sharks included studies of shark brains, eyes, and teeth.[17] Among paleontologists, Steno is famous for spotting the similarities between the teeth of White Sharks (fig. 1.18) and *glossopetrae,* or "tongue stones." Glossopetrae were in reality the teeth of extinct megatoothed sharks like *Otodus megalodon,* but until Steno deduced the common origins of shark teeth and "tongue stones," people of the Middle Ages thought these fossils were the stone tongues of biblical serpents who were cursed by Saint Paul (although the name "tongue stones" appears to have come from Pliny the Elder, who noted the teeth's similarity in shape to that of human tongues and speculated that they fell to Earth during lunar eclipses.[18])

As time marched on, naturalists continued to be fascinated by sharks, although public awareness of sharks had not reached the levels we see today. In the tenth edition of his book *Systema Naturae,* Carl von Linné (1707–1778), now referred to by his academic *nom de plume* Linnaeus, laid out the modern system of taxonomic classification and began to formally describe elasmobranchs in a manner approaching modern taxonomic standards, although he did not refrain from offering some commentary.[19]

By the nineteenth century, naturalists like Charles Darwin (1809–1882) and Thomas Huxley (1825–1895) occasionally referenced sharks in their communications and scholarly works. Their interest in cartilagenous fishes focused predominantly on anatomy and development. In fact, Huxley was the naturalist who formally named the clade Chondrichthyes and assigned elasmobranchs and holocephalans (ratfish or chimeras) to the taxonomic class on the basis of their skeletonal composition. At that time, detailed studies of shark behavior and ecology were still almost a century into the future.

This brings us to the twentieth century, the century in which shark science began to flourish, thanks in part to advances in the fields of physiology and animal behavior. Physiologists took an interest in the sensory biology of sharks, and as the world became more globalized, international shark fisheries grew. Perhaps some of the most influential events that put sharks in the public consciousness were a

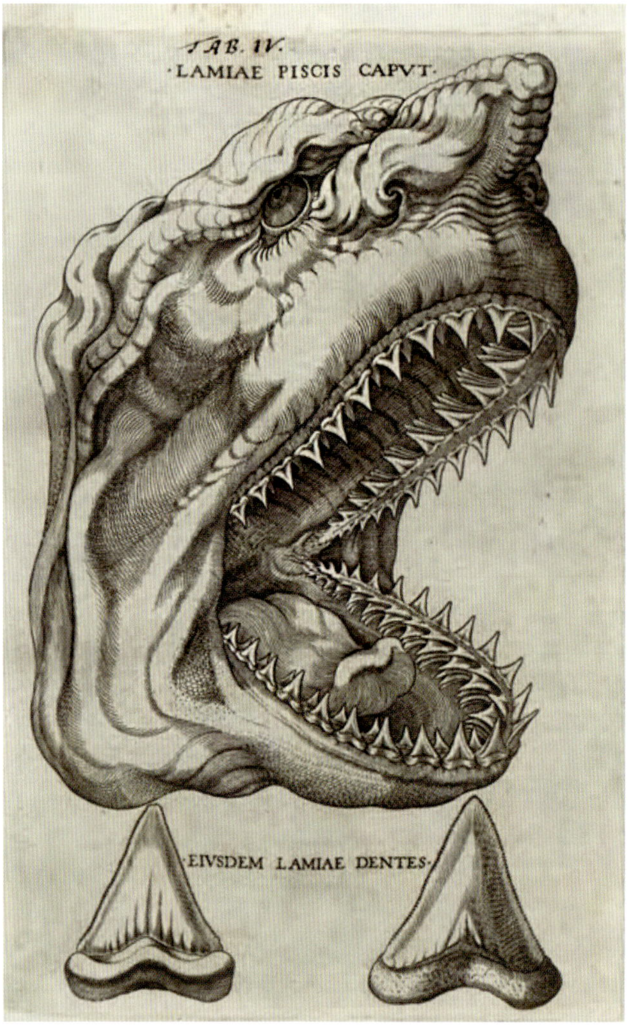

Figure 1.18. A stylized illustration of a White Shark's head and teeth that appeared in Steno's work comparing tongue stones to the teeth of modern sharks (From Steno, N. 1667. *Canis carcariae dissectum caput*).

string of shark attacks in the summer of 1916 in New Jersey, USA, which made the national news. Prior to that, awareness of sharks was largely relegated to fishers and academics, and their potential danger to humans was rarely considered.

Then came World War II, during which surviving members of Allied and Axis navies found themselves adrift when their ships were torpedoed. Needless to say, being torpedoed can dampen morale. The prospect of being bitten by a curious shark drawn to the loud, smelly, and unfamiliar site of a sinking ship did nothing to help. Thus, the US Navy devoted considerable resources to the study of shark behavior and physiology, trying to assess the actual risk of shark attack and to develop an effective shark repellent. By the 1950s, numerous marine labs were staffed by researchers focusing on the anatomy, physiology, and behavior of elasmobranch fishes. Add to that the emerging economic significance of sharks as targets of international fisheries and a newfound public fascination, and the seeds of modern shark biology in Europe and North America started to take root and grow.[20]

In the Media

It is difficult to avoid sharks in the media. Discovery Channel's *Shark Week* and National Geographic's *Sharkfest* saturate cable airways for a week each summer and in countless replays, with recent annual viewership exceeding 50 million. Feature films[21] like *Finding Nemo, Shark Tale, Open Water, 47 Meters Down, The Shallows,* and *The Meg* include sharks as central characters. Other shark movies include those with campy[22] titles like *Santa Jaws, Atomic Shark, Toxic Shark, Sharkenstein, Ice Sharks, Ozark Sharks,* and *Sky Sharks.* These may be representative of the genre, but the apex of shark camp belongs to the Sharknado franchise, of which there are an original and seemingly endless sequels. In these cult flicks, hungry sharks are entrained in a waterspout and then plunked down where they terrorize a cast of unknown actors trying to start, or aging celebrities trying to resurrect, a career. Some films that also pass themselves off as documentaries are, in fact, camp. Paramount among these is *Megalodon: The Monster Shark Lives.* It is difficult enough to educate the public about believable but incorrect assumptions or conclusions about sharks without the tidal wave of pseudoscience that some shark documentaries purvey. Moreover, some of these shows have incited harassment of government officials.

Shark Tank on cable TV features five *sharks,* millionaire venture capitalists, who vet ideas of erstwhile entrepreneurs for commercial success and savage them—apparently the eponymous reference of the title—if their proposals are unlikely to make money. Press coverage by TV, print, and Internet media continues unabated, especially if a shark bites a swimmer. And the ubiquity of sharks on social media? Some informative sites with stunning photographs and videos, but many more with misinformation or naïve opinions.

In Hemingway's masterpiece *The Old Man and the Sea,* sharks represent one of nature's forces that the aged fisher Santiago battles in his epic quest first for triumph and ultimately survival. Steven Spielberg's summer blockbuster *Jaws,* based on Peter Benchley's novel, petrified beachgoers in 1975, and endures as one of the most identifiable sources of irrational and unjustified panic that has insinuated itself into the psyche of hundreds of millions of people to this day. Beachgoers are more threatened by errant surfboards, jet skis, rip currents, jellyfish, bacteria in the water, SUVs in the beach parking lot, and even toasters[23] as they prepare breakfast, than they are by sharks, but foremost in their heads as their toes sink into the hot sand is the foreboding, ominous *Jaws* theme. Author Benchley was so disturbed that his book had emotionally tattooed generations with the stereotype of sharks as consummate killers, he spent the remainder of his life working for shark conservation. In December 2022, filmmaker Steven Spielberg apologized for the negative impact the Jaws franchise (four movies in all, but Spielberg directed only the original) had on our perception and persecution of sharks.

Speaking of the theme from *Jaws,* a very cool study by Andrew Nosal and colleagues from Scripps Institution of Oceanography, the University of California San

Diego, and Harvard, entitled *The Effect of Background Music in Shark Documentaries on Viewers' Perceptions of Sharks*,[24] concluded that attitudes toward sharks are influenced or reinforced by how ominous the background music is. Don't you just love science?

Although author Benchley denied any connection, the plot of *Jaws* is linked to a series of shark attacks in 1916 along the Jersey Shore.[25] During a 12-day period that summer, four people were killed and one was injured by a series of shark attacks.

Discerning the perpetrator or perpetrators of these attacks makes an interesting scientific detective story. At first, that a shark or sharks could be responsible for these attacks was quickly dismissed, since many ichthyologists (fish biologists) and other marine scientists were fairly certain, even in the early twentieth century, that sharks did not attack swimmers at the beach, at least along the US East Coast. This seems extraordinary, even unbelievable, today, but 100 years ago there were fewer beachgoers and thus fewer interactions with sharks other than on ships at sea or at shark-bite hotspots, and, of course, no Internet to communicate and sensationalize the events.

The list of causes more likely than sharks of the Jersey Shore attacks put forth at the time by experts and others, almost hilariously if not for the seriousness of the attacks, included German U-boats, sea turtles, and even Orcas. The panic that ensued from these events, and the response to it, presaged the unrelenting fear of sharks that persists today, as well as the resulting vengeful persecution of them that continues.

From a biological perspective, the most interesting question, and over 100 years later one still somewhat unsettled, is what kind of shark was responsible for the 1916 attacks. The issue was seemingly resolved at the time when a 7.5 ft (2.3 m), 325-lb (147 kg) juvenile White Shark, caught in a bay only a few miles away from the site of two of the attacks, was reported to have human remains in its digestive tract. Although the presence of human remains found in a White Shark would seem to conclusively settle the case (since the concept of *fake news* was still a century in the future), we may never know with certainty whether the Jersey Shore attacks of 1916 were caused exclusively by that species, or perhaps by two different species, the second being a Bull Shark, but we can be fairly certain that it was not an Orca, sea turtle, U-boat, or toaster.

Why so much interest in sharks? As biologists and conservationists, we are happy that a growing number of people are interested in sharks exclusively because of the vital ecological role that sharks play in maintaining marine biodiversity, or because of the suite of fascinating adaptations they possess; for example, the streamlined shape of a Shortfin Mako that makes it one of the speediest beasts in the sea. However, it is equally clear that, for perhaps the majority, the fascination with sharks owes more to fear and the perception of the danger they pose than to their ecological roles, biological attributes, or threats to their survival.

Another example of sharks in *our* lives: On May 9, 2017, we received this e-mail from a staff reporter for the *Palm Beach Post* that started:

> *Hi Dr. Abel / Dr. Grubbs,*
> *I'm sure you've seen the video of the adult film star allegedly getting bit by the shark . . .*

After we quickly recovered from the shock of a reporter we barely knew expressing certainty of our familiarity with adult film stars, we, along with other shark specialists, immediately deduced that the wound was human-inflicted as a very successful quest for web fame. The bite of one of the sharks in the video would likely have either removed a chunk of flesh or left a series of tooth punctures in an arc corresponding to the arched shape of a shark jaw, but not a clean laceration.

Why We Fear Sharks

If you are a predator, especially an apex predator, you begin life saddled with an image problem: the *perceived* threat you pose invokes a fear disproportionate to the *real* threat and may replace the awe we should have for you with irrational terror. We thus are numbed to the very real challenges to your survival. We root for the underdog (e.g., the old, young, or otherwise infirm) wildebeest that lions have separated from its herd, or the impala being chased by the stealthy, speedy leopard. And when you step into the water at the shore, your sympathy for the plight of sharks is supplanted by the comfort of knowing that, because of overfishing, habitat destruction, and so on, there are fewer of these toothy beasts lurking nearby in the shallows awaiting your entry. Yes, we love Shark Week, but for most of us, watching sharks from the safety of our screens or from a boat is just fine, if not preferred.

Dolphins are predators as well, but cute and cuddly ones according to our perceptions. In the 1980s, when people in the US saw graphic images of dolphins killed in the capture of tuna for the canned seafood industry in the eastern tropical Pacific Ocean, consumers united to demand action, and the *Dolphin-Safe* movement arose. No movement of equivalent scale exists for sharks. Moreover, the net effect (no pun intended) of the Dolphin-Safe campaign was taking the lives of sharks, sea turtles, small fish, etc. in numbers that dwarf the mortality of dolphins, as well as causing the Mexican tuna industry to crash. We discuss this further in chapter 11.

Why do we care less about sharks than we should? Ignorance of their ecological role or threats to them may explain some of our indifference to their plight, but not all. The other factor is fear, even though many know that a shark bite constitutes a low risk. Humans fear the dark, spiders, snakes, heights, public speaking, a visit to the dentist, etc.

Our brains conjure up reasons for our fears: What evil could be lurking in the dark? What if I stepped on a Copperhead Snake or Gaboon Viper that I did not see? What if I fell from this ninth story balcony? And the more we contemplate these unlikely occurrences, the more we nurture our fear. Fear evolved in humans as a survival mechanism, and there are times when it is rational to be fearful, but much of our fear borders on panic and is not based on high likelihoods of the occurrence of harmful things.

Fear of sharks is most like fear of the dark, which really is fear of the unknown. For all typical beachgoers know, there could be hundreds of hungry sharks anticipating their entry into the surf, especially in murky water. And layered on top of this fear are the gruesome descriptions and images of shark attack victims that populate social media. That death by shark attack would be a particularly horrifying way to die, or to lose a limb, is the icing on the cake for fearing sharks.

Shark Bites, Repellents, and Deterrents

In the new millennium, the years 2001 and 2015 have stood apart from the other years, at least in the eyes of the American media, as *Years of the Shark*, because of the perception of a significant increase in the number of shark bites and attacks.

Prior to the horrific events of 9/11, the summer of 2001 was a slow news period. Beginning on that summer's 4th of July weekend, several gruesome human-shark interactions occurred, including serious injuries to an 8-year-old boy and others, and fatalities in North Carolina and Virginia. News footage of migrating Blacktip Sharks off the coast of Florida, a common annual occurrence, fueled the hysteria (fig. 1.19). Although in terms of bites, the summer shark season started out furiously in 2001, but its end that year was not a particularly unusual one. Although regrettably there were fatalities, what really differed about the year 2001 was the increased focus on even minor shark bites, a trend that has only mushroomed since. The year 2015 started out similarly to 2001 along the US East Coast, with numerous bites, although no fatalities, and ended as a slightly atypical year.

Explanations for localized and/or perceived upticks in shark bites include improved dissemination of reports of interactions, more beach visitors and thus more opportunity for interactions, and two events that could be associated with global climate change: elevated water temperatures, and wind and current patterns that may have pushed shark prey, and thus sharks as well, closer to shore. Increases in shark bites have also been attributed to habitat alteration, changes in water quality, and decreases in shark prey due to overfishing. Since the number of shark bites is low and there is significant annual variation, accurately cataloging the incidence of shark bites and explaining their causes are for now elusive goals.

Hand in hand with the widely held fear of sharks (remember from the Preface that we call this particular kind of fear *galeophobia* or *selachophobia*) is the wrong-headed idea that we would face fewer risks in a world without sharks. This is bad reasoning in large part because of the low likelihood of being bitten by a shark; the number of documented, unprovoked shark bites hovers around 100 annually,

Figure 1.19. Aerial photographs of Blacktip Shark aggregations off the east coast of Florida, which are annual occurrences (Courtesy of Stephen Kajiura).

according to the Florida Museum of Natural History's International Shark Attack File[26] (the one-stop shop for the best-available information on the subject). In 2023, 69 unprovoked bites were reported, almost exactly the decadal average of 70 per year. Along the US east coast, where Blacktip Sharks have been responsible for numerous bites in Florida, recent increases in seawater temperatures associated with climate change are thought to have delayed or reduced migrations of these sharks from the north, and thus resulted in substantially fewer bites, although the number of bites varies annually.

Being bitten by a shark is thus a statistical possibility, but the number of shark-related fatalities and serious but non-life-threatening injuries is negligibly low compared to the threats we face every day and consider routine. We shorten our lives by smoking, drinking sugary sodas, eating too many carbs, and texting while driving. In the US, air pollution causes as many as 200,000 premature deaths annually. Breathing the dust emitted by our tires and running shoes as they wear away is a bigger risk to our health than are sharks. We know far more people who have experienced direct lightning strikes than people who have been attacked by a shark, and we are in the shark business.

One of our students wrote in an essay that you are more likely to be struck by *lighting* than bitten by a shark. We are pretty sure she meant to write *lightning*, but she was correct as the word was written: More people are injured by falling lighting fixtures than by sharks annually. A few years ago, a meme circulated on the Internet that you were more likely to be killed in the tropics by strikes from random coconut falls. Being killed by an errant coconut may be one of the few risks less likely than shark attack, although no International Falling Coconut Database collects data on coconut strikes.

On a somewhat serious note, as field biologists who work at sea often in bad weather, we are aware that the odds of being struck by lightning are increased from those while ashore. So, the authors try to avoid thunderstorms while we work on the water, in part because we fear the headline that might result: *Shark Biologists Were Correct: Struck by Lightning!*

So, it boils down to these two conclusions: First, the most dangerous activity you engage in daily is waking up and conducting your daily affairs. Second, as we stated above, given enough time, even extremely unlikely events will happen. That is why people buy lottery tickets.

Note that we have used the word *bite* in describing these shark-human interactions. Some shark biologists distinguish between *bites* and *attacks,* the former typically causing only minor injuries after *bite-and-release* behavior associated with mistaken identity.[27] A spate of these often occur in the summer months along the US southeast coast, particularly in naturally murky waters, where Blacktip Sharks, streamlined, swift-swimming eaters of small schooling fish, are thought to confuse human hands and feet with these fish and may bite. However, within fractions of a second, they apparently recognize the mistake and release, often leaving only a series of shallow puncture marks.

The term *attack,* a word that many shark biologists think should be replaced with *serious* or *fatal bites,* is reserved for cases involving more serious injury, most often involving White Sharks, Tiger Sharks, or Bull Sharks. In some of these cases, there are repeated bites after the first. The damage of even a single bite from these species may be extensive, even if they are considered mistaken identity interactions.

Ways to Avoid Shark Bites

Common sense and good fortune are the best deterrents to getting bitten by a shark. When we are in the water and see sharks, our instinct is to swim *toward* them. However, do not try this yourself (our liability lawyers made us write that). Seriously, the sensible action to take if you are in the water and see sharks, except perhaps small harmless species like horn and bamboo sharks, is to calmly but deliberately return to your boat or the shore and leave the water. And do not swim into a school of small fish (sometimes called a *bait ball*), or other areas where sharks might be attracted, like a fishing pier. Some sharks cruise closer to shore at dawn and dusk, so the time of day should also be a consideration. Is there an increased risk of being bitten by a shark if you are menstruating?

According to the International Shark Attack File, there is "no obvious pattern of increased shark encounters"[28] during menstruation. However, prudence should be practiced given that sharks have an array of senses that include a high sensitivity to blood.

Are there repellents (sometimes also referred to as *deterrents*, although technically the terms have different definitions) effective against sharks? The answer is yes, but none are foolproof, and their efficacy depends on the application. Below is a listing of some of the repellents used historically and/or currently. For a comprehensive summary, we refer you to Dudley and Cliff's *Shark Control: Methods, Efficacy, and Ecological Impact*;[29] the 2015 report by McPhee, Blount, and Lincoln-Smith,[30] (whose categories of deterrents we use below); and the 2021 report by McPhee et al.[31] Since these reports were published, research on deterrents has accelerated; we urge you to conduct your own searches for up-to-date information. The focus of many of these studies is the efficacy of deterrents on the sharks considered the *big three*, that is White Sharks, Bull Sharks, and Tiger Sharks.[32]

Large Scale

Shark Surveillance

The use of lifeguards or shark spotters on the beach or in boats to locate sharks and warn bathers has had mixed success. Major drawbacks of surveillance methods involving humans to detect sharks include the needs for an appropriate vantage point, relatively clear and calm water, sharks to stay near the surface, spotters to be alert for long durations, communities to support such efforts, and appropriate temporal and spatial coverage.

One very successful effort is the *Shark Spotters* program, which employs people equipped with binoculars to detect White Sharks in False Bay and other areas along the South Peninsula of Cape Town, South Africa. Since 2004, they have reported over 2000 shark sightings. Warnings are conveyed via sirens, a system of colored flags, and social media.

The use of aerial drone surveys, which have been called *game changers* among shark early warning systems, is expanding. Drones, also referred to as *unmanned aerial vehicles*, have also elucidated aspects of shark behavior useful in understanding interactions with humans. Butcher et al.[33] provide a comprehensive review of the use of drones, both aerial and underwater, in shark science. Other methods of shark surveillance include high-resolution sonar (e.g., *Cleverbuoy*), telemetry (acoustic or satellite tags on sharks), and airplanes/helicopters, all of which also provide information on shark movement ecology that also informs scientists about human–shark interactions.

Shark Barriers

Barriers to exclude sharks date back as far as 1907, when a steel structure was erected at a beach in Durban, South Africa. A variety of other materials have been tested but the mainstay has been cheap and effective mesh beach nets.

Unfortunately, these nets entrap and kill numerous species of sharks, and they have been called *walls of death*. One long-term study[34] in South Africa reported that 44 km (27 miles) of nets caught an average of 1470 sharks annually (in addition to 536 other large marine animals). Newer nets are constructed of more rigid plastics and thus ensnare fewer or even no sharks and other marine life, are more durable, and have reduced operating costs.

Researchers have also explored alternative barrier systems, including bubbles (largely ineffective), electricity, magnets, and other physical barrier systems. Electrical/magnetic fields in the range of 3–7 volts per meter have shown some efficacy as deterrents, but the response varies by species.

Another device proposed as an alternative to shark nets is the *Shark Exclusion Barrier* (an improved version of the Sharksafe Barrier, which allowed sharks to enter at the surface), which consists of both a visual and magnetic barrier. Powerful rare earth magnets, which have been demonstrated to repel sharks, are used along with a visual barrier of PVC pipes anchored to the bottom such that they can move with waves and currents, and an additional surface barrier. The exclusion barrier was recently tested on Bull Sharks,[35] with promising results.

Culling

The last large-scale method we discuss is culling, which is a category of shark persecution. Culling traditionally involves catching and killing of sharks on both small and large scales, either as a preventive measure or in response to shark bites.

While it is undeniable that culling removes sharks from an area, at least temporarily, the long-term success as a shark deterrent is not known. What is known is that any practice that removes predators, particularly top predators, which are typically the least abundant animals in any ecosystem, from their environment likely will have reverberations on populations of other organisms throughout that system.

A hybrid approach to culling involves catch-and-release fishing in which captured sharks are translocated, that is, released at some distance from their capture location. A problem facing this type of fishing is post-release mortality, a phenomenon we discuss in chapter 9. A relatively new system, the Shark-Management-Alert-in-Real-Time (SMART) drumline[36] (essentially a float to which from one to several baited hooks are attached), addresses this mortality by quickly removing sharks from the drumlines (having been alerted electronically of their presence) and tracking them post-release with satellite and acoustic tags.

Personal Deterrents

In recent years, numerous personal deterrents employing different repellent techniques (physical screen, chemical, acoustic, visual/camouflage, electric, magnetic, chainmail wetsuits, and cages) have become available in the marketplace. Prior to this, personal repellents in the US were developed mainly by the Navy for use in shipwreck/plane crash scenarios. Currently, most personal deterrents are proprietary and many have not undergone meaningful, robust scientific scrutiny that assesses both their efficacy and, ideally, their underlying physiological basis.[37]

Websites for commercial personal deterrents may feature impressive in-house tests, often with equally impressive videos, demonstrating that sharks avoid their repellent. There may also be testimonials in support of the products. Our experience with the companies that produce and/or market these devices is that they are earnest, and they encourage independent tests of their products. And many, but certainly not all, of their repellents show some promise as repellents in some situations.

Finally, we remind you that shark bites are extremely low-likelihood events, so in the overwhelming majority of situations, personal repellents are redundant; that is, you are already protected by virtue of not being among the regular prey of any sharks. If you *still* want to purchase a personal shark repellent, we suggest you ask a resident of Western Australia (WA) to make the purchase for you, since the WA government has offered a shark deterrent rebate on selected repellents.

We conclude this section with the following breaking news, which should ease the anxiety of readers in the UK:

> *You may rest assured that the British Government is entirely opposed to sharks.*
>
> Prime Minister Winston Churchill, February 20, 1945, in the House of Commons

Biomedical and Other Non-Food Uses of Sharks

Before we discuss the ways in which humans use sharks, we remind you that the single best *use* of sharks is allowing sharks the unfettered ability to perform their ecological roles by preserving their biodiversity. This serves sharks and humans both, as well as the entire planet.

That said, sharks have played and will continue to have roles in human health. You may have heard that *sharks don't get cancer.* Is the claim true? Unfortunately, no. Cancer refers to a disorder in which cells undergo genetic changes and lose their growth constraints and thus reproduce unchecked and invade other tissues, often leading to death. It may surprise you to learn that virtually all kinds of multicellular animals, from insects to whales, experience cancer, although the prevalence varies considerably.

In 2004, cancer researcher Gary Ostrander and colleagues published a thorough analysis[38] of this claim. We quote from the paper:

1 / Sharks do get cancer.

2 / The rate of shark cancer is not known from present data.

3 / Even if the incidence of shark cancer were low, cancer incidence is irrelevant to the use of crude extracts for cancer treatment.

Irrespective of the degree to which sharks are protected from cancer, there is *something* special about their primitive but powerful immune system. One characteristic that has particular clinical promise is the presence of a class of heavy antibodies, proteins produced by the immune response system to fight invasive foreign substances or pathogens (called *antigens*). Typical vertebrate antibodies possess two light and two heavy protein chains, and sharks are no exception. However, sharks, and inexplicably camelids (camels and their relatives), have antibodies composed only of heavy chains which, despite the name, are very small and are in a class known as *nanobodies*. Clinically, nanobodies from sharks and other sources may prove extremely useful in the field known as *immunotherapeutics* (fighting viruses, tumors, and diseases), as well as in developing diagnostic technology. Australian scientists are testing a drug based on the shark antibodies to treat the disease idiopathic pulmonary fibrosis (IPF). Additionally, researchers have found that sharks and their relatives possess more genes related to wound healing than are found in bony fishes.

Examples of other biomedical uses of sharks include:

- *Skin grafts/wound dressing*—Diabetic foot ulcers are treated with a mixture of silicone, cow collagen, and shark cartilage.
- *Antibacterial coverings*—A commercial surface coating/plastic sheet designed to prevent bacteria from sticking to a surface uses a *ridge* and *ravine* micropattern inspired by the skin of sharks.
- *Antibiotic/Antiviral agents*—The compound squalamine, which is found in the liver and stomach of some sharks, may have broad antibacterial and antiviral properties
- *Bone regeneration*—Apatite bioceramics from sharks may be useful in repairing, replacing, and/or treating bone defects or in other maxillofacial surgical applications
- *Therapy for neurodegenerative disorders*—The compound squalamine also appears to prevent the buildup of a lethal protein associated with Parkinson's disease and dementia.
- *Therapy for arthritis*—The compound chondroitin sulfate, derived from both shark and cow cartilage, in combination with the compound glucosamine, may slow the progression of osteoarthritis. In combination with other compounds, chondroitin may also have applications in cataract surgery and reduction of urinary tract infections in women.
- *Medical creams*—Shark liver oil is one of the ingredients in the antihemorrhoidal cream *Preparation H*.
- *Use as models for biomedical research*—Marine biomedicine is the discipline that uses marine organisms, sharks among them, to study biological problems relevant to human health.

Concluding Comments: Why Sharks Matter

We conclude this chapter by answering the question *why do sharks matter?*[39] Arguments in favor of caring about the plight of sharks include ecological, economic, spiritual, and ethical reasons.

Let us examine the ecological argument first. Sharks, even filter-feeding species, are predators. Some (but not most) are top, or apex, predators. Although an oversimplification, apex predators are defined as the animals at the top of the food chain, typically large-bodied, with no or extremely few natural predators as adults. In addition to some sharks like Whites, Tigers, Bulls, Oceanic Whitetips, Shortfin Makos, and Reef Sharks, marine apex predators include Orcas, Polar Bears, Saltwater Crocodiles, and Swordfish. Smaller sharks (e.g., Atlantic Sharpnose, Blacknose, and Bonnethead) and some rays (e.g., Cownose) that are one or two feeding levels below apex predators are known as *mesopredators.*

Some sharks may be *keystone species* in their ecosystems; that is, they are of disproportionately large importance. Removing the keystone species from an ecosystem is analogous to removing the keystone from an architectural arch, which causes the entire structure to collapse. The impact of the former, however, is often less apparent and not always easy to quantify. Still, the normal function, stability, and resilience of marine ecosystems are threatened if apex predators, particularly if they are keystone species, are removed or decreased in numbers. Tiger Sharks are both top predators and possibly a keystone species in Australian seagrass ecosystems, where they prey on sea turtles and dugongs. In doing so, they play a role in preventing these species from overconsuming seagrass.

Predatory sharks, whether apex or mesopredators, help preserve an ecosystem's biodiversity. At the same time, apex predators are particularly vulnerable to threats like overfishing, habitat destruction, and pollution because they are the least abundant organisms in an ecosystem.

The next argument for shark conservation is economic, namely preserving *ecosystem services* (goods and services that are required for human well-being and that ecosystems provide for free), for which there are not sufficient technological replacements.

The list of ecosystem services provided by the world's biodiversity is extensive and includes food, clean air, clean water, flood and erosion control, building materials, pollination, medicines, etc. We depend on these ecosystem services, and global health and the global economy would both most assuredly collapse in their absence.

In addition to the biomedical uses and ecotourism we listed above, the kinds of economic services provided by sharks include maintaining biodiversity of marine ecosystems such that populations of commercially and recreationally important fishes continue to exist. Products of economic value from sharks include shark meat, fins for the Asian delicacy shark fin soup, leather from shark skin, cartilage and collagen for the alternative health care industry, livers for omega-3 fatty acids and other products, and teeth and jaws.

Similarly, the recreational fisher, scuba diver, snorkeler, photographer, or ecotourist contributes to local and national economies by buying gear and permits, traveling, room and board, and other purchases.

That brings us to the final reasons, both intangibles, as to why we need sharks. First, sharks provide spiritual and aesthetic refreshment and add an ineffable value to our lives. Ecologist E.O. Wilson identified an innate human need known as *biophilia* (in his eponymous book) as *the urge to affiliate with other forms of life.* If you have ever marveled at seeing a shark in nature, or even in an aquarium, then you understand the value that a shark adds to your life.

Second, sharks, like every other species on the planet, have intrinsic value; that is, they have a right to exist based on their heritage, and this right is irrevocable.

Given all of the benefits, tangible or not, that sharks and the ecosystems in which they live provide, we ignore their plight at our own risk. Sharks are simply essential for our survival. It thus is incomprehensible that anyone would want to live in a world without sharks.

NOTES

1. Like poet Henry Reed's eponymous poem, which you might consider reading.

2. For a refresher on the hierarchy of zoological nomenclature, refer to Box 1.1.

3. *Inspiring Sediment* would make a great band name. Feel free to use it or any of the numerous other awkward word combinations you will find in this book.

4. In some taxonomic circles, the class to which the sharks, batoids, and chimaeras belong is listed as Elasmobranchiomorpha rather than Chondrichthyes, a distinction that acknowledges that the preceding group members are not the only cartilaginous fishes. The chondrosteans, a primitive bony fish group of sturgeons and paddlefish, possess an internal skeleton that is mostly cartilage (hence the "chondr" in their name). The cartilage-bone dichotomy embodied in the Chondrichthyes–Osteichthyes nomenclature is so well-established and descriptive that *Chondrichthyes* has persisted as the name of the class.

5. Density is a measure of heaviness, and it is expressed as mass per unit volume; for example, grams per ml (sometimes written as $g{\cdot}ml^{-1}$).

6. Porter, M.E. et al. 2006. J. Exp. Biol. 209: 2920–2928.

7. Cloaca (pronounced *klo-aý kah*) refers to the common and only opening for the reproductive, digestive, and urinary tracts in chondrichthyans, bony fishes, amphibians, reptiles, birds, and some mammals.

8. *Freak! Malaysian Fisherman Reels in Shark with 'Webbed Feet'.* 2007. http://www.underwatertimes.com/news.php?article_id=73210864950. (Accessed 9/19/25).

9. Lucifora, L.O. et al. 2011. PLoS One 6: e19356. https://doi.org/10.1371/journal.pone.0019356 (Accessed 9/19/25).

10. Mojetta et al. 2018. Reg. Stud. Mar. Sci. 21: 30–38.

11. In this section, we describe two examples of sharks assuming spiritual significance in ancient native cultures. Other fascinating examples not discussed here due to the limitations of space include mythologies of Aborigines of Australia, as well as the Aztecs and Mayans of Central America. Each of these cultures included sharks in their mythologies

dating back to the first century BC or earlier, and we encourage interested readers to learn about these cultures.

12. Polynesian peoples are those of the Polynesian Triangle, a geographic region between Hawaii, Rapa Nui (Easter Island), and Aotearoa (New Zealand).

13. The impressiveness of this maritime feat cannot be overstated. Polynesian wayfinders memorized sun, star, and current positions to navigate outriggers or double-hulled canoes more than 4800 km (3,000 mi)!

14. Taylor, L. 1993. *Sharks of Hawai'i: Their biology and cultural significance.* University of Hawaii Press.

15. Stewart, H. 1979. *Looking at Indian art of the Northwest Coast.* University of Washington Press.

16. Ganias et al. 2017. Fish. Fish. 18: 1038–1055.

17. Cuvier, G. 1995. *Historical portrait of the progress of ichthyology: From its origins to our own time.* Pietsch, T.W. (ed). Johns Hopkins University Press.

18. Forli, M. and Guerrini, A. 2022. *The history of fossils over centuries: from folklore to science.* Springer.

19. In *Systema Naturae* (10th ed) Linnaeus observed that the biblical prophet Jonah was consumed by a White Shark—not a whale—which was standing interpretation at the time among some scholars. In fact, tracing the linguistic descriptions of the mythological leviathan across translations, any mention of a whale came after many iterations of the story and was likely not in the original text.

20. Castro, J.I. 2016. Mar. Fish. Rev. 78: 14–33.

21. See Brooks, H. 2025. *The Best Shark Movies Of All Time.* https://www.ranker.com/list/the-best-shark-movies/ranker-film for a list of 49 films featuring sharks. (Accessed 9/19/25).

22. "Consciously artificial, exaggerated, vulgar, or mannered; self-parodying, esp. when in dubious taste," according to Dictionary.com.

23. Yes, toasters. Do not think so? Watch this: https://www.youtube.com/watch?v=ht-pxGXOlh3U (Accessed 9/19/25).

24. https://doi.org/10.1371/journal.pone.0159279. (Accessed 9/19/25).

25. The events' history is recounted in the books *Close to Shore* by Michael Capuzzo (2002; Broadway Books) and *Twelve Days of Terror: Inside the Shocking 1916 New Jersey Shark Attacks* by Richard Fernicola (2016; Lyons Press).

26. https://www.floridamuseum.ufl.edu/shark-attacks/. (Accessed 9/9/25).

27. Neff, C. and Hueter, R. 2013. J. Envir. Stud. Sci. 3: 65–73.

28. https://www.floridamuseum.ufl.edu/shark-attacks/reduce-risk/menstruation/. (Accessed 9/19/25).

29. Dudley, S.F.J. and Cliff, G. 2010. In: Carrier, J.C. et al. Sharks and their relatives II: Biodiversity, adaptive physiology, and conservation. CRC Marine Biology Series. CRC Press.

30. McPhee, D. et al. 2015. NSW Department of Primary Industries: 59916026.

31. McPhee, D. et al., 2021. Ocean Coast. Mgmt. 201: 105492.

32. Clarke, T. M. et al. 2024. Sci. Rep., 14(1): 16307.

33. Butcher, P.A. et al. 2021. Drones, 5(1): 8.

34. Cliff, G. and Dudley, S.F. 1992. Mar. Freshwater Res. 43: 263–272.

35. O'Connell, C.P. et al. 2024. Aquat. Conserv.: Mar. Freshw. Ecosys. 34(6), e4209.

36. Guyomard, D. et al. 2019. Fish. Res. 216: 6–17.

37. Clarke, T. M. et al. 2024. Sci. Rep., 14(1): 16307.

38. Ostrander, G.K. et al. 2004. Cancer Res. 64: 8485–8491.

39. Between editions of *Shark Biology and Conservation*, a new book of the same name as this section's subtitle (Why Sharks Matter) by colleague David Shiffman was published. We refer readers to Shiffman's publication for his take on the topic.

2 / Evolution of Sharks

CHAPTER OUTLINE

Introduction: Nature's Perfect Predators?

Many authors from Ernest Hemmingway to Peter Benchley have used sharks to symbolize a relentless, unyielding natural force. Perhaps it is unsurprising that by the end of the twentieth century, the popular media emphasized that sharks were nature's perfect killing machines. The proof? Sharks have remained virtually unchanged for millions of years, and, as the old saying goes, "if it ain't broke, don't fix it."

The problems with this interpretation are many. For one thing, evolution does not represent a march toward perfection. Rather, it is about what works given the circumstances, and circumstances change, sometimes rapidly or sometimes over millions of years. Another issue with that line of reasoning is that it assumes that sharks have not changed and are considered *living fossils* by many. But how accurate is that?

To assess how much sharks, or any group of organisms for that matter, have changed and to develop scientific hypotheses that explain the biodiversity we see today (which we will review in chapter 3), we must consider their evolutionary past. We need to compare extinct and modern species and establish how species are related. Only then can we deduce sharks' evolutionary origins and the extent to which they have diversified. We are, in a real sense, evolutionary detectives, and our clues are hidden in the sharks themselves and the fossil record.

How Do Scientists Study the Evolution of Sharks?

An unfortunate reality for paleontologists studying sharks is that cartilage does not preserve well. Thus, the fossil record of sharks is sparse compared to that of bony fishes, but it is not without useful information. That is because imprints of some shark and shark-like fossils have been preserved in anoxic (devoid of oxygen) fine sediments, where there is little bacterial decomposition.

Moreover, the information from fossil shark teeth (one of the most numerous vertebrate fossils), spines, and scales is not insignificant. Paleontologists thus know much less about the evolutionary history of sharks than we would like, but they know a lot about shark dentition and have made inferences from this knowledge.

Paleontologists know, for example, that among those sharks swimming today, the most *primitive*, that is, the ones whose lineage we can trace back the farthest, are the cow sharks (order Hexanchiformes; fig. 2.1A), whose morphologically distinct teeth have been found in sediment 180 to 200 million years old (fig. 2.1B).

A

B

Figure 2.1. (A) Bluntnose Sixgill Shark, a representative cow shark, which is among the most primitive known extant sharks. Note the distinct cockscomb-shaped teeth. (B) Teeth from fossil sixgill sharks (From the Natural History Museum, Bonn University; Courtesy of Ghedoghedo).

The newest and one of the most powerful methods for understanding relationships of sharks, indeed all organisms, is using the structure of their genetic material, specifically their mitochondrial and/or nuclear DNA or RNA, to work out a group's phylogeny, or evolutionary history and relationships. This molecular biology technique has revolutionized the fields of systematics and taxonomy.

How do molecular systematics and phylogenetics work? Recall that DNA is composed of nucleotide building blocks. The more closely related organisms are, the higher percentage of common nucleotide sequences they possess. The more geological time that has elapsed since species diverged from each other, the more some nucleotides have been deleted and others have been inserted, and as natural selection has acted on the new combinations, some have been retained and others eliminated. Typically, meaningful information can be achieved by sequencing around 1000 pairs of nucleotides. Once the nucleotide sequence has been obtained, the analysis and evaluation take place (fig. 2.2).

Using all of these methods, it is known that the evolutionary lineages leading to modern sharks and their relatives diverged very early on the vertebrate evolutionary tree (nearly 420 million years ago). They are among the oldest branches of jawed vertebrates for which there are still living representatives (fig. 2.3).

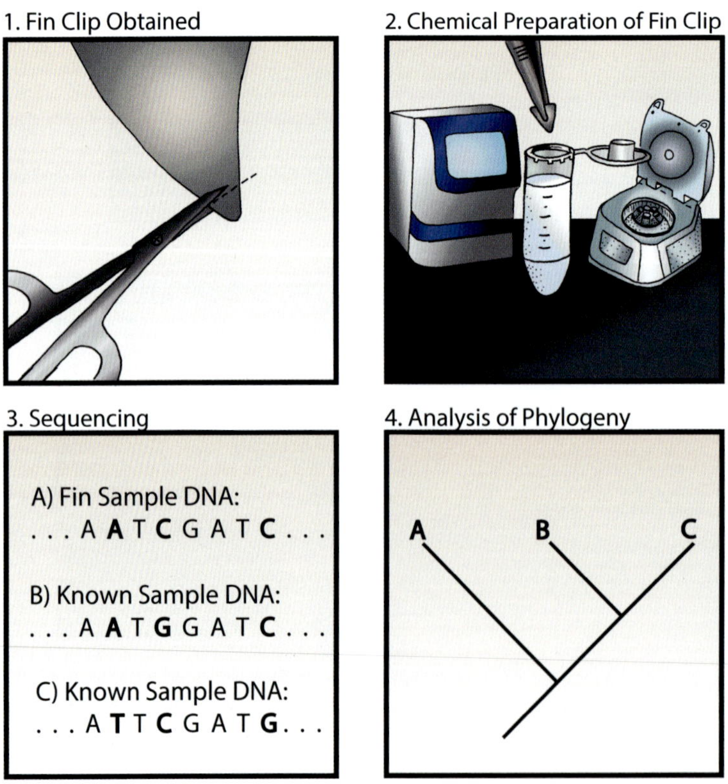

Figure 2.2. Overview of how scientists construct phylogenies based on DNA sequences.

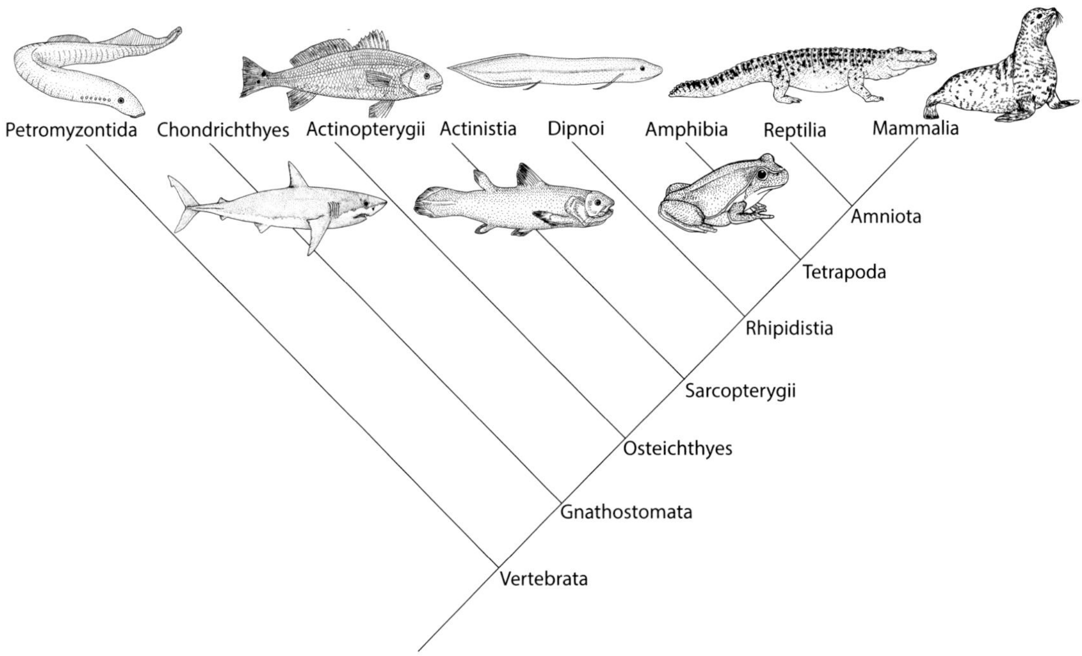

Figure 2.3. Evolutionary tree of vertebrates (Redrawn from fig. 23.1, Cracraft, J. and Donoghue, M. 2006. *Assembling the Tree of Life.* Oxford University Press).

Relationships Among Extant (Living) Forms

Let us start with a quick activity to illustrate how systematists and taxonomists—scientists who classify and name organisms—might classify sharks. Look at Figure 2.4, which is a composite of outlines of 15 different kinds of sharks and rays.

1 / Examine the outlines, and remember they are not drawn to scale. Look for a characteristic that differs among them, that is, a characteristic that some individuals possess and others do not. This characteristic could be qualitative (descriptive) or quantitative (measured or calculated using numbers).

Write down the characteristic, being as anatomically accurate as you can. If you do not know the correct term, then be as descriptive and detailed as you can. Now divide all the organisms into two groups on the basis of that characteristic.

2 / Now identify a new characteristic within each group that you could use to further divide the two initial groups into smaller groups. Recording the characteristics or anatomical basis of your divisions may be helpful.

3 / Repeat this sequence of division, characteristic identification, and subdivision until you have separated the original stack into 15 individuals.

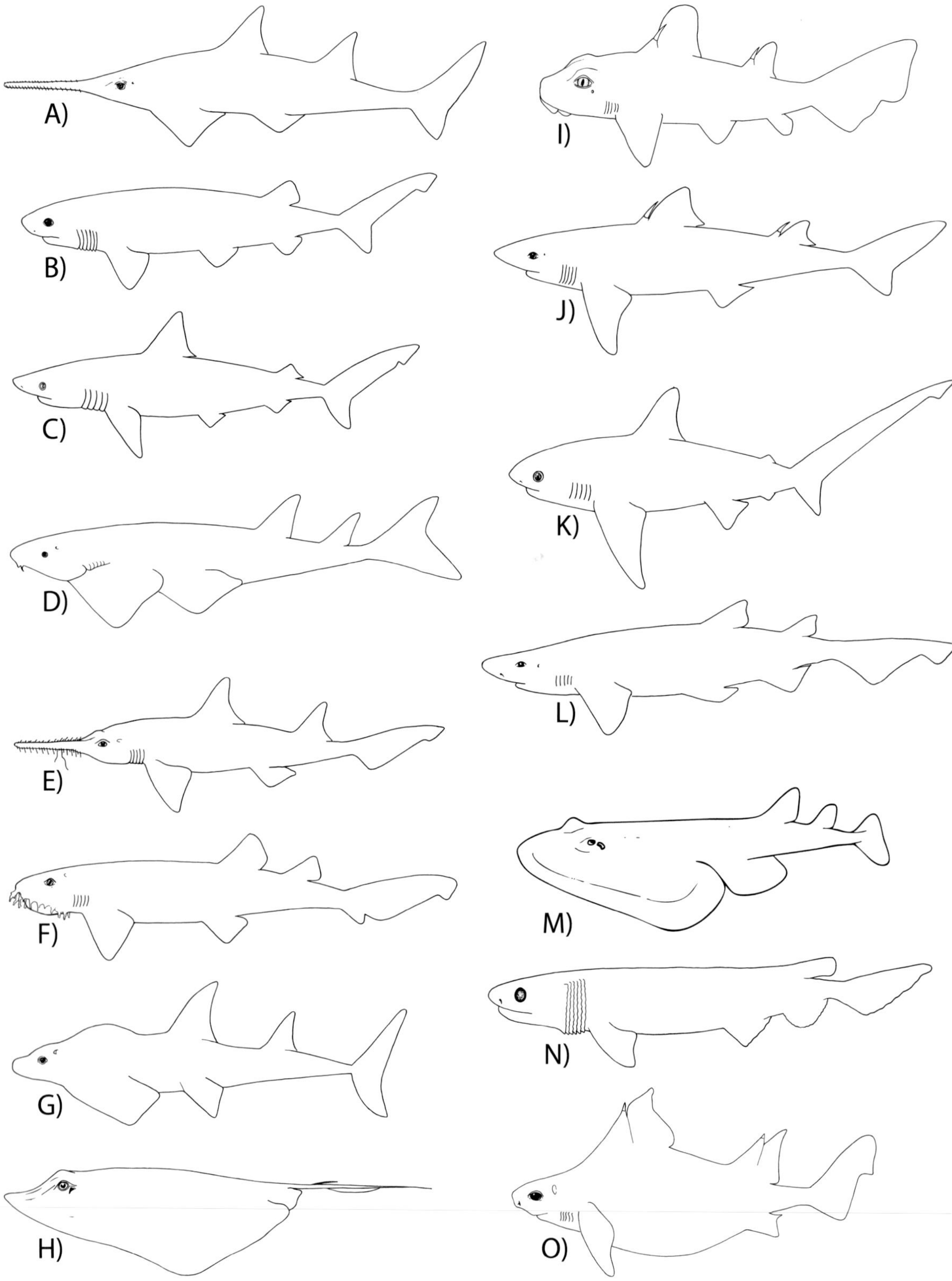

Figure 2.4. Outlines of 15 different kinds of sharks and rays (Note that M is portrayed as slightly tilted toward you). Refer to the text for instructions on how to conduct this quick taxonomic activity.

This exercise typically results in groups within groups. The deeper two groups are nested within a larger group, the more closely related you might hypothesize they are.

Over 25 years of doing this activity with our students, we have noticed some trends. Many students start by dividing the species they consider *sharks* into one group and *rays* into another, likely based on general body shape. The group you might call *sharks* was *compressed* (pushed in from side-to-side) and rays were *depressed* (flattened).

If you did use this method, then your rationale would be a good one and generally accurate, except that some sharks have evolved to live ecologically like rays, and thus to look more like a ray, and vice versa. Accordingly, you may have placed two batoids, letter A, the Smalltooth Sawfish *(Pristis pectinata)*, and letter G, the Bowmouth Guitarfish (*Rhina ancylostoma*), more commonly called the Shark Ray, in the *shark* group, and letter D, the Atlantic Angel Shark (*Squatina dumeril*), in the *batoid* group.

This raises the question: *What is the difference between sharks and rays?* According to estimates based on the rate of change in genetic sequences, batoids probably diverged from sharks nearly 230 million years ago (mya). Until recently this divergence was known as the *Great Jurassic Split* (which sounds more like an extraordinary if painful gymnastic move than a pivotal point in the evolution of elasmobranchs) because it was thought to have happened during the Jurassic, a little more than 200 mya.

It is tempting to distinguish between sharks and batoids based on a depressed body shape in batoids and laterally compressed, more spindle-shaped bodies in sharks. This difference is generally true, but it is secondary to the location of the gill slits. In sharks, the 5–7 bilateral gill slits are *in all cases* located above, or dorsal to, the well-defined pectoral fins. In rays, the 5 gill slits (6 in one species) are located on the underside of the body, below where the broadly expanded pectoral fins, which we call the wings, are attached to the body.

If you divided the organisms as we suspected, then you incorrectly grouped the Smalltooth Sawfish and Bowmouth Guitarfish with the sharks on the basis of their more *derived* body plan, and the Angel Shark with the batoids because it is flattened.

From this point, there are several different ways that you could have proceeded. Here are some of the anatomical characteristics that you may have used in creating your taxonomic tree:

1 / one or two dorsal fins

2 / presence or absence of an anal fin

3 / terminal (at the end of the snout) or subterminal (underslung) mouth

4 / head shape (tapered or hammerhead)

The identification of each of the organisms is in this note.[1] If you want to check the accuracy of your attempt to classify the elasmobranchs in this activity, jump to chapter 3 or visit the informative *Chondrichthyan Tree of Life* website.[2]

Cladistics

Grouping organisms according to external anatomy is only a first step, and it is fraught with potential wrong turns that may not group organisms together according to their *phylogeny*, or common ancestry, but rather on their superficial similarities, which may or may not reflect their phylogeny.

The taxonomic system that focuses on evolutionary relationships is known as *cladistics*, or *phylogenetic systematics*.[3] Cladistics distinguishes between characteristics that are evolutionarily older, that is, ancestral (or original), and those that are more recently evolved, or changed (or derived).

The former are known as *plesiomorphic* characteristics and the latter as *apomorphic*. An easy way to recall these terms is that the first *p* in *plesiomorphic* stands for *primitive* and the *a* in *apomorphic* stands for *advanced*. You should be cautioned that this does not imply that an advanced state is necessarily more adaptive or of higher value than a primitive one.

You may be more familiar with the terms *homology/homolog* and *analogy/analog*. Traditionally, *homologous* traits are the traits due to common ancestry, and *analogous* traits are the results of convergent evolution.

Here is one example. Shortfin Makos (*Isurus oxyrinchus*) and Albacore Tuna share numerous adaptations that make them high-performance predators (e.g., fusiform, or torpedo-shaped body; keeled caudal fin; elevated body temperature), but since the former is a chondrichthyan and the latter an osteichthyan, these common characteristics are the result of convergent evolution and are thus *analogous* traits. In other words, these species did not inherit their similar traits from a common ancestor. Rather, their evolution independently converged on a similar body type.

Within the elasmobranchs, plesiomorphies include a simple jaw structure, tribasic fins (with three rigid cartilages at the base), and claspers on the inner margins of males' pelvic fins. These characteristics, in other words, have been retained from ancestors.

One example of an elasmobranch apomorphy is something that is absent: an anal fin in the shark order Squaliformes, a group of seven families that includes the dogfish and bramble sharks (fig. 2.5). Another elasmobranch apomorphy, from the batoids, is the presence of a tail spine, which is a modified scale, in members of the order Myliobatiformes, informally called stingrays.

When constructing phylogenies, it is critical to know what traits are shared by group members because of inheritance from an ancestor common to all the group members. Apomorphies that are shared by taxa within a group and their most recent common ancestor are called *synapomorphies* (*syn* = together, with).

Figure 2.5. An apomorphic (more recently evolved, or derived) characteristic of elasmobranchs, loss of the anal fin in the Little Gulper Shark (order Squaliformes).

Thus, synapomorphies are derived traits shared by organisms in the same taxonomic group and are therefore very useful in establishing phylogenies and relationships among extant organisms.

Synapomorphies of chondrichthyan fishes (elasmobranchs + chimaeras, or ratfish) include (see fig. 1.4):

- Skeleton composed of cartilage, some of which is prismatically calcified (also known as *tesserated*[4] cartilage). According to a 2012 review paper[5] on the subject, this characteristic is ***the*** (emphasis theirs) *critical defining character of this group.*
- Pelvic claspers as intromittent (copulatory) organs
- Teeth not firmly affixed to jaws, with serial replacement
- Chondrocranium (braincase) without sutures (seams)
- Soft fin rays (ceratotrichia) that are unsegmented

Taxa united by synapomorphies form *monophyletic* groups called *clades.* In other words, a monophyletic clade consists of a single ancestral species and *all* its descendants. It is often helpful to diagram the relationship between different clades because some clades are more closely related to one another than others. Such a diagram is called a *cladogram.*

For a group, such as a taxonomic family or order, to be *monophyletic,* all of the species *belonging* in that group should be *classified* within that group and not be mistakenly classified in another group (fig. 2.6). Additionally, no species from other groups can be spuriously included within the group in question. Establishing interrelated monophyletic groups supported by synapomorphies is the goal of cladistics.

If one or more of the descendants of an ancestor are missing from the group, say, because we simply do not know all of the details yet, the group is not considered monophyletic but rather is *paraphyletic* (*para* = near; fig. 2.6). Reptiles are considered a paraphyletic group if birds are erroneously excluded (as they were until recently), because birds and reptiles share a common ancestor and should have been classified together.

If a group is *polyphyletic* (*poly* = many; fig. 2.6), it means that it wrongly does not include the *common* ancestor but rather ancestors common to *some* of the members of the group. An example of polyphyly is classifying mammals and birds in the same group separate from the reptiles, as was also done in early taxonomies, because both are endothermic (warm-bodied).

Thus, paraphyletic and polyphyletic taxonomies are wrong. They usually arise as educated best guesses based on incomplete information.

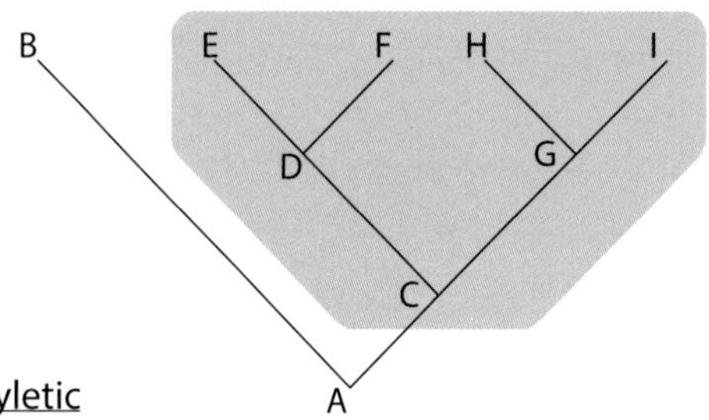

Monophyletic
A single ancestor and all descendants.

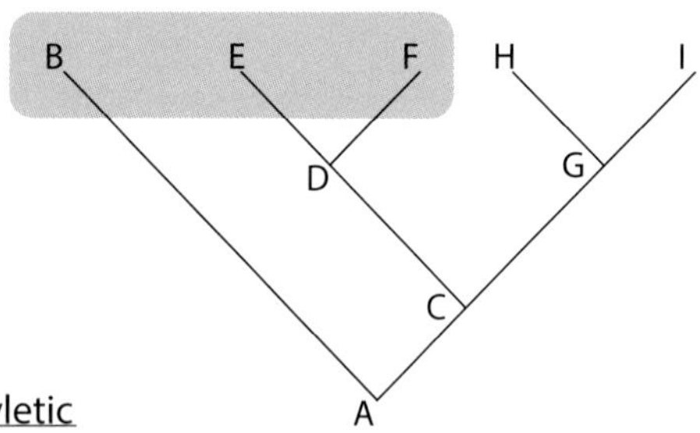

Polyphyletic
Descendants that are related, but are derived from two or more ancestors.

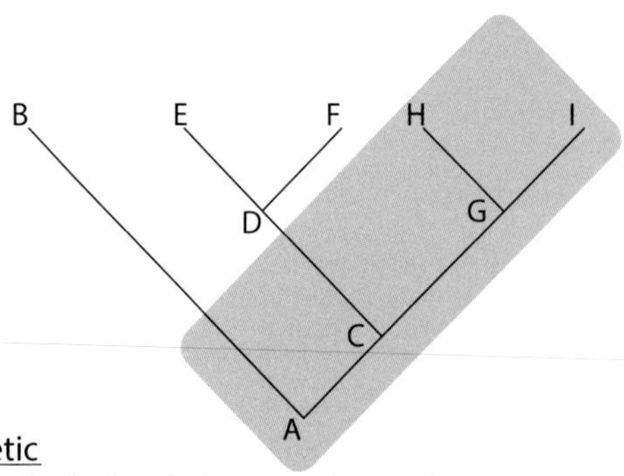

Paraphyletic
Single ancestor, but excludes some descendants.

Figure 2.6. Cladograms showing monophyletic, paraphyletic, and polyphyletic groups.

While these terms may seem confusing, once you begin observing and applying the concepts, you will become more facile with them.

Here is an example. As long as we can remember, the scientific name for the most common coastal stingray of the SE Atlantic Coast and Caribbean Sea, the Southern Stingray (fig. 2.7), has been *Dasyatis americana*. The 97 known species of whip-tail stingrays, including the Southern Stingray, are in the family Dasyatidae, which has been thought to be a monophyletic group. That means that these 97 species would be more closely related to one another than to other rays, because they all have the same common ancestor.

However, a recent study[6] employed a combination of morphology and molecular analyses and concluded that most of the genera were *not* monophyletic, and that the genus *Dasyatis* was polyphyletic, requiring the revision. In other words, these species were all related but not as closely as once thought and were derived from more than a single common ancestor. As a result, an entirely new classification was created, and the Southern Stingray was renamed *Hypanus americanus* to distinguish it from other rays that we now know are not as closely related as previously thought.

Gnathostome Phylogeny: The Evolution of Jaws

Before we describe the evolution of the chondrichthyan fishes, we want to emphasize a preceding evolutionary development of singular significance: the evolution of jaws.

Figure 2.7. The Southern Stingray, formerly known as *Dasyatis americanus*, but now called *Hypanus americanus* based on new morphological and molecular analyses (Courtesy of Trey P. Spearman).

Five hundred million years ago, the oceans were dominated by jawless fishes, or *agnathans* (*a* = without + *gnathos* = jaws). One evolutionary lineage of agnathans still has living representatives, the lampreys and hagfish. Today, there are about 124 species of lampreys and hagfish (fig. 2.8A). They have persisted for about 500 million years, making their clade, the cyclostomes (*cyclo* = round + *stome* = opening), ancient but not nearly as biodiverse as the approximate 34,000 bony fishes and 1250 elasmobranchs seen today.

Early agnathan fishes that are collectively called "*ostracoderms*" (meaning *shelled skin*; fig. 2.8B) represent important characters in the evolutionary story of jaws. "Ostracoderms" got their name from the plates of dermal bone that covered

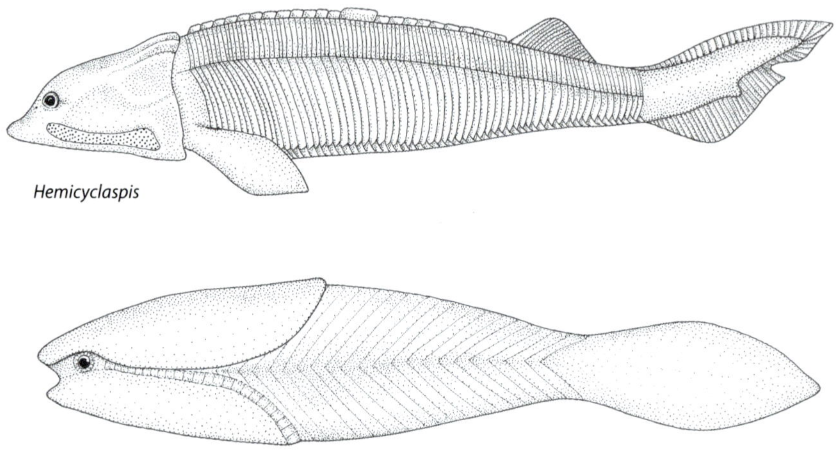

A

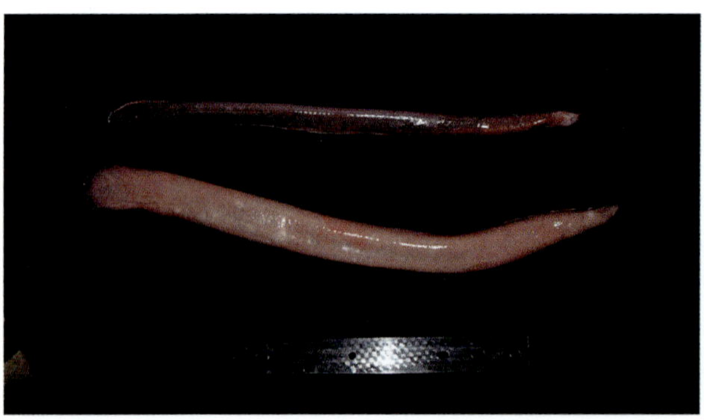

B

C

Figure 2.8. (A) Ostracoderms, early jawless fishes with dermal bone covering their heads. (B) Two species of hagfish. (C) *Cardipeltis bryanti*, a fossil ostracoderm from the Devonian of Wyoming (Courtesy of James St. John).

their heads (fig. 2.8C). Dermal bone, like many of the bones of our skulls, grows between layers of tissue and not from a template of cartilage. The quotation marks around the name "ostracoderms" indicate that it is a poly- or paraphyletic group. Fossil evidence tells us that first "ostracoderms" likely arose about 480 mya, in the early Ordovician Period. By the Silurian Period, 416 to 443 mya, multiple lineages of "shelled skin" jawless fishes had evolved and diversified.

Although "ostracoderms" lacked jaws, they did have mouths, which they used to consume small, soft-bodied prey. Without jaws, though, they were probably limited in the size and type of prey they could consume. For researchers studying the evolution of jaws, one group of "ostracoderms" stands out, the *osteostracans.* These fishes had heterocercal caudal fins, paired pectoral and pelvic fins, and a well-defined head, which likely made osteostracans effective and efficient swimmers. Presumably, osteostracans inherited these traits from an ancestor. If only that ancestor had jaws, imagine the benefit! Osteostracan "ostracoderms" are represented in Figure 2.8A by *Hemicyclaspis.*

As luck would have it, an evolutionary offshoot of that ancestral species did give rise to the first fishes with jaws. Jaws are essentially highly modified skeletal arches like the series of arches that support gills. When genes associated with segmentation of a developing vertebrate head were not expressed (that is, translated into proteins) during development, *mandibular arches* formed in the first *gnathostome* fishes (*gnath* = jaws; *stome* = opening). Mandibular arches are the developmental template for jaws, and in early gnathostomes and sharks, those mandibular arches were structurally supported by modifications of the second arch in the series, the *hyoid arch* (fig. 2.9).

Why would natural selection favor the modification of mandibular arches into jaws? It helps to consider the advantages of jaws. First, grasping became possible. Second, once grasped, prey or parts of prey could be handled or manipulated, cut, and further processed. This expanded the menu of our early gnathostome ancestors considerably, opening up new sources of nutrients and ecological niches. Once the first jaws evolved, natural selection went to work, modifying jaws into the myriad forms seen in cartilaginous and bony fishes as well as tetrapods. In many ways, the evolution of jaws marks a turning point not just in the story of sharks but in the story of vertebrates, including humans!

Overview of the Evolution of Chondrichthyans

While jawed fishes, including the species that gave rise to sharks and their relatives, existed before the Devonian Period (360 to 420 mya), it was during this 60-million-year period that bony fishes diversified, so much so that the Devonian is known as the *Age of Fishes.* It is thought that the increased abundance of warm, shallow seas led to a profusion of biogenic reefs (formed by animals such as sponges and extinct coral relatives), which in turn led to diversification of bony fishes by providing innumerable living spaces and abundant food, which are ecological

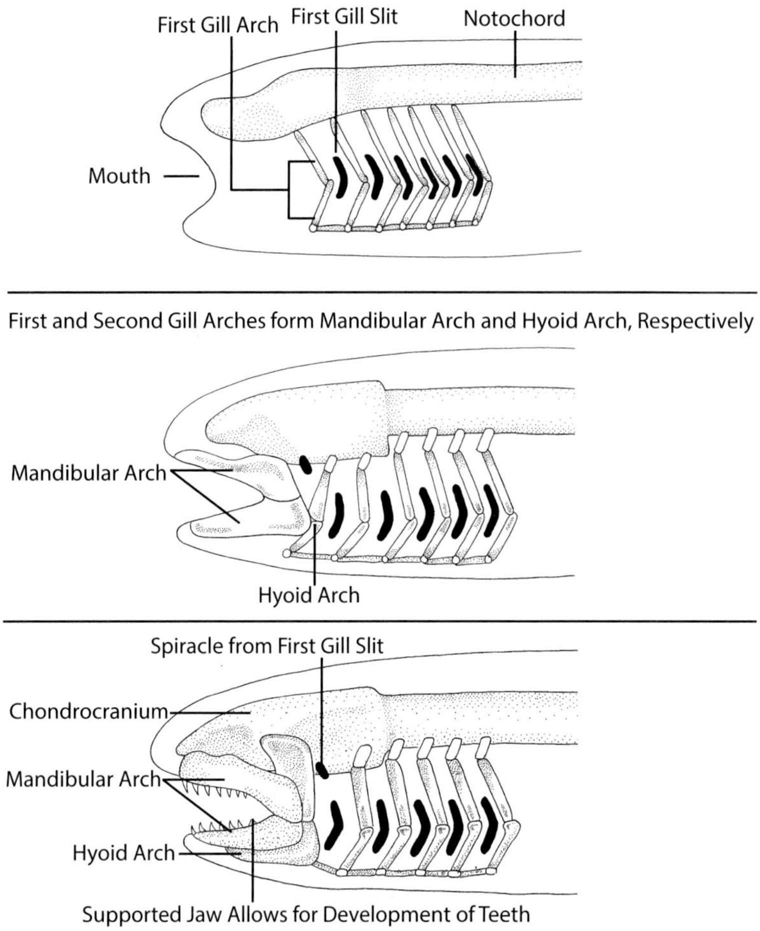

Figure 2.9. The evolution of jaws from a set of skeletal arches in the same series as the arches that support the gills.

roles that current coral reefs also perform. All major clades of fishes arose during this period.

Early sharks also swam in Devonian seas (fig. 2.10). Today, their fossils are found in Devonian deposits, but they were not nearly as diverse or numerous a group as the bony fishes in the clade Osteichthyes. Two other groups of interest flourished during the Devonian: the "placoderms" and "acanthodians." Both of these groups have at one time or another been considered possible ancestors of the chondrichthyans (see "Who are the Ancestors of Sharks?"). The Devonian ended with a mass extinction event (or, more likely, a string of extinction events over 20–25 million years) that killed between 70–80% of marine species, especially those in shallow, tropical waters, likely due to extensive hypoxia and anoxia (low or no dissolved oxygen).

The Carboniferous Period (300–360 mya) followed the Devonian. It is often referred to as the *Golden Age of Sharks*, and for good reason. Chondrichthyans diversified considerably. If we expand our understanding of the term shark to

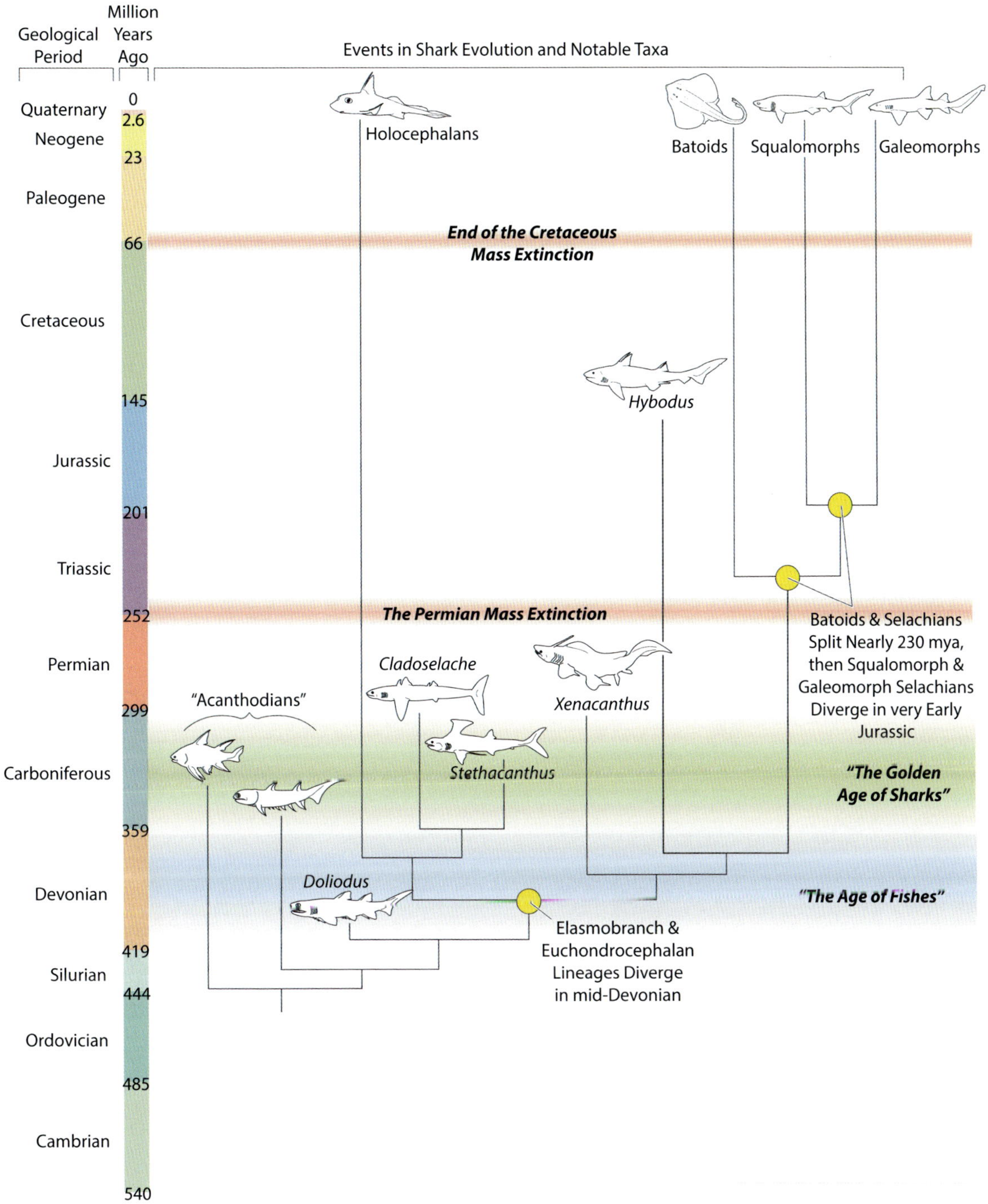

Figure 2.10. Major events in chondrichthyan evolution and the phylogenetic relationships between major clades. Estimates of clade divergences are based on Coates *et al.* 2018. Proc. R. Soc. B. 285: 20172418, Achilman *et al.* 2012. Mol. Phylogenetics Evol. 63: 28-42., and Maisey, J.G. 2012. J. Fish. Biol. 80: 917-955.

include their extinct and extant sister taxa in the clade Euchondrocephali (*eu* = true + *chondro* = cartilage + *cephali* = head), then we see some truly remarkable and bizarre sharks (or, more broadly, chondrichthyans) unlike any in the ocean today. For example, the Damocles Shark (*Damocles serratus*), which modern researchers hypothesize was more closely related to ratfish (see an overview of holocephalans in chapter 3) than modern sharks, had a large, curved spine with bristles that hung over the heads of male specimens (fig. 2.11A). The placement of this structure is reminiscent of the mythological sword of Damocles, which hung over the ancient Greek's head as he sat upon the throne.

Other fascinating examples of bizarre prehistoric sharks of the Carboniferous are those of the order Iniopterygiformes, also called iniopterygians. These small euchondrochephalans had long, broad pectoral fins with spiny denticles on the leading edges (fig. 2.11B). It is unclear how these small sharks eked out a living. Were they benthic (associated with the sea floor) or pelagic (living higher in the water column)? Some researchers point out that modern fishes with similar pectoral fins can leap out of the water and glide for considerable distances. We call them flying fish, and in the Carboniferous, it is possible that iniopterygians were flying sharks.

Damocles sharks, iniopterygians, and their relatives have all been unearthed in the Bear Gulch Limestone of Montana. This formation represents a fossil

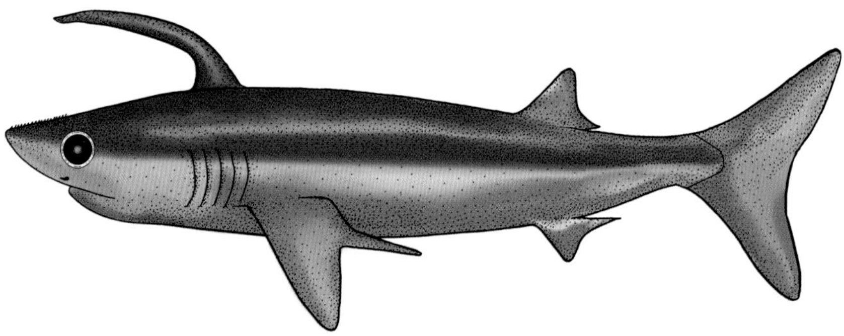

A *Damocles*

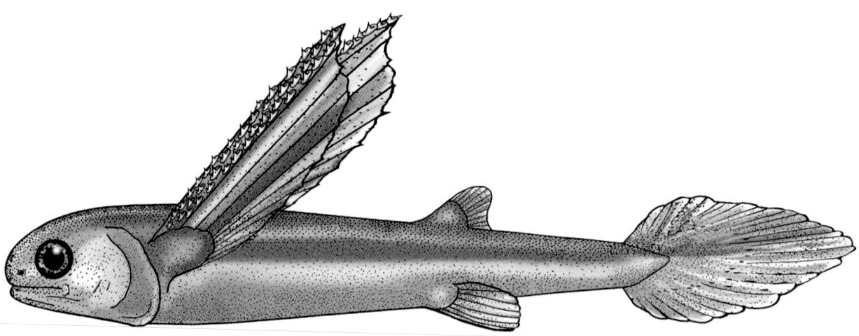

B Iniopterygiformes

Figure 2.11. (A) Damocles Shark (*Damocles serratus*) with a bristly spine that hung over its head. (B) Iniopterygiformes had long, wing-like pectoral fins.

lagerstätte, a German word that means "storage place." Fossil *lagerstätten* (plural) are sites of exceptional fossil preservation. The fossil deposits of Bear Gulch date back to about 323 mya. The former bay, some of whose inhabitants are preserved as fossils in the limestone, existed for only about 1000 years. Within the approximately 27 m (90 ft) thick system of sediments, over 65 species of sharks were discovered. By the end of the Carboniferous and throughout the following Permian Period, sharks diversified, and several groups of sharks occupied saltwater and freshwater habitats. It is our pleasure to refer you to *Research, Conservation, and Education Focus* Box 2.1, in which paleontologists Richard Lund and Eileen Grogan discuss the importance of the Bear Gulch *lagerstätte* and what it tells us about shark evolution based on their decades of painstaking scientific exploration.

The Permian Period ended about 250 mya, with one of the worst mass extinctions reflected in the fossil record. During the *Permian Mass Extinction*, which is also called the *Great Dying*, 90–96% of all marine species (including chondrichthyan fishes) were wiped out, as well as 70% of terrestrial life.

What caused the Great Dying? Many experts point to a runaway greenhouse effect that altered Earth's atmosphere. As massive volcanic eruptions in modern day Siberia released volcanic magma into oil shales, enormous forest fires and lava fields burned out of control for nearly 180,000 years. Levels of the greenhouse gases CO_2 and CH_4 (methane) spiked. For marine organisms, that meant ocean acidification, increases in ocean temperature, and widespread anoxic conditions. The ocean thus did not contain enough oxygen to meet animals' metabolic demands. With this in mind, it is not surprising that many of the chondrichthyan lineages that survived the Permian extinction were represented by species that lived in deeper waters. There, the temperature was cooler, and the water could carry more dissolved oxygen.

Following the Permian Mass Extinction, the Triassic Period (approximately 200 to 250 mya) saw the proliferation of sharks in the clade Neoselachii (*neo* = new + *selachii* = sharks), which includes sharks, skates, and rays (fig. 2.10). Neoselachians are often called "modern sharks" because they share many of the same anatomical features of sharks we see today. For example, neoselachians, apart from skates and rays, have three layers of mineralized tissue called enameloid covering the crowns of their teeth. Neoselachians also have hyostylic jaw suspension (see chapter 4), and notochords constricted by vertebral centra. Neoselachians diversified considerably throughout the Triassic, Jurassic, and Cretaceous Periods. At the end of the Cretaceous nearly 66 mya, another mass extinction event occurred, one with which dinosaur lovers are well-acquainted.

The end-Cretaceous extinction was likely caused by a 10 km (6.2 mi) asteroid slamming into the northern coast of what is now Mexico's Yucatán Peninsula, as well as widespread volcanic activity. Over 40% of elasmobranch genera and 45% of shark species became extinct (and 75% of all marine species). The ecological impact was severe. After this extinction, movement of the Earth's crustal plates created new continental margins in relatively warm climates in temperate and tropical

BOX 2.1

Research, Conservation, and Education Focus

Relations of Extant and Paleozoic Chondrichthyes

Richard Lund and Eileen D. Grogan

Carnegie Museum of Natural History & Saint Joseph's University
Philadelphia, PA, US

The research team of Richard (Dick) Lund and Eileen D. Grogan have collected 5966 fish specimens from more than 100 quarry sites within the Bear Gulch lagerstätte (sedimentary deposit). This deposit in current day Montana preserves the remains of a Paleozoic (Mississippian) tropical marine bay. The quarry sites sample areas from the headwaters to the mouth and from shore to shore of the original Paleozoic bay. Virtually all of the recovered fish fossils are new taxa, illustrating a level of diversity and specialization otherwise not evident from other deposits of this age, in great part due to the high quality and fine detailed preservation of the fossilized organisms. This preservation *and* intensive sampling has provided data relating to ontogenetic change, sexual dimorphism, mating behavior and reproductive strategies for multiple taxa.[1] The resulting dataset of laboratory and field information (collected in context for each specimen) allows detailed analyses of the fish, their habitats, community structure and their co-variations around the bay.

What made the Bear Gulch Limestone deposit a lagerstätte? Conditions at the time of deposition and infilling of the bay often favored retention of soft-tissues, skin pigment, and gut content information! The highest quality preservation resulted from storm-raged monsoonal seasons prevailing upon the Paleozoic bay and its inhabitants.[2] During such a season, fish were trapped in a sheet-wash of fresh water and sediments which, simultaneously, suffocated and buried them. In the process, lack of oxygen and lack of predation helped preserve the entire fish and, in some cases, also preserved information on fish behavior and mating habits.

The majority of Bear Gulch fossil fish are either bony (Osteichthyes) or cartilaginous (Chondrichthyes) fish. Of these, the chondrichthyans were the most morphologically diverse. The extent to which Bear Gulch chondrichthyans experimented in their morphologies challenges one to recognize some taxa as cartilaginous fishes given our images of extant forms. Therein lies a great challenge working on the classification of extinct fishes relative to extant forms. The range of Bear Gulch chondrichthyan diversity precludes many taxa from assignment to either Elasmobranchii (sharks, skates and rays, or Holocephali [chimaeras], the subclasses into which all living chondrichthyans are assigned).

Bear Gulch chondrichthyans *do* possess the morphologies diagnostic of the Class Chondrichthyes (principal dentitions on the palatoquadrate and Meckel's cartilage, internal fertilization via intromittent organs derived from pelvic fin axes, and cartilage endoskeleton). However, Bear Gulch and other Paleozoic chondrichthyans present an assortment of transitional morphologies between the elasmobranch and holocephalan subclasses.

Modern elasmobranchs are characterized, in part, by separate external gill openings for each branchial chamber, a palatoquadrate not fused to the neurocranium and supported by a dorsal hyoid cartilage (hyomandibular) derived from the following gill arch, and multiple tooth families. Modern holocephalans are characterized by a single opercular gill chamber cover, a palatoquadrate fused to the neurocranium, a hyoid arch not involved in jaw suspension, and few ever-growing dental plates. So, how are these morphologically disparate lineages united? Are there Paleozoic forms that bridge the gap and retain a monophyletic Chondrichthyes? Do Bear Gulch chondrichthyans provide some resolution?

The higher taxonomic groups of Bear Gulch chondrichthyans are currently identified as Elasmobranchii and Euchondrocephali, with the unique Iniopterygii positioned as an early stem offshoot of the Class Chondrichthyes.[3] Fundamental *suites of characters* establish a firm divide between the two clades. Like modern elasmobranchs, the Paleozoic forms are characterized by jaws stabilized against the cranium via a hyomandibula and separate gills positioned in succession behind the cranium and anterior to the pectoral girdle. In contrast, the Euchondrocephali have their gill supports arranged in a nested fashion, positioned slightly ventral to the hind region of the cranium, and covered by a soft operculum. The upper jaws

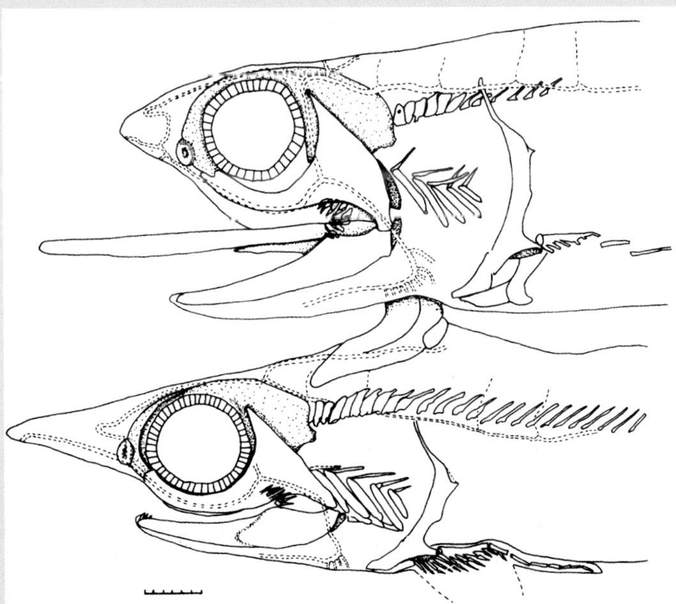

These male and female *Falcatus* represent the quality of many of the fossils recovered by Dick Lund and Eileen Grogan in Bear Gulch. The male is recognizable by his large, modified fin spine. The position of the animals has led many to speculate that these shark-relatives died as they prepared to pass their genes on to a new generation.

are either autodiastylic and, thereby, suspended from the cranium by two processes or, alternatively and in a derived condition, with the upper jaws fused with the cranium. The latter, holostyly, is a feature found in extant chimaeroid chondrichthyans. Whether possessing autodiastylic or holostylic jaws, the hyoid arch of euchondrocephalans remains independent of the jaws, instead charged with supporting the soft opercular covering of the nested gills. The euchondrocephalan pectoral girdle is positioned close to the rear of the cranium, in contrast to the more posteriorly positioned elasmobranch girdle.

While about 50% of Bear Gulch chondrichthyan taxa have been formally published to date, the collective data of *all taxa* reveals the majority of chondrichthyans to be euchondrocephalan. The plesiomorphic autodiastylic forms provide a closer relationship to the more derived Holocephali. The Euchondrocephali is the sister-group to the Elasmobranchii and, thus, provides support for a monophyletic fossil and extant Chondrichthyes.

1. Lund, R. 1990. Environ. Biol. Fishes 27: 1–19.

2. Grogan, E.D. and R. Lund. 2002. Geobios. 24(2): 295–315.

3. Grogan, E.D. *et al.* 2014. In: Carrier, J. *et al.* (eds.) The biology of sharks and their relatives. 2nd Edition. CRC Press.

latitudes. These shallow, warm waters provided conditions ideal for the development of seagrass, salt marshes, mangroves, and coral reef communities. Coincident with this was a similar proliferation of bony fishes, which took over marine systems.

Who Are the Ancestors of Sharks?

By now you may have noticed that the word shark has been applied to cartilaginous fishes in multiple clades. One reason this word has been applied so broadly is that many fossil fishes exhibit characteristics that we associate with sharks. Examples of such traits include streamlined bodies with heterocercal tails, cartilaginous skeletons, pelvic claspers, fin spines on dorsal fins, rows of teeth, and prominent jaws. Some fishes exhibited enough of these shark-like characteristics that the shark moniker was applied and stuck.

Fossil fishes called "acanthodians" (*acanth* = spine) first appeared in the Silurian and diversified in the Devonian. They are often referred to as *spiny sharks*. They did bear a superficial resemblance to sharks, with prominent dorsal fin spines; they also had cartilaginous skeletons, so the name seems justified. But are we right to call them sharks?

Unlike sharks, "acanthodians" also had spines associated with their pectoral and pelvic fins. Some "acanthodians," like *Climatius* (fig. 2.12), had pre-pelvic spines on their bodies between their pectoral and pelvic fins. Many "acanthodians" lacked the files of teeth we typically associate with sharks, although some did have shark-like dentitions.

Researchers have long debated where "acanthodians" fit in the vertebrate tree of life. At one point, "acanthodians" were proposed as the sister taxon of bony fishes. Now, they are widely considered a stem group of chondrichthyans, meaning they are very early representatives of that evolutionary lineage. If every fish that has ever been labeled an "acanthodian" were included in this group, it would almost certainly be paraphyletic.[7] Recent technological advances in the fields of biomedical imaging and histology have helped sort out closely related "acantho-

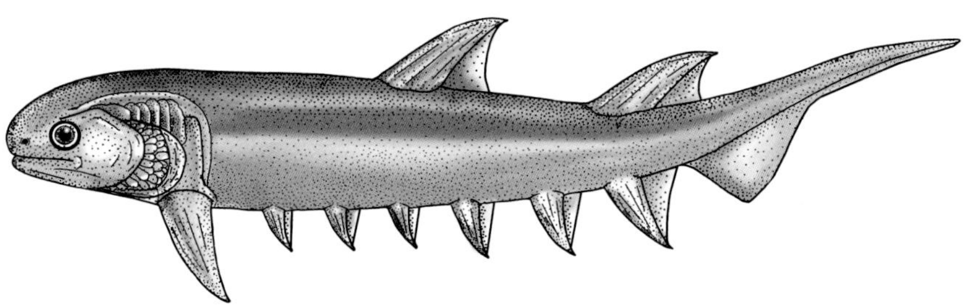

Climatius

Figure 2.12. Acanthodian fishes bore prominent prepelvic spines as seen in the acanthodian genus *Climatius*.

dians" and support their position as stem chondrichthyans with evolutionary origins dating back more than 440 million years.

At this point it is worth mentioning another group of early gnathostomes that was once thought to have given rise to sharks, the "placoderms" (*plac* = plate; *derm* = skin). Similar to the jawless and paraphyletic ostracoderms, "placoderms" had large plates of dermal bone around their heads, but unlike the ostracoderms, "placoderms" had jaws. This is probably one reason "placoderms" flourished and diversified between their origins 440 mya in the Silurian and their extinction approximately 359 mya at the end of the Devonian or early Carboniferous (fig. 2.13).

For a long time, "placoderms" were considered monophyletic, united by their bony headshields. More recently, researchers have suggested that "placoderms" represent paraphyletic clusters of species that branched off the gnathostome lineage. Just how many clusters of placoderm species there are is an active area of research and debate in paleontological communities, but their paraphyly is a widely accepted hypothesis.

Why did paleontologists once hypothesize that "placoderms" where the immediate ancestors of sharks? It helps to consider two groups of "placoderms," the *arthrodires* and *ptyctodonts*. Male representatives of both groups had claspers, though the claspers of "placoderms" were not attached to the pelvic fins the way shark claspers are. This exemplifies convergent evolution. Ptyctodonts had reduced bone, and fossils of the ptyctodont *Materpiscis* (*mater* = mother +

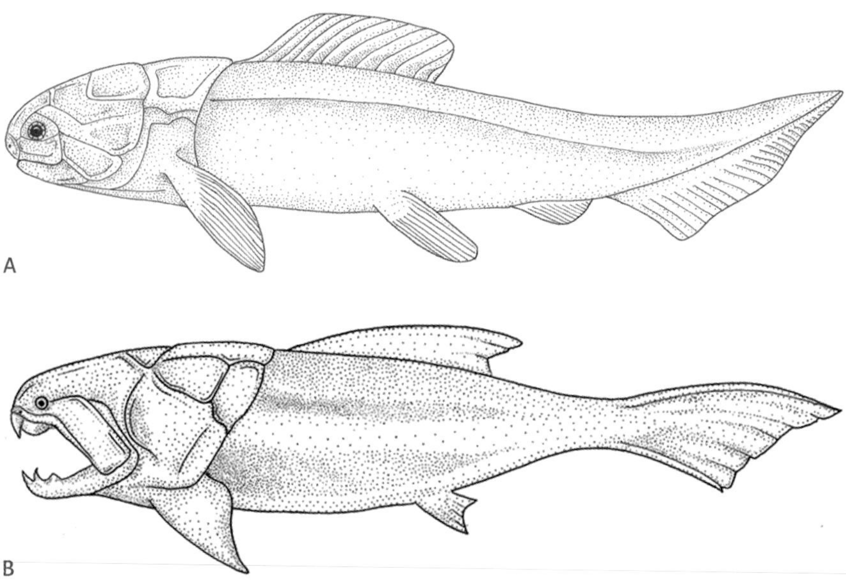

Figure 2.13. (A) A representative placoderm. (B) *Dunkleosteus*, illustrated to show size and body shape (Both redrawn from Maisey, J. 1996. *Discovering Fossil Fishes*. Holt).

piscis = fish) even show preserved embryos inside the mother, suggesting viviparity (live birth).

Arthrodire placoderms included the six-meter-long (about 20 feet) *Dunkleosteus*. These placoderms showed many shark-like traits (fig. 2.13B) including large jaws, though the jaws of *Dunkleosteus* were covered in dermal bone forming *gnathal plates*. *Dunkleosteus* also had a heterocercal tail and a cartilaginous skeleton that resembled the skeletons of modern sharks. Despite these superficial similarities, "placoderms" do not represent the most immediate ancestors of sharks, and the discovery of transitional fossils bridging the gap between "acanthodians" and sharks help to rebut the placoderm-to-chondrichthyan hypothesis.

Doliodus latispinosus: A Transitional Fossil

In 1997, a 23 cm (9 in) section of a fossil fish was unearthed from mudstone in Canada. When the paper describing the specimen was published in 2003,[8] it turned the chondrichthyan paleontology world upside down. The fish was assigned to the genus *Doliodus*, a taxon previously known exclusively for teeth and portions of fin spines. What distinguished this specimen from congeners (members of the same genus) and indeed most other chondrichthyan fossils of that age was the relative completeness of the articulated specimen. Most of the hard anatomical parts were present and preserved close to their original orientation in life, an extremely unusual and lucky find.

The 1997 specimen of *Doliodus* was originally assigned to the species *problematicus*, though subsequent study revealed that fin spines from the same species were discovered earlier and attributed to *Ctenacanthus latispinosus*,[9] which resembled an acanthodian fish, with a cylindrical body and terminal mouth and was so classified. According to the rules of taxonomy, the revised genus identification sticks, but the older species name takes priority. Thus, the remarkable specimen is properly referred to as *Doliodus latispinosus*. While this may seem like a trivial matter, the naming history highlights the significance of the specimen.

Doliodus latispinosus was also first identified as an acanthodian, hence its original, but later revised, assignment to the acanthodian genus *Ctenacanthus*. At the time, it was perfectly reasonable because the fin spines resembled those of other acanthodians. When the more complete specimen was found in 1997, researchers noted that the species also had shark-like characteristics that differed from many acanthodians, such as jaws full of teeth arranged in discrete tooth files and a pectoral girdle, the skeletal structure supporting the pectoral fins, that more closely resembled the more robust pectoral girdle of a shark (fig. 2.14). Therefore, *Doliodus latispinosus* is a *transitional fossil*, demonstrating a link between the "acanthodians" at the stem of the chondrichthyan lineage and the more derived sharks and their relatives.[10]

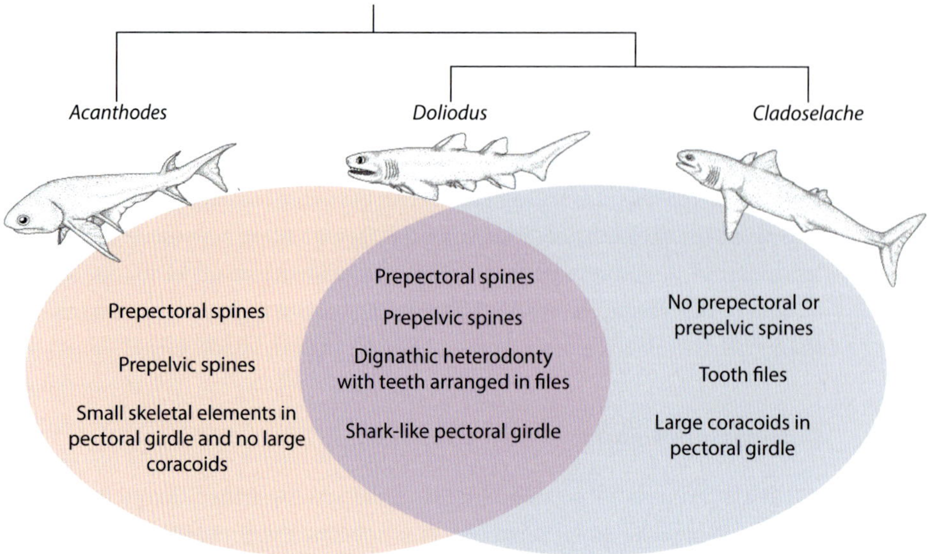

Figure 2.14. *Doliodus* exhibited some acanthodian characteristics and some characteristics associated with more derived chondrichthyans. This Venn Diagram identifies acanthodian and shark-like traits of *Acanthodes*, *Doliodus*, and *Cladoselache*.

Notable Fossil Sharks

Piecing together the evolutionary history of chondrichthyans and resolving the relationships between extinct sharks is no easy task. A fossil record that is relatively poor compared to that of bony fishes and tetrapods complicated matters. Why is the shark fossil record so poor by comparison? Cartilage, the defining characteristic of chondrichthyan skeletons, does not fossilize well under most conditions. This unfortunate reality has several major implications. The first is that most of the fossil record for chondrichthyans is limited to scales, fin spines, and teeth. The second implication is that fossil *lagerstätten* like Bear Gulch, in which conditions did favor more complete fossilization, take on important roles in our understanding of shark evolution. However, they only represent a snapshot in a long history. The third implication is that species represented by complete or unusual fossils tend to factor more prominently in accounts of shark evolution. Let us familiarize ourselves with some of those noteworthy species and consider the current state of knowledge for these taxa. As we go, it may be helpful to track our progress on Figure 2.10 which shows the divergence of many of these taxa.

Cladoselache and the Euchondrocephalans

One genus of extinct sharks is widely hailed as containing the first "true" shark, *Cladoselache*. These sharks first appeared in the Devonian Period more than 370 mya and likely went extinct by the early Carboniferous Period. *Cladoselache* grew

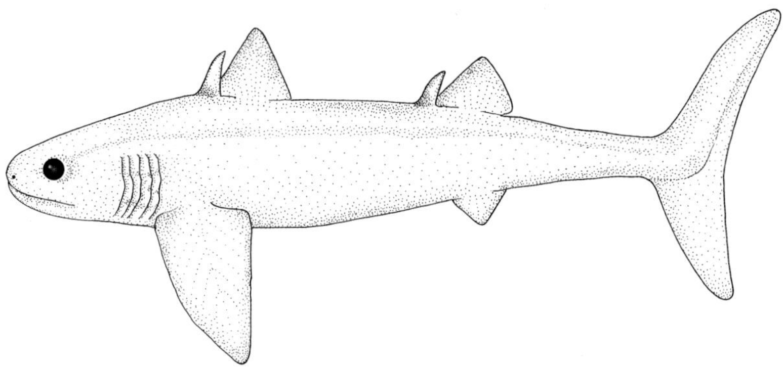

Figure 2.15. *Cladoselache*, an early euchondrocephalan (Redrawn from Maisey, J. 1996. *Discovering Fossil Fishes*. Holt).

to about 2 m (6.6 feet) in length and had a streamlined, shark-like body with two dorsal fins, a broad caudal fin, and robust pectoral fins, all of which suggest that it was a fast-swimming pelagic predator (fig. 2.15). Chronologically, its existence overlapped with that of the bus-sized placoderm *Dunkelosteus*.

The jaws of *Cladoselache* were suspended under their braincases in close association with the hyoid arch and had limited jaw mobility (termed *amphistylic* jaw suspension; see chapter 4). Unlike most modern sharks, *Cladoselache* had a terminal mouth, meaning it was positioned at the front end of the snout rather than under it. Today, only a handful of sharks have terminal mouths (e.g., the frilled sharks in the genus *Chlamydoselachus* and the Megamouth Shark *Megachasma pelagios*).

In 2023, a team of researchers described a new genus and species of Devonian shark from the deserts of Morocco.[11] This shark had many of the same anatomical traits as *Cladoselache*, including pectoral fins with broad, flat internal structures called radials supporting the fin, a cleaver-shaped upper jaw, and teeth with multiple cusps on either side of a large, pointed cusp. This new fish was given the name *Maghriboselache mohamezanei*, and it is considered the sister taxon to *Cladoselache* within the family Cladoselachidae.

Were these really the first true sharks? Yes and no. For much of the twentieth century, *Cladoselache* was widely considered an elasmobranch, a classification that all modern sharks and batoids share today. By the end of the 1990s and into the early 2000s, an alternate hypothesis gained traction, based on the braincase. The proportions, processes (in anatomy, a *process* is an outgrowth of a structure), and the distribution of holes through which nerves and blood vessels could enter the braincase of *Cladoselache* were very similar to other chondrichthyans in the order Symmoriiformes.[12] The catch is that symmoriiforms are grouped into the larger clade Euchondrocephali, meaning that *Cladoselache* would be more closely related to ratfish than any modern shark. But as we have seen, the word shark has been applied broadly across a variety of species. One thing is certain, *Cladoselache* was more shark-like than *Doliodus* and the "acanthodians."

Despite its prominence in the story of shark evolution, there are still mysteries to be solved about *Cladoselache*. For example, why have researchers never found a specimen with claspers? Did *Cladoselache* lack claspers? Or perhaps *Cladoselache* sexually segregated, with males and females frequenting different areas. We see this today in adults of many species, such as the Scalloped Hammerhead (*Sphyrna lewini*) and Blacktip Shark (*Carcharhinus limbatus*). If *Cladoselache* sexually segregated, perhaps the females frequented shallow marine habitats where the conditions required for quality chondrichthyan fossils (i.e., loose, shifting mud that could rapidly burry a body before it decomposed) were more common.

The euchondrocephalan sister taxa of *Cladoselache* have also captured the attention of paleoichthyologists. We have already encountered one such species, the Damocles Shark (*Damocles serratus*) in our discussion of Bear Gulch, but that was not the only euchondrocephalan Bear Gulch had in store.

Falcatus (see figure in *Research, Conservation, and Education Focus* Box 2.1), which has been called the Unicorn Shark because of its large, curved, scaled spine just posterior to the head on the dorsal surface of males of this group. *Falcatus* was a very small (0.15 to 0.25 m, or 6–9 in) schooling shark that preyed on small invertebrates, like shrimp. It was first described in 1883 from the St. Louis Limestone and later rediscovered by paleontologists Dick Lund and Eileen Grogan in Montana's Bear Gulch, a limestone laid down in tropical fresh and brackish waters about 320 mya. Lund also made a serendipitous[13] discovery in the Bear Gulch of a pair of *Falcatus* in which the female appears to be grasping the head appendage of her partner. Among living chondrichthyans, only male chimaeras possess a cephalic clasper, or *tenaculum*, reminiscent of the heavily modified fin and spine on *Falcatus*. Cranium-associated tenaculum-like, sexually dimorphic structures occur in Bear Gulch chondrichthyans, and they are specifically found in the euchondrocephalan taxa.

One of the more bizarre clades of euchondrocephalans is the superfamily Edestoidae, which includes the families Helicoprionidae and Edestidae (fig. 2.16A). These were the "whorl tooth" and "scissor tooth" that swam the seas between the end of the Devonian and the dawn of the Triassic. Both families of sharks had symphyseal biting teeth, meaning that the teeth sat on the midline of the jaw, or *symphysis*, where the left and right halves meet. *Helicoprion*, sometimes called the "buzzsaw" shark thanks to its symphyseal dentition forming a spiral of teeth (fig 2.16B), likely grew to about the size of a modern White Shark (*Carcharodon carcharias*). Paleontologists hypothesize that these sharks used their tooth blades to cut through the shells of cephalopods like ammonites and nautiluses, which were abundant and potentially valuable food sources.[14]

We have only just scratched the surface in our survey of euchondrocephalans. In addition to *Cladoselache, Damocles, Falcatus*, and the edestoids, there are many other euchondrocephalan ancestors and sister taxa to modern ratfish and elasmobranchs. Ancient sharks called petalodonts, for example, had compressed

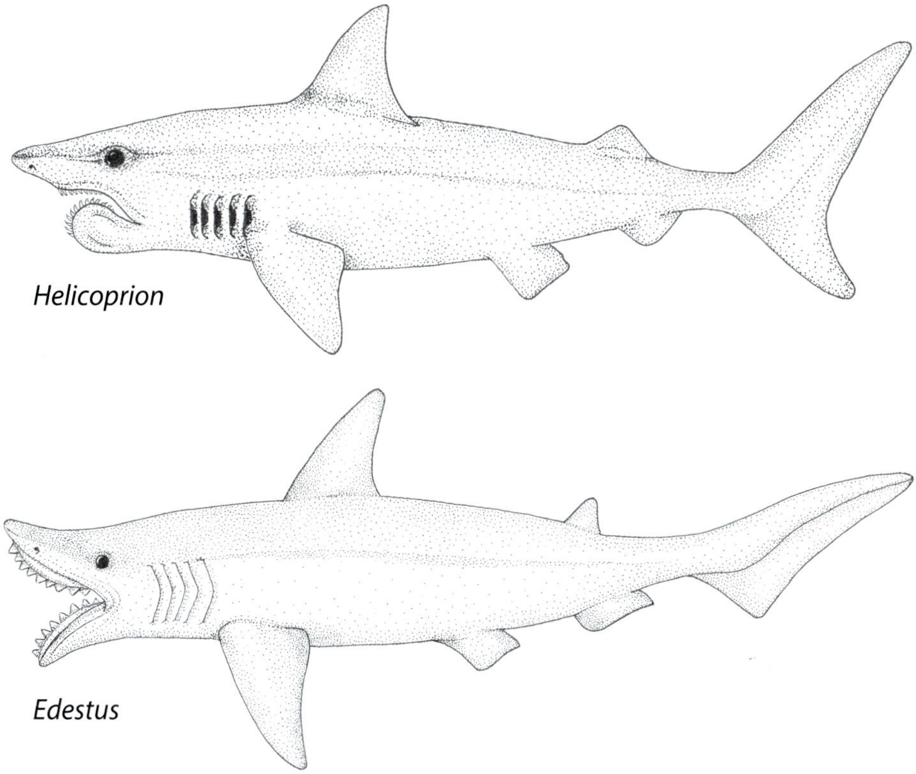

A

B

Figure 2.16. (A) Euchondrocephalans with bizarre jaws. *Helicoprion* and *Edestus* each require some artistic interpretation to reconstruct fully, but fossils of their unique teeth and portions of their jaws give researchers clues as to what their heads may have looked like. (B) Fossil tooth whorl of *Helicoprion* from the Permian of Idaho (Courtesy of James St. John).

bodies that were vaguely leaf-shaped and teeth suited for scrapping bryozoans off rocks for a meal.[15] The bizarre *Stethacanthus* (a member of the same family as *Damocles* and *Falcatus*) exhibited amazing sexual dimorphism as it prowled Devonian and Carboniferous seas—males had a modified dorsal fin that was enlarged and flattened with rows of denticles on top. The result was a chondrichthyan that looked like it was wearing a Roman helmet on his dorsal fin (fig. 2.17). The point made clear by this diversity is that anyone who claims that sharks have remained virtually unchanged for millions of years, clearly is not considering the whole chondrichthyan story.

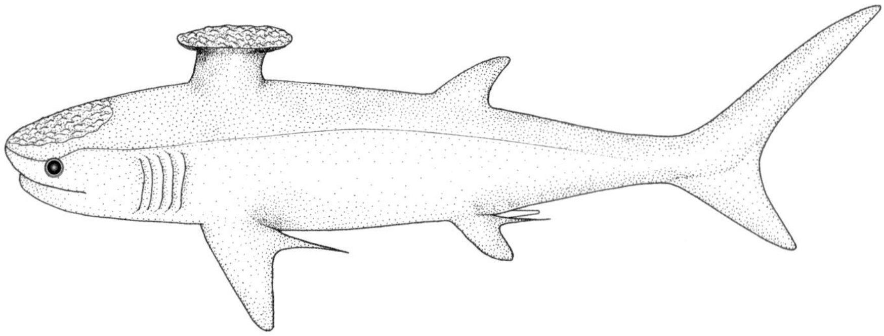

Figure 2.17. A male *Stethacanthus*, a euchondrocephalan with a highly modified dorsal fin.

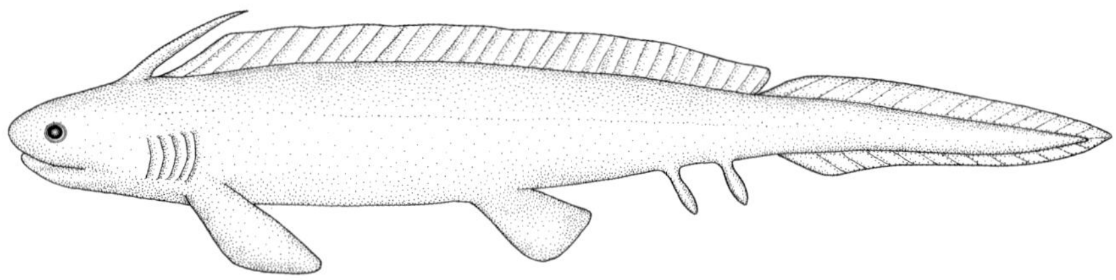

Figure 2.18. *Xenacanthus* was an early genus of elasmobranch and gets its genus name from the large spine on its head.

Xenacanthiforms

Another extinct clade of chondrichthyans that has captured human interest is the order Xenacanthiformes. These chondrichthyans are among the first we have encountered in our overview of notable fossil sharks that are elasmobranchs[16] (as opposed to the euchondrocephalans of the preceding section). This means that Xenacanthiforms were more closely related to the lineage that gave rise to modern sharks and rays than to the euchondrocephalan lineage in which we find modern ratfish.

The order Xenacanthiformes reached its peak in the Carboniferous and Permian Periods. Two hundred sixty mya, nearly 160 million years before the first crocodilians, early elasmobranchs like the eel-like *Orthacanthus* stalked the swamps of what is now Europe and North America. Although *Orthacanthus* reached lengths of 3 m (almost 10 ft), its sister *Xenacanthus* (fig. 2.18) seldom grew larger than 2 m (about 6 ft). These xenacanthiforms, along with earlier Carboniferous members of the clade, such as *Diplodoselache*, were predominantly freshwater sharks with double-cusped teeth. Their ordinal name (*xena* = strange or foreign + *acanthus* = spine) is an allusion to the long cephalic spine at the base of their heads. A xenacanthiform shark is illustrated in figure 2.18 representing an early offshoot of the elasmobranch lineage. Although the heyday of Xenacanthiformes was in the Permian, a few of these sharks, such as *Triodus*, barely survived the Permian Mass Extinction and did not go extinct until about 212 mya during the early Triassic.

Hybodonts

Hybodont sharks belonged to the order Hybodontiformes (*hybo* = hump + *odont* = tooth), and these were the dominant sharks during the Triassic, Jurassic, and Cretaceous Periods. Hybodontiformes includes some of the most common sharks during what humans typically consider the Age of Dinosaurs, and like the non-avian dinosaurs, hybodonts did not make it past the mass extinction at the end of the Cretaceous.

Hybodont sharks included freshwater and marine species, and although some hybodonts reached respectable lengths of 3 m (about 10 ft), others grew to only a little more than 30 cm (1 ft). The Jurassic sharks of the genus *Hybodus* exemplified many of the characteristics researchers associate with hybodonts (fig. 2.19). Each dorsal fin had a long spine. In well-preserved spines, small denticles are seen

Figure 2.19. (A) Hybodont shark. Illustrations of hybodont sharks frequently bear a striking resemblance to modern horn sharks. Why is that? Since hybodont sharks had the same kind of heterodont dentition, that is, with two different tooth morphologies, as modern horn sharks, artists interpreted that they must also look like them (Redrawn from Maisey, J. 1996. *Discovering Fossil Fishes*. Holt). (B) The hybodont shark *Hybodus fraasi* from Eichstätt, Germany (Courtesy of Haplochromis. https://commons.wikimedia.org/wiki/File:Hybodus_fraasi_(fossil).jpg. Accessed 9/16/24).

on the apical end. Males had one or more pairs of prominent cephalic spines on either side of their head just above and behind the eyes. Were these used to ward off revival males, or perhaps to coax a mate? In the absence of a time machine or an incredibly fortunate fossil yet to be discovered, we may never know.

The bodies of most hybodonts were streamlined and shark-like, and their jaws were suspended under their braincases in much the same way most modern sharks' jaws are suspended.[17] The hyoid arch, the skeletal structure between the first gill arch and the jaws, supported the jaws, which hybodonts could probably protrude a modest distance. This form of jaw suspension is called hyostyly (see chapter 4).

Some fossils have given researchers glimpses into the trophic ecologies of these extinct sharks. Fossils of the small hybodont *Tribodus* have been recovered with the shells of benthic shellfish detectable in the shark's stomach. Another fossil of a Jurassic ammonite bears the tooth of the hybodont shark *Planohybodus*, attesting to a predator-prey relationship between the species.[18]

Hybodont tooth morphology also helps inform our understanding of their trophic ecology. Hybodonts exhibited varying degrees of *monognathic heterodonty*, meaning teeth in one region of the jaw were shaped differently than teeth in another region of the same jaw. In several hybodont species, teeth in the front of mouth had pointed cusps while teeth in the back of the mouth were flat, suitable for crushing shelled prey.

The Rise of Modern Sharks

By the end of the Jurassic, about 144 mya, two new clades of elasmobranchs had evolved, the *batoidea* (skates and rays) and the *selachimorpha* (sharks as we would recognize them today). Together, batoids and selachimorphs are sister taxa and make up the monophyletic *neoselachians* (*neo* = new + *selachian* = sharks). These are the modern sharks. The rise of neoselachians coincides with the breakup of the supercontinent Pangea into Laurasia in the north and Gondwana in the south. By the late Jurassic, the outlines of North America and South America/Africa were becoming evident, separated by a young Atlantic Ocean and the tropical Tethys Sea, parts of which were destined to become the Indian Ocean. The breakup of the continents provided shallow water habitat over the now-increased continental shelves—an ideal habitat for elasmobranch fishes to exploit new niches and to diversify.

Neoselachians share anatomical features that unite them in a monophyletic clade. Many of these features represent changes in fins and mineralized structures, such as teeth and spines.[19] These traits include:

- *A notochord segmented by vertebral centra.* Until the neoselachians, sharks relied on the notochord, a stiff rod of glycoproteins and collagen running the length of the body, to lend structural support during swimming. Today, modern ratfish still have a continuous notochord, and only its external sheath is mineralized and vertebral elements form around (not through) the notochord.[20] In neoselachians, the centrum of each vertebra segments the notochord. The result is a sturdier vertebral column suitable as an anchor for trunk muscles that power swimming.

- *Pectoral fins attached to a pectoral girdle with a coracoid bar.* In ancestral sharks, a skeletal element connecting the left and right pectoral fins was not always present. Formation of a complete, robust pectoral girdle provides more stability as well as anchoring for muscles. In modern sharks, the pelvic girdle connects both pelvic fins, which ancestors also lacked.
- *Triple layer enameloid in the teeth.* Although batoids appear to have secondarily lost this feature, the current consensus is that neoselachians are united by the presence of three layers of enameloid covering the crowns of their teeth. Each layer is discernable from the other two by the orientation of the mineral crystals that make up that layer.
- *Fin spines covered by smooth enameloid.* In the modern sharks that have retained their dorsal fin spines (e.g., the Spiny Dogfish [*Squalus acanthias*]), the spine is covered by a thin layer of smooth enameloid. This may be linked to the evolution of triple layer enameloid in the teeth.
- *No more than three tiny cartilaginous elements connecting the claspers to the skeleton of the pelvic fin.* In modern male sharks and batoids, there are only one to three intermediate cartilaginous elements that connect the skeleton of the clasper to the skeleton of the pelvic fin. Interestingly, ratfish also have only a single cartilaginous element there, but we infer from their phylogenetic distance and the presence of multiple intermediate cartilages in the pelvic fins of *Hybodus* that the reduction of intermediate pelvic cartilages occurred independently in neoselachians.

The traits listed above are all helpful for phylogeneticists trying to organize the chondrichthyan tree of life. However, it is worth noting that other characteristics, which may not strictly be phylogenetically informative, do pop up on multiple occasions in many modern sharks, suggesting a few reoccurring evolutionary themes, such as the imperative to efficiently capture food and move throughout your environment. For example, modifications in how loosely jaws are suspended under the braincase and a reduction in the size of radial cartilages in the fins that make fins more flexible seem to be common in neoselachians. The functional implications of each of these will be discussed more in chapter 4.

By "modern sharks," we mean anatomically modern sharks, not that the species we have today first appeared in the Triassic or Jurassic. There are many extinct "modern sharks," such as *Paleospinax* (fig. 2.20A), which was an early, or stem, neoselachian from the early Jurassic. Stem neoselachians were present in the Triassic, but the group diversified in the Jurassic. *Paleospinax* superficially resembled a modern carpet shark (family Orectolobidae), although it possessed a spine on both dorsal fins that carpet sharks lack, leading some to conclude that it

A

B

Figure 2.20. (A) *Paleospinax*, an early neoselachian shark (Redrawn from Maisey, J. 1996. *Discovering Fossil Fishes*. Holt). (B) Fossil *Protospinax*, which shows its body shape and solid spinal column with vertebral centra (Courtesy of High Contrast).

more closely resembled a modern dogfish (family Squalidae). Like many of today's sharks, *Paleospinax* had a subterminal mouth; shortened, loosely suspended jaws; a vertebral column complete with centra segmenting the notochord; enameloid-coated teeth; and an elongated snout.

Other extinct neoselachians include *Protospinax*, which was a bottom-dwelling shark that likely resembled a slightly flattened dogfish with broad, rounded pectoral fins and a blunted head (fig. 2.20B). Recent work by paleoichthyiologists suggests that this taxon was closely related to modern angel sharks (family Squatinidae) and sawsharks (family Pristiophoridae).[21] *Protospinax* and its sister taxon of the Jurrasic, *Synechodus*, probably never grew larger than 1.5 m (almost 5 ft) in length.

Neoselachians are divided into three major monophyletic superorders: *Squalomorphii, Galeomorphii*, and the *Batoidea (sometimes referred to as Batomorphii)*, whose extant taxa will be described in chapter 3.

Recall that batoids diverged from the elasmobranch line about 230 mya.[22, 23] Among the sharks swimming today, those with some of the most primitive features (indicating we should expect to trace their lineages back very far) form the group of sharks known as the cow sharks (see fig. 2.1). These include the sixgill and sevengill sharks as well as two extant species of frilled shark, all in the order Hexanchiformes.

The divergence of hexanchiform sharks and the other squalomorph sharks, occurred relatively early in the squalomorph lineage. Fossilized teeth of frilled sharks and sevengill sharks (order Hexanchiformes) date back to the Cretaceous, so we know that they had already diverged from the rest of the squalomorphs almost 85 mya. This is not far off (by the standards of geological time) from the divergence estimates based on genetic sequences that suggest that Hexanchiformes diverged about 130 mya.[24] That means that the main squalomorph lineage was established before the Cretaceous, likely in the Jurassic. This interpretation is supported by the divergence estimates of other squalomorphs like dogfish (order Squaliformes), which diverged about 163 mya, and the angel sharks (Squatiniformes), which diverged about 155 mya.

The superorder Galeomorphii is often regarded as a more recent taxon than Squalomorphii, but that is not necessarily true. This perception is likely due to the presence of many species in the clade exhibiting what we consider derived anatomical traits, such as hyostylic jaw suspension and a loss of fin spines. Among galeomorph sharks, the horn sharks (order Heterodontiformes) are the oldest in the fossil record. According to fossil evidence, that order diverged in the early Jurassic.[25] Extant horn sharks (in the genus *Heterodontus*) date back as far as 42.5 mya.[26] The next oldest galeomorph group after the horn sharks are the carpet sharks (order Orectolobiformes; 166 mya), followed by the mackerel sharks (Lamniformes; 162 mya), and the requiem sharks (Carcharhiniformes; 179 mya).

Megalodon and the Origins of the White Shark

We conclude this chapter by discussing an extinct neoselachian that has received a lot of attention thanks to its size and large teeth, *Otodus megalodon*[27] (fig. 2.21A). The story of Megalodon sheds little light on the major questions of shark evolution. Out of more than 420 million years of chondrichthyan evolution dating back to the Silurian, Megalodon prowled the seas for only about 20 million years of the most recent geological period, the Neogene. And like many chondrichthyans' before it, Megalodon's lineage terminates in an evolutionary dead end. Still, the species' size (topping out close to 22.9 m, or 75 ft, in length according to one recent study[28]) have captured the attention of shark enthusiasts, so it merits our attention here.

The popular media often draw comparisons between Megalodon and the modern White Shark (*Carcharodon carcharias*). There were similarities to be sure. Like the White Shark, Megalodon was probably endothermic (warm-blooded) and regularly consumed marine mammals. Megalodon teeth bear a superficial similarity to those of White Sharks. Both are large, triangular, and serrated (fig. 2.21B). This led researchers throughout the twentieth century to assign Megalodon to the same genus as the modern White Shark, *Carcharodon*, and accept it as a sort

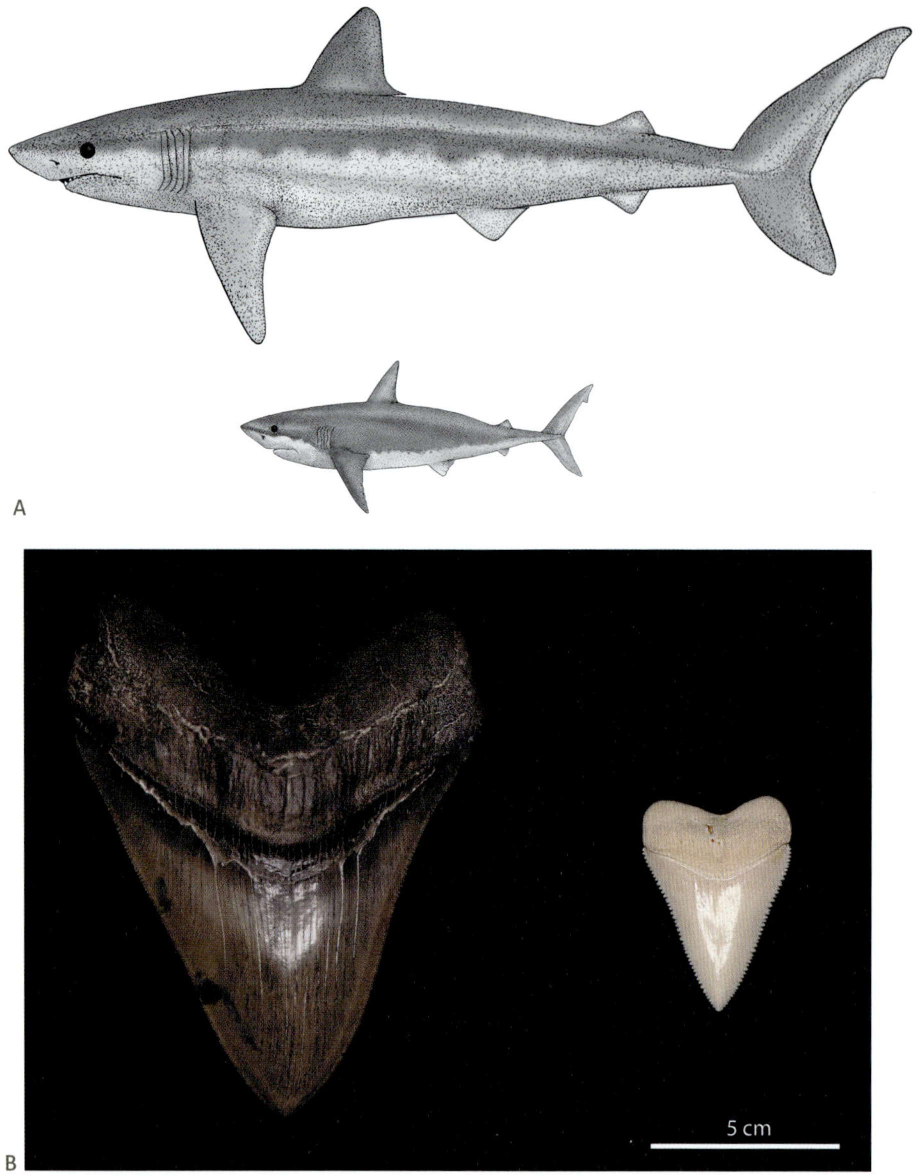

Figure 2.21. (A) Megalodon compared to an extant White Shark. (B) Megalodon tooth compared to a modern White Shark tooth.

of evolutionary "grandfather" of the modern White Shark. However, several observations cast doubt on this interpretation of Megalodon's relation to the White Shark.

Similarities in the dimensions of tooth roots and crowns between the White Shark and the extinct mako shark *Cosmopolitodus* (sometimes classified as *Carcharodon*) *hastalis* suggest close kinship between the species. These similarities are not observed between White Shark and Megalodon teeth. The description of a new transitional fossil, *Carcharodon hubbelli*, in 2012[29] linking extinct makos and White Sharks made it clear that the modern *Carcharodon carcharias* most likely evolved from a lineage of mako sharks in the late Miocene Epoch, not more than 6 to 8 mya.

Was Megalodon a super-predator? Yes. Is it a fascinating example of gigantism among sharks and a major player in marine food webs of the Miocene? Of course. Does its story explain the evolutionary trajectories and diversity of chondrichthyans today? No, not to an appreciable degree given that the taxonomic orders of modern neoselachians existed tens of millions of years before Megalodon.

Concluding Comments

This chapter covered over 400 million years of chondrichthyan evolution, and we only scratched the surface of the many strange and fascinating extinct sharks that swam the seas of bygone ages.

Are sharks still evolving? Of course! Evolution is not static. The questions are, how will sharks evolve, and will all the evolutionary lineages we see represented today survive the next extinction event? The answers to both those questions depend in large measure on human activity. How we continue to remake the planet will shape the selective pressures and extinction rates facing all organisms, including sharks.

NOTES

1. A) Smalltooth Sawfish B) Bluntnose Sixgill Shark C) Atlantic Sharpnose Shark D) Angelshark E) Longnose Sawshark F) Ornate Wobbegong G) Bowmouth Guitarfish H) Southern Stingray I) Horn Shark J) Spiny Dogfish K) Bigeye Thresher L) Mouse Catshark M) Common Torpedo N) Frilled Shark O) Prickly Dogfish.

2. https://sharksrays.org. (Accessed 9/9/25).

3. *Taxonomy* is the theory and practice of naming things. *Systematics* is the process of classifying organisms typically according to their phylogeny (i.e., based on common ancestry). The words are typically used incorrectly as synonyms. Most current taxonomies of organisms are systematic taxonomies.

4. A *tessera* (pl. *tesserae*) is a tile, usually in the shape of a cube, that comprises a mosaic.

5. Grogan, E.D. et al. 2012. In: Carrier, J.C. et al (eds.). Biology of sharks and their relatives. CRC Marine Biology Series. CRC Press. 3–31 pp.

6. Last, P.R. et al. 2016. Zootaxa 4139: 345–368.

7. Coates, M.I. et al. 2018. Proc. R. Soc. B. 285: 20172418.

8. Miller, R.F. et al. 2003. Nature. 425: 501–504.

9. Burrow, C.J. et al. 2017. Can. J. Earth Sci. 54: 1248–1262.

10. Maisey, J.G. et al. 2017. Am. Mus. Novit. 3875: 1–16.

11. Glug, C. et al. 2023. Swiss. J. Paleo. 142. https://doi.org/10.1186/s13358-023-00266-6. (Accessed 9/19/2).

12. Maisey, J. G. 2007. Bull. Am. Mus. Nat. Hist. 307: 1–122.

13. Serendipity, meaning *a fortunate surprise*, often is associated with hours, days, week, years, or even decades of meticulous hard work, which is clearly the case here.

14. Tapanila, L. et al. 2020. Anat. Rec. 303: 363–376.

15. Lund, R. 1989. J. Vert. Paleo. 9: 350–368.

16. Luccisano, V. et al. 2021. J. Syst. Palaeontol.19: 1623–1642.

17. Maisey, J. G. 2008 J. Morph. 269:1022–1040.

18. Vullo, R. 2011. Sci. Nat. 98: 545–549.

19. Maisey J.G. et al. 2019. In: Johanson et al. Evolution and Development of Fishes. Cambridge University Press. 87–109 pp.

20. Didier, D. A. 1995. Am. Mus. Nat. Hist. Novit. 3119: 1–86.

21. Jambura, P.L. et al. 2023. Diversity. 15. https://doi.org/10.3390/d15030311 (Accessed 9/19/25).

22. Aschliman, N.C. et al. 2012. Mol. Phylogenet. Evol. 63: 28–42.

23. Divergence times between clades is usually estimated by calculating the rate of change in genes and then calibrating that based on the earliest known fossils of the clades of interest. The result is a range, and sometimes those ranges can differ quite a bit. One study (Heinicke et al. 2009. In: Hedges, S. B and Kumar, S. The Timetree of Life. Oxford University Press. 320–327 pp.) found that molecular divergence estimates can differ by as much as hundreds of millions of years from what the fossil record leads us to believe.

24. Sorenson, L. et al. 2014. J. Evol. Biol. 27: 1536–1548.

25. Maisey, J. G. 2012. J. Fish. Biol. 80: 918–951.

26. Slater, T.S. et al. 2020. Pap. in Paleonto. 6: 425–437.

27. Many people refer to *Otodus megalodon* simply as Megalodon. Taxonomically, this is tantamount to calling *Tyrannosaurus rex* simply Rex. That is not proper taxonomy, but for the sake of efficiency, we will hold our noses and refer to this shark as Megalodon. It is also worth noting that some authors still assign Megalodon to the genus of extinct sharks *Carcharocles,* but this does not change the information presented here.

28. Shimada et al. 2025. Palaeontol. Electronica. 28.1.a12. https://doi.org/10.26879/1502

29. Ehret, D.J. et al. 2012. Palaeontol. 55: 1139–1153.

3 / Diversity of Sharks

CHAPTER OUTLINE

Introduction: Why So Few Kinds of Sharks?

Exactly how many different species of sharks are there? The truthful answer is that we do not know, in part because the use of molecular techniques combined with increased sampling in the deep-sea and remote coastal regions make this a moving target as new species are discovered and some species are deemed invalid, for example if two described species turn out to be the same and are synonymized. On average, about 300 new species of bony and cartilaginous fishes are described each year. In the five years between writing the first and second editions of this book, 46 new species of chondrichthyans have been described, but 18 species have been deemed invalid, for a net increase of 28 species.

New species of extant sharks are typically not discovered in Eureka! moments when a specimen with unique coloration or other physical traits is first observed in its environment or is captured. This sort of lucky confluence of a species new to science and the presence of a knowledgeable fisher or shark scientist does indeed happen, but less so than you might expect. More frequently, new species are "discovered" through painstaking examination of specimens that may involve x-ray or more advanced imaging techniques and, frequently, genetic analysis. Consider the Painted Horn Shark, *Heterodontus marshallae,* from NW Australia. The shark had long been known to science, but was confused with its very similar New Zealand cousin, the Zebra Bullhead Shark, *Heterodontus zebra.* A combination of differences in gene sequence, external morphology, coloration, and egg cases led to the reclassification as a new species.[1] Finally, the elucidation of a new species of hammerhead shark, *Sphyrna alleni,* which was previously considered the more broadly distributed Bonnethead, *Sphyrna tiburo,* was published in 2024.[2] The new species is found along the West Atlantic coast from Belize to Brazil.

In 1984, the authoritative book on sharks, Compagno's *Sharks of the World*[3] listed 342 species of sharks. As of this writing, an annually updated checklist[4] as well as the *Chondrichthyan Tree of Life* website,[5] documented 550 kinds of sharks among 1263 species of chondrichthyans, which also included 660 batoids and 53 holocephalans. This breaks down to 14 orders, 65 families, and about 211 genera. We have used a figure of about 550 described species of sharks and 1250 total species of sharks, skates, rays, and chimaeras. This number will certainly increase in the future, assuming deep-sea research continues.

In this chapter, we survey the diversity of chondrichthyan fishes, with greater detail provided for sharks. We cover all orders and most families within them, as well as species that are iconic, ecologically important, commercially valuable, overfished, and/or are otherwise interesting. If the diversity of shapes, sizes, and lifestyles of sharks and other chondrichthyan fishes fascinates you, this chapter will be most interesting.

Let us consider why there are so few kinds of sharks compared to the dominant aquatic vertebrates, the bony fishes, of which there are over 35,000 kinds.

Bony Fishes Have More Offspring

First, all sharks use internal fertilization, and the overwhelming majority of bony fishes fertilize externally. Bony fishes typically invest reproductive resources into a large number of eggs (e.g., several million in the case of cod). These eggs or resulting hatchlings have a high degree of *vagility,* or ability to disperse, if they are pelagic and are entrained in surface currents that carry them perhaps thousands of miles. The wider their dispersal, the greater the odds of isolating and diversifying into new species.

Sharks use an entirely different life-history strategy of investing resources into a limited number of large offspring. Early development of the embryo occurs inside the female for the sharks that lay eggs, like horn and bamboo sharks, and complete development occurs internally for sharks like White Sharks (*Carcharodon carcharias*) or Blacktip Sharks (*Carcharhinus limbatus*). The advantage for sharks is that each individual has a greater chance of survival than a single bony fish hatchling. The disadvantage is decreased vagility (large numbers of offspring dispersing). Thus, there is a far greater opportunity for speciation in bony fishes.

Most Sharks Are Small

In addition to eggs and neonates (newborns), adults are also capable of dispersal. However, many sharks are also handicapped by their size. Surprisingly, most species of shark (over 300 species) are small (< 100 cm, or 3.3 ft). Migrating across the distances required to achieve the geographic isolation needed for speciation is very risky for small sharks. Moreover, about 80% of shark species are bottom-associated, and bottom-associated species are less likely to disperse great distances. Most bony fishes are also not large, but recall that they often produce pelagic eggs and larvae (hatchlings) that can readily disperse.

Bony Fishes Have a Swim Bladder

Swim bladders are gas-filled spaces (think *internal balloon)* in the body cavity of most bony fishes which act as flotation devices and allow fish that possess them to adjust their buoyancy, and hence their vertical position in the water column. Sharks are heavier than water and, without a swim bladder, must swim continuously using their fins to provide lift or live on the seafloor.

The evolution of swim bladders provided two critical advantages. First, bony fishes could save energy by avoiding having to continuously move, and saving energy is a major driving force for an organism's success, enabling it to devote energy to other survival tasks. Second, fish with swim bladders could expand their *niches*; that is, occupy myriad habitats with more or better food, fewer predators, or appropriate environmental conditions, by simply adjusting their buoyancy and using their fins for fine scale movement instead of providing lift.

A swim bladder enables bony fishes to exploit the many tight spaces on coral reefs, which account for most of the living space in these ecosystems. In other words, a swim bladder facilitates being small. No sharks on coral reefs are as small as the smallest and/or most common bony fish species.

Elasmobranchs had about 300 million years during which they could have evolved small forms before teleosts did; however, they did not. The teleosts, with their fancy swim bladders, came along and literally took over the marine and freshwater aquatic world, aided by the evolution of complex habitats like reef-building corals, seagrasses, and mangroves.

Sharks Possess Relatively Rigid Pectoral Fins with a Small Range of Motion

A major evolutionary advancement for sharks was replacing a rigid pectoral fin base with three cartilages (see fig. 1.9), endowing the pectoral fin with greater flexibility; however, the pectoral fins of sharks are still less flexible than those of most bony fishes. Absent the restrictions imposed by these three cartilages, many bony fishes (e.g., damselfish) have acquired the ability for fine-scale movement. Thus, no sharks capable of darting in and out of crevices exist, due largely to their inflexible pectoral fins.

Bony Fishes Have More Complex Heads and Jaws

Finally, bony fishes have evolved seemingly unlimited types of head shapes and mouths and thus different feeding styles, whereas sharks have stuck with the same basic plan with a few exceptions.

What allows this variation in bony fishes is the complexity of the head (see fig. 1.8). The number of bones in the skulls of bony fish varies by species and can number more than 100. Each part of the head of bony fishes, especially mobile maxillae and premaxillae, can be molded through evolution to create jigsaw patterns culminating in a variety of mouth and head shapes. With every different mouth position and shape in bony fishes came the opportunity to exploit a new habitat or food source, and these structural and functional variations led to the evolution of new species.

In contrast, the skull of sharks is fairly simple. There is a suture-less (one solid piece) *chondrocranium*, and a single upper jaw (the *palatoquadrate*) and a single lower jaw (*Meckel's cartilage*), each composed of two cartilages (see fig. 1.8). There is little raw material for evolution to work on. To be sure, the basic shark head plan has been an unqualified success and the variation in feeding morphology achieved is impressive. Compare the crushing jaws of a Horn Shark (*Heterodontus francisci*), the melon-ball scooping jaws of the Cookiecutter Shark (*Isistius brasiliensis*), and the grasping jaws of the Goblin Shark (*Mitsukurina owstoni*). However, along with the other features described, this simple head plan deprived sharks of the ability to diversify to the same extent as bony fishes. There will never be a shark with a mouth like a seahorse, wrasse, houndfish, or butterflyfish.

Overview of Shark Taxonomy

To start, if you need a taxonomic refresher, consult Box 1.1, which provides an overview of the Linnaean system of classification along with some helpful conventions used in naming organisms. Box 3.1 contains definitions for some of the terms that appear frequently as distinguishing features.

Chondrichthyans and humans are both in the phylum Chordata. So are 3000 species of tunicates (basically sea squirts and their relatives—Did you know that you are more closely related to blobs known as sea porks or sea livers than you are to, say, a squid?), 30 species of lancelets (small, obscure, sediment-dwelling fish-like creatures), and about 70,000 other vertebrates (also known as *craniates*).

We will start our taxonomic journey with the superclass Gnathostomata, the jawed fishes. One widely accepted scheme has the gnathostomes as a monophyletic[6] group of five clades: the extinct Placodermi and Acanthodii, the latter sharing a common ancestor with the extant Chondrichthyes (sharks, skates, rays, and chimaeras) and two extant clades of osteichthyan fishes, the Actinopterygii (ray-finned fishes) and Sarcopterygii (lobe-finned fishes as well as the tetrapods, e.g., amphibian, reptiles, birds, and mammals).

The class Chondrichthyes includes two extant subclasses, the Holocephali and the Elasmobranchii, the latter of which is the dominant living chondrichthyan group. Consult Figure 3.1, a cladogram of the class Chondrichthyes, for an overview of relationships among the 14 orders.

BOX 3.1

Frequently Encountered Terminology Used in Descriptions of Distinguishing Features of Chondrichthyans

Jaw Suspension—refers to how the upper jaw (palatoquadrate) and lower jaw (Meckel's cartilage) connect to the skull (chondrocranium) and other supporting structures, particularly the hyomandibular cartilage, part of the hyoid arch (the second gill arch that moved forward in evolution to support the jaws to some degree; see fig. 2.10). Consult chapter 4 for more details. The following types are in descending order from primitive to most derived, and from least to most mobile.

Amphistylic: The upper jaw is attached by ligaments (strong, fibrous material that connects skeletal elements) at two points to the braincase. Posteriorly, the jaw is provided with very limited support by the hyomandibular cartilage.

Orbitostylic: A projection from the upper jaw called the *orbital process* articulates with the skull, and the hyomandibular arch, while still buttressing (support) the jaw, is shorter.

Hyostylic: There is a ligamentous connection (the *ethmopalatine* ligament) between the palatoquadrate and chondrocranium anterior to the orbit, and there is buttressing (support) by the hyomandibular cartilage.

Euhyostylic: There are no connections (no ligaments or articulations) between the upper jaw and braincase, and the jaws are supported exclusively by the hyomandibular cartilage at the corners of the jaws.

Modes of Embryological Development—refers to whether there is live birth or hatching from an egg, or some intermediate form, and includes the mode of nutrition for the developing embryo. This topic is discussed in detail in chapter 6. The three main categories of embryonic development in elasmobranchs are:

Oviparity: egg-laying

Yolk-sac Viviparity (formerly Aplacental Viviparity): live birth without a placenta, a structure formed from maternal and embryonic tissues that connects the embryo to the wall of the uterus and provides nutrition to the developing embryo

Placental Viviparity: live birth with a placenta, although different from the mammalian placenta

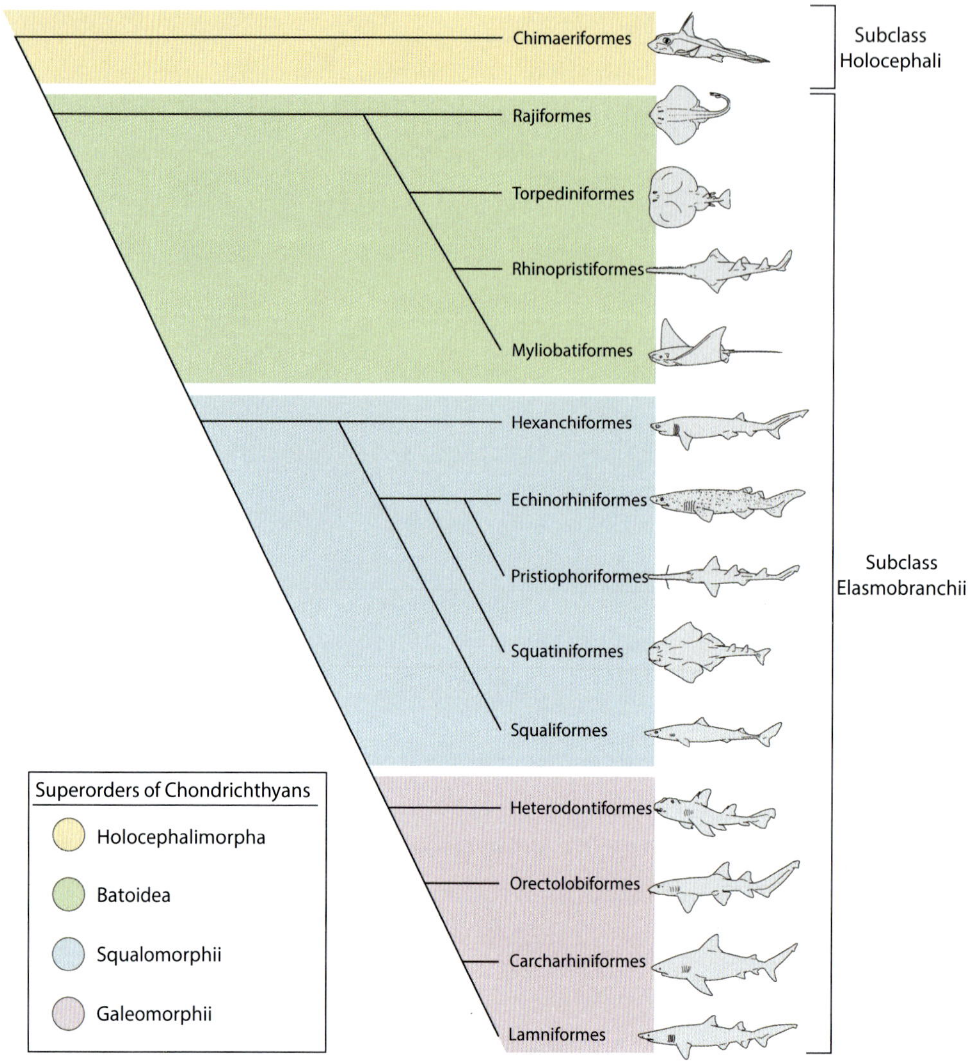

Figure 3.1. Cladogram of the Class Chondrichthyes (Redrawn from fig. 23.1, Cracraft, J. and Donoghue, M. 2006. *Assembling the Tree of Life*. Oxford University Press).

Subclass Holocephali

Superorder Holocephalomorpha

ORDER CHIMAERIFORMES

Extant members of the Holocephali (fig. 3.2) are a monotypic subclass (i.e., there is a single order, Chimaeriformes [*chimaer* = monster]), and they are commonly known as the *chimaeras* (or *ghost sharks* in Australia). The > 50 described species of Chimaeriformes comprise three families: Chimaeridae (42 sp.), Callorhinchidae (3 sp.), and Rhinochimaeridae (8 sp.). Holocephalans represent a monophyletic sister-group to the Elasmobranchii.

Most chimaeras live in deep water but the best-known species, often displayed in public aquaria, is the spotted ratfish, *Hydrolagus colliei*, which is found

Figure 3.2. Juvenile chimaera (Courtesy of NOAA).

along the US West Coast. Chimaeras are the only living chondrichthyans with holostylic jaw suspension (the upper jaw is fused to the skull) and thus have no jaw producibility. Other notable features include large eyes, a large mildly venomous spine on the first dorsal fin, and in addition to the pelvic claspers typical of all male chondrichthyans, male chimaeras possess a tooth-bearing clasper (or tenaculum) on their heads.

Subclass Elasmobranchii

The most speciose and diverse members of the Chondrichthyes are the elasmobranchs. Most possess dermal denticles (placoid scales), protrusible jaws of varying extents depending on superorder, serially replaced teeth not embedded firmly in the jaws, and tribasic pectoral and pelvic fins. Recall that the group is named for their gills, which are described as naked, strapped, or plate gills. These all refer to their gills lacking the protection afforded by the operculum of bony fishes.

Superorder Batoidea (Skates and Rays)

Batoids are the most speciose chondrichthyans and elasmobranchs, with 660 species in 26 families identified,[7] with species yet to be described. One explanation for this high diversity of the group is the very protrusible jaws, which allow for a variety of feeding modes.

Batoids are a monophyletic group united by many *apomorphic* (derived, or evolutionarily advanced) characters that include[8] (fig. 3.3):

- Pectoral fins fused to the head above the gill openings.
- No orbital articulation of the jaw (upper jaw does not connect with chondrocranium), which allows for *extremely* protrusible jaws in most taxa (in some, there is less producibility because the jaw muscles have become hypertrophied and are used for crushing). This is known as *euhyostylic* jaw suspension.
- A synarcual cartilage, essentially a series of fused cervical vertebrae and vertebral elements which anchor the enlarged pectoral fins. Interestingly, placoderms possessed a similar synarcual.

In addition to these derived characteristics, batoids have a flattened (depressed) body, a reduced or absent anal fin, five pairs of gill slits on the ventral surface

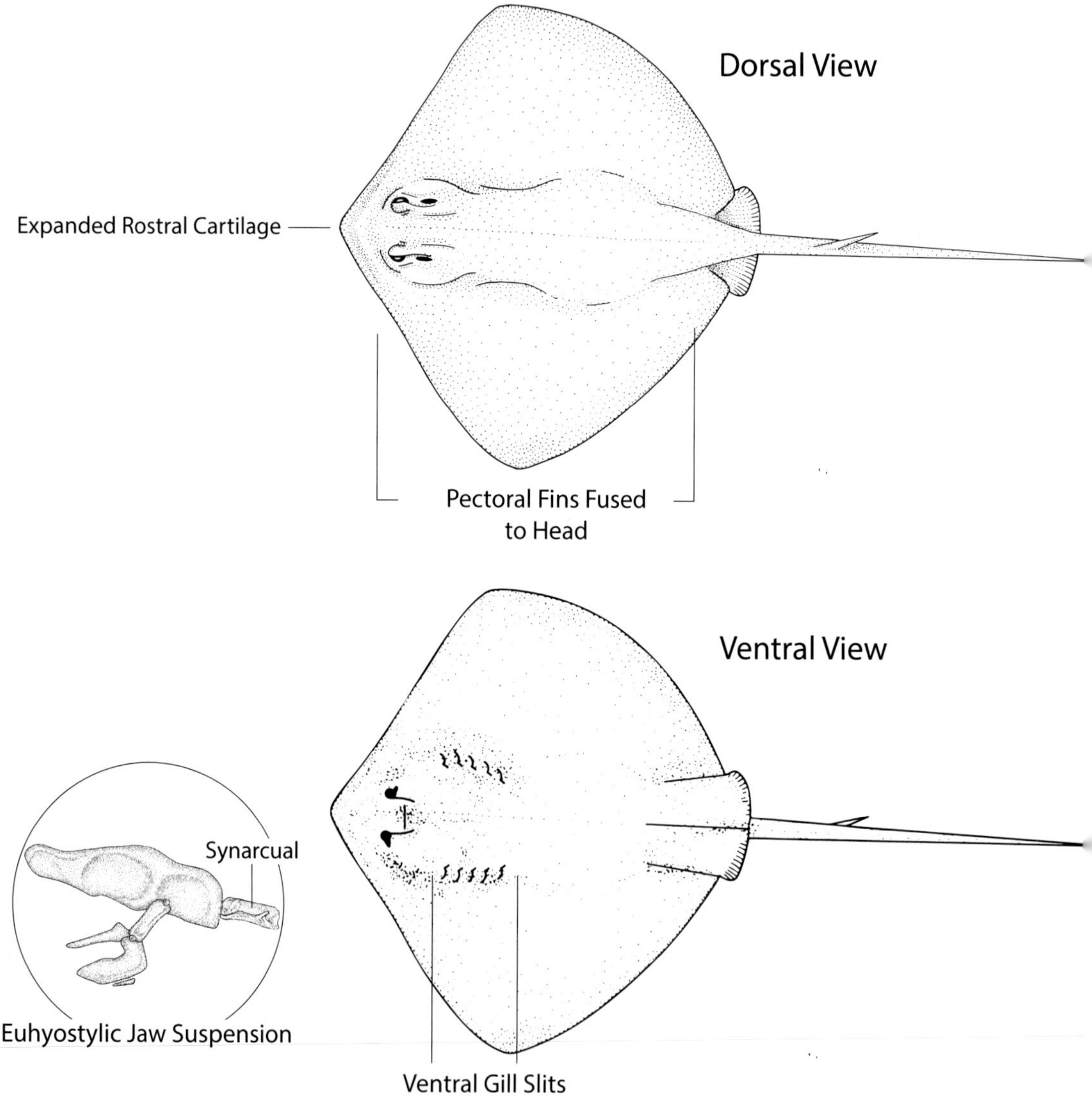

Figure 3.3. Apomorphic characters of batoids (Redrawn from fig. 41, Yokota, L. and De Carvalho, M.R., 2017. Zootaxa, 4332: 1–74).

(except for a single species with six pairs), and a pair of enlarged spiracles on the dorsal surface behind the head.

Less-enlightened shark biologists often describe batoids as *pancake* or *flat sharks*. Underlying this light-hearted condescension is the outdated idea that batoids are merely highly derived sharks. In fact, batoids diverged from sharks between 240 and 290 mya and evolved independently of sharks over that period. Recently, the tables have been turned, with emboldened batoid specialists referring to sharks as *sausage rays*.

Ecologically, most batoids are benthic, but there are pelagic forms as well, and whereas most rays you are familiar with are coastal, nearly half (mostly skates) live in the deep-sea. Many batoids feed on invertebrates and some are *durophagous* (eating hard-bodied prey), but some are *piscivorous* (fish-eaters) and others *planktivorous* (filter feeding plankton-eaters). Many have small teeth that are used for grasping or crushing. Members of three families, Pristidae, Dasyatidae, and Potamotrygonidae, inhabit low-salinity brackish waters or even fresh water.

Batoids are threatened by habitat destruction/alteration (especially freshwater forms) as well as growing fisheries. Like sharks, rays have life-history patterns that make them extremely vulnerable to commercial fisheries. Highly publicized misinformation about Cownose Rays (*Rhinoptera bonasus*) in Chesapeake Bay led to the development of a fishery that may critically threaten their population along the US East Coast (see chapter 11).

The superorder Batoidea comprises four orders: skates (order Rajiformes), electric rays (Torpediniformes), guitarfish and sawfish (Rhinopristiformes), and stingrays (Myliobatiformes) (fig. 3.4). Batoids have also been called *hypotremates* (*hypo* = under; *tremo* = opening), a term dating back to the early 1900s that refers to placement of the gill openings on the ventral surface beneath the pectoral fins. Correspondingly, sharks have been called *pleurotremates* (gills on the side). An abbreviated overview of batoids follows. For more information, we recommend the 2016 exquisitely illustrated reference *Rays of the World*.[9]

ORDER RAJIFORMES (SKATES)

The rajiforms are the skates (fig. 3.4), and at 293 species in four families, they represent nearly half of the diversity of batoids. Skates are benthic organisms, and occur in all oceans, from Arctic to Antarctic, in shallow coastal shelves to abyssal regions. Skates are most common in temperate cold or deep waters, where they are the dominant batoid. Deep-sea skates are mostly brownish in color.

Skates are characterized by a prominent rostral cartilage that is more rigid than that in rays, a quadrangular or rhomboidal body disc, two reduced dorsal fins (or one or both are absent) placed posteriorly on a stocky but spineless tail, and no caudal fin. Skates do not possess venomous spines.

Skates have a peculiar modification of the inner margin of their pelvic fins, an additional lobe (fig. 3.5). This bilobed pelvic fin creates a mobile pelvic girdle that

A

B

C

D

Figure 3.4. Representatives of each of the four orders of batoids. (A) Rajiformes: Clearnose Skate. (B) Rhinopristiformes: Smalltooth Sawfish. (C) Torpediniformes: Marbled Electric Ray (Aquapix/Shutterstock.com). (D) Myliobatiformes: Spotted Eagle Ray (Courtesy of Annie Guttridge).

Figure 3.5. The unique bi-lobed pelvic fin of a skate. The anterior lobes rotate and are used for movement on the bottom.

is used to crawl or scull on the bottom, a phenomenon known as *punting*. Each fin can move independently of the other. In members of the family Anacanthobatidae, the terminal lobes of the mobile pelvic fins are extended such that they are commonly called "leg skates." Most elasmobranchs are capable of only limited pelvic mobility, so this represents a big deviation from the norm, and a very cool adaptation.

Skates possess small cusped or flat, plate-like teeth. Many species possess *bucklers*, enlarged, thorn-like scales along the midline of the dorsum and tail. Males possess rows of enlarged scales on the cheeks near the eyes and also on the wingtips (termed *malar* and *alar* spines, respectively) that are used to maintain contact with females during copulation (see chapter 6). Skates are *oviparous*, with their *mermaid's purse* egg cases frequently seen by beachgoers.

Finally, some skates have a weak caudal electric organ (discharges of <4 V) whose function is enigmatic, but which may be used for communicating with each other. Skates are considered the most primitive among the living batoids and diverged from the rest of the group more than 130 mya.

Rajiform families include Anacanthobatidae (Smooth or Leg Skates; 13 sp.), Arhynchobatidae (Softnose Skates; 109 sp.), Gurgesiellidae (Pygmy Skates; 19 sp.), and Rajidae (Skates; 152 sp.; the most speciose family of skates).

ORDER TORPEDINIFORMES (ELECTRIC RAYS)

The order Torpediniformes (*Torp* = numb, numbness), the electric rays, are a group of about 63 species in five families (fig. 3.4). They are small to moderately large (total length to about 1.8 m, or 5.9 ft, but most species are less than 1 m, or 3.3 ft, long) with no dermal denticles (naked), except the Family Platyrhinidae, and with large dorsal spiracles. Most torpediniform rays have two well-developed dorsal fins, a well-developed caudal fin, and small mouths with rounded to slightly cusped teeth. Their large caudal fin is atypical of batoids outside of the rhinopristiforms. Their caudal fin is used for propulsion more so than undulations of the pectorals.

The group is named for its huge pectoral-cephalic (anterior) electric organ derived from honeycomb-shaped muscle cells aligned like batteries in a series. The electric organ in some species is capable of 200-volt, high amperage discharges capable of rendering victims temporarily paralyzed or even dead.

Torpediniform families include Torpedinidae (Torpedo Rays; 18 sp.), Narcinidae (Numbfishes; 30 sp.), Narkidae (Sleeper Rays; 9 sp.), Hypnidae (Coffin Rays; 1 sp.), and Platyrhinidae (Fanrays; 5 sp.). The platyrhinids are the only members of the torpediniform rays that do not possess electric organs and the only to possess thornlike denticles.

ORDER RHINOPRISTIFORMES (GUITARFISH AND SAWFISH)

The order Rhinopristiformes, was formerly two separate orders, the Rhinobatiformes (guitarfish and wedgefishes) and the Pristiformes (sawfish) until molecular data suggested that sawfishes are simply wedgefishes with a highly specialized rostrum, and the orders were combined.

This Rhinopristiformes consists of about 70 species of guitarfish, wedgefish, panrays, and sawfish (fig. 3.4). Most of these species superficially resemble sharks, at least posteriorly, more so than rays. Like the torpedinids, most members of this order have a well-developed caudal fin, which they use for propulsion. They possess an elongated, stiffened rostrum (like the skates), two well-developed spineless dorsal fins (which also aid in propulsion), and 65–75 rows of small, pavement-like teeth. They are bottom feeders, preferring small crustaceans and fish, except the sawfishes which use their toothed rostrum to immobilize large fishes swimming above them. The rhinopristiforms are yolk-sac viviparous. This group includes the most endangered of all chondrichthyan fishes, the wedgefishes, giant guitarfishes, and sawfishes. The rhinopristiform families include Glaucostegidae (Giant Guitarfishes; 9 sp.; all critically endangered), Pristidae (Sawfishes; 5 sp.; all critically endangered), Rhinidae (Wedgefishes; 11 sp.; nine of which are critically endangered), Rhinobatidae (Guitarfishes; 40 sp.), Trygonorrhinidae (Banjo Rays; 3 sp.), and Zanobatidae (Panrays; 2 sp.; two species).

ORDER MYLIOBATIFORMES (STINGRAYS)

The Myliobatiformes (*Mylio* = mill, molar) are the stingrays, whose scientific name is based on their grinding teeth. This is a big, diverse group, containing 234 species in 11 families. Myliobatiform rays exhibit a variety of color patterns and body shapes—ovoid, rhomboid, and trapezoid. In stark contrast to the Rhinopristiformes, the myliobatiform rays have no rostral cartilage. All have five gill slits except the one species in the family Hexatrygonidae. Their skin is naked except for patches of dermal denticles. They have a slender tail with or without spines. Most possess a dorsal fin-fold, whereas others have a single dorsal fin. This group is viviparous, employing lipid-rich uterine milk. They dominate coastal shallow ecosystems in warmer climates, include the only pelagic and planktivorous batoids, and members of two families live in fresh water.

Ever since Crocodile Hunter Steve Irwin died from a stingray spine penetrating his chest, there has been renewed interest in this group. Stingrays are not aggressive, but as many as 2000 injuries a year in the US are caused by stingrays. Fatalities, like Mr. Irwin's, are rare and are due to penetration or infection more so than the direct effect of the toxin, which is produced by goblet cells in living epithelial tissue that surrounds the spine, which is a modified dermal denticle. Stingray injuries are described as extremely painful and may be accompanied by nausea, vomiting, and cramps. Treatment includes bathing the wound in water as hot as can be tolerated without burning tissue (< 46°C, or 115°F).[10] If the barb is removed improperly, the epithelium may remain in the wound and continue to produce or release the venom for a short time, which can lead to permanent, serious tissue damage.

Myliobatiform families include Hexatrygonidae (Sixgill Stingray; 1 sp.), Urolophidae (Round Rays; 28 sp.), Urotrygonidae (Smalleyed Round Rays; 16 sp.), Plesiobatidae (Deepwater Stingray; 1 sp.), Dasyatidae (Stingrays; 98 sp.; the largest myliobatiform family), Potamotrygonidae (River Rays; 38 sp.; unique among the myliobatiforms in that the group has secondarily invaded freshwater, all in the Amazon basin, and one species has reinvaded marine environments; see chapter 8), Gymnuridae (Butterfly Rays; 12 sp.), Myliobatidae (Eagle Rays; 18 sp.), Aetobatidae (Pelagic Eagle Rays, 5 sp.), Rhinopteridae (Cownose Rays; 8 sp.), and Mobulidae (Manta and Devil Rays; 9 sp.).

Is it a Skate or a Ray: A Quick Review

Here is a quick guide to distinguish the orders of batoids. First, rule out the torpediniforms and rhinopristiforms, based on their large caudal and dorsal fins. That leaves the orders Rajiformes (the skates) and Myliobatiformes (the stingrays). If the individual in question has thorns (also called *bucklers*) along its dorsal midline, has a bilobed pelvic fin, rigid rostrum, and a relatively stout tail without spines, and it lays eggs, then it is a skate. If instead, the individual lacks bucklers, has a fleshy snout, single-lobed pelvic fin, and a whip-like tail (or very short tail), often with one or more spines, then it is a ray.

Superorder Squalomorphii (Dogfish Sharks)

The superorder Squalomorphii (*squal*= dogfish) is a diverse group of 183 species in five orders and eleven families. They are typically found in colder water (e.g., at mid- to high latitudes and/or in deep water).

Squalomorph sharks are characterized by two synapomorphies (common derived characteristics): loss of an anal fin in all but the most primitive group and orbitostylic jaw suspension (box 3.1), an intermediate form that allows a moderate level of jaw mobility (that is, *cranial kinesis*; see chapter 4). Many have evolved enlarged spiracles. All squalomorph sharks are yolk-sac viviparous, with some employing limited histotrophy through mucoid uterine secretions.

ORDER HEXANCHIFORMES (COW SHARKS)

The order Hexanchiformes consists of two primitive families, Chlamydoselachidae (frilled sharks) and Hexanchidae (cow sharks). The morphological similarities between the two families include having extra gill arches (6 in 5 species, 7 in 2 species), loss of the first dorsal fin, and these are the only squalomorph sharks with an anal fin. However, there are also major morphological differences between chlamydoselachids and hexanchids (see below).

Both families are predominantly deepwater, although one hexanchid is found in shallow waters in cold temperate climates, and others are likely diurnal vertical migrators.

Families in the order Hexanchiformes:

- Frilled Sharks (Family Chlamydoselachidae; *chlamy*=cloak or garment; fig. 3.6)
    - » two species, the Frilled Shark (*Chlamydoselachus anguineus*) and the Southern African Frilled Shark (*C. africana*), the latter described in 2009 and found from southern Angola to southern Namibia. Both have a head like a reptile, a terminal mouth, big green eyes, an eel-like body, gills with frilly edges that extend beyond the gill slits (hence the common name), the first slit continuous across the throat, distinctive three-pronged pitchfork-type teeth swept backward similar to a python's and unlike those of any modern shark, short rounded pectoral fins, a small dorsal set far back, and an anal fin that is larger than the dorsal. They likely feed on soft-bodied prey. The lineage to which this group belongs dates back as

Figure 3.6. (A) Frilled Shark (Paulo de Oliveira/ Biosphoto). (B) Unique pitchfork-shaped teeth found in frilled sharks.

long as 150 mya. Frilled Sharks have a prominent notochord, but no distinct vertebral centra. They are found in deep water (100–1200 m, or 330–3940 ft) and have small litters (average of six). One study estimated their gestation period at an extraordinary 3.5 years.

Representative Species

Frilled Shark (*Chlamydoselachus anguineus*; fig. 3.6). This is the best-known frilled shark. In both species, dentition is alike in upper and lower jaw and the teeth are distinct from those of all extant sharks, and more similar to the Carboniferous shark *Xenacanthus*. Each tooth is pitchfork-like with three fang-like cusps and two smaller cusps between them. These form interlocking rows and point inward. There are about 300 teeth in about 25 rows. If a prey item is unfortunate enough to be grasped by a Frilled Shark, there is no way it can escape.

The absence of a first dorsal fin may facilitate anguilliform (eel-like) swimming and rolling when feeding, similar to an alligator death roll. This species is commonly seen in Japanese trawls, but is far more widespread, including both sides of the Atlantic.

Chlamydoselachus anguineus (Frilled Shark)

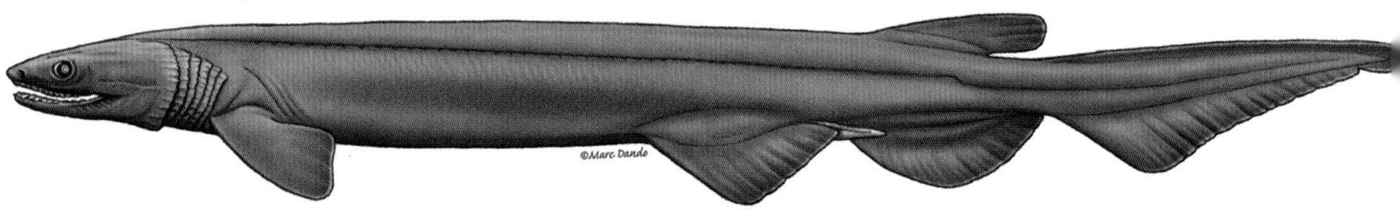

- Cow Sharks (F. Hexanchidae; *Hex*=six; *anch*=gill; fig. 3.7)
  - » five species in 3 genera, *Hexanchus, Notorynchus,* and *Heptranchias.* Hexanchids are primitive and are the sister-group to the Chlamydoselachidae. Some have large litters (up to 108 have been reported). Most are deepwater. They have big green eyes capable of harvesting dim light. They also have odd, cockscomb-shaped teeth similar to ancient sixgill sharks from >200 mya (see fig. 2.1B). These teeth are capable of sawing large chunks of flesh from large prey and carrion. This group, along with most other squalomorphs, has retractable eyes. Finally, their vertebral column can be called *notochordal* (the vertebral centra are not fused or calcified), and the soft column is so flexible that hexanchiform sharks can bite their own tails.

Representative Species

Bluntnose Sixgill Shark (*Hexanchus griseus*; fig. 3.7). The most well-known hexanchid is *H. griseus* (*hexanchus*=sixgill; *griseus*=gray). These are widely distributed, big (to ≥5 m, or 16 ft), stout-bodied predators and scavengers but with surprisingly weakly-calcified jaws and notochord, which may be explained by problems depositing calcium salts at the depth at which they spend much of their

B

Figure 3.7. Sharks of the family Hexanchidae (A) Atlantic Sixgill Shark (Courtesy of Katie Grudecki). (B) Bluntnose Sixgill Shark with eye partially withdrawn into its socket.

time. Hexanchid teeth (see fig. 2.1A) are like mini-saw blades, and the additional flexibility of weakly-calcified jaws allows the jaws to bend as they encounter large food items, such as whale falls (dead whales that sink to the bottom). This flexibility brings more of the serrated teeth in contact with the whale's flesh, allowing the shark to saw off chunks of blubber and meat.

Because of the lack of calcified skeletal structures and a large liver, this species is lighter and is close to neutrally buoyant. Most are deepwater (200–1000 m, or 650–3300 ft), a section known as the oceanic *twilight zone* where their large green eyes harvest the remaining blue-green light, but they do occur in shallow water in some cold temperate locations. They have a single, spineless dorsal fin set far back along the body, which reduces friction and allows them to spin more easily when sawing chunks of flesh from prey.

Hexanchus griseus (Bluntnose Sixgill Shark)

Atlantic Sixgill Shark (*Hexanchus vitulus*; fig. 3.7). In 2013, we captured a 1.5 m (4.9 ft) Atlantic Sixgill Shark (*Hexanchus vitulus*) on a longline in approximately 250 m (820 ft) of water near Bimini, The Bahamas. Because the shark was dead when retrieved, we dissected the specimen for educational purposes, and discovered that the rectal gland, typically finger-shaped, looked like a kidney bean, and was much smaller than expected. One explanation for the different morphology is that the genus *Hexanchus*, one of the most primitive shark genera, with fossils from the early Jurassic (> 190 mya), largely has occupied the deep sea throughout its evolution, and thus it is possible that the evolutionary age and degree of isolation of the genus has resulted in different anatomical and physiological systems.

ORDER ECHINORHINIFORMES (BRAMBLE OR PRICKLY SHARKS)

The entire order Echinorhiniformes, which is also a very primitive group, consists of only two species in one family.

The single family in the order Echinorhiniformes:

- Bramble or Prickly sharks (Family Echinorhinidae; *echino* = spiny; *rhinus* = nose or snout, due to large denticles on snout; fig. 3.8)
  - » two species of primitive, large, sluggish, deepwater, bottom-dwelling, poorly known sharks. Echinorhinids are distinguished by large, thorn-like denticles and two small, spineless dorsal fins far posterior on the body. They also lack an anal fin. They have distinct teeth with one main cusp and 3–4 smaller cusplets in both the upper and lower jaws (fig. 3.8). Bramble sharks may reach > 3 m (10 ft) and, although stout, they are relatively soft-bodied with flabby musculature.

Representative Species

Prickly Shark (*Echinorhinus cookei*; fig. 3.8). *E. cookei* is found in the Pacific Ocean.

Echinorhinus cookei (Prickly Shark)

A

B

Figure 3.8. (A) Teeth of Prickly Shark (Courtesy of D. Ross Robertson) (B) Prickly Shark.

Figure 3.9. Common Angel Shark (LuisMiguelEstevez/Shutterstock.com).

ORDER SQUATINIFORMES (ANGEL SHARKS)

The order Squatiniformes consists of one family of 23 species of ray-like sharks. They are coastal in temperate seas and deeper in the tropics. Although they resemble batoids, with pectoral fins expanded (but not connected to the head) and they in fact have some ecological similarities to batoids, they are true sharks.

The single family in the order Squatiniformes:

- Angel Sharks (Family Squatinidae; fig. 3.9)
  - » twenty three mostly shallow-water species (one-third are deep sea species), all in the genus *Squatina*, with prominent spiracles without valves. Like most squalomorphs, they are cold-water sharks. They are the only elasmobranchs with an inverse heterocercal tail (one in which the lower lobe is larger). Also called a *hypocercal* caudal fin, this type of tail fin was found in some Devonian placoderms. A hypocercal tail elevates the head, which benefits them since angel sharks are lie-and-wait predators with huge gapes on their terminal mouths, and they ambush prey from the bottom. They have been commercially fished since the 1980s and may be served as *monkfish* in some locations.

Representative Species

Angel Shark (*Squatina squatina*; fig. 3.9). Once very common, this species is now listed as critically endangered by the International Union for Conservation of Nature (IUCN) in the NE Atlantic, Mediterranean, and Black Seas, and overfishing continues.[11] It has been extirpated from the North Sea and parts of the Northern Mediterranean Sea. The only relatively healthy population occurs in the Canary Islands. In 2019, this species and two other species of angel sharks were listed as in danger of extinction by the Spanish government, prohibiting all harvest.

Squatina
(Angel Shark)

ORDER PRISTIOPHORIFORMES (SAWSHARKS)

The order Pristiophoriformes (*prist* = comb; *phor* = possessing, or bearing) are the sawsharks (not to be confused with the sawfish, which are batoids). There are 10 species, all deepwater, with two long barbels midway on the tapered rostrum, and two spineless dorsal fins. The long barbels were hypothesized to have a chemosensory or electroreceptive function but are now known to be mechanosensory or tactile in function. Like most squalomorphs, sawsharks lack anal fins. They grow to 1.7 m (5.6 ft).

It is interesting to note that the sawed rostrum, *pristification*, evolved five times in evolutionary history of chondrichthyans: the pristids (sawfish), the Pristiophoriformes (this group), the Carboniferous species *Bandringa rayi*, a group of skates that went extinct about 150 mya, the sclerorhynchids, and in at least two species of extinct holocephalans.

The single family in the order Pristiophoriformes:

- Sawsharks (Family Pristiophoridae; fig. 3.10)
    - ten species in 2 genera, *Pliotrema* (3 sp.) and *Pristiophorus* (7 sp.). The teeth of the saw typically alternate between large and small and are conical, not bladelike as in the batoid pristids. Unlike sawfish, pristiophorids are capable of regenerating rostral teeth that may be lost and adding teeth as the rostrum grows. Pristiophorids typically feed on fish, squid, and crustaceans, using their cusped, non-flattened teeth. They cruise the bottom, using the tactile barbels and ampullae of Lorenzini on the saw to detect prey buried in the sediment. They then hit victims with side-to-side swipes of the saw, crippling them. The sawsharks are the only family of modern

Figure 3.10. Eastern Sawshark (marinethemes.com/Kelvin Aitken).

elasmobranchs that includes both five-gilled and six-gilled species and are the only non-hexanchiform sharks with more than five gill slits (one species of ray, *Hexatrygon bickelli,* also has six gills).

Representative Species

Sixgill Sawshark (*Pliotrema warreni*; see fig. 10.8). This is one of three six-gilled species of sawsharks, the other two only recently described. It is vulnerable to deep benthic trawls but is listed as *Least Concern* by the IUCN. It is found off South Africa and Mozambique at depths ranging from 10 to 915 m (33–3000 ft).

Pliotrema warreni (Sixgill Sawshark)

ORDER SQUALIFORMES (DOGFISH SHARKS)

The order Squaliformes consists of 141 species in 6 families. This diverse order of sharks includes the smallest species of shark (the Dwarf Lantern Shark [*Etmopterus perryi*] reaches only 20 cm [8 in]) and among the largest sharks (the Greenland Shark [*Somniosus microcephalus*] reaches 7 m [23 ft]). All have five gill slits, two dorsal fins (most with spines, which is not characteristic of the other squalomorphs), no anal fin, a more protrusible jaw than other squalomorphs, and large spiracles. They are typically cold water or deepwater sharks. The Spiny Dogfish (*Squalus acanthias*) and the Pacific Spiny Dogfish (*S. suckleyi*) are shallow, cold-temperate species in this order. Though two species of the large sleeper sharks, the Greenland Shark (*Somniosus microcephalus*) and the Pacific Sleeper Shark (*S. pacificus*), can be found to depths of 2000 m (6700 ft) or more, they can also be found in shallow estuaries in the boreal parts of their ranges. With these exceptions, the members of this order are deep water species.

Families in the order Squaliformes:

- Dogfish Sharks (Family Squalidae; fig. 3.11)
    - » forty species in 2 genera, *Squalus* and *Cirrhigaleus.* These true dogfish are found globally in tropical, temperate, and boreal seas, in nearshore to 800 m (2624 ft) or greater. Cylindrical in cross section, they possess two dorsal fins with strong ungrooved but mildly venomous spines. Teeth in both jaws are small, similar in size, and are arranged in a single overlapping functional row like a knife blade (fig. 3.11). Their life-history characteristics make them vulnerable to overfishing (see chapter 11). We include the Spiny Dogfish (*Squalus acanthias*) as the representative species, but it is in fact the oddball since it is one of the few coastal members of the family. The Hawaiian Spurdog (*Squalus hawaiiensis*; fig. 3.11) is a more typical squalid shark.

Figure 3.11. (A) Dentition of sharks in the family Squalidae. (B) Hawaiian Shortspine Shortspur.

A

B

Representative Species

Spiny Dogfish, Spurdog, Piked Dogfish (*Squalus acanthias*). This species is named for the presence of spines preceding both dorsal fins. It is intertidal to 900 m (2950 ft) and is the only *Squalus* in the Atlantic which is also coastal. The sister-species in the Pacific, the Pacific Spiny Dogfish (*S. suckleyi*), also has a coastal and deep-sea presence. The Spiny Dogfish age of maturity is about 20 years, and the gestation period of 22–24 months is among the longest known in animals. The IUCN lists it as *Vulnerable*, with the Northeast Atlantic population considered *Critically Endangered*.

- Gulper Sharks (F. Centrophoridae; *centro* = spine, or point; *phor* = possessing, or bearing)
    - » sixteen species of deepwater sharks in 2 genera, *Centrophorus* (12 sp.) and *Deania* (4 sp.). The family is named for the short, stout, pointed, grooved dorsal spines. Gulper sharks are deepwater sharks found throughout the world's oceans (except in polar regions) with cylindrical bodies; low and long dorsal fins; huge green to yellowish eyes; and long, narrow rear lobes of the pectoral fins. They reach 1–2 m (3.3–6.6 ft). Gulper sharks have erect, cusped teeth on the upper jaw and bladelike teeth, like the squalids, on the lower jaw. Their bodies are *very* slimy.

Representative Species

Gulper Shark (*Centrophorus granulosus*). This is a slender shark with green eyes and a long snout. It is among the largest gulpers and is frequently caught on deepwater longlines. The liver of Gulper Sharks has a large volume of oil. Their conservative life history makes them very susceptible to overfishing. Gulper Sharks are listed as *Endangered* by the IUCN.

Figure 3.12. Green Lantern Shark (adults and juveniles). Note the photophores (bioluminescent organs) on the underside of the head.

- Lantern Sharks (F. Etmopteridae; fig. 3.12)
  - » fifty-three species in 4 genera, including among the smallest sharks, named for their bioluminescent photophores, which essentially mimic the frequency of downwelling light and make the silhouette disappear when viewed from below. They range in size from 20 to about 100 cm (8 to 39 in). They are found globally, either in the bathypelagic (1000–4000 m, or 3280–13,120 ft) or mesopelagic (200–1000 m, or 650–3280 ft) zones. Etmopterids are also distinguished by two dorsal fins with grooved spines, the second fin and spine usually larger than the first, and abundant photophores in species specific patterns mostly on the ventral surface.

Representative Species

Dwarf Lantern Shark (*Etmopterus perryi*). This species is noteworthy because it may be the world's smallest shark, reaching only about 0.2 m (8 in).

Etmopterus perryi (Dwarf Lantern Shark)

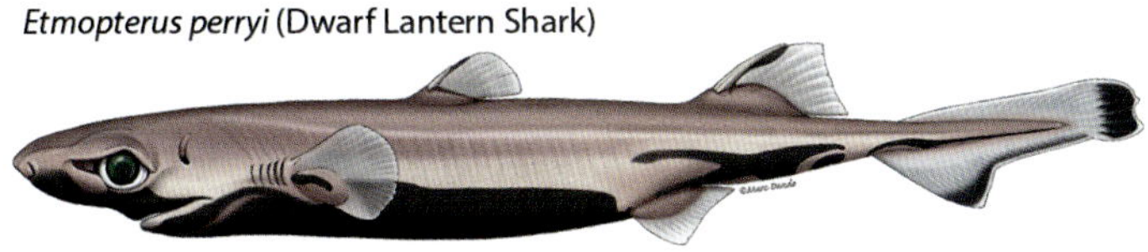

Viper Dogfish (*Trigonognathus kabeyai*). This shark is known only from a few localities, including off Japan, Taiwan, and the northwest Hawaiian Islands. It has long, dagger-like teeth. Its morphology is similar to that of the osteichthyan dragonfishes (Family Stomiidae), an example of convergent evolution.

Trigonognathus kabeyai (Viper Dogfish)

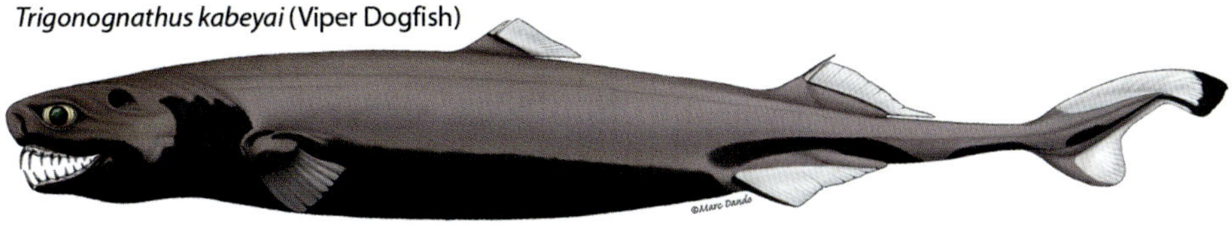

Figure 3.13. Greenland Shark from the North Atlantic (Courtesy of Charles Cotton and the MAR-ECO project, www.mar-eco.no).

- Sleeper Sharks (F. Somniosidae; fig. 3.13)
  - » (*Yawn*). Seventeen species in 6 genera of deepwater benthic and oceanic species that are dark brown to pitch black, ranging from small (40 cm, or 1.3 ft) to gigantic (> 6 m, or 20 ft). All have a broad head; flat snout; large spiracles; two small dorsal fins; very small, deep green eyes; and low angular or rounded pectoral fins.

Representative Species

Greenland Shark (*Somniosus microcephalus*; *microcephalus* = small head; fig. 3.13). This is the largest high-latitude shark, reaching 7 m (23 ft) or longer. Although it appears slow moving, its diet consists of active prey (seals, moose, and polar bears have been found in its digestive tract). Two scientific discoveries on this species have occurred recently. First, it was observed on camera and ultimately caught at about 1750 m (5740 ft) in the Gulf of Mexico. Given that water temperature at that depth in the Gulf, about 4°C (39°F), is very similar to those in its known high-latitude home, this discovery was not too surprising. Second, a 2016 study used radiocarbon dating to estimate a mean age of 392 (± 120) years for one specimen. If valid, this would make the Greenland Shark the oldest known living vertebrate.

Somniosus microcephalus (Greenland Shark)

- Rough Sharks or Prickly Dogfish (F. Oxynotidae; fig. 3.14)
  - five species, all in the genus *Oxynotus*, found on continental and insular shelves and slopes in the Eastern Atlantic (including the Mediterranean), Western Atlantic, and Western Pacific. They are easily distinguished by their very high and compressed body, which is triangular in cross section, with high, sail-like dorsal fins, each with a large spine that may be obscured by the fin. There is a lateral ridge on the abdomen between the pectoral and pelvic fins. Their skin is very rough (hence, the name), with a luminous organ. They remain lightly studied but diet studies suggest they specialize in eating embryos of oviparous chondrichthyans, making a small incision in the egg cases and sucking the contents out. Molecular data suggest *Oxynotus* are nested within the family Somniosidae and thus this family is paraphyletic.

Representative Species

Caribbean Rough Shark (*Oxynotus caribbaeus*; fig. 3.14).

Oxynotus caribbaeus (Caribbean Rough Shark)

Figure 3.14. Caribbean Rough Shark (Courtesy of NOAA).

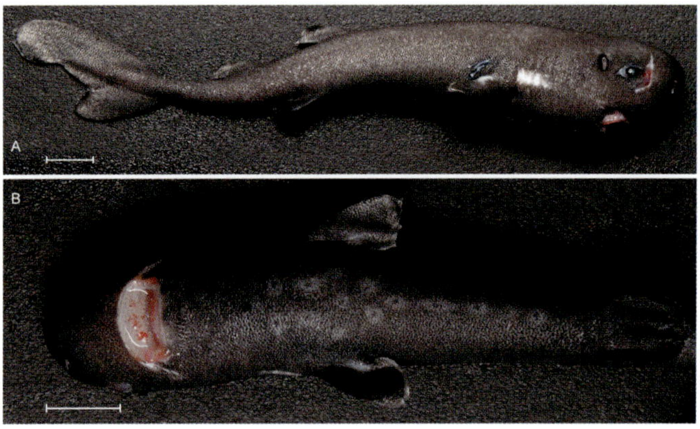

Figure 3.15. American Pocket Shark (Courtesy of NOAA).

- Kitefin Sharks (F. Dalatiidae; figs. 3.15 and 3.16)
  - » ten species in 7 genera of dwarf to medium-sized (< 2 m, or 6.6 ft) deep-water sharks, with a narrow head and conical, short snout. Their dorsal fins (or first fin only) are without spines. They are found globally. This family includes the pocket sharks, a rare group known only from two species, the Pocket Shark (*Mollisquama parini*) from the South Pacific, and the American Pocket Shark (*Mollisquama mississippiensis*), recently described from a midwater trawl in the Gulf of Mexico.[12] While these are noteworthy because of their rarity and, well, the cuteness of their bulbous head, what stands out about them is their eponymous pockets, bilateral pouches located dorsal to the insertions of the pectoral fins. Unique to this genus, these muscular glands are thought to secrete a luminous fluid (fig. 3.15) that could function to attract prey or fool predators.

Representative Species

Cookiecutter or Cigar Shark (*Isistius brasiliensis*; fig. 3.16). This iconic species is known for its unusual lifestyle. Cookiecutter Sharks are neutrally buoyant vertical migrators with luminous organs that attract larger fish and mammals.[13] They possess photophores (bioluminescent organs) on the underside except for a darkly pigmented collar. The light emitted from these photophores matches the downwelling light, so the silhouette of the shark disappears from beneath, except for the darkly pigmented collar, which stands out and may trick a potential predatory fish beneath into interpreting that it is a small fish. When the fish investigates, the Cookiecutter Shark accelerates toward the erstwhile predator stealthily and sometimes at great speed using its relatively large and powerful caudal fin and attacks, scooping out a chunk of flesh . Despite their small size (up to 0.6 m, or 2 ft), Cookiecutter Sharks have powerful jaws and large teeth (fig. 3.16A), which are more heavily calcified than in other squaliforms. In addition to fish and marine mammals (fig. 3.16B), they have been known to attack nuclear submarines, and there are multiple cases of attacks on people over deep water at night off Hawaii.

Isistius brasiliensis (Cookiecutter or Cigar Shark)

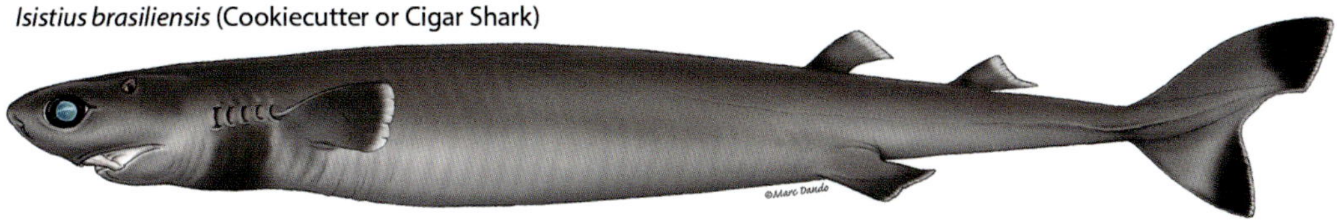

A

B

Figure 3.16. (A) Mouth of Cookiecutter Shark showing fleshly lips and teeth. (B) Cookiecutter Shark scars on an Opah at a fish auction in Hawaii.

Kitefin or Seal Shark (*Dalatias licha*; fig. 3.17). This deepwater shark (usually 200–600 m, or 660–1970 ft) is included here because the morphology of its mouth so closely resembles that of the Cookiecutter Shark that it very likely shares the same mode of feeding, although the Cookiecutter Shark is the only shark recognized for this method of feeding.

Pygmy Shark (*Euprotomicrus bispinatus*). At about 25 cm (10 in), the Pygmy Shark is the second smallest known shark.

A

B

Figure 3.17. (A) Kitefin, or Seal, Shark. (B) Mouth and teeth of this species, which show striking similarities to those of the Cookiecutter Shark, and which likely are used to feed in a manner similar to that of the Cookiecutter Shark.

Euprotomicrus bispinatus (Pygmy Shark)

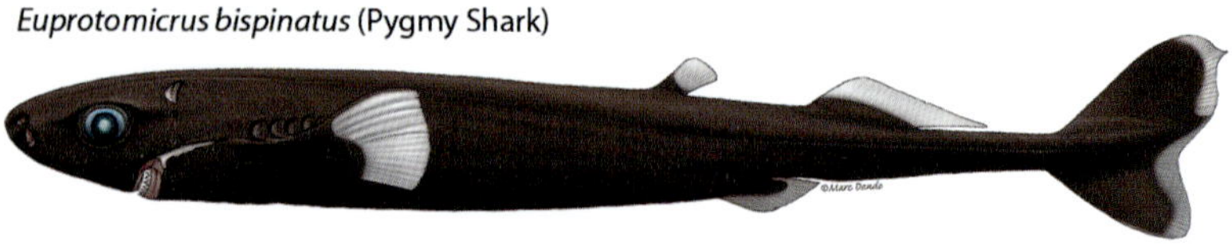

Superorder Galeomorphii (Galea or Galeomorph Sharks)

The galeomorph sharks are a diverse, monophyletic group ranging from small demersal sharks with tiny, cusped teeth to large, pelagic predators with bladelike teeth. These include the beasts that come to mind when people think of sharks, but the group also includes those whose form has diverged. There are about 367 species in 24 families and 4 orders. They are mostly coastal and pelagic in warmer climates, but the largest family is found in cold deep waters. All have anal fins as well as two well-developed dorsal fins that lack spines (except in the horn sharks). They have hyostylic jaws which affords high producibility in many species, and a variety of rostrum shapes. The four orders fall into two clades, with the Heterodontiformes and Orectolobiformes as one sister-group, and the Lamniformes and Carcharhiniformes as another (see fig. 3.1). A major difference between them is the morphology of the snout, with the former group possessing short snouts and no expanded rostral cartilage and the latter possessing elongated rostral cartilages and larger snouts.

ORDER HETERODONTIFORMES (BULLHEAD OR HORN SHARKS)

The Heterodontiformes (*hetero* = different; *dont* = teeth) include a single family, the Heterodontidae, and ten widely distributed Pacific and Indo-Pacific, primarily nocturnal, bottom-dwelling species (fig. 3.18). All have heterodont dentition, with cusped teeth in front for grasping and molariform teeth in the rear for crushing prey like sea urchins, using hypertrophied jaw muscles (see figs. 4.19 and 4.21). Occasionally the sea urchin spines will pierce the heart, which would likely be fatal were it not for a tough heart and the pericardio-peritoneal canal, which drains blood from around the heart before it fatally compresses it (see chapters 1 and 7). They have spines on both dorsal fins, a plesiomorphic character not found in other galeomorph sharks. Some have ocular spines (spines over their eyes). They also have prominent supraorbital crests; hence the name *horn shark*. Heterodontiform sharks also possess an *oronasal groove* (see fig. 5.7), which directs seawater from the nares to the mouth, where it can be sensed a second time. They have tiny spiracles and no nictitating membrane. The group is entirely oviparous. They are typically < 1 m (3.3 ft) long.

Among galeomorph sharks, horn sharks are considered basal and are most closely related to the order Orectolobiformes, with whom they share a common ancestor about 200 mya, during the second major radiation of sharks. One exam-

Figure 3.18. Port Jackson Shark (Dirk van der Heide/Shutterstock.com).

ple of their close relationship is the coloration of *Heterodontus* and the Nurse Shark, *Ginglymostoma*. Young of both species are vividly spotted, whereas adults are darker brown. Heterodontids superficially resemble the hybodonts, which lived at the end of Carboniferous through the Jurassic and are believed to include the ancestors to modern elasmobranchs.

The single family in the order Heterodontiformes:

- Bullhead or Horn Sharks (Family Heterodontidae; fig. 3.18)
    - ten species of nocturnally active, sluggish, bottom-dwellers in the Eastern Pacific and Western Indian Oceans, all with a truncated snout with reduced or absent rostral cartilage, which makes the snout appear attenuated (or squished). This shape facilitates eating invertebrates, which they suck into their mouth.

Representative Species

Port Jackson Shark (*Heterodontus portusjacksoni*; fig. 3.18). This species is perhaps the best-known member of its family. It is found along the coast of southern Australia.

Galapagos Bullhead Shark (*Heterodontus quoyi*). This species is found in shallow waters (to 40 m, or 131 ft) in the Eastern Tropical Pacific.

ORDER ORECTOLOBIFORMES (CARPET SHARKS)

The Orectolobiformes (*orecto* = stretched out, spread out; *lob* = lobe, fin), or carpet sharks, consist of about 45 species in seven families, which represents low diversity at the species level but higher diversity at the family level (numerous families with relatively few species). There is strong morphological and molecular support, however, to unite the nurse, zebra, and whale sharks, currently in three separate families, into a single family. There is a wide array of body shapes in the order. This group includes the Nurse Shark (*Ginglymostoma cirratum*), Whale Shark (*Rhincodon typus*), and wobbegongs. All except one species are demersal (bottom-associated). They have a short snout, with only a short central rostral cartilage, two spineless dorsal fins, and a very short mouth well in front of the eyes. Like the Heterodontiformes, most have prominent nasoral grooves. Their spiracles are moderate to large and below the eye, except in the Whale Shark. Many have small gill slits, with the fourth and fifth slits overlapping. All except two species are Indo-Pacific. Reproduction ranges from oviparity to yolk-sac viviparity.

Families in the order Orectolobiformes:

- Collared Carpetsharks (Family Parascyllidae; fig. 3.19)
  - » eight species in 2 genera, all of which are intertidal to high subtidal in the Western Pacific. These are among the few sharks with an eel-like niche, scuttling and slithering using their pectorals. Parascyliids are oviparous.

Representative Species

Collared Carpet Shark (*Parascyllium collare*; fig. 3.19).

Figure 3.19. Collared Carpet Shark (marinethemes.com/Kelvin Aitken).

- Blind Sharks (F. Brachaeluridae)
  - » two species, both coastal off Australia. Despite its common name, blind sharks have typical orectolobiform vision. These demersal sharks are yolk-sac viviparous retaining eggs internally through hatching.

Representative Species

Blind Shark (*Brachaelurus waddi*). This species is nocturnal and is found in caves and tidepools.

Brachaelurus waddi (Blind Shark)

- Wobbegongs (F. Orectolobidae; fig. 3.20)
  - » twelve species of demersal, shallow water sharks, most of which occur in the Western Pacific and Indo-Pacific from Japan to Australia. They have a wide head, flattened body, a mouth that is close to terminal, spiracles larger than eyes, and prominent nasal barbels. They are sluggish and eat benthic invertebrates but also employ ambush predation to capture fish using very strong jaws. Most are small, but they can exceed 3 m (9.9 ft). They are yolk-sac viviparous.

Representative Species

Ornate Wobbegong (*Orectolobus ornatus*; fig. 3.20). The Ornate Wobbegong is a nocturnal ambush predator.

Orectolobus ornatus (Ornate Wobbegong)

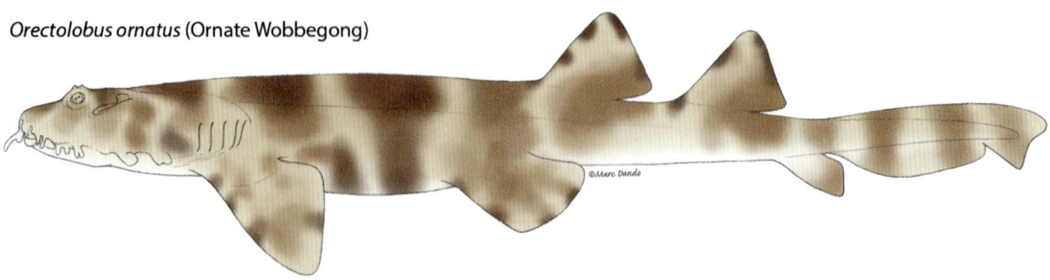

Figure 3.20. Ornate Wobbegong.

- Bamboo or Longtailed Sharks (F. Hemiscylliidae; *hemi* = half; *scyll* = sea monster)
    - » seventeen species of small (most less than 0.5 m, or 1.6 ft), sluggish, tropical Indo-Pacific sharks mostly associated with coral reefs, with tails that may be longer than the rest of their body. They are popular aquarium sharks. Most are oviparous.

Representative Species

Epaulette Shark (*Hemiscyllium ocellatum*). Named because of a black spot outlined in white behind the pectoral fins, which resembles military epaulettes, this species is capable of crawling over emergent rocks and coral.

Hemiscyllium ocellatum (Epaulette Shark)

Grey Bamboo Shark (*Chiloscyllium griseum*). This is a common aquarium species and is also eaten by people.

Chiloscyllium griseum (Gray Bamboo Shark)

- Nurse Sharks (F. Ginglymostomatidae; *ginglymo* = hinge; *stoma* = mouth; fig. 3.21)
    - » four species, one Atlantic and three Pacific, in shallow, tropical and subtropical waters. They are durophagous (eat hard-bodied prey) and have jaw structure and smaller mouths similar to those of horn sharks but with less hypertrophied muscles and lack heterodont teeth. They employ a *suck, crush, spit, repeat* mode of feeding on bottom invertebrates (mostly mollusks and crustaceans).

      They are chocolate brown, with two nearly equal-sized dorsal fins set far back on the body. Their mouth is near-terminal, with long protruding tactile barbels. They can reach 3–4.3 m (10–14 ft) and are yolk-sac viviparous retaining egg cases internally.

Representative Species

Nurse Shark (*Ginglymostoma cirratum*; fig. 3.21). This inshore, bottom-dwelling, largely nocturnal species is robust and reaches about 3 m (10 ft).

Figure 3.21. Juvenile Nurse Shark (Courtesy of Chelle Blais).

- Zebra Sharks (F. Stegostomatidae; *stego* = cover; *stoma* = mouth)
  » monotypic, one species in the Indo-Pacific, which occupies a niche very similar to that of the ginglymostomatids. The common name applies to the somewhat zebra-like pattern of juveniles. Adults are spotted, with five prominent longitudinal ridges and with a caudal fin almost as long as the rest of the body. They are common in public aquaria. They are oviparous.

Representative Species

Zebra Shark (*Stegostoma tigrinum*).

- Whale Shark (F. Rhincodontidae; fig. 3.22)
  » monotypic, iconic family. The Whale Shark is a huge, planktivorous, filter-feeding shark with a terminal mouth. It is the largest fish (up to 18 m, or 59 ft) and the only pelagic orectolobiform. It is cosmopolitan in tropical and warm-temperate waters. Large aggregations of Whale Sharks have been observed in about 20 so-called hotspots, including off the coasts of Australia, Belize, the Maldives, and the Gulf of Mexico, where they gather to feed (fig. 3.23).

  In some parts of their range, Whale Sharks are killed for meat, liver oil, cartilage, and especially their large fins, which are exported. The 2001 documentary *Shores of Silence: Whale Sharks in India* exposed the unregulated artisanal Whale Shark fishery along NW India. The film led the Indian government to list the species on the Indian Wild Life (Protection) Act, with additional conservation measures by non-governmental organizations like the Wildlife Trust of India. The efforts have drastically reduced but not eliminated the Whale Shark catch there.

  Whale Sharks are yolk-sac viviparous, retaining egg cases internally; one specimen harpooned by fishers contained 300 embryos. Surprisingly, Whale Sharks are maintained successfully in captivity, even after being shipped 8000 miles.

Representative Species

Whale Shark (*Rhincodon typus*; fig. 3.22).

Rhincodon typus (Whale Shark)

Figure 3.22. Whale Shark (Chainarong Phrammanee/Shutterfly.com).

Figure 3.23. Low-altitude aerial photograph of Whale Shark aggregation from 2009 showing 68, along with tourist boats (From de la Parra Venegas, R. *et al.* 2011. PLoS One, 6(4): p.e18994).

ORDER LAMNIFORMES (MACKEREL SHARKS)

The order Lamniformes includes only 15 species in seven families (some monotypic); it is diverse only at higher taxonomic levels. This group includes the ultimate predators as well as planktivores. All have two spineless dorsal fins, the second typically much smaller than the first, and usually a more conical snout without any lateral expansion of rostral cartilage (an adaptation for speed). They are mostly pelagic, inhabiting cool waters, but some are deep sea. They have a variety of tooth morphologies. Lamniform sharks all lack a nictitating membrane. All have anal fins. All species with known reproductive modes are viviparous with oophagy (egg-eating); that is, during development the embryos consume unfertilized ova continuously produced by the mother during her pregnancy.

Families in the order Lamniformes:

- Sand tigers (Two families: Carchariidae, *kárcharos* = saw-like; and Odontaspididae, *odont* = teeth, *aspid* = shielded)
  - » Three sand tiger species historically comprised the family Odontaspididae based on similarities in tooth and body shape. Today, the poorly known deep-water (300–500 m, or 985–1640 ft) species *Odontaspis ferox* and *O. noronhai* remain in the family Odontaspididae. The better-known Sand Tiger, *Carcharias taurus*, represents Carchariidae.

Representative Species

The Sand Tiger (*Carcharias taurus*), also known as the Grey Nurse Shark in Australia or the Ragged Tooth Shark in South Africa, is the best-known of the sand tigers. Common in aquariums, Sand Tigers are slow swimming, occasionally gulping air into their stomachs to hover. Like all sand tigers, *C. taurus* is moderately large (up to 3.5 m, or 11.5 ft) with a sizable mouth; long, pointed teeth with one or more cusplets on either side; and dorsal fins set back on the body. Unlike odontaspidid sand tigers, the carchariid Sand Tiger is a coastal species of the Mediterranean, Atlantic, and Indo-Pacific. They are oophagous and embryophagous, meaning that the embryos will consume each other in utero until only one remains in each uterus. The Sand Tiger, *C. taurus*, is listed by the IUCN as Critically Endangered.

- Crocodile Shark (F. Pseudocarcharidae; fig. 3.24)
  - monotypic, one species, the Crocodile Shark (*Pseudocarcharias kamoharai*), which grows to 1 m (3.3 ft), and is a mesopelagic, diurnal vertical migrator found in the tropical Atlantic and Indo-Pacific. It has a pointed snout, with large eyes, and dagger-shaped teeth, which may grab squid before swallowing them whole. They often have a white spot on their cheeks.

Representative Species

Crocodile Shark (*Pseudocarcharias kamoharai*; fig. 3.24).

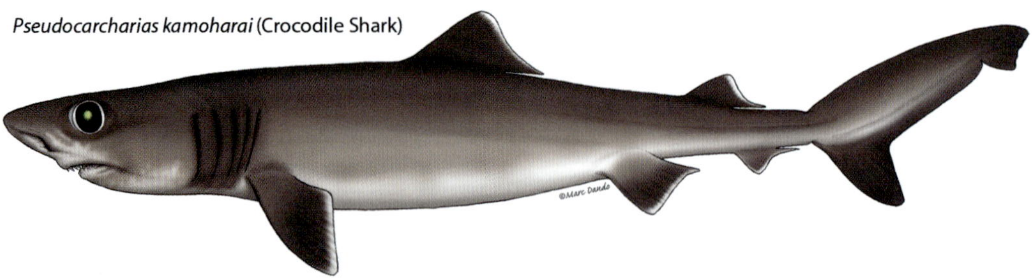

- Goblin Shark (F. Mitsukurinidae)
  - » This is a monotypic family whose single member, the Goblin Shark (*Mitsukurina owstoni*; fig. 3.25), has a flabby, elongate body, tube-like nostrils,

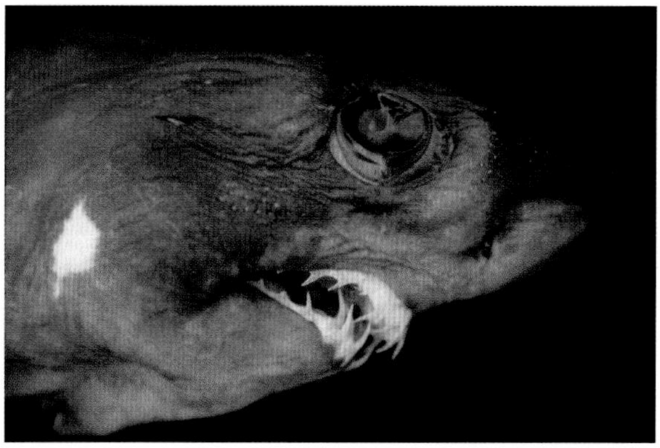

Figure 3.24. Head of Crocodile Shark.

blade-shaped snout (unique among lamniforms), and extremely protrusible jaws with dagger teeth. Like other sharks with dagger or bladelike teeth, its jaws are relatively weak, but highly protrusible. The jaws are drawn back stretching ligaments that then rapidly shoot the jaws forward like a slingshot to grab prey. The Goblin Shark lives in deep waters (550 m, or 1800 ft) in the Atlantic, Indian and Pacific Oceans, where it feeds on fish and squid.

This species has an extremely large liver, which makes them nearly neutrally buoyant. It possibly grows to 3.5 m (11 ft) and is probably yolk-sac viviparous with oophagy. Pictures of the Goblin Shark often depict it with the jaws protruded, which is not the case when they are swimming and is more likely the result of photographing or drawing a dead specimen.

Representative Species

Goblin Shark (*Mitsukurina owstoni*; fig. 3.25).

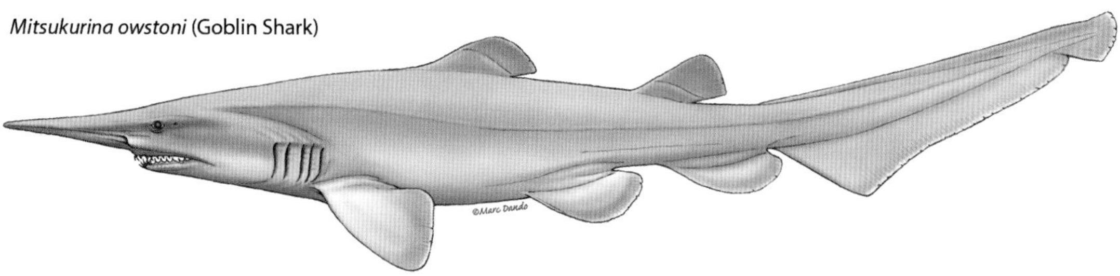

- Megamouth Shark (F. Megachasmidae; fig. 3.26)
  - » Another monotypic family, the Megachasmidae was unknown to science until 1976, when a Megamouth Shark (*Megachasma pelagios*), then unnamed, was caught enwrapped in a sea anchor off Hawaii. The fact that a shark of that size and unique appearance had eluded the notice of the marine biology community was considered astounding. The event is used to this day to support the absurd idea that a *Megalodon* could also be

Figure 3.25. Goblin Shark (Paulo de Oliveira/Biosphoto).

lurking in the deep, undetected by science. (In fact, a more defensible argument is that greatly expanded exploration of the deep sea would undoubtedly add new sharks, but they would be small and brown). The next specimens were not caught until 1984 and 1987. As of 2016, there were about 67 documented catches. That number exploded the next year when the captures of 32 specimens in 2013 and 2014 alone off Taiwan were reported. As of this writing, the number of confirmed specimens is nearing 300, with more than half of these occurring off Taiwan! In 2022, for the first time two free-swimming Megamouth Sharks were filmed off the coast of California. The species is probably found globally in the tropics, but fewer in the Atlantic (only one has been caught in the North Atlantic, in Puerto Rico in 2017)

Among the largest sharks, measured Megamouths have ranged from 177 to 570 cm long (5.8 to 18.7 ft). The Megamouth Shark is a planktivore, with fleshy lips that were hypothesized to possess luminescent photophores to attract its food source. This would be the only shark outside of the Squaliformes to possess photophores and alas—this theory has not been validated. However, the Megamouth is thought to feed differently than other elasmobranch planktivores, by swallowing water and then forcing it forward through the gill rakers, which strain out the plankton.

Representative species:

Megamouth Shark (*Megaschasma pelagios*; fig. 3.26).

Figure 3.26. Megamouth Shark (Bruce Rasner/Jeffrey Rotman/Biosphoto).

- Threshers (F. Alopiidae; *Alop* = fox; fig. 3.27)
   » three species, all large (up to 6 m, or 20 ft) with large eyes and small mouths. Their most identifiable feature is their curved, elongate, whip-like caudal fin, whose upper lobe may be one-half the total body length. They reside inshore to pelagic waters (surface to > 350 m, or 1150 ft) in all warm tropical seas. The mouths of the Common Thresher (*Alopias vulpinus*) and the Pelagic Thresher (*A. pelagicus*) are small with weak jaws, all that is needed for consuming small pelagic shoaling fish like anchovies and sardines. The elongated tail is used to first cause schools of their prey to tighten before swatting and stunning members prior to eating them. The Bigeye Thresher (*A. superciliosus*) eats larger prey, has a somewhat larger mouth and larger teeth and is the only pelagic shark that has dorsal binocular vision, allowing them to target with the tail individual prey swimming above them near the surface. This feeding strategy is likely why Bigeye Threshers are frequently hooked in the tail by commercial fishers. They are all yolk-sac viviparous with oophagy.

   The presence of the elongate caudal fin comes with a hydrodynamic cost. You will see in chapter 4 that the extended upper lobe of the caudal fin causes the net force exerted by the fin to be upward at about a 45° angle. Thus, to compensate for the elevation of the rear of the body, thresher sharks typically have greatly expanded pectoral fins that act as planing surfaces and provide lift.

   The Pelagic Thresher is listed as *Endangered* and the other two species are listed as *Vulnerable* by the IUCN.

Representative Species

Common Thresher (*Alopias vulpinus*). This is the most common thresher along the US East Coast. They have tiny teeth. *A. vulpinus* gives birth to two to six live pups born at 1.5 m (5 ft) long.

Alopias vulpinus (Common Thresher Shark)

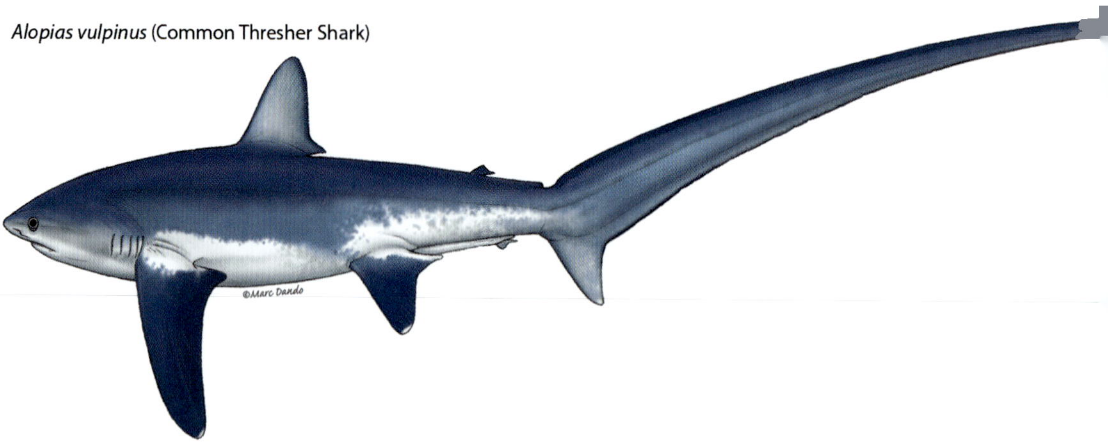

Bigeye Thresher (*Alopias superciliosus*; fig. 3.27A). The Bigeye Thresher is aptly named. Not only is the eye larger, it also is oriented for binocular vision over its head, the only non-demersal shark so adapted. This ability enables them to attack prey above them.

Pelagic Thresher (*Alopias pelagicus*; fig. 3.27B. The Pelagic Thresher is found only in the Pacific and also has huge pectoral fins, though not as large as those of the Bigeye Thresher.

- Basking Shark (F. Cetorhinidae; *cet* = whale; *rhin* = nose; fig. 3.28)
  - » A monotypic family, containing only the Basking Shark (*Cetorhinus maximus*), the second largest shark (and fish as well), which is located globally in cool water, but has been found to migrate from New England to the Amazon River plume at a depth of about 500 m (1640 ft). They grow to as large as 10 m (33 ft). They are planktivorous, with huge gill slits almost encircling their head. They filter plankton using gill rakers, which are deciduous (they fall out and are replaced annually).

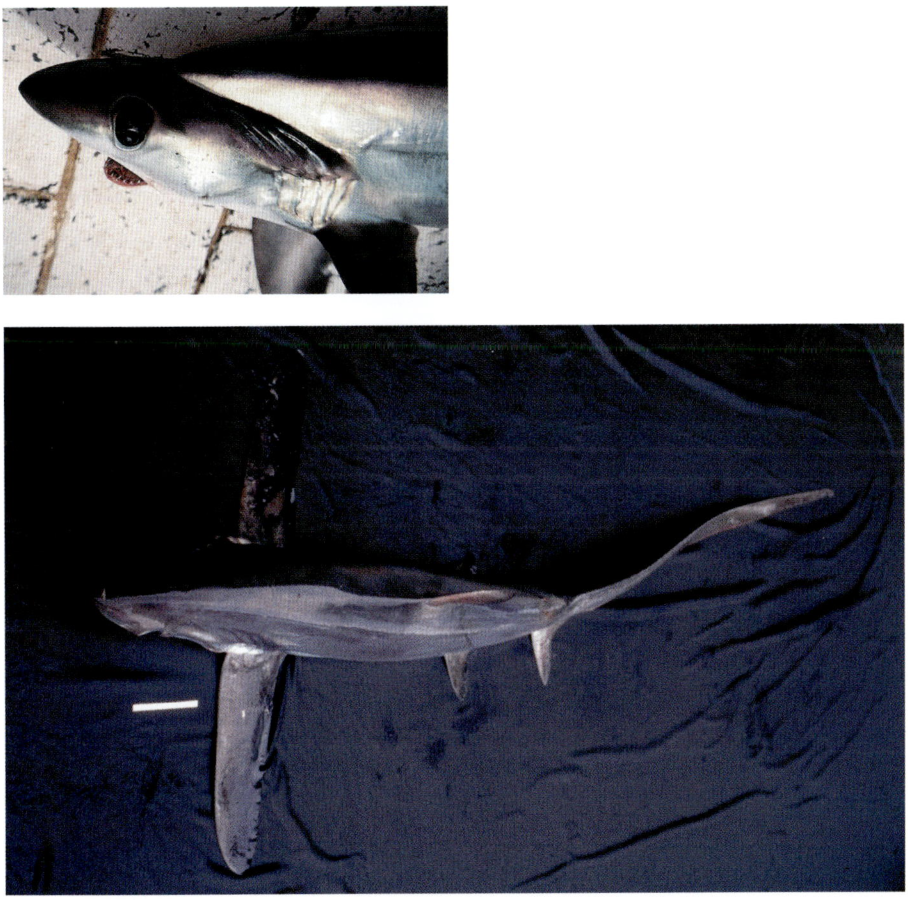

A

B

Figure 3.27. (A) Head of Bigeye Thresher, showing upward-focusing eye (NOAA Fisheries/Apex Predators Program). (B) Pelagic Thresher. Note the expansiveness of the pectoral fins.

Basking Sharks supported a large fishery in the North Atlantic until the mid- twentieth century when populations crashed (see chapter 11). Populations have stabilized or began to recover but globally they are listed by the IUCN as *Endangered.* Aerial surveys along the NE US coast beginning in 1980 revealed aggregations as large as 1400 individuals.

The size of the brain of the Basking Shark is small relative to its body weight,[14] which is attributed to its planktonic feeding, a mode that involves weak swimming and thus lower sensory requirements. They recently have been observed leaping from the water along the coast of Ireland.

Representative Species

Basking Shark (*Cetorhinus maximus*; fig. 3.28).

Cetorhinus maximus (Basking Shark)

- Mackerel Sharks (F. Lamnidae; figs. 3.29–3.31)
  - » five species in 3 genera, characterized by a large fusiform body (as big as 4–7 m, or 13–23 ft), lunate (superficially homocercal) caudal fin, and pointed snout. They also maintain their body temperature above ambient, which allows them to develop more muscle power and faster processing times for sensory information (see chapter 8). All are pelagic, though some species can occur coastally (typically in 150 to 1000 m, or 490 to 3280 ft) in most temperate and some tropical seas, and they feed on fish, cephalopods, and marine mammals. They are yolk-sac viviparous with oophagy.

Representative Species

Shortfin Mako (*Isurus oxyrinchus*; fig. 3.29). This species is perhaps the most magnificent fish in the sea, with its beautiful coloration (brilliant blue or purple on top, white on the bottom) and its near perfect streamlining. The species is aptly

Figure 3.28. Basking Shark filter-feeding (Martin Prochazkacz/Shutterfly.com).

Isurus oxyrinchus (Shortfin Mako)

named: *isurus* = equal tail; *oxyrinchus* = sharp snout. It is found globally in temperate and tropical waters.

Distinguishing features include a conical snout, moderately short pectoral fins, origin of first dorsal posterior to the free rear tips of pectoral, lunate tail, dagger teeth, and huge gill slits. They grow to > 4 m (13 ft). They swallow some of their prey whole (bony fishes and cephalopods, predominantly) and are considered opportunistic apex predators. As they age, their teeth become broader and flatter, enabling them to widen their prey options to include organisms too large to swallow whole but from which they can remove a chunk of flesh (e.g., swordfish, tuna, sharks, sea turtles, and marine mammals). Both the Shortfin and the Longfin Mako (*I. paucus*) have been captured or observed impaled with swordfish bills in their head regions. They are yolk-sac viviparous with oophagous embryos. As a result, the developing embryos, as many as 18, have enlarged guts known as egg (or yolk) stomachs (see fig. 6.16A).

Porbeagle (*Lamna nasus*). The Porbeagle resembles the Salmon Shark (*L. ditropis*; fig. 3.30), including the second caudal keel. However, the Porbeagle has three-cusped teeth (the Salmon Shark has only one additional, small cusplet), and it has a white blotch at the posterior of its first dorsal fin. It is found globally in cold temperate latitudes except in the north Pacific where the Salmon Shark replaces it. The Porbeagle typifies overfished sharks (see chapter 11).

Figure 3.29. Shortfin Mako (Wildestanimal/Shutterstock.com).

Lamna nasus (Porbeagle)

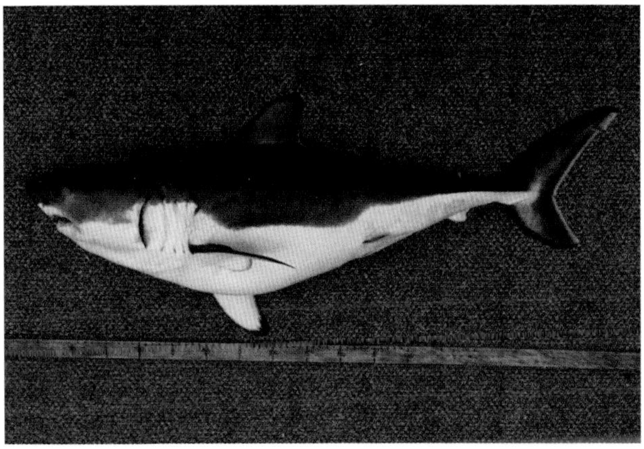

Figure 3.30. Juvenile Salmon Shark, *Lamna ditropis* (Courtesy of Dave Itano).

White Shark (*Carcharodon carcharias*; fig. 3.31). What can we say about one of the most iconic species on the planet? Are they overrated? Well, if overrated means that the species receives attention disproportionate to its ecological importance or conservation status, often to the exclusion of more interesting and endangered sharks, then yes. On the other hand, if caring about this species leads to awareness of the plight of other sharks and inhabitants of the planet in general, and thus valuing our natural environment, then no.

Characteristics include its large size (up to 6.1 m or 20 ft and 1900 kg or 4200 lb), a jaw full of triangular serrated teeth for cutting, long gill slits, very black eyes, vivid color changes on its sides, and black tips under the pectoral fins. It is globally distributed in both coastal and oceanic waters of between 12°C and 24°C (54°F and 75°F). Contrary to the public's perception, some populations are not declining and in fact have been increasing for 20 or 30 years, though globally the species is listed as *Vulnerable* by the IUCN.

The White Shark is perpetually in the news, but recent reports merit attention. The first are reports near SE Farallon Island, south of San Francisco, and more recently off South Africa and Australia, of White Sharks being killed by Orcas,

Figure 3.31. White Shark (Courtesy of Craig O'Connell).

which apparently removed the liver and possibly other internal organs. In South Africa, the behavior has been linked to specific Orcas in a region where prior to 2015 a shark cage diving industry was built around White Shark sightings. By 2020, White Shark sightings decreased to near zero. In 2024, White Sharks had returned to some locations but the frequency of sightings was well below levels prior to predation by Orcas.

Carcharodon carcharias (White Shark)

The order Carcharhiniformes contains 297 species currently in 10 families. The order includes the Tiger (*Galeocerdo cuvier*), Bull (*Carcharhinus leucas*), and other sharks with which you are likely familiar. All have two spineless dorsal fins, and an anal fin. They lack nasoral grooves, possess nictitating membranes, and have intestines with a spiral or scroll valve. All three major modes of embryonic development are present, including oviparity, yolk-sac viviparity, and placental viviparity.

Families in the order Carcharhiniformes:

- Deepwater Catsharks (Family Pentanchidae; fig. 3.32)
  - » one hundred twenty-four species in 15 genera, making this the most speciose family in the order. Deepwater catsharks are mostly small (< 1 m, or 3.3 ft), and are found predominantly in cold and deep waters globally to about 2,000 m (6,500 ft) deep. Nearly all are oviparous. They have elongated bodies with two very small dorsal fins positioned far back on the body. Many are dark brown in coloration, but some are strikingly patterned. The members of this family were recently separated from the Scyliorhinidae (Catsharks). Unlike scyliorhinids, Deepwater Catsharks lack orbital crests on the chondrocranium. Despite their diversity, you are unlikely to encounter many of these unless you study deep-sea sharks, so we will mention only a couple of species.

Roughtail Catshark (*Galeus arae*; fig. 3.32). The Roughtail Catshark is a small catshark common in the NW Atlantic, Gulf of Mexico and Caribbean Seas. The genus *Galeus* are called sawtailed catsharks because they possess enlarged scales on the dorsal edge of the upper lobe of the caudal fin.

Panama Ghost Catshark (*Apristurus stenseni*). The genus *Apristurus* includes about 37 species (with more still likely to be discovered) of deepwater catsharks and is also the most common deepwater genus of sharks. They are small, dark black to brown sharks with tiny eyes, which slowly cruise or, depending on the species, sit on the seafloor.

Apristurus stenseni (Panama Ghost Catshark)

Figure 3.32. Roughtail Catshark, with close-up showing enlarged denticles on the upper lobe of the caudal fin.

- Catsharks (Family Scyliorhinidae; fig. 3.33)
  - » thirty-seven species of small (0.25 to 1.6 m, or 0.8 to 5.2 ft) tropical to temperate sharks occurring from nearshore to more than 600 m (2,000 ft) deep. Most are oviparous. They have elongated, catlike eyes (hence the name) and two small dorsal fins positioned far back on the body. Orbital crests above the eyes are present on the skull.

Representative Species

Small-spotted Catshark (*Scyliorhinus canicula*). This species is a shallow water inhabitant and is one of the most abundant sharks in the North Sea, Mediterranean, and adjacent areas in the Atlantic. It has a very slender body and is oviparous.

Scyliorhinus canicula (Small-spotted Catshark)

Chain Catshark (often referred to as Chain Dogfish [*Scyliorhinus retifer*; fig. 3.33]). The Chain Catshark is found from 36 to 750 m (118–2461 ft) deep in the Western North Atlantic, Gulf of Mexico, and Caribbean Sea. It has a reticulated pattern and is commonly displayed in aquaria. Despite the common name, the species is not a dogfish.

Figure 3.33. Chain Catshark at Norfolk Canyon off the coast of Virginia. Note the abandoned fishing net in the background (Courtesy of NOAA).

Striped Catshark or Pajama Shark (*Poroderma africanum*). The Pajama Shark is found off South Africa and is included here because it is the only shark with horizontal stripes. It is commonly displayed in aquaria.

Poroderma africanum (Striped Catshark, Pajama Shark)

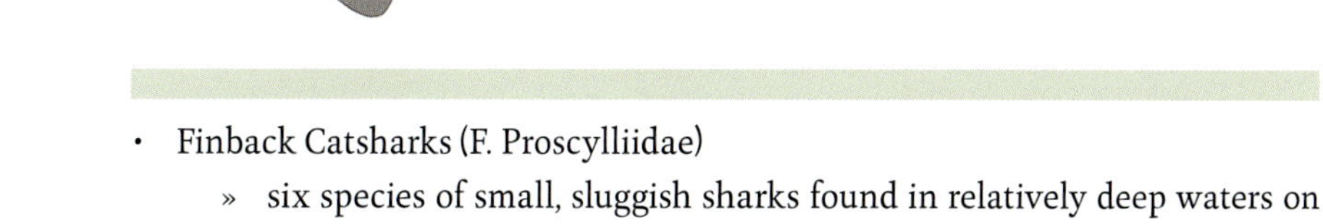

- Finback Catsharks (F. Proscylliidae)
    - » six species of small, sluggish sharks found in relatively deep waters on the edge of continental and insular shelves and slopes (50–700 m or 160–2,300 ft). They can sometimes be quite common in tropical areas. Some proscylliids are oviparous whereas others are yolk-sac viviparous.

Representative Species

Harlequin Catshark (*Ctenacis fehlmanni*). The Harlequin Catshark is only known from the type specimen caught off Somalia in the western Indian Ocean.

Ctenacis fehlmanni (Harlequin Catshark)

- False Catsharks (F. Pseudotriakidae; fig. 3.34)
    - » six species of small- to relatively large (0.6 to 2.7 m, 2 to 9 ft) deepwater sharks, found from 200 to over 1500 m (656–4921 ft) deep.

Representative Species

False Catshark or Sofa Shark (*Pseudotriakis microdon*; fig. 3.34). Superficially these flabby sharks look like a gulper shark (black with big green eyes), and they possess large spiracles. However, the gulpers are squalomorph sharks with a tighter jaw suspension and no anal fin, whereas the pseudotriakids are galeomorph sharks, which have a looser jaw suspension and an anal fin. Thus, the resemblance is an example of convergent evolution and is not phylogenetically based. This is the only shark outside the lamniforms known to reproduce using a form of oophagy.

Pseudotriakis microdon (False Catshark or Sofa Shark)

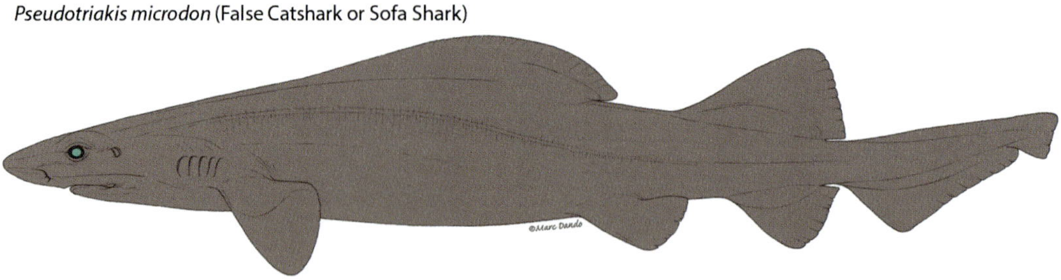

- Barbeled Houndshark (F. Leptochariidae)
    - » a monotypic family, the single species being the Barbeled Houndshark (*Leptocharias smithii*).

Representative Species

Barbeled Houndshark (*Leptocharias smithii*). Found in the Eastern Atlantic, this demersal species is found to depths of 75 m (246 ft). It is yolk-sac placental

Figure 3.34. False Catshark, or Sofa Shark.

viviparous. The evolutionary relationship of this species among the other carcharhiniforms has been difficult to resolve, hence it was placed in its own family.

Leptocharias smithii
(Barbeled Houndshark)

- Houndsharks (F. Triakidae; fig. 3.35)
  - » forty-eight species in 9 genera of mostly coastal sharks, including the Leopard Shark (*Triakis semifasciata*; fig. 3.35A) and the Dusky Smoothhound (*Mustelus canis*; fig. 3.35B). They all have a long, pointed snout, and arched mouth. Dentition varies from pavement like to sharply cusped to bladed. They are demersal to coastal pelagic in cool temperate to tropical waters from the surface to > 450 m (1475 ft), where they feed primarily on crustaceans, molluscs, fish, and squid. Triakids are found in the Atlantic, Pacific, and Indian Oceans. Interestingly, about half the species are yolk-sac viviparous and half are placental viviparous. Triakids range from 50 to 200 cm (1.5–6.5 ft).

Representative Species

Spotted Houndshark (*Triakis maculata*). The Spotted Houndshark is a small, harmless coastal shark found in the Pacific off the coast of South America.

Triakis maculata (Spotted Houndshark)

School Shark or Soupfin Shark (*Galeorhinus galeus*). Also known as the Vitamin Shark, this species is easily identified by the large lobe on the upper caudal fin. They have a slender body, long snout, and small second dorsal. Their vitamin A-rich liver led to their being the biggest shark fishery in California in the 1930s and 1940s.

Galeorhinus galeus (School Shark, Soupfin Shark)

Dusky Smoothhound or Smooth Dogfish (*Mustelus canis*; fig. 3.35). The Dusky Smoothhound (Smooth Dogfish) is a common placental viviparous species in the NW Atlantic. There are two subspecies, *M. canis canis* and *M. canis insularis*. The former is most common along the shallow continental shelf from New York to South Carolina, and it pups in shallow estuaries. Oddly, it was recently reported pupping at 400 m in the Gulf of Mexico, which means that the same subspecies utilized two very different depth regimes for giving birth. In both habitats, they specialize on primarily eating crabs. *Mustelus canis insularis* is the island form of the species and is found in the tropics in deeper waters.

Mustelus canis
(Dusky Smoothhound, Smooth Dogfish)

A

B

Figure 3.35. (A) Leopard Shark, commonly found off the California coast, at the Birch Aquarium in La Jolla, California (Courtesy of Matthew Field, www.photography.mattfield.com). (B) Dusky Smoothhound.

- Weasel Sharks (F. Hemigalidae)
    - » eight species, with a robust body and elongate, rounded snout. They live in inshore waters from the surface to 130 m (425 ft), where they feed on fish. They have unusual, highly serrated teeth for tearing apart octopus, bony fish and small sharks. The group is placental viviparous and reaches 2.4 m (8 ft). They are all found in the Indo-Pacific.

Representative Species

Weasel Shark *(Hemigaleus microstoma)*. This rarely encountered species is found in southern India, southern China, and parts of SE Asia.

- Requiem Sharks (F. Carcharhinidae)
    - » about 57 species, 36 of which are in the largest genus, *Carcharhinus*. They are the second largest galeomorph family and are the dominant sharks in tropical and subtropical waters, and some even live in freshwater. Carcharhinids have a long, arched mouth often with bladelike upper teeth and narrow lowers, a nictitating membrane, and no spiracle. They are found in both inshore and pelagic waters from the surface to 800 m (2625 ft), where they feed on fish, cephalopods, crustaceans, turtles, and mammals. They are all placental viviparous, occupy all oceans, and some are potentially dangerous. Over 24 species are in the range of 1.7–5 m (5–16 ft).

      Species in the genus *Carcharhinus* can be superficially divided into two broad groups based on the presence or absence of a slightly elevated ridge on the dorsum between the first and second dorsal fins (fig. 3.36). This feature, the interdorsal ridge, is a very useful first character to use for identifying many of the species.

Representative Species

RIDGEBACK SHARKS

Caribbean Reef Shark (*Carcharhinus perezi*; fig. 3.37A). The Caribbean Reef Shark is the most common shark of Caribbean coral reefs, and it reaches a maximum

Figure 3.36. Conspicuous interdorsal ridge characteristic of one group of carcharhinid sharks, in this case a Sandbar Shark. A small ectoparasitic copepod is also shown.

size of 3 m (3.3 ft). This species eats bony fishes. It occupies a similar ecological niche to the Indo-Pacific Grey Reef Shark (*C. amblyrhynchos*).

Carcharhinus perezi
(Caribbean Reef Shark)

Sandbar Shark or Brown Shark (*Carcharhinus plumbeus*; fig. 3.37B). The Sandbar Shark is a coastal water species distributed globally in temperate to subtropical waters shallower than 100 m (330 ft). It reaches a maximum length of 2.4 m (8 ft). In addition to having a prominent interdorsal ridge, it can be identified by its oversized first dorsal fin far forward on the body. The Sandbar Shark is the dominant large coastal shark along the US East Coast as well as in Hawaii, but it occupies a deeper water habitat in the latter. The biggest Sandbar Shark nursery (the area where they are born and/or spend their early years) in the world is Chesapeake

A 

B 

Figure 3.37. Ridgeback sharks. (A) Caribbean Reef Shark. (B) Sandbar Shark.

Bay. In part because of their large first dorsal fin's value to the shark fin soup industry, Sandbar Sharks drove the US East Coast directed shark fishery until they became overfished.

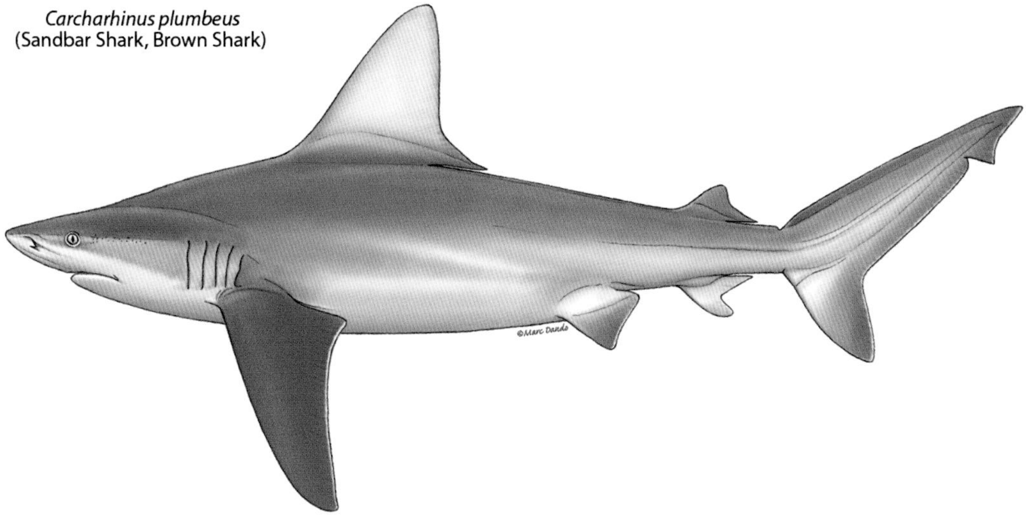

Carcharhinus plumbeus
(Sandbar Shark, Brown Shark)

Dusky Shark (*Carcharhinus obscurus*; fig. 3.38). The Dusky and Galapagos Sharks (*Carcharhinus galapagensis*) are so closely related that some geneticists considered them as a single species. The Dusky Shark remains one of the most overfished sharks on the US East Coast, and its full recovery is not expected until after 2100.

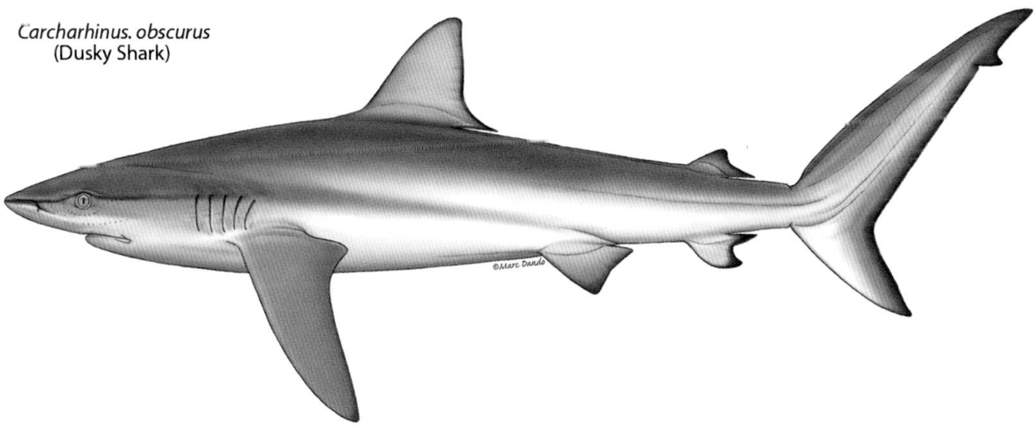

Carcharhinus. obscurus
(Dusky Shark)

Silky Shark (*Carcharhinus falciformis*). Named because of its small dermal denticles that are smooth to the touch, the Silky Shark can be identified by the convex rear margin of its first dorsal combined with an elongated free tip on the second dorsal fin and long, sickle-shaped pectoral fins. It is one of the most common tropical pelagic sharks in the world. In some areas, Silky Sharks are overfished because of purse seine fisheries for tunas.

Figure 3.38. Dusky Shark (Courtesy of Andrew Raak).

Oceanic Whitetip Shark (*Carcharhinus longimanus*; fig. 3.39). The specific epithet *longimanus* means long hands, which refers to its elongate, paddle-shaped pectorals, and the common name describes the prominent edges of the fins, which appear to have been sprayed with white paint. Interestingly, newborns have black tipped anal and pelvic fins. The Oceanic Whitetip Shark is a stocky, large (around 2 m, or 6.6 ft), pelagic shark found in temperate and tropical oceans. It has a reputation of being an aggressive and dangerous shark. The Oceanic Whitetip Shark is inquisitive and may bump people it encounters.

Carcharhinus longimanus
(Oceanic Whitetip Shark)

Figure 3.39. Oceanic Whitetip Shark, another ridgeback species (Courtesy of Lance Jordan).

Blacktip Reef Shark (*Carcharhinus melanopterus*; fig. 3.40A). The Blacktip Reef Shark, whose fins look as if they have been dipped in ink, is often confused with the Blacktip Shark (*C. limbatus*), especially in the media. The Blacktip Reef Shark is found only on Indo-Pacific coral reefs, whereas the Blacktip Shark is distributed globally in coastal tropical and temperate waters, and the eastern Mediterranean.

Blacktip Shark (*Carcharhinus limbatus*; fig. 3.40B). The Blacktip Shark is widely distributed and is one of the most common non-ridgeback sharks. It is found both inshore and offshore, in the upper 100 m (330 ft). Along the US East Coast, the species is frequently observed in dense schools of thousands along its migratory route every spring and fall (see fig. 1.19. These are streamlined, swift-moving, fish and squid-eaters, and are often blamed for shark bites along the US Atlantic Coast in the summer.

Bull Shark (*Carcharhinus leucas*; fig. 3.41). This is a stout shark with a robust, blunt, rounded snout. It has a large first dorsal fin situated far forward on the body, but without an interdorsal ridge. Its eyes are relatively small, but its teeth are broad and heavily serrated for shearing. The Bull Shark is predominantly found in shallow tropical and temperate waters less than 30 m (100 ft) deep, but it can be found shallower and as deep as 150 m (492 ft).

The Bull Shark is considered dangerous, especially in the developing world where the daily lives of inhabitants find them in Bull Shark habitat, particularly brackish and fresh waters. When Bull Sharks are encountered by a boat in shallow, clear tropical waters, they may go into an agonistic (threat) display, lowering the fins and hunching their back, and when that occurs, they may charge and strike the boat.

Figure 3.40. Two non-ridgeback sharks that are often confused with each other. (A) Blacktip Reef Shark (Courtesy of Michael Scholl). (B) Blacktip Shark.

Carcharhinus leucas (Bull Shark)

Figure 3.41. Adult Bull Shark, a non-ridgeback species (Nicolas-SB/Shutterstock.com).

Lemon Shark (*Negaprion brevisostris*; fig. 3.42A). Lemon Sharks are a large (up to 3.4 m or 11 ft), migratory, shallow water, temperate and tropical shark found along the Atlantic and Eastern Tropical Pacific Coasts. Juveniles inhabit mangrove-lined lagoons and adults are continental shelf, estuarine, and reef associated. They are a muted lemon color with two dorsal fins of approximately the same size. They are philopatric; that is, they return to their birthplace to give birth. Lemon Sharks are a species whose ecology, behavior, life history, and physiology have been extensively studied for over 30 years by Dr. Samuel Gruber and his graduate students and associates at the Bimini Biological Field Station in the Bahamas.

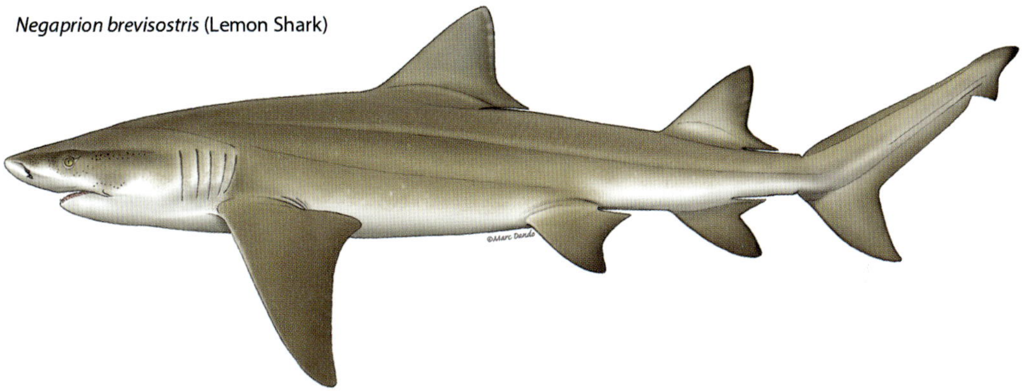

Whitetip Reef Shark (*Triaenodon obesus*; fig. 3.42B). The Whitetip Reef Shark, not to be confused with the Oceanic Whitetip Shark, is a common, small (1.6 m, or 5.2 ft), slender, Indo-Pacific shark. Named for its white-fringed fins, the Whitetip Reef Shark sleeps during the day and actively forages on reef fishes in the evening, an activity facilitated by the flexibility of this species.

Blue Shark (*Prionace glauca*; fig. 3.43). Blue Sharks are large (to 3.3 m, or 10.8 ft), pelagic, cold-temperate sharks found in all of the world's oceans. They are one of the most abundant large sharks. They are slender, show countershading (blue on top), and have long pectoral fins. Like other carcharhinids, they are placental viviparous. They have one of the largest reproduction potentials of the group and are capable of having litters as big as 100.

Atlantic Sharpnose Shark (*Rhizoprionodon terraenovae*; fig. 3.44). The Atlantic Sharpnose Shark is a small (1 m or 3.3 ft), hugely abundant Atlantic coastal shark, characterized by having a small, black second dorsal fin, a small ridge in front of the anal fin (a preanal ridge), long labial furrows, and elongated ampullary pores (like oval slits) on their cheeks behind their eyes. Beginning in May and throughout the summer, in temperate estuaries and nearshore environments, the neonates (newborns) are nearly ubiquitous.

A

B

Figure 3.42. (A) Lemon Shark (Courtesy of Annie Guttridge). (B) The ecologically similar Whitetip Reef Shark (Courtesy of Lesley Rochat).

Triaenodon obesus (Whitetip Reef Shark)

Prionace glauca (Blue Shark)

Figure 3.43. Blue Shark.

Figure 3.44. Atlantic Sharpnose Shark. Note the black second dorsal and white spots, distinguishing features of this species.

- Tiger Shark (F. Galeocerdidae)
    - » one species, Tiger Shark (*Galeocerdo cuvier*; fig. 3.45). Found globally in tropical and temperate coastal waters, this iconic, large shark (5 m or 16.5 ft), is the only member of its family and genus. It is thought Tiger Sharks split from placental carcharhinids nearly 100 million years ago and subsequently evolved embryotrophic viviparity, a unique form of reproduction. They possess unique homodont dentition (same in both jaws) with teeth resembling a can opener, a predorsal ridge, a very wide, broad head and a keeled caudal peduncle (fig. 3.45). It is easily identified by its markings, which are most vivid when it is a juvenile, and by its long caudal fin. Like the cow sharks, Tiger Sharks surprisingly have relatively weak jaws, an adaptation for sawing pieces from sea turtles, among their favorite prey.

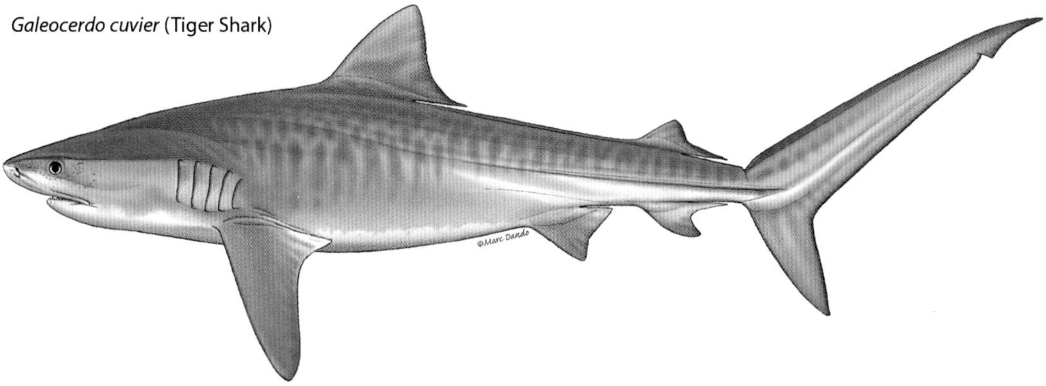

Figure 3.45. Tiger Shark showing its distinguishing features: predorsal ridge, long lateral keel, and wide head (Courtesy of Annie Guttridge).

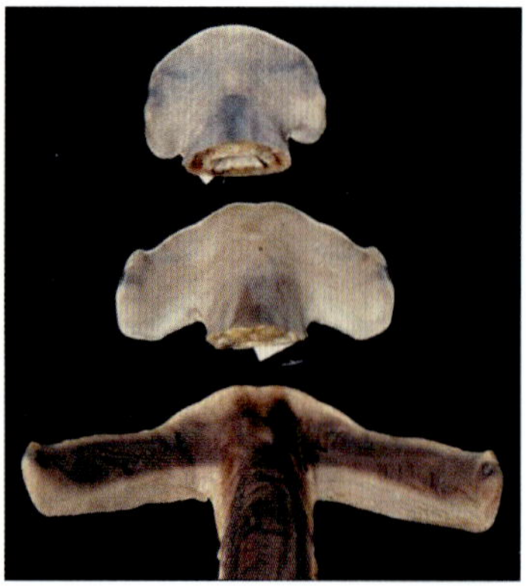

Figure 3.46. Cephalofoils of selected members of the family Sphyrnidae. From top to bottom: Bonnethead, juvenile Scalloped, and Winghead Hammerheads.

- Hammerheads (F. Sphyrnidae; *sphyrna* = hammer)
    - » nine species, but it is likely that more will be added to it. The hammer-shaped head, technically known as a *cephalofoil*, ranges in size from the relatively small Bonnethead Shark (*Sphyrna tiburo*) to the bizarrely wide head of the Winghead Shark (*Eusphyra blochii*), an Indonesian species whose head is half as wide as its body length (fig. 3.46). Hammerheads are found in all tropical and warm temperate seas in both inshore and pelagic environments, with animals having been tracked from the surface to nearly 1000 m (3280 ft). They feed on fish, including batoids, squid, and crustaceans. Members of the family range in size from the 1.2 m (3.9 ft) Bonnethead to the 6.2 m (20.3 ft) Great Hammerhead (*S. mokarran*). All hammerheads are placental viviparous.

Representative Species

Bonnethead (*Sphyrna tiburo*; fig. 3.47). This small, beautifully speckled shark is extremely common in the Gulf of Mexico and along the US Atlantic coastline. *S. tiburo* is also the only shark with an obvious sexually dimorphic head shape. Males exhibit an anterior bulge on the cephalofoil whereas the head is more

Sphyrna tiburo (Bonnethead)

Figure 3.47. Bonnethead.

rounded in females. Bonnetheads eat crustaceans, especially blue crabs, along with fish and other invertebrates.

Great Hammerhead (*Sphyrna mokarran*; fig. 3.48A). There are no projections on the head of this species, but there is a hump instead of scallops or smooth curve. The first dorsal fin is enormous and there are large pelvic fins as well as a huge upper caudal fin.

Sphyrna mokarran
(Great Hammerhead)

Slender Hammerhead or Winghead Shark (*Eusphyra blochii*; fig. 3.48B). This exclusively Indo-Pacific, small (1.5 m, or 4.9 ft) species has an extremely wide cephalofoil and is found in murky estuaries.

A

B

Figure 3.48. (A) Great Hammerhead. (B) Winghead Shark.

Concluding Comments

Although not exhaustive, this chapter's overview of the diversity of sharks, which includes iconic species as well as those of ecological and commercial importance, we hope fed your sense of wonder at these magnificent beasts. At the same time, it should have acquainted you with sharks with which you may have been unfamiliar and the characteristics and adaptations that distinguish them from each other. Both of these become important as we discuss the physiology, ecology, behavior, fisheries, and conservation of sharks in future chapters.

NOTES

1. White, W.T. et al. 2023. Diversity 15(7): 849.
2. Gonzalez, C. et al. 2024. Zootaxa 5512(4): 491–511.
3. Compagno, L.J. 1984. Sharks of the world: an annotated and illustrated catalogue of shark species known to date. Food & Agriculture Org. (No. QL 638.9. C65).
4. Weigmann, S. 2016. J. Fish Biol. 88: 837–1037.

5. https://sharksrays.org. (Accessed 9/9/25).

6. Recall that a *monophyletic* group simply means that all members of the group are most closely related to each other and are not placed in other groups.

7. Last, P.R. et al. (eds.) 2016. *Rays of the World*: supplemental information 40: 1–10. CSIRO Publishing.

8. Batoids in fact have eight synapomorphies. For details, see: Aschliman, N., "The batoid tree of life: recovering the patterns and timing of the evolution of skates, rays, and allies (Chondrichthyes: Batoidea)," PhD diss., (Florida State University, 2011).

9. Last, P. et al. (eds.). 2016. *Rays of the World*. CSIRO Publishing.

10. Or, if hot water is not available, a hot oat milk latte ~~reluctantly~~ enthusiastically transferred from one of the author's thermos tumblers into a plastic bag worked on one occasion.

11. Morey, G. et al. 2019. *The IUCN Red List of Threatened Species* 2019: e.T39332A117498371. https://dx.doi.org/10.2305/IUCN.UK.2019-1.RLTS.T39332A117498371.en. (Accessed 9/9/25).

12. Grace, M.A. et al. 2019. Zootaxa 4619: 109–120.

13. Widder, E.A. 1998. Envir. Biol. Fish. 53: 267–273.

14. Kruska, D.C. 1988. Brain Behav. Evolut. 32: 353–363.

PART II

Adaptational Biology: How Sharks Work

4 / Functional Anatomy of Sharks

CHAPTER OUTLINE

Introduction: Shark Skin and Adaptations

In 1996, the swimsuit company Speedo developed a competition, full body swimsuit called the *Fastskin,* and in 2004 they introduced the advanced *Fastskin FSII.* These were both advertised as revolutionary, performance-enhancing body coverings made from the *fastest fabric in the world* that significantly reduced drag, the major frictional force that slows moving objects and the avowed enemy of elite swimmers. Promotional literature from the company emphasized that the design of this line of swimsuits was inspired by sharks.

In advertising campaigns, the company juxtaposed images of swimming sharks and humans that highlighted superficial similarities in form between the two groups. The real design breakthrough, however, according to Speedo, was creating a fabric that channeled water smoothly from front-to-back along the swimmer's

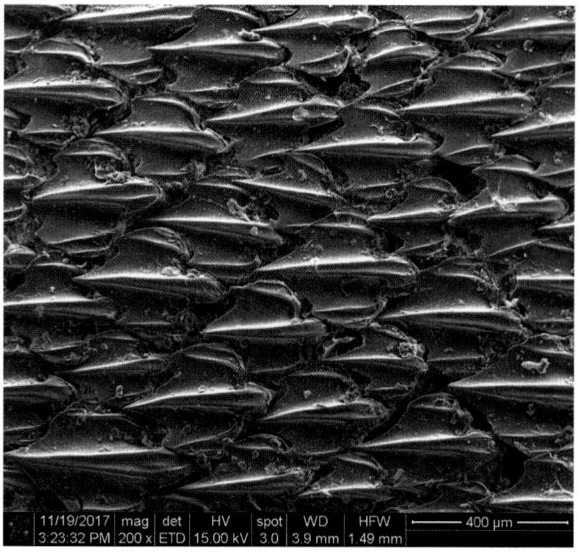

Figure 4.1. Riblets (ridges and grooves) on dermal denticles of Genie's Dogfish shown on a scanning electron photomicrograph (Courtesy of Charles F. Cotton).

body, much in the same way that a shark's dermal denticles were thought to do thanks to their ridges and grooves (fig. 4.1).

A high-performance swimsuit inspired by sharks would be a cool story, and an exquisite example of *biomimicry* (the act of copying nature's best ideas for human uses), except for the fact that humans swim differently than sharks. In fact, a 2012 paper[1] found no increase in speed in tests using Speedo fabric compared to an average increase of 12.3% using membranes made from shark skin, when each was compared to controls using a *robotic flapping device* (got yours yet?). Humans do not swim like robotic flapping devices either.

Although the Speedo swimsuits may not have allowed wearers to swim like sharks, the company's engineers were correct in their understanding of the drag-reducing role of the dermal denticles of real shark skin. According to the research referenced above, the dermal denticles both decreased drag and increased thrust.

Dermal denticles, and the skin of sharks, have additional functions as well. First, they protect sharks. We frequently catch sharks on our experimental fishing gear with bites from other sharks that either do not penetrate the skin or only poke holes but do not tear the skin (fig. 4.2), which is due to the protection offered by the dermal denticles, as well as the thickness and inherent toughness of the underlying skin.

The skin likely also plays a role in reducing adhesion of fouling organisms (like algae, barnacles, etc.) and ectoparasites (see fig. 3.36), whose presence might create eddies of swirling water against the shark's body, thereby increasing friction and slowing the shark or requiring more energy to swim, not to mention potentially affecting the health of the shark.

The functions that scientists attribute to shark dermal denticles and skin are *adaptations*. Adaptations refer to any features of an organism that enhance its

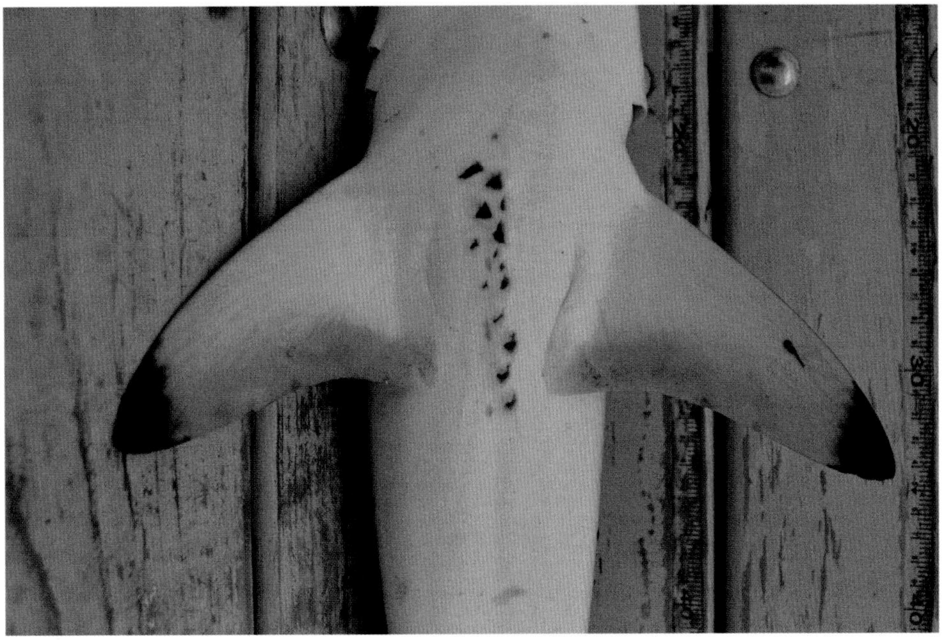

Figure 4.2. Blacktip Shark from experimental longline with series of puncture marks from the bite of another shark. This shark was successfully released.

ability to survive and reproduce in its environment *and* which are hereditary, meaning it has the potential to be inherited by offspring.

Adaptations can be anatomical (structural), physiological (internal function), or behavioral. Though the value of an adaptation to an organism is not always obvious, nature selects for adaptations that promote survival and at the same time, if possible, conserve energy.

Adaptations may entail evolutionary trade-offs; for example, the extended rostrum, or saw, of a sawfish functions in part to detect the presence of prey and is remarkably effective at swiftly slashing small prey, either stunning them or impaling them on the sharp rostral teeth. It is also an effective defense mechanism. At the same time, swimming with such an awkward structure protruding from its head must be energetically inefficient for a sawfish, although this has not been studied. Prominent physiological ecologist George Bartholomew put it best: *The organism is a compromise. The result of natural selection is adequacy and not perfection.*[2]

A reoccurring theme in the study of adaptation is whether an adaptation increases an organism's evolutionary *fitness.* Does the alleged adaptation directly or indirectly enhance the organism's ability to get its genes into the next generation? The crucial act of reproduction requires energy in the production of gametes (sperm and eggs) as well as finding and mating with another member of your species. But that is not the only way in which animals use energy. All animals (including sharks) use energy for growth; movement; maintenance of stable internal conditions (a state called *homeostasis*; see chapter 5); ventilating the gills; balancing internal water and salt concentrations; pumping blood through the circulatory system; and numerous other processes at the organ, tissue, and cellular levels. These energy-dependent functions ultimately enhance the odds that a shark will live to maturity and reproduce. Given the importance of energy, it is no wonder that many adaptations conserve energy and aid in consuming energy, which we refer to as food. And what we call food for a predator are organisms with their own adaptations to avoid being food. Energy-efficient survival is a key step toward evolutionary fitness.

In this chapter, we will examine some anatomical adaptations that make sharks such good survivors, or, as fictional ichthyologist Matt Hooper said in the blockbuster film *Jaws*, "miracles of evolution." In subsequent chapters, we will examine physiological and behavioral adaptations. Collectively, anatomy, physiology, and behavior determine the ecological roles sharks play in their respective ecosystems, and they give us insight into the diversification of sharks over evolutionary time. Moreover, these fields of study all allow us to answer the question "How do sharks work?"

Before engrossing yourselves in the magical world of shark functional anatomy, we urge you to read the enlightening essay on how to know a shark by coauthor Joshua Moyer in *Research, Conservation, and Education Focus* Box 4.1.

Ecomorphology and Body Shape

Modern sharks come in a wide variety of shapes and sizes (recall the sampling of chondrichthyan biodiversity in chapter 3; fig. 4.3). How did this diversity evolve? Why have some morphological traits evolved convergently in distantly related species? Answers to these questions lie in the *functional significance* of morphological traits under ecological conditions. In other words, how do the sizes and shapes of anatomical structures affect the way an organism completes an ecologically relevant task, such as feeding or reproducing, that contributes to its evolutionary fitness? Studying links between morphological form, functional significance, and an organism's ecology is called *ecomorphology*, and it can help researchers develop evolutionary hypotheses to explain morphological variation in sharks and other organisms. A wonderful example of shark ecomorphology concerns body shape and locomotion.

How and where you move affects your ability to find suitable habitats, mates, and food, and to avoid predators. There is no doubt that movement is ecologically relevant, and movement through water can be energetically expensive. Water is about 800 times heavier and 50 times more viscous, or resistant to flowing, than air. But water has one major advantage over air: it physically supports organisms better. That is why objects that are denser than the water they displace (*negatively buoyant*) sink more slowly than they would fall in air. As far as sharks are concerned, the trick is moving through the water in an energy-efficient manner. As

BOX 4.1

Research, Conservation, and Education Focus Old Lessons and New Technologies in Shark Biology

Joshua K. Moyer

Department of Ecology & Evolutionary Biology
Yale University
New Haven, CT, US

It is wonderful to behold a student getting a hands-on chance to learn about sharks, whether that is by observing sharks at a field site or aquarium or by dissecting specimens to learn about the literal ins and outs of sharks. The first reaction of most students to these experiences is one of amusement and curiosity. An animal that has occupied a space in the student's consciousness can now be explored, and a new understanding is waiting to be developed. For that student, sharks have just become a little less abstract.

Then, as I have seen many times while teaching courses in comparative vertebrate anatomy, ecological morphology, and marine biology, a new question dawns on the student and is usually accompanied by a furrowed brow: "What do I need to know about this shark?" In most cases, the student has remembered the reality of course grades or the prospect of needing to establish professional credibility. There is so much to know about sharks, how could I ever make a comprehensive list of the things a student could or should learn from a specimen?

To recapture my students' enthusiasm and to help them get the most from their shark-centered experience, I tell them that there is no magic number of observations to be made. Rather, I encourage them as my academic advisors encouraged me, "Look at your fish" (or shark as the case may be). In the spirit of many nineteenth-century naturalists who advocated what would become a modern comparative specimen-based approach, the goal is to see how little you saw before and then to contextualize your observations in a broader framework. Each new vantage point will allow for new observations and inferences. Each organ system examined, each tissue layer peeled back will enable you to make new statements of fact. But facts by themselves are just trivia until they are used as a foundation for a more expansive understanding of the organism in question.

These are the reasons that my research in the fields of the comparative and functional anatomy of sharks involve the use of technologies such as computed tomography (often abbreviated CT) as well as polarized light microscopy and histology. I have applied these methodologies many times in my study of the microstructure and development of shark teeth. They provide new ways to look at the shark in front of me, especially their teeth and jaws. I can see things in micro-CT scan or under polarized light through a microscope that I could not see with my naked eyes, such as delineations in the tissues that make up a shark tooth. As a result, I am able to make observations of one taxon or specimen that often lead me to ask, "How does this compare to other specimens?" When I can answer that question, I know that I have successfully built on my initial observations.

The same is true when I venture into the field to study the feeding behaviors of wild sharks, documenting how they use the teeth and jaws that I became familiar with in the lab. In recent years, advancements in camera technology have bolstered our understanding of shark behavior. With durable, high-definition cameras that fit in the palm of my hand, I can film feeding sharks to replay in ultra-slow motion. This may not sound like a game changing advance but consider this: many shark bites happen in a fraction of a second, as do the changes in body position and movement that accompany a shark bite. If we had to rely on traditional camera technology or our own vision, we would literally blink and miss many of the nuances that vary from one shark to another, especially when it comes to large sharks in their natural environment. In the absence of field-portable video technology, we could only make educated guesses as to what variation in feeding performance may exist in the wild.

The common thread uniting the CT scanning, microscopy, and new camera technology I use, also unites advancements in tagging technology, social network modeling, and even genomics. They provide researchers new ways to look at their sharks. Though the technology we use may be figuratively, and in cases such as satellite telemetry, literally space-age compared to what was available to our naturalist predecessors, we are collectively doing what they spent much of their careers doing: finding new ways to look at their fish.

That is why as a scientist and an educator, I am elated when a student abandons the question "What do I need to know" and instead asks, "What other sharks are like this and why?" When I hear this, I know we are about to find a new way to look at our sharks.

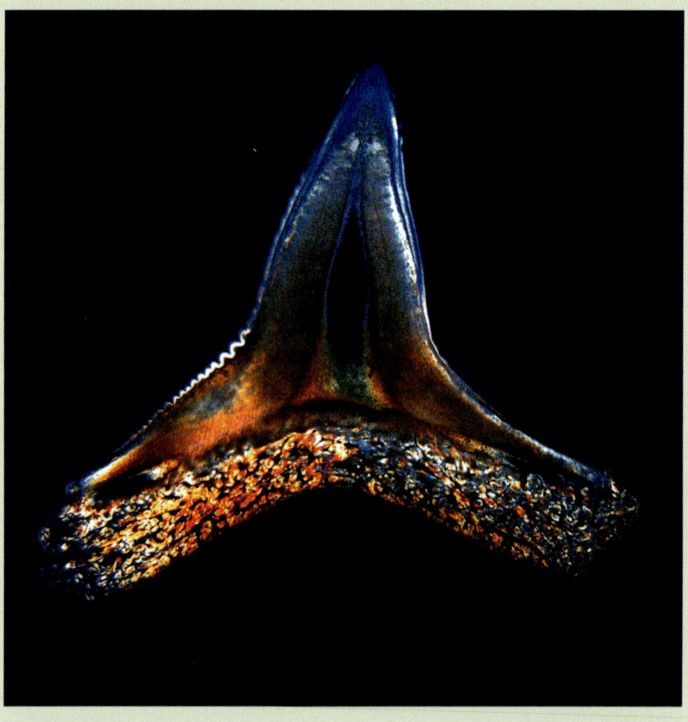

Polarized light microscopy provides another way to examine and visualize specimens. By sectioning shark teeth, such as this tooth of a Bull Shark (*Carcharhinus leucas*) to approximately the thickness of a piece of cardstock and then visualizing it through polarizing filters, Joshua Moyer differentiated boundaries between tissue layers that would not be as apparent to the naked eye.

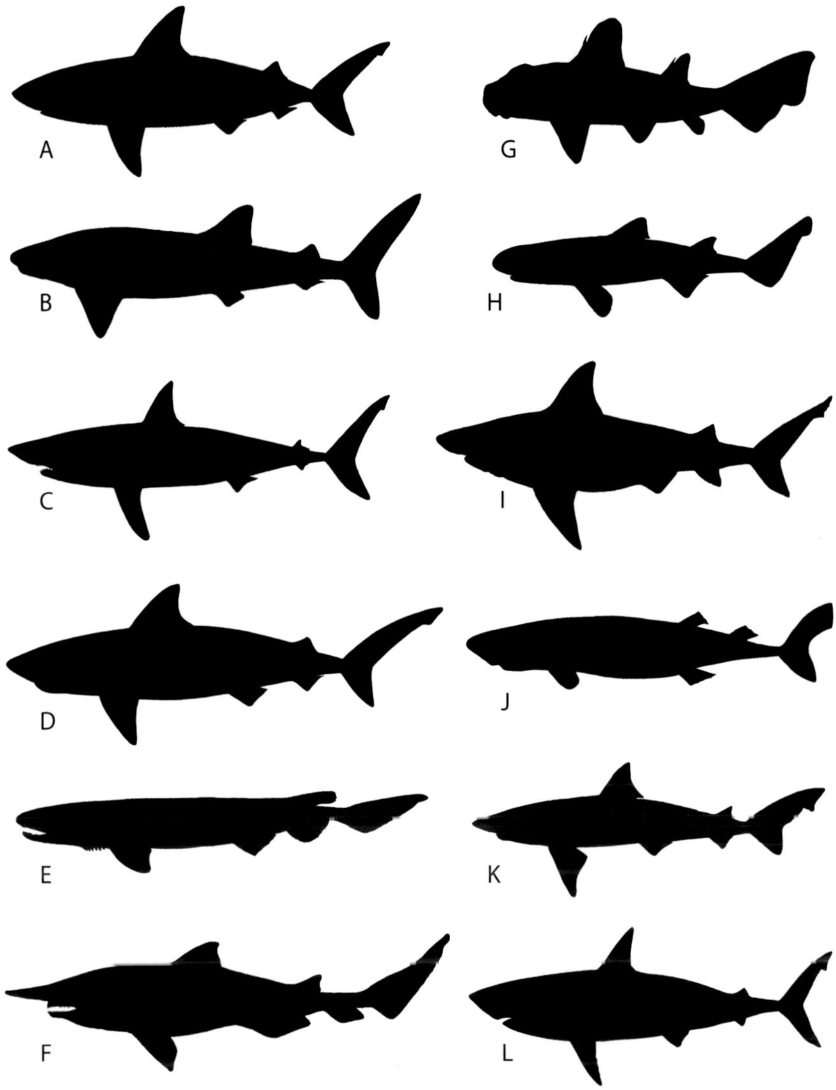

Figure 4.3. Silhouettes of sharks showing diversity of body, head, and caudal fin shapes (Body sizes not drawn to scale). Key: (A) Salmon Shark; (B) Whale Shark; (C) Shortfin Mako; (D) Tiger Shark; (E) Frilled Shark; (F) Goblin Shark; (G) Horn Shark; (H) Kitefin Shark; (I) Bull Shark; (J) Cookiecutter Shark; (K) Soupfin Shark; (L) White Shark.

you might imagine, the energetic demands of this task differ depending on how far and fast a shark needs to move. Unsurprisingly, sharks known for speed and distance swimming share morphological similarities in body shape.

As a shark (or any organism living in water) moves through the water, it must contend with two types of drag: surface drag and pressure drag. *Surface* or *frictional drag* occurs when there is friction between the water molecules and the external surface of the shark. *Pressure drag* occurs when the *boundary layer* of water flowing over the shark is disrupted. The more rapidly a shark undulates its

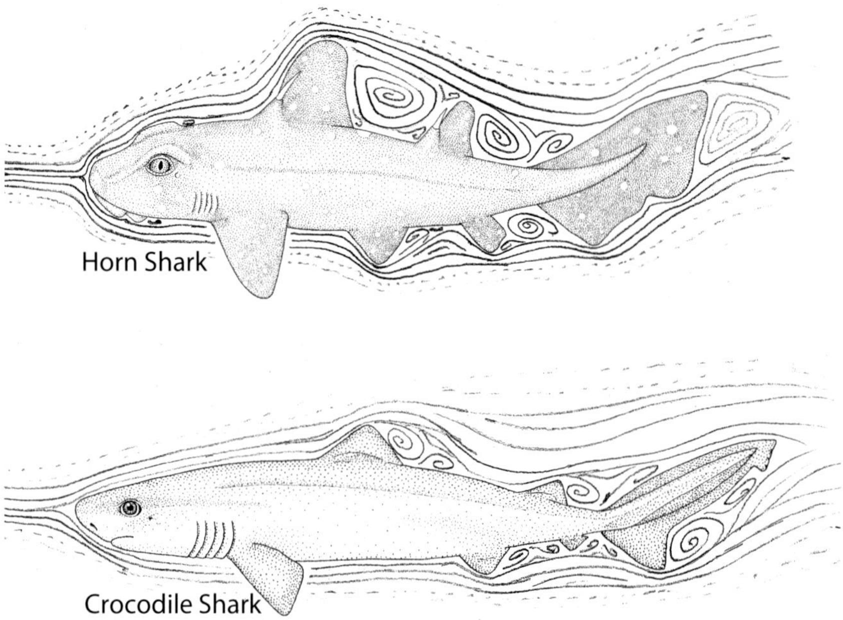

Figure 4.4. Body shape and form drag. Streamlines on a Crocodile and Horn Shark (Roi Gurka).

body and tail, the more the boundary layer is disrupted. This creates vortices in the water displaced by the shark. As the vortices spin backward, they pull against the shark's forward movement. This increases the energetic demand on the shark if it is to keep moving forward. These pesky vortices also occur when the boundary layer must flow over or around blocky or bumpy surfaces. It should come as no surprise then that pressure drag is minimized by a tapered, or *fusiform*, body shape, which is streamlined and less disruptive to the boundary layer of water (fig. 4.4). Minimizing drag and the energy expenditure required to overcome it is crucial if you are a shark that swims long distances or engages in high-speed pursuits of prey! Thus, body shape has a *hydrodynamic* functional significance in sharks with different ecologies.

In a 1977 study, researchers organized 56 species of sharks into four categories based on body types and tail form.[3] Today, these categories are called *ecomorphotypes*. Ecomorphotypes are groupings of organisms that share morphological and ecological characteristics. The four main body-type categories identified in the 1977 study are *macropelagic* (Type 1), *littoral* (2), *benthic* (3), and *micropelagic/bathic* (*bathic* = deepwater) (4).

- Macropelagic sharks (Type 1), include members of the families Cetorhinidae (Basking Shark) and Lamnidae (mackerel sharks, e.g., the Shortfin Mako) as well as the Whale Shark (*Rhincodon typus*). All are oceanic cruisers with more-or-less conical heads (the Whale Shark being a rare exception), deep bodies, and caudal fins with high *aspect ratios* (see *Thrust*). These traits are associated with sustaining increased power and efficiency at high swimming speeds. At lower speeds, they still increase efficiency.

- Littoral sharks (Type 2), include members of the families Mitsukurinidae (Goblin Shark), Odontaspididae and Carchariidae (sand tigers), Alopiidae (threshers), Carcharhinidae (requiem sharks), and Sphyrnidae (hammerheads). Here, *littoral* refers to continental shelf cruisers. Some of these may seem superficially similar to macropelagic sharks, but littoral shark bodies are less deep, the head is blunter, and the caudal fin is more asymmetrical with a lower aspect ratio. Swimming speeds of sharks in this category are wide-ranging.
- Benthic sharks (Type 3), include members of the families Heterodontidae (horn sharks), Orectolobiformes (carpet sharks), Hexanchiformes (cow sharks), Pristiophoriformes (saw sharks), and members of the order Carcharhiniformes except those in the family Carcharhinidae. Sharks in this group have a very low tail with a reduced or absent ventral lobe; a large, blunt head; anterior pelvic fins; and more posterior first dorsal fins. Needless to say, sharks thus described are slow-swimming.
- Micropelagic/bathic sharks (Type 4), include members of the order Squaliformes. All of these lack an anal fin, have slightly elevated pectoral fin insertions, and possess a large upper caudal fin lobe. These are found in the littoral zone, in the deep sea, and in shallower waters as well.

Figure 4.5 illustrates examples of each ecomorphotype. If we expand the list of body types to include dorsoventrally flattened forms, including the remaining batoids and the angel sharks, we could add a fifth ecomorphotype for elasmobranchs.

Be careful not to confuse ecomorphotypes with phylogenetic clades. Shared ecology does not necessarily equal shared evolutionary lineage. Sawfish (Rhinopristiformes) and sawsharks (Pristiophoriformes) serve as a cautionary example. Both share the benthic ecomorphotype, but sawfish are batoids and sawsharks are sharks.

Since the publication of the four original shark body types, the study of elasmobranch ecomorphotypes has grown with the inclusion of other morphological characteristics as well as habitat and behavioral traits in the original descriptions.[4, 5] Cramming more than 1200 living species of elasmobranchs into four categories has made for some counterintuitive bedfellows at times. Some biologists have reconciled this by introducing more than a dozen subtypes, grouping sharks first by habitat and then subdividing the groups based on morphology.[6] Other studies measure and quantify morphological traits (an approach called *morphometrics*) and then determine whether ecological similarities exist within morphometric groups.[7] Regardless of the approach, an ecomorphological form–function relationship between body shape and ecology is the basis for understanding much of the morphological diversity in modern sharks.

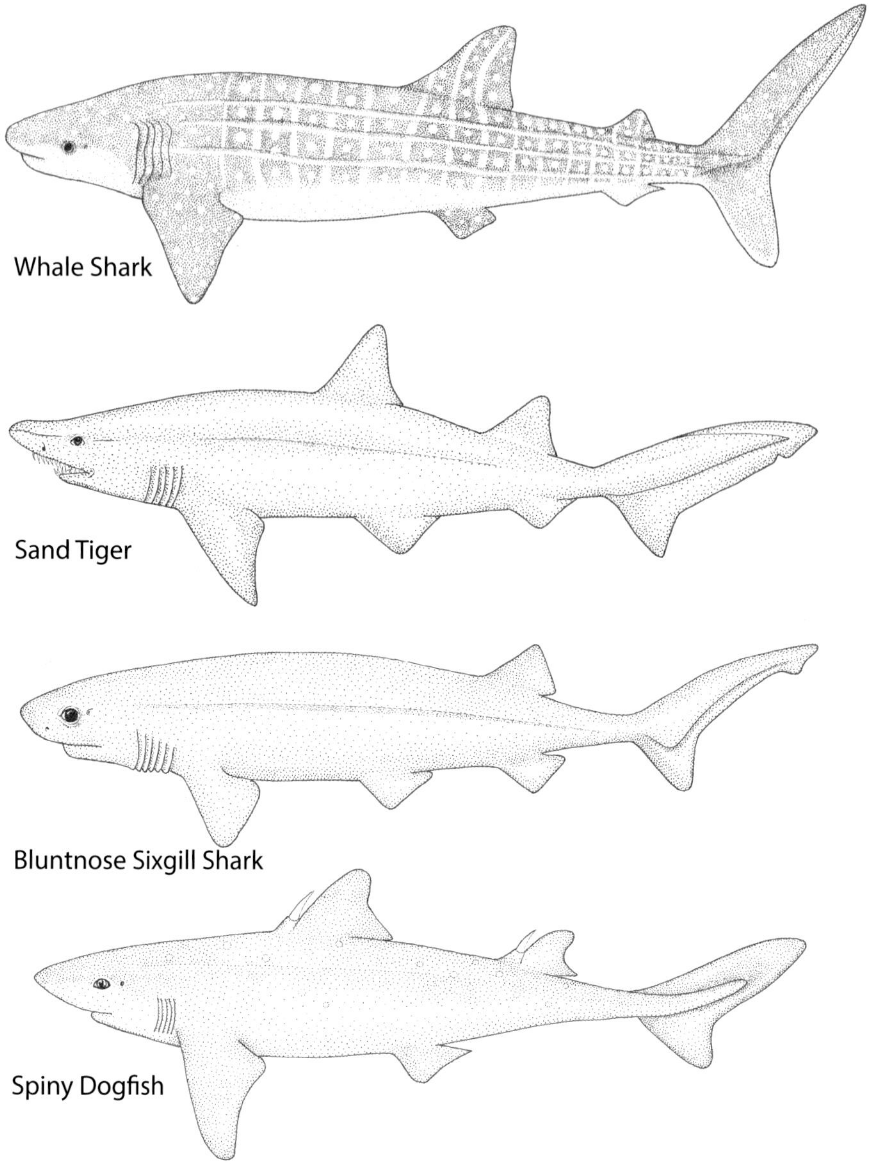

Figure 4.5. Four basic ecomorphotypes of sharks based on Thomson and Simanek (1977) and Dolce and Wilga (2013).

Fins

Like overall body form, the morphology of a shark's fins is correlated with the species' ecology. Most sharks spend their lives somewhere between the floor and ceiling of their environment, and fins facilitate three-dimensional movement in that space. To better understand how, it is helpful to first consider the position and structure of those fins.

The fins of sharks and rays differ structurally from those of modern bony fishes (figs. 4.6 and 4.7). The fins of bony fishes are supported by hard spines and softer

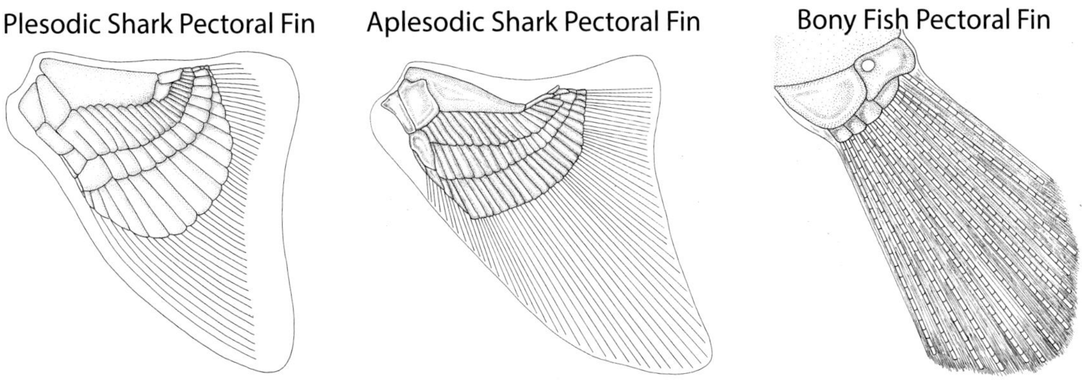

Figure 4.6. Fins of bony fishes and sharks, compared (Redrawn from fig. 5.10, Wilga, C.A. and Lauder, G.V. 2004. In: Carrier, J.C. *et al.* (eds.). *Biology of Sharks and Their Relatives*. CRC Press. 139–164).

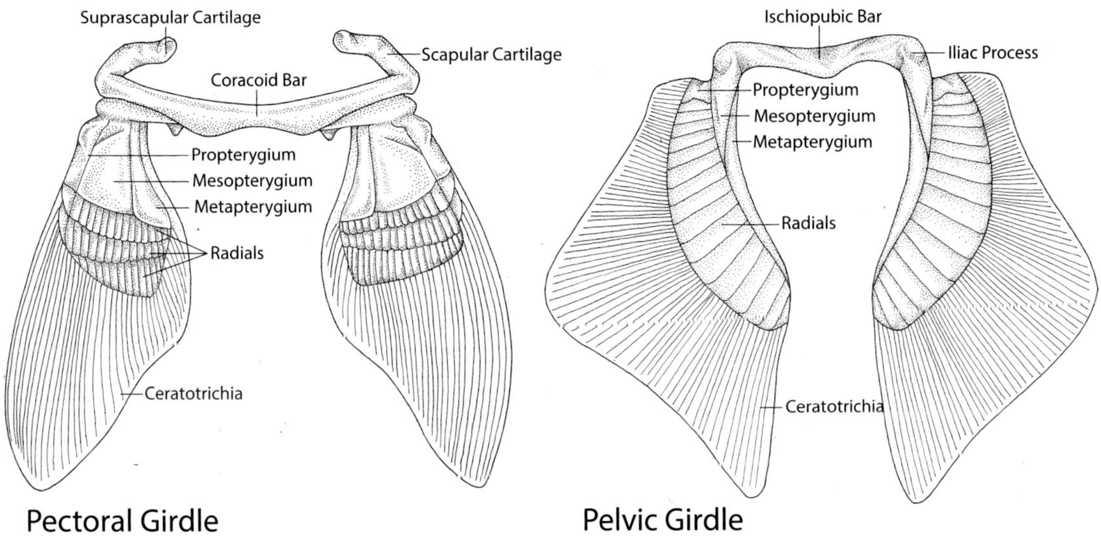

Figure 4.7. Pectoral and pelvic girdles of sharks (Redrawn from Gilbert, S. 1973. Pictorial Anatomy of the Dogfish. University of Washington Press).

segmented rays called *lepidotrichia,* which consist of segments of bone that give each fin ray a jointed appearance when examined closely. The functional result is that a bony fish's fins are compressible, and the *lepidotrichia* enable conformational changes in the surface of the fin. At the base of the fins, bony skeletal elements called *pterygiophores* lend extra support.

In contrast to the fins of bony fishes, shark fins are given their shape by fibers of the protein keratin called *ceratotrichia.* Ceratotrichia are unsegmented, and consequently sharks cannot fold or undulate their fins to the same extent as the average bony fish. Incidentally, ceratotrichia are the primary ingredient in shark fin soup. In addition to fibrous ceratotrichia, shark fins are supported by cartilaginous skeletal elements called *radial cartilages* (sometimes called *radial pterygiophores*), but unlike ceratotrichia, radials do not extend to the edges of the fins. At the base of the fin where it joins the body, a shark has another series of cartilaginous skeletal elements called *basal pterygiophores* (fig. 4.7; see figs. 1.9 and 1.10). The only exception to this general anatomical scheme of shark fin anatomy is the caudal fin (tail) which is supported by a portion of the vertebral column in the upper lobe (fig. 4.8).

Median Fins

Sharks have two median fins, the dorsal(s) and anal. The anal fins of many fast-swimming sharks (e.g., macropelagic sharks like the Shortfin Mako), are reduced in size, possibly reducing drag. In littoral sharks, a group called *continental cruisers* that includes many carcharhinid sharks, the anal fin is moderately sized. These sharks are characterized by being highly maneuverable, so it is not a stretch to assume that the anal fin plays a role in maneuverability. Sharks in the order Squaliformes, which are typically benthic sharks, lack an anal fin, the functional significance of which is not yet known.

The first dorsal fin is perhaps the most well-known of all shark fins. It is this fin that is depicted in company logos, advertisements, and movies to signal the sensationalized presence of a shark, specifically, a large pelagic shark. In pelagic sharks (species that spend most of their time in the water column above the sea floor), the first dorsal fin tends to be large and triangular and located anteriorly, above or just behind the pectoral fins. This is due to the fist dorsal fin's role in maintaining stability while swimming. Smaller benthic sharks, such as cat and carpet sharks, typically have reduced dorsal fins located more posteriorly on the body, since their bottom-associated lifestyle means they are not continually swimming for long periods.

The second dorsal fins of high-speed cruisers like lamnids are greatly reduced. The second dorsal fins of continental cruisers are more moderately sized. This is consistent with the idea that speed and drag reduction are paramount to the former group of sharks, with speed and maneuverability both being important to the latter.

Some sharks, such as the frilled and cow sharks, in the order Hexanchiformes, have lost one of their dorsal fins and have their lone dorsal fin far back along the body. The absence of an anterior dorsal fin allows them to roll with less drag when they are feeding and as they roll, they shear pieces of flesh from their prey with their saw-like teeth.

Many fast-swimming bony fishes (e.g., tunas and some billfish) can fold their dorsal fins (and pectoral and anal fins as well) flat against their bodies. In doing so, they become more streamlined, and decrease drag at higher swimming speeds. These fins can be extended when needed for maneuverability or stability. Because of the differences in fin anatomy, sharks are not capable of folding their fins in the same way.

Caudal Fin

The caudal fins of ray-finned bony fishes and cartilaginous fishes alike provide thrust, but there are fundamental anatomical differences between the two. Recall that the caudal fins of most derived bony fishes, whether it is a goldfish or a salmon, are mostly symmetrical and referred as *homocercal* caudal fins. The *dorsal* (upper) and *ventral* (lower) lobes of the fin are essentially the same with respect to the skeletal elements that support the surface area of the fin. A shark's *heterocercal* tail also has a dorsal and ventral lobe, but the vertebral column extends into the dorsal lobe. *Hypochordal rays*, which are expanded arches on the underside of caudal vertebrate, lend further support to the dorsal lobe of the caudal fin (fig. 4.8). The ventral lobe of a shark's heterocercal tail consists of ceratotrichia and lacks any vertebral elements.

There is considerable variability in the shape of caudal fins. At one extreme are the threshers (see fig. 3.27), whose greatly expanded upper lobe is about half of the length of the body and is used to herd and stun small schooling fish. Sharks that rest on the bottom have reduced lower caudal lobes. Sharks like White Sharks (*Carcharodon carcharias*) and Shortfin Makos (*Isurus oxyrinchus*) have tails that are superficially symmetrical (but still have the vertebrae extending to the tip of the upper lobe), an adaptation that allows them to generate greater thrust, which is an advantage if you are an apex predator that needs to pursue or ambush prey.

Finally, let us consider the caudal fin of an angel shark (see fig. 3.9). It lives on and over the bottom, and it feeds by surprising its prey from below (not from behind or above, the importance of which will be evident when we discuss thrust below). Over evolutionary time, the relative sizes of caudal fin lobes switched in

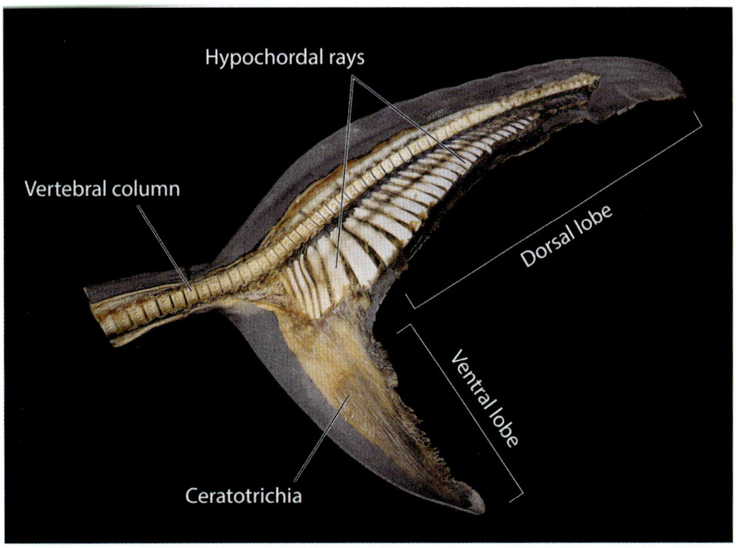

Figure 4.8. Skeletal anatomy of a heterocercal tail of a Porbeagle depicting vertebral column and hypochordal rays.

angel sharks. The lower lobe of an angle shark is the longer of the two (although the vertebrae still extend to the end of the upper lobe). This is known as a *hypocercal* caudal fin.

Pectoral and Pelvic Fins

Sharks exhibit two different types of pectoral fins, *plesodic* (*ples* = near) and *aplesodic* (*a* = without) (fig. 4.6). In the former, the radial cartilages extend outward more than 50% of the way toward the fin's tip, which provides more support and stiffens the fin. In the latter, radials extend less than 50% of the way, which makes the fin more flexible. Plesodic fins are the more derived type and are found in sharks of the order Lamniformes as well as the carcharhiniform families Hemigaleidae (weasel and snaggletooth sharks), Carcharhinidae, and Sphyrnidae, as well as all batoids except those in the families Pristidae (sawfish) and other members of the order Rhinopristiformes (guitarfishes).

In sharks, the pectoral fins are ventrolateral (lower on the flanks) and are relatively stiff. The pectoral fins of sharks are supported by sturdy internal structures: the pectoral girdle is formed by the *scapulacoracoid* cartilage (actually, two fused structures, the single *coracoid* bar and paired *suprascapular* cartilages; fig. 4.7). Three enlarged basal cartilages and more distal radial cartilages inside the fins lend further structural support (fig. 4.7; see fig. 1.9). The three enlarged basal cartilages of sharks are (in anterior to posterior order) the *propterygium*, the *mesopterygium*, and the *metapterygium*.

For a fascinating example of how evolution has tweaked the skeletal anatomy and morphology of pectoral fins, meet the Epaulette Shark (*Hemiscyllium ocellatum*). Native to the waters of New Guinea and northern Australia, the Epaulette Shark forages on small fishes and benthic invertebrates. Usually less than 100 cm in length, it ventures into tide pools in search of prey. When the tide goes out, an Epaulette Shark can walk across the exposed reef flats to the next tide pool by undulating its body and lifting itself up on modified pectoral and pelvic fins. How can it do this? The species has lost one of the ancestrally three basal pterygiophores at the base of its pectoral fins.[8] The remaining mesopterygium and metapterygium are widely spaced, which allows a wider range of motion for the pectoral fins of the Epaulette Shark (fig. 4.9).

Like the pectoral fins, sharks' pelvic fins also attach to a sturdy internal structure. This structure is the pelvic girdle. A cartilaginous structure called the *puboischiadic bar* joins the left and right halves of the pelvic girdle, and the pelvic fins themselves are supported by basal pterygiophores, radial cartilages, and ceratotrichia. In male sharks, additional cartilaginous skeletal structures called *mixopterygii* extend posteriorly from the basal pterygiophores. These form the skeletal support for the males' claspers. While these extensions of the pelvic fins do not serve a locomotory function, they have undergone extensive modification in elasmobranchs to facilitate the delivery of sperm into a female shark's reproductive tract.

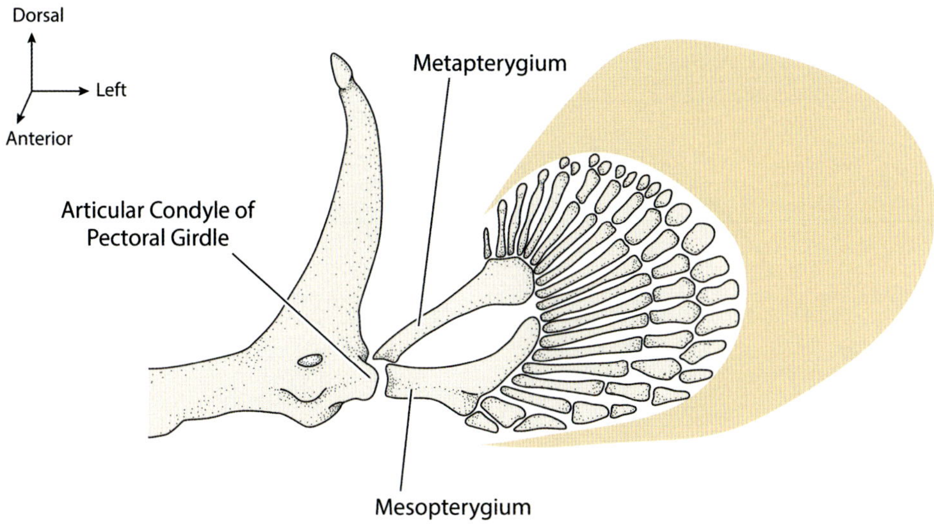

Figure 4.9. Modified fin skeletal structures of the Epaulette Shark.

How Sharks Swim

When it comes to locomotion, a shark is faced with two challenges: moving its body and stabilizing it. Movement involves a change in position in one or more dimensions: left to right, forward and backward, or up and down. Stabilization requires the shark to fine tune its body position as it moves. Because a shark moves in three dimensions, there are three orthogonal axes (axes at right angles to each other) on which a shark may adjust its body position and stabilize itself: *pitch* determines whether a forward moving body goes up or down. A shark with a positive pitch is angled "nose up." A negative pitch angles a shark "nose down." *Yaw* is a side-to-side movement around the vertical axis. *Roll*, which can make a shark go belly-up, is rotation of the body around its longitudinal axis. Refer to Figure 4.10. The primary functional significance of shark fins is to enhance movement and enable stabilization.

Some of the original work that uncovered the roles of the various shark fins was done in what some may consider a gruesome manner, by removing one or more fins of a living shark and observing the resulting motion. Other studies have used models of sharks in flumes (water tunnels) or have inferred function based on the underlying structure of the fins or from behavioral observations. These have all contributed to the knowledge base, but until appropriate measurements can be made on sharks swimming naturally in their environment, the best information has come from studying sharks swimming in controlled laboratory conditions. The major drawback to this has been the limited number of species studied, which has included principally those most amenable to swimming in captivity, specifically Leopard Sharks (*Triakis semifasciata*), Spiny Dogfish (*Squalus acanthias*), and Bamboo Sharks (*Chiloscyllium plagiosum*).

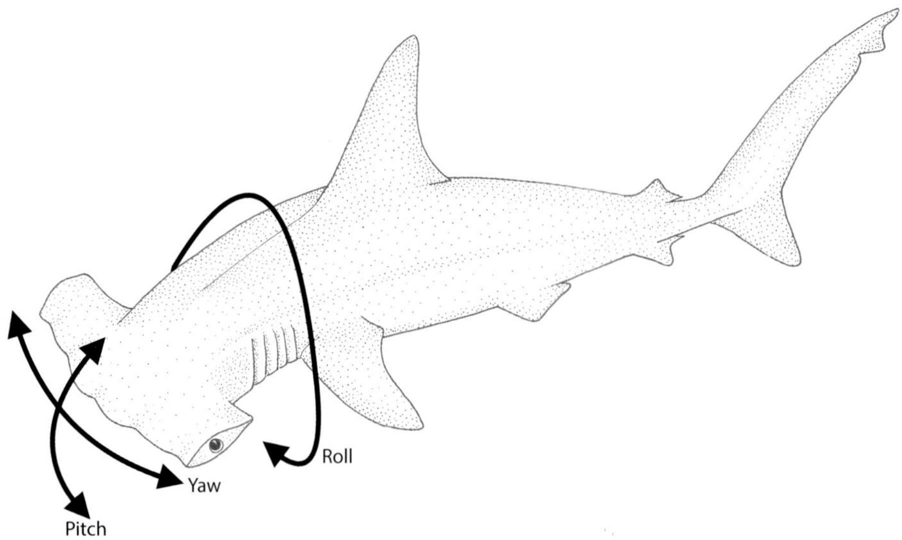

Figure 4.10. Responses to forces acting on moving objects—pitch, roll, yaw.

Thrust

Thrust is the *propulsive* force that causes a shark to move through the water. Like most fishes, sharks swim by contracting *myomeres* (muscle segments) along the length of the body. When myomeres on one side of the body contract, they pull connected body segments to that side while myomeres on the opposite side relax. By alternating myomere contraction and relaxation on each side of the body, the shark laterally undulates its body, passing propulsive waves toward the caudal fin. As the undulations reach the caudal fin, the fin oscillates back and forth. Recall what our friend Isaac Newton taught us in his Third Law of Motion: if one object exerts a force on a second object, the second object exerts an equal and opposite force on the first. As the caudal fin moves, it exerts an angled force on the water (angled because of the movement of the caudal fin). These forces are *vector forces,* meaning that they can be resolved into forward and lateral components. The forward component of the force *pushes* the shark through the water (fig. 4.11). Understanding this general mechanism allows us to spot another form–function relationship, this time in the aspect ratios of caudal fins (fig. 4.12).

The *aspect ratio* of a shark's caudal fin gives us an idea of how important thrust and acceleration are to the shark. Higher aspect ratios provide more thrust relative to drag and are found in species with higher aerobic metabolisms; that is, high-performance sharks that include lamnids and some carcharhinids. The tradeoff is that stiff, high aspect ratio caudal fins are not capable of the same level of maneuverability. Caudal fin aspect ratio is calculated as follows:

$$A = h^2/s$$

Where A = aspect ratio, h = straight line vertical span, and s = surface area of the fin.

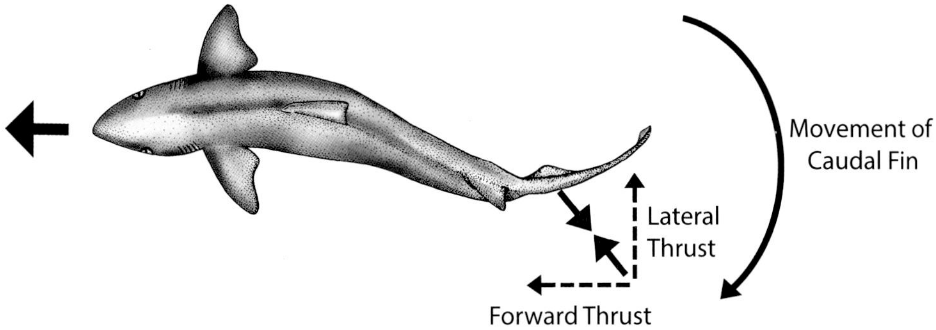

Figure 4.11. Diagram of thrust generation and vector forces in lateral undulation.

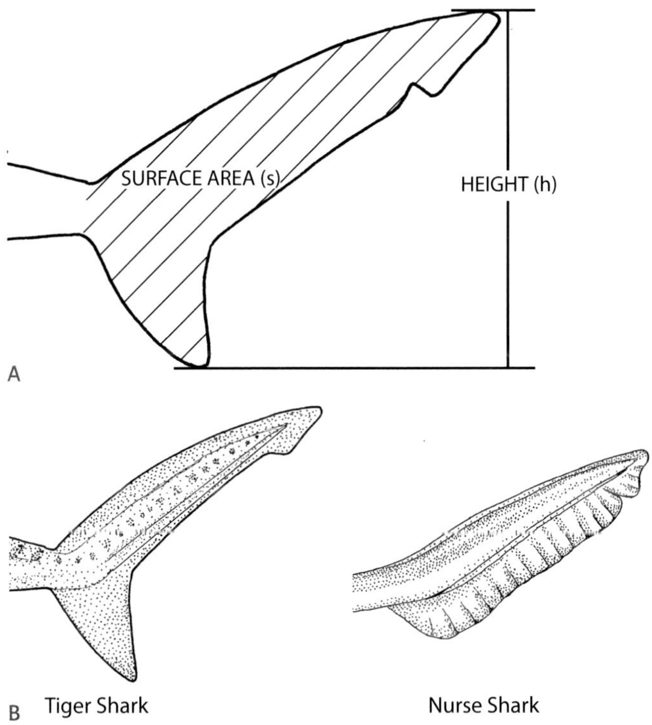

Figure 4.12. (A) Diagram showing how an aspect ratio is determined for a generalized caudal fin. (B) Nurse and Tiger Shark caudal fins.

Among all fishes, the caudal fins of tunas have the highest known aspect ratios, up to about 7.5 in the Bigeye Tuna (*Thunnus obesus*). Among sharks, aspect ratios vary from 0.76 in the Nurse Shark[9] to > 14 in the Tiger Shark[10] (*Galeocerdo cuvier*; fig. 4.12B).

If we compare patterns in how much lateral undulation of the body is used to generate thrust across different species of sharks, we see a continuum of *swimming modes* (fig. 4.13).

- *Anguilliform (eel-like).* A form of axial (along the body) undulation found in slim, elongate sharks like those in the order Orectolobiformes and families Chlamydoselachidae and Scyliorhinidae, in which a wave moves down along the entire body. An entire wave is typically present in the body when swimming fast, with significant lateral head movement (yaw) as speed increases.
- *Carangiform and Subcarangiform (jack-like).* Here, the axial undulations occur in the posterior half of the body, with less than a full wave present. The wave amplitude (the lateral body motion) diminishes as it moves posterior, and there is some head yaw. Subcarangiform swimming is a subdivision and involves more than one half of the body but less than a full body wave, as well as slightly more head yaw than in carangiform swimming. Most sharks are either carangiform or subcarangiform swimmers. Examples include sharks of the orders Squaliformes, Carcharhiniformes, and Lamniformes exclusive of the family Lamnidae. Most are pelagic.
- *Thunniform (tuna-like).* Thunniform swimming represents the swimming mode on the opposite end of the continuum from anguilliform swimming. Like a windup fish toy, only the caudal fin and peduncle oscillate in thunniform swimmers, and there is little head yaw. Among sharks, thunniform swimming is restricted to members of the family Lamnidae.

For reasons that are self-evident, most batoids are incapable of these swimming modes. Rather, batoids rely more on *undulation* (e.g., the Southern Stingray *Hypanus americanus*) or *oscillation* of their pectoral fins (e.g., the Cownose Ray *Rhinoptera bonasus*).[11] The exceptions to this trend in batoids are members in the orders Rhinopristiformes and Torpediniformes. Both are more basal batoid groups

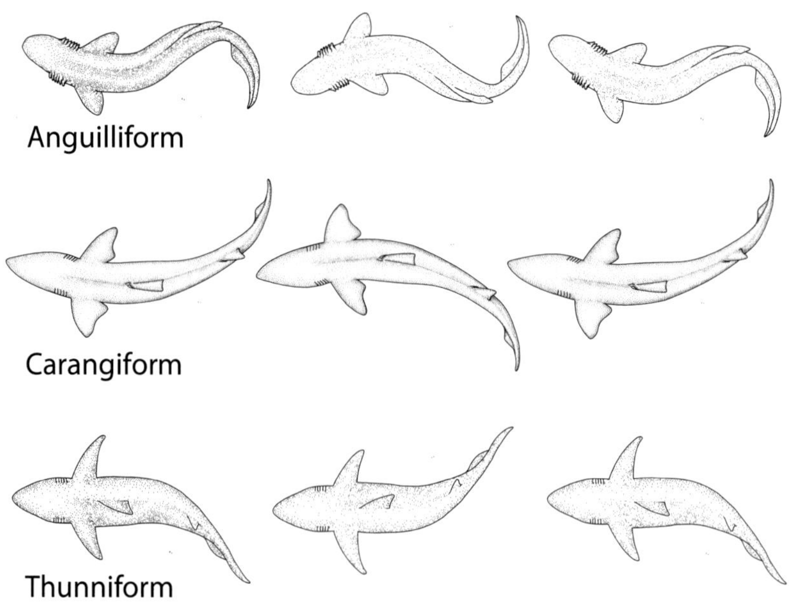

Figure 4.13. Major modes of swimming in sharks and batoids (Redrawn from fig. 7, Sfakiotakis, M. *et al.* 1999. J. Ocean. Eng. 24: 237–252).

and are axial swimmers, using both their body and tail. Likewise, angel sharks swim more like a torpedinid ray than a shark.

In undulatory (also called *rajiform*) swimming, a wave passes down their pectoral fins (wings). Most batoids (all skates, most stingrays) are undulatory swimmers, and most undulators are benthic. In oscillatory (also called *mobuliform*) swimming, batoids flap their "wings" up and down like birds. This small group includes some members of the order Myliobatiformes (cownose, eagle, and manta rays). All are pelagic.

Lift

Lift is a force produced when a fluid, like a gas or liquid, flows around an object. We typically associate lift with an upward movement. In reality, lift is a force that is perpendicular (90°) to a flowing liquid. We are used to thinking about lift in flying or swimming, during which we move forward and use lift to overcome gravity. For a shark swimming horizontally into seawater, lift is generated as water flows over its pectoral fins and body and is displaced by an asymmetrical heterocercal tail.

To better understand lift, it is helpful to picture a macropelagic shark, like the Blue Shark (*Prionace glauca*; see fig. 3.43), which must continually generate lift to maintain its position in the water column. As the shark swims forward, water encounters the *leading edges* of the shark's pectoral fins. The water can flow in boundary layers around the fin on the dorsal (upper) side or the ventral (lower) side (fig. 4.14). Because the dorsal surface is more curved, water flowing on the dorsal side must move faster than the water on the ventral side if it is to meet the water flowing along the ventral side again at the back edge of the fin as the shark moves forward. As the 18th-century physicist Daniel Bernoulli told us, when fluid flows over a surface quickly, pressure drops. Thus, for the shark, there is less pressure in the boundary layers above the fin than below, which results in lift. In this regard, the pectoral fins of a Blue Shark and the wings of an airplane are very similar.

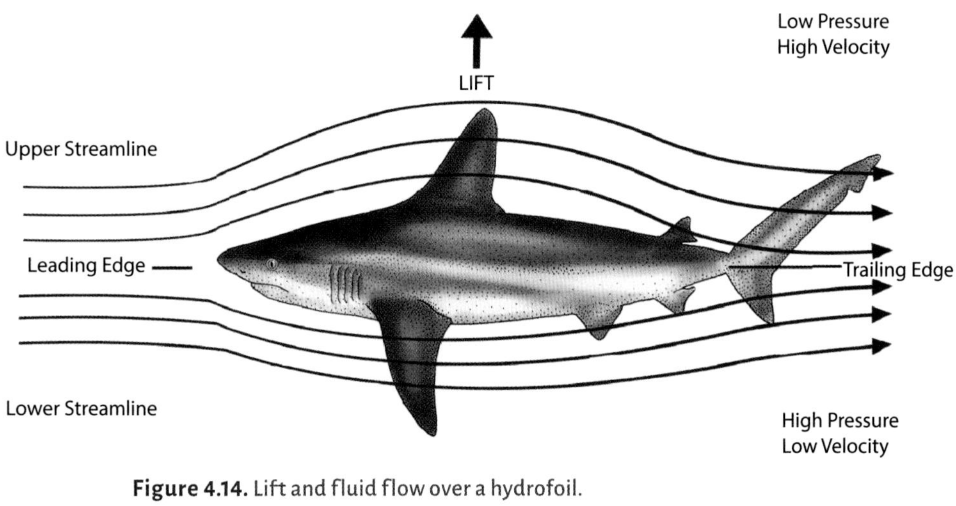

Figure 4.14. Lift and fluid flow over a hydrofoil.

Another often overlooked source of vertical lift for a swimming shark is its asymmetrical tail. As the caudal fin generates forward thrust, the dorsal lobe of the fin also pushes water downward and backward. Thanks to Newton's Third Law, we know that the backward component of the water's movement can help push the shark forward. Likewise, the downward component contributes to maintaining the sharks' position in the water column and can contribute to a positive *pitch*.

Water flow over the body, the pectoral fins, and the caudal fin collectively contribute to maintaining or changing a shark's position in the water column. The lift generated by these structures is complemented by a large, oily liver. The low-density oil found in shark livers makes the organ positively *buoyant* (meaning it floats), which helps offset the negative buoyancy of the sharks' musculoskeletal structures. *Buoyancy* is determined by the density of an object and the density of the water that the object displaces. If the density of the object is less than that of the displaced water, it floats. If it is greater than the density of the displaced water, it sinks. This phenomenon is not the same as lift as far as physics is concerned, but to a shark, lift and buoyancy both affect its vertical position in the water column and the energy required to maintain it.

Stabilization

Moving forward with a level path is one thing, but how do sharks keep from rolling, and how do they maintain or alter their headings? Pectoral fins capable of being adjusted independently and dorsal fins play key roles.

To alter its *yaw* (to bear left or right), a swimming shark uses the muscles at the base of the pectoral fins to change the angle of the fins as they cut through the water. By changing the angle at which the fin on the inside of the turn meets the oncoming water, a shark can increase the amount of fin surface area that is meeting the water perpendicularly. The result is increased drag on that side of the shark but not the other, causing the shark to veer into the turn as it swims.

Veering into a turn due to a difference in drag on the left and right sides of the body isn't the only way a shark can change its heading. Many smaller sharks, like those in the family Hemiscyllidae, can bend their bodies quite a bit. For example, Epaulette Sharks (*Hemiscyllium ocellatum*) can almost touch their noses to the tips of their tails.[12] However, larger sharks do not show the same flexibility. The structural stiffness of the vertebral column appears to be a limiting factor as do the length and width of the body. Long and skinny sharks appear to be the most flexible.

Because there is considerable variability in the sizes, shapes, and placement of dorsal fins in sharks, generalizing their importance in stabilizing *roll* is tricky. In some species with flexible dorsal fins positioned far back on the body, the dorsal fins may help contribute to thrust. However, in active, faster swimming species, including the Spiny Dogfish (*Squalus acanthias*)[13] and White Shark (*Carcharodon carcharias*),[14] dorsal fins appear to stabilize the shark. If such a shark were to begin rolling, the surface area of a rigid dorsal fin would encounter the water, and an opposing force would dampen the roll as the shark began to spin. Newton's Third Law strikes again!

Scales and Skin

Shark skin is covered in scales called *placoid scales*. Rays and skates also have placoid scales, but their distribution tends to be more regionalized on the body. Here we opt to refer to these scales by a different name to reflect just how unique they are: *dermal denticles*. The name dermal denticles means tiny skin teeth, and that's exactly what they are. Like oral teeth, shark scales have a pulp cavity surrounded by dentine and covered in an enamel-like coating. These tiny teeth are embedded in the skin and differ in size and shape depending on where they are on the body and from species to species.

Hydrodynamics of Skin Teeth

The dermal denticles of sharks have multiple functions. Here, we will consider their hydrodynamic importance. As we do, bear in mind that in swimming, the name of the game is achieving the correct balance between thrust and drag while maintaining sufficient lift to stay buoyant. Denticles play a central role in that game, along with body shape, metabolic performance, etc.

The main hydrodynamic role of dermal denticles is reducing frictional drag caused by *surface shear*. Surface shear is a stress that causes friction and generates vortices that twist, rotate, and interact with one another near the surface of the shark. These vortices may be shed into the boundary layer leading to a more chaotic flow and motion-slowing turbulence. The *sine qua non*[15] of frictional drag reduction in sharks are the riblets, or longitudinal ridges creating grooves, of the dermal denticles (see fig. 4.1).

One way that the riblets on a denticle can reduce frictional drag is by directing the vortices shed as water flows over the shark. Remember that sharks and their denticles are three-dimensional, and that means that vortices are shed to the sides of a denticle as well as off its back. Vortices that are shed to the side of a denticle still represent water flowing against an object, so friction will result, and friction resists the movement of the shark through the water. By channeling the vortices backward and minimizing their lateral movement over the surface of a denticle, denticle riblets reduce friction between the skin and water.

It also appears that riblets perform two other functions. First, in the posterior part of the body, the dermal denticles actually create directed turbulence that, counterintuitively, reduces the drag generated by a shark's overall body. This happens because the total surface area of the body in contact with a given drag-inducing vortex is functionally reduced if the vortex lifts off from the riblet, away from the body of the shark. As a result, the boundary layer flowing over the shark does not separate as much as it would if the vortex were to stay in close contact with the skin. Thus, shedding a vortex through directed turbulence helps maintain a smoother boundary layer of water. Secondly, dermal denticles may also enhance lift and thrust as their textured surface forces water to flow quickly over them.[16]

We should bear in mind that the size and shape of denticles vary dramatically across sharks, as do swimming speeds, body movements, and habitats. Therefore, we must be cautious about overgeneralizing the functional significance of denticles across species. Because dermal denticle morphology varies, assessing the full range of dermal denticle functions is a daunting challenge. However, that same variability also represents an opportunity to look for ecomorphological correlations between denticle morphology and the habitat and lifestyle of selected species.

Studies conducted in 1984[17] and 1986[18] examined dimensions of dermal denticles in 15 species of sharks separated into three groups based on swimming modes: fast-swimming neritic, intermediate swimmers, and slow-swimming epibenthic.

The dermal denticles of the Nurse Shark (fig. 4.15), a slow-moving species that frequently rests at the bottom, were smaller and lighter than those of continuously swimming species, and the denticles were less-densely packed and did not form the lengthwise channels created by the scale pattern on the other species. Fast-swimming sharks (Shortfin Mako, Scalloped Hammerhead, and Blacktip Shark) possessed smaller, thinner, and lighter, but more densely packed, dermal denticles. Characteristics of dermal denticles of the intermediate-speed group (Bull, Sandbar, and Blue Shark) placed between the other two.

Finally, scales along the flank of the Shortfin Mako were found to be highly flexible, an adaptation the authors of the study that discovered this thought would enhance burst swimming.[19]

Skin as an Exotendon

When muscles contract, force is transmitted through *tendons* to move a part of the body. We often think of conventional tendons like our Achilles tendon that connects muscles in the calf to our heel bone. Sharks do have conventional tendons throughout their bodies, but their skin displays properties of another type of tendon: an exotendon. As you might infer from the prefix *exo*, exotendons are structures that act like tendons in transmitting force, but are *external* to the muscles, tendons, and skeletal structures primarily responsible for movement.

When a shark laterally undulates its body (or swings its tail from side to side in the case of thunniform locomotion), the myosepta that separate myomeres act as tendons, connecting the vertebral column to the skin. In the *dermis*, or inner layer of the skin, layers of the fibrous connective tissue collagen form a cross-helical pattern, wrapping around the shark. The faster a shark swims, the more

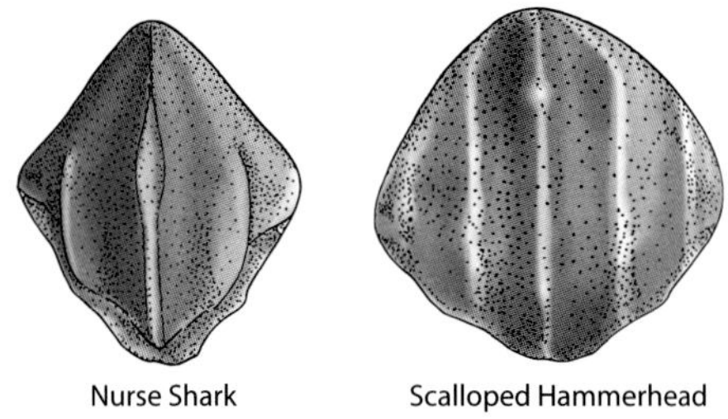

Figure 4.15. Diversity of scale types: Nurse and Scalloped Hammerhead.

Figure 4.16. Helices of collagen fibers in the skin up and down the body.

contracting myomeres and flowing blood increase the internal pressure on tissues, including the skin. Consequently, the skin stiffens, and the cross-helical collagen layers stretch on the convex (outward bending) side of the body. Energy is stored in these stretched collagen layers, and as the shark begins to bend in the opposite direction (remember, this is a laterally undulating shark), those stretched collagen layers release their energy in an *elastic recoil.* Because they are arranged in a cross-helical pattern, the energy of the recoil is channeled toward the tail (fig. 4.16). The result is that the shark expends less energy unbending its tail as it prepares to swing it in the opposite direction. Once again, we see that energy efficiency and conservation is a reoccurring theme in shark adaptations.

Functional Anatomy of Feeding

The anterior ends of sharks have fascinated humans for centuries. That is where we find the majority of a shark's sensory structures and the brain that processes the sensory input. It is also where we find sharks' jaws and teeth. As you might imagine, sensing and reacting to your environment and finding, procuring, and consuming food are important ecological tasks. So, it should come as no surprise that sharks show a wide range of adaptations in their heads. We will discuss sensory biology in detail in chapter 5. Here, we will focus on the functional anatomy of the head and feeding apparatus, specifically the jaws and teeth.

Head Shape

The shape of a shark's head is related to both locomotion and diet. But how do we know which selective pressures best explain the shape of a shark's head? Do the benefits of efficient, hydrodynamic motion shape the head, or perhaps the distribution of sensory structures and jaw muscles offer advantages to a predatory shark? The truth is it is not a dichotomy. A given head shape can offer different benefits under different conditions. To illustrate this, let us consider a group of sharks that represent an extreme in head shape: the hammerheads (family Sphyrnidae).

The outline of the hammerhead *cephalofoil* resembles the wing of an airplane in that it has *camber* (an asymmetry between the top and bottom surfaces creating lift), and its placement brings to mind a canard wing, or forewing (a planing surface anterior to the pectoral fins or a set of smaller wings on an airplane in front of the main wings). On airplanes, canard wings enhance maneuverability as the plane rolls. Perhaps counterintuitively, a 2003 study[20] studying the maneuverability of Bonnetheads and Scalloped Hammerheads (*Sphyrna tiburo* and *S. lewini*) found that these species, while highly maneuverable, did not roll as they reoriented toward an odor source or electric stimulus. In those circumstances, the cephalofoil may have helped stabilize the shark, but it was not acting as a canard wing.

Does that mean that it has no hydrodynamic function? Not necessarily. In 2016, a study[21] was published detailing how Great Hammerheads (*Sphyrna mokarran*; fig. 4.17) spent much of their time cruising at a roll angle between 50° and 70°. Researchers calculated that under those conditions, when the shark was not actively reorienting toward a stimulus, the lift afforded by the canted cephalofoil likely reduced the energetic cost of locomotion by nearly 10%.

When hunting or navigating, the cephalofoil of a hammerhead likely offers other benefits. In a 2009 study,[22] binocular vision, which results in better depth perception as the brain processes and integrates the slightly different view from each eye, was greater on two of three hammerheads (Scalloped [*Sphyrna lewini*] and Winghead [*Eusphyra blochii*]) than that in two carcharhinid sharks. However, the wide separation of the eyes among sphyrnids creates a large anterior blind area close to the head in most species. To compensate, hammerheads swing their heads from side to side as they swim.

A third function of the cephalofoil, prey handling, is based on two published anecdotal observations of this phenomenon in Great Hammerheads (*Sphyrna mokarran*) in which individuals used their heads to pin a Southern Stingray (*Hypanus americanus*)[23] and Spotted Eagle Ray (*Aetobatus narinari*)[24] to the bottom before eating them.

Figure 4.17. Great Hammerhead in Bimini, Bahamas. Note the tight turning radius, enabled in part by using the laterally expanded head as a stabilizing rudder (Courtesy of Annie Guttridge Photography).

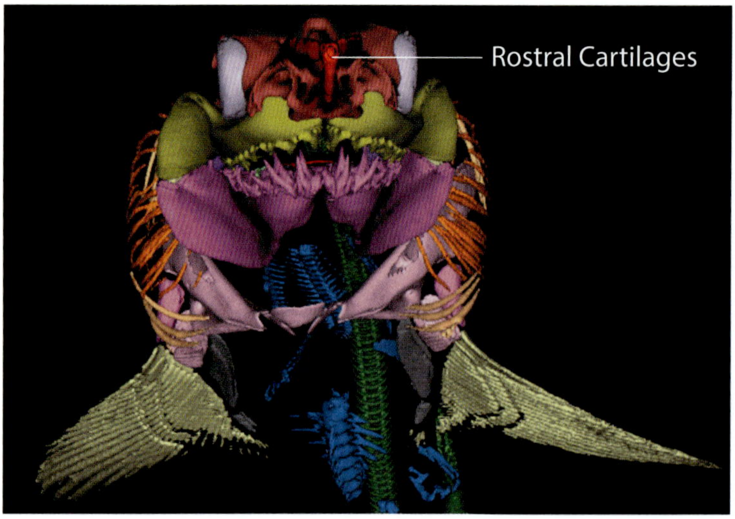

Figure 4.18. Tribasic snout cartilage of the Shortfin Mako shown in a CT scan (Courtesy of Gavin Naylor).

Among sharks that display less exaggerated head morphologies, carcharhinid sharks such as the Bull Shark (*Carcharhinus leucas*) and the Tiger Shark (*Galeocerdo cuvier*) typically have blunter snouts and heads that are wider than they are high. Exceptions to this trend include the Blue Shark (*Prionace glauca*) and Spinner Shark (*Carcharhinus brevipinna*), which have slightly pointier snouts than other carcharhinids.

In contrast to carcharhinids with blunt snouts, lamnid sharks have conical snouts supported by pointed rostral cartilages (Fig. 4.18). The Shortfin Mako (*Isurus oxyrinchus*) exemplifies this morphology, and its head is roughly bullet-shaped, the hydrodynamic benefits of which are likely evident by this point.

One group of sharks that merits special attention for head shape is the family Heterodontidae. This family is comprised of the horn sharks, and due to their blunt snouts and ridged heads, they have collectively been given the name bullhead sharks. These slow-moving, bottom-dwelling sharks eat sedentary or sluggish benthic invertebrates like sea urchins. They do not require streamlining to be happy and well-adjusted close to the seafloor, but powerful jaws capable of crunching shells are a must for these sharks.

All of the above sharks have *subterminal* mouths; that is, the mouths are on the lower side of the head (also called underslung jaws). Frilled sharks, angel sharks, the Whale Shark, and the infrequently encountered Megamouth Shark (*Megachasma pelagios*) have *terminal* mouths; that is, at the tip of the snout. The last two of these are filter-feeders in which a large gape allows them to maximize the volume of seawater they filter as they swim.

Regardless of the overall shape of the head, it is important to recognize that its morphology is the result of underlying musculoskeletal structures. These structures include the braincase, rostral cartilages, jaws, and associated soft tissues (e.g., the muscles, tendons, and ligaments powering and supporting the jaws).

Jaw Suspension

Jaw suspension refers to how the upper and lower jaws connect to the skull and hyoid arch (see chapter 2 for an overview of jaw evolution). Try to move your lower jaw without moving the rest of your head. Most of us can do this easily. Now try to move your upper jaw without moving the rest of your head. If you can do this, you should probably stop reading and proceed to the hospital. As a human your upper jaw is (or should be) fused to your braincase and is considered *autostylic* (*auto* = self; *styly* = types or styles) or more specifically, *craniostylic*, meaning that the upper jaw is fused to the cranium, but sharks exhibit different types of jaw suspension.

To understand the types of jaw suspension of sharks, we first need to review some anatomical terminology (see fig. 2.9). The upper and lower jaws of sharks are known, respectively, as the *palatoquadrate* and *Meckel's cartilage*. Together, these structures are called the *mandibular arch* (in recognition of their evolution from a series of skeletal arches that include the branchial arches), but we can simply call them *jaws*. Immediately posterior to the jaws is the *hyoid arch*, which represents the second branchial arch that moved forward in evolution to support the jaws to some degree. It is important to understand that the palatoquadrate is only connected to the shark's braincase, or *chondrocranium*, by suspensory ligaments—not fusion of skeletal structures. This arrangement allows for a greater degree of jaw protrusion, or *protraction*, and is why a shark filmed and replayed in slow motion may look as if it is losing its dentures when it bites.

The extent to which the jaws are supported by the hyoid arch and the way in which they rest under the chondrocranium differs between groups of sharks and rays (fig. 4.19). Here is a brief summary of jaw suspension types in modern sharks and rays, along with the functional significance of each type:

- Amphistylic Jaw Suspension

  Amphistylic (*amphi* = both) jaw suspension is found in sharks displaying more primitive or ancestral features, including the frilled and cownose sharks, as well as many of the older fossil species.

  In amphistylic suspension, the palatoquadrate is attached by two sets of ligaments (strong, fibrous material that connects skeletal elements to each other)—the *ethmopalatine ligament* just anterior to the *orbit* (the region that houses the eye) and aptly named the *postorbital ligament*, which is found just posterior to the orbit. The palatoquadrate is not tightly fixed to the chondrocranium. Posteriorly, the jaws are provided with very limited support by the hyomandibular cartilage, part of the hyoid arch. This anatomical arrangement does not allow for nearly as much jaw protraction as is observed in more derived species.

- Orbitostylic Jaw Suspension

  Researchers hypothesize that orbitostylic jaw suspension evolved from an amphistylic ancestor. Today, we see orbitostylic jaw suspension in dogfishes of the order Squaliformes. Here, the orbital process of the palatoquadrate articulates via a ligament associated with the orbital region of the braincase (hence the name *orbito*stylic). The ethmoid and postorbital articulations seen in amphistylic suspension have been lost in orbitostylic species.

Orbitostylic jaw suspension typically involves a longer suspensory ligament relative to the suspensory ligaments of comparably sized amphistylic jaws. The functional significance of this is that the orbitostylic jaws can be protracted farther away from the chondrocranium. At this point, it is important to recognize that when comparing measures of force, distance, or speed, one should consider that larger animals can cover more distance or exert greater force simply because of their size. That is why comparative studies that quantify jaw protraction typically do so as a percentage of another measurement, such as head length.

Orbitostyly (the condition of being orbitostylic) does have one drawback: when the jaw is fully protruded, the maximum gape is substantially reduced by nearly two-thirds compared to the gapes of batoids and galeomorphs. Thus, it should come as no surprise that many species of squaliform sharks scrape, twist, and suck their food to ingest it. Perhaps the best-known example of this is the Cookiecutter Shark (*Isistius brasiliensis*), which latches on to larger prey items, such as tuna or marine mammals, and twists its body until it removes a chunk of flesh.

- Hyostylic jaw suspension

Hyostyly is characterized by articulation of the jaws, particularly Meckel's cartilage, with the hyoid arch, which helps support the jaws. The portion of the hyoid arch called the *hyomandibular cartilage* articulates with the chondrocranium at one end and the jaws at the other.

Historically, researchers classified all modern sharks as hyostylic because they all have a hyoid arch that articulates with the jaws, but as more functional and comparative anatomical studies distinguished between types of hyostyly, the term has come to be applied most commonly to the jaw suspensions of Carcharhiniformes, Orectolobiformes, Heterodontiformes, and Lamniformes.

In these sharks, evolution has tinkered with the arrangement of suspensory ligaments. The *ethmopalatine* ligament is present between the palatoquadrate and chondrocranium anterior to the orbit in carcharhiniforms, orectolobiforms, and heterodontiforms. In basal lamniforms, the ethmopalatine ligament is complemented by a single new ligament that connects the midpoint of the palatoquadrate to the underside of the chondrocranium between the nasal capsules. This is the *palatonasal ligament.* Interestingly, the modern lamnids (White Shark, makos, Porbeagle, and Salmon Shark) evolutionarily lost the ethmopalatine ligaments.

Compared to amphistylic and orbitostylic jaws, hyostylic jaws are highly mobile, and the position of the hyomandibula allows for considerable protraction to facilitate biting, grasping, and manipulation of prey. The Goblin Shark (*Mitsukurina owstoni*) is a good example of just how mobile hyostylic jaws can be. A 2016 study found that Goblin Sharks could protrude their jaws 32.3% of the length of its head![25] This extreme feeding performance was given the name "slingshot feeding," as the jaws move downward and backward and then shoot outward to snag prey.

- Euhyostylic Jaw Suspension

  In the batoids, jaw protraction reaches its most extreme case. In euhyostylic (*eu*=true) jaw suspension, there are no ligamentous connections between the palatoquadrate and the braincase, and the skeletal support for the jaws is provided exclusively by the hyomandibular cartilage at the corners of the jaws.

  Euhyostyly equals freedom of jaw movement. A 2014 study[26] showed that batoids exhibit a high level of trophic diversity (different feeding mechanisms). Apparently, not restricting jaw mobility allowed the evolution of muscles to accommodate feeding styles ranging from *suctorial feeding* (suctioning or "slurping" up soft prey like worms) to *durophagy* (feeding on hard prey like

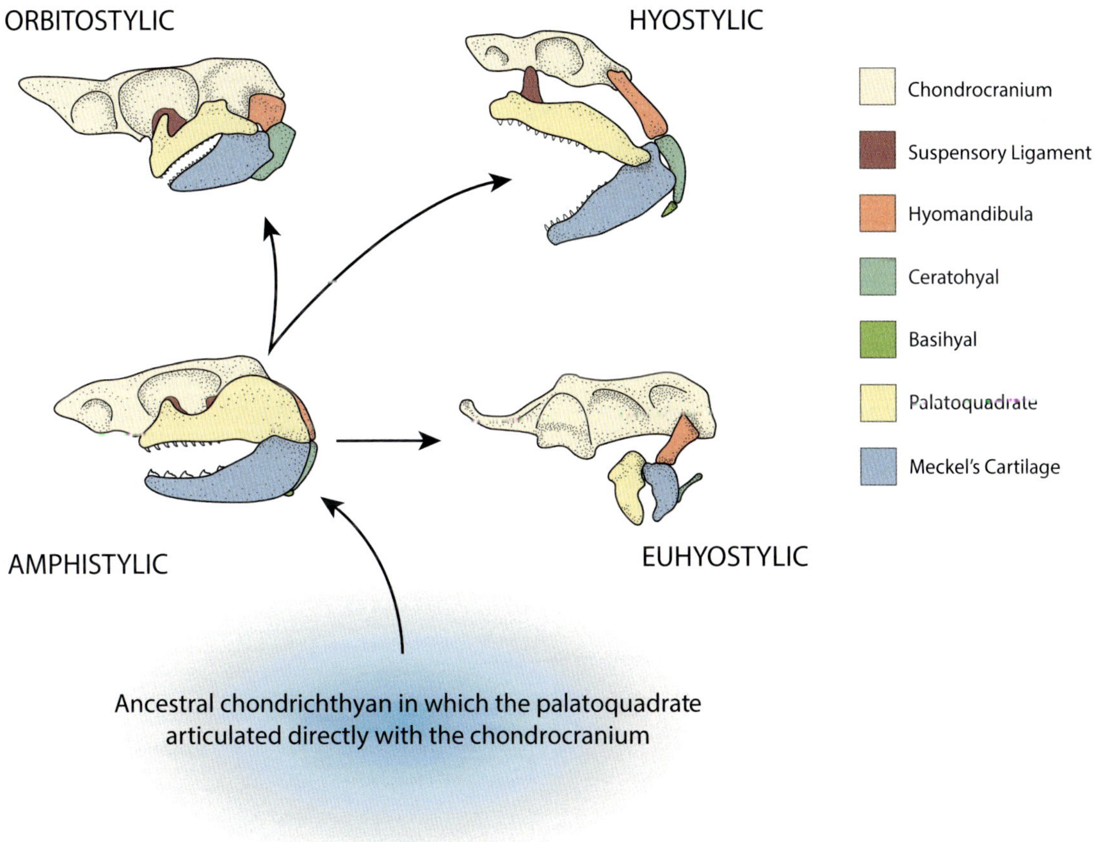

Figure 4.19. Types of jaw suspension in chondrichthyans. The amphistylic jaw suspension of hexanchiforms differs from the orbitostylic jaw suspension of squaliforms, hyostylic jaw suspension of many galeomorphs (e.g., Carcharhiniformes and Lamniformes), and the euhyostylic jaw suspension of batoids. Note the position of the jaws with respect to the chondrocranium and hyoid arch.

clams). When we catch batoids on our experimental longlines, we are careful not to pull larger specimens onboard by the leader, lest we damage the rear connections of their feeding apparatus. Imagine being lifted by your tongue or from behind your jaws, to fathom this situation.

Teeth

Except in filter-feeding species like the Basking Shark (*Cetorhinus maximus*), shark teeth are essentially force applicators. As the jaws close, the force exerted by the jaw musculature is transferred through teeth to whatever food item a shark has in its mouth. Two key questions are: what does the shark have in its mouth, and what does it need to do with it? In other words, what are the physical properties of the food item, and is the goal to pierce and clutch the food item, cut it, or crunch it? These factors will determine how force needs to be applied by the teeth during a bite, influencing the evolution of tooth morphology as well as the arrangement of teeth in the dentition.

Shark teeth are often categorized by their assumed function during a bite. Heavily serrated teeth, like those of an adult White Shark (*Carcharodon carcharias*) are regarded as "cutting type" teeth (fig. 4.20A). When a White Shark tooth is drawn across a material, in this case we will assume it is animal flesh, the tooth applies force on that flesh, and the flesh pushes back against the tooth. Serrations concentrate force applied by the tooth and apply it to a small area. The result is that concentrated force overcomes whatever resistance it encounters from the flesh, splitting the flesh along the serration. As the tooth is drawn across the flesh, it is pushed deeper, and the series of serrations part the flesh.[27]

Some sharks eat smaller fishes that can be swallowed whole but may be difficult to catch or hold on to. In these species, "piercing type" or "clutching type" teeth would be advantageous. The crown of a piercing type of tooth is long and slender and typically lacks serrations. The teeth of a Shortfin Mako (*Isurus oxyrinchus*) exemplify this tooth type. Clutching type teeth, like those of the Nurse Shark (*Ginglymostoma cirratum*) or Porbeagle Shark (*Lamna nasus*), typically have a crown with a pronounced central cusp and tiny cusplets on either side (fig. 4.20B).

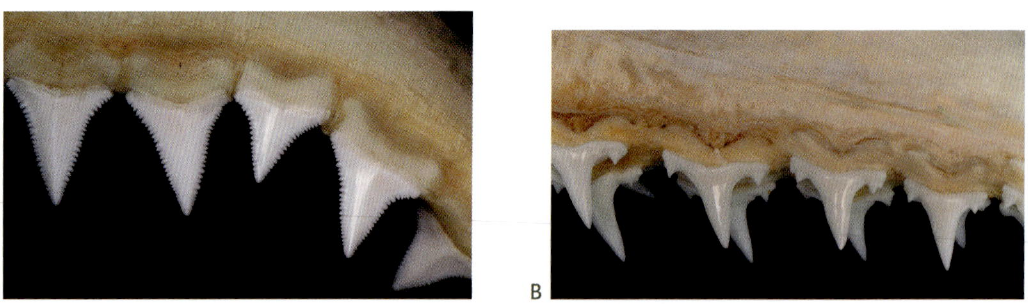

A B

Figure 4.20. (A) White Shark teeth with prominent serrations on a broad, triangular tooth crown. (B) Porbeagle teeth with smooth edges and small cusplets on either side.

Many species of sharks and rays, including the Horn Shark and Cownose Ray, feed on hard-bodied prey like scallops, clams, snails, crabs, and sea urchins. Eating hard bodied prey (*durophagy*) does not require teeth that shear or snag, rather teeth that are broader and resemble molars or crushing plates. Horn sharks are fascinating examples of a durophagous species. The teeth in the distal regions of their jaws (toward the corners of their mouths) are molariform, meaning they resemble molars and are "crushing type" teeth. Teeth toward the middle or central region of the jaw are much smaller and pointed, representing clutching type teeth. These teeth are used when a horn shark has grasped an urchin or shellfish but has not yet positioned the food in its mouth for a sufficient crunch (fig. 4.21). Having a dentition made of teeth of different morphologies in the same jaw is called *monognathic* (*mono* = one + *gnathic* = jaws) *heterodonty* (*hetero* = different + *odont* = tooth). The monognathic heterodonty of horn sharks is so striking that their taxonomic order is named for it, Heterodontiformes.

Carcharhinid sharks typically exhibit another type of heterodonty, *dignathic* (*di* = two + *gnathic* = jaws) *heterodonty*, in which teeth of the upper jaw have different morphologies than teeth on the lower jaw. Examine the dentition of a Caribbean Reef Shark (*Carcharhinus perezi*) or Snaggletooth Shark (*Hemipristis*

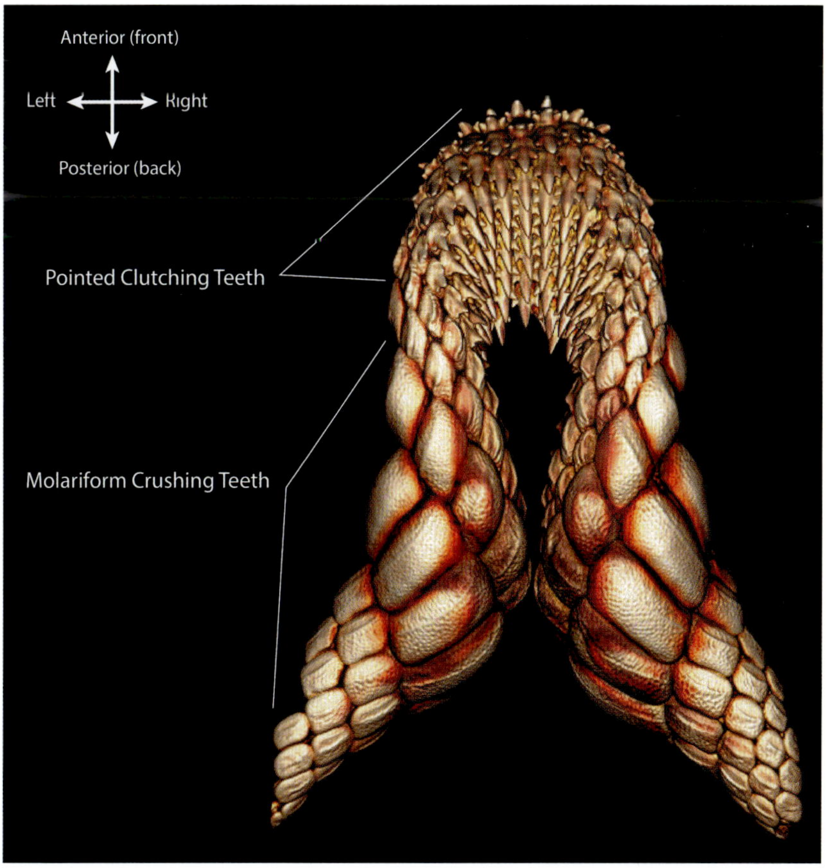

Figure 4.21. CT scan of a Horn Shark's lower jaw showing heterodonty between the forward-most teeth and those in the rear of the dentition.

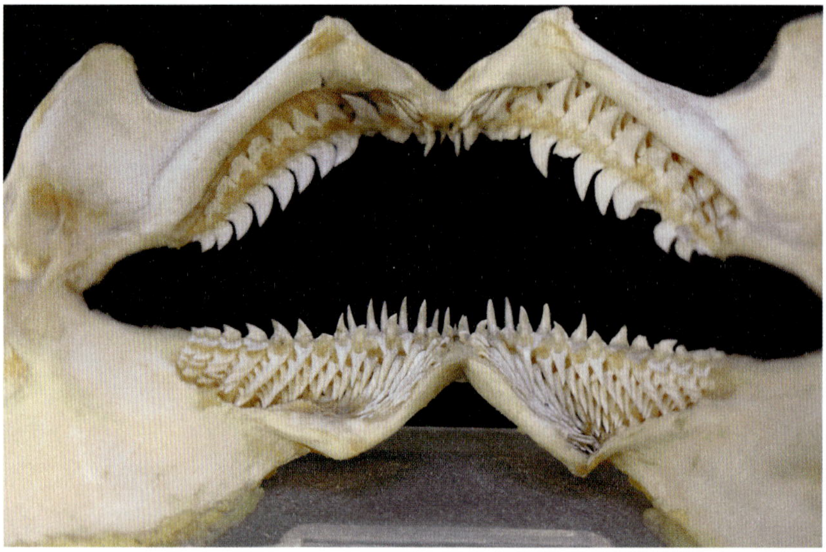

Figure 4.22. Variation in tooth morphology between teeth of the upper and lower jaws of a shark, in this case a Snaggletooth Shark.

elongate; fig. 4.22) and you will see what we informally call a "knife and fork dentition." The teeth of the palatoquadrate are broad and serrated (think cutting) while the teeth of Meckel's cartilage are longer and pointed like the tines of a fork.

All sharks and rays exhibit *polyphyodont* (*poly*=multiple+*phy*=growth or generation+*odont*=teeth) dentitions in which teeth are replaced continuously throughout the animal's lifetime. In contrast, humans are *diphyodont,* meaning we only have two sets of teeth. Another distinction between sharks' teeth and human teeth is how they are anchored in the jaw. While we have teeth in sockets within our jaws, shark teeth are held in *tooth files* by a *dental ligament* in the gum tissue.

In sharks, each tooth begins as a hollow tooth crown made of *enameloid* at the back of a tooth file. Enameloid is very similar to the enamel covering the crowns of human teeth, but it has a few structural and biochemical differences, including the presence of fluoride in the mineral crystals that make up the enameloid. Incidentally, fluoride is the same substance dentists use to strengthen our enamel.

As a tooth develops, the crown fills with *dentine,* and a root forms. All the while, the developing tooth is moving from the back of the tooth file to the front (fig. 4.23). When a developing tooth erupts from the gum tissue at the front of a tooth file, it is referred to as a *functional tooth* because it could be used in prey capture or handling. From one corner of the mouth to the other, sharks' functional teeth form rows. Eventually, functional teeth will fall out (or be yanked out by struggling prey). Theoretically, a shark could produce tens of thousands of teeth

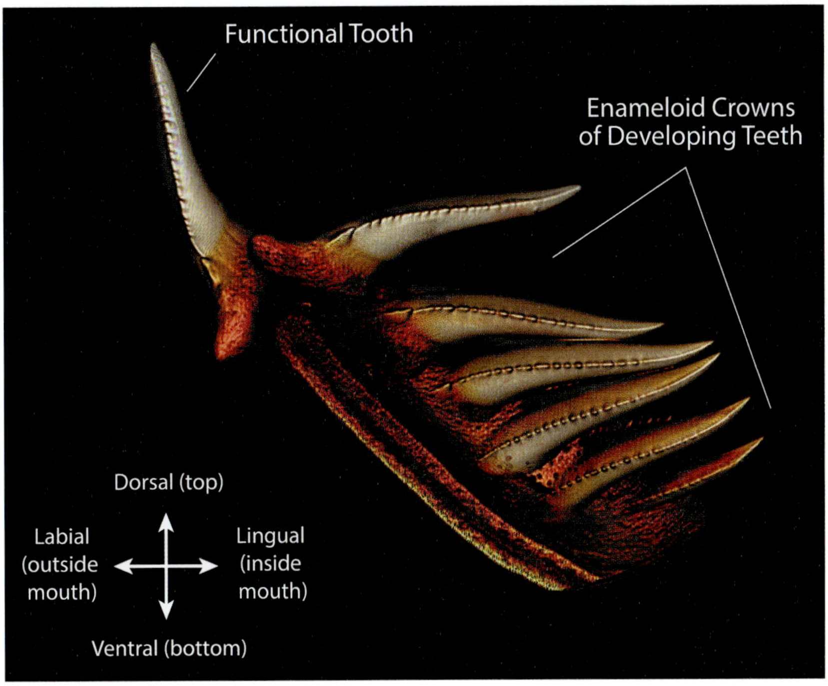

Figure 4.23. CT scan of a tooth file from the lower jaw of a White Shark.

in its lifetime, but verifying those estimates is tricky. One 1990 study of captive Nurse Sharks found that the rate of tooth replacement varied with water temperature. In warmer summer months, the sharks replaced an entire row of teeth in 9 to 21 days. In the winter, the rate slowed to one row replaced every 51 to 70 days.[28]

Feeding Performance

To a functional anatomist, *performance* relates to how an animal uses its anatomical structures to do something. In sharks, feeding performance is broadly described based on the degree to which a shark relies on its own forward movement or suction to bring food into its mouth and what the shark does with the food once it is in its mouth.

Ram vs. Suction Feeders

When a shark opens its mouth and depresses the lower elements of its hyoid arch, the volume of the oral cavity increases, which means the pressure inside the mouth momentarily drops. Water rushes into the mouth to fill the increased volume, carrying food with it. This is *suction feeding* in action, and many sharks use it to obtain prey. Perhaps the best-known example of a suction feeding shark is the Nurse Shark (*Ginglymostoma cirratum*), which can generate a pressure inside its oral cavity of -1.1 atmospheres. That is like getting sucked into space. Other sharks that rely heavily on suction feeding include angel sharks (family Squantinidae) and carpet sharks of the order Orectolobiformes, such as the Whitespotted Bamboo shark (*Chiloscyllium plagiosum*).

On the opposite end of the feeding performance spectrum, we find sharks that rely heavily on *ram feeding*. Ram feeding sharks use their forward movement to overtake prey rather than sucking water and prey into their mouths. For example, Basking Sharks (*Cetorhinus maximus*) employ ram feeding as they propel themselves into clouds of zooplankton. The shark engulfs the plankton, which is filtered out of the water via *gill rakers*. Many macrophagous (*macro* = large + *phagous* = eat) sharks that consume larger prey also use ram feeding to close the distance between themselves and their food prior to a bite. Anyone who has ever watched (either in person or on television) a White Shark launch itself out of the water with a seal or sealion in its jaws has witnessed a ram feeding event.

Bite Force

How do you quantify the feeding performance of sharks? For filter-feeding species, researchers typically calculate the number of liters of water filtered per hour or the mass of food particles removed. For sharks that rely on biting to capture and handle prey, the performance metrics are a little different.

The force exerted by the jaws during a bite is a widely used measure of performance. Shark bites are among the strongest in the animal kingdom, right? Actually, no and yes. The force of shark bites closely matches that needed to grasp and hold onto their prey, or to take a chunk of flesh from prey, and these forces may or may not be as high as expected. Our expectation of extremely high bite forces likely is based on continuous media reinforcement of that perception on the one hand, or our belief that sharks simply look like they should have high bite forces.

A 1967 scientific study[29] on bite force in sharks used a device called a *gnathodynamometer*, which was encased in approximately 1.4 kg (3 lb) of Bonito flesh, to measure bite force in the adult Tiger Shark and five other species in pens around the Bahamian island of Bimini. The highest force the researchers found applied by a single tooth was 60 kg (132 lb) over an area of 2 mm^2, which is equivalent to 3000 kg (or more ominously, 3 tonnes) per cm^2 (42,700 lb per in^2). Although these numbers alone reflect significant force, when they were extrapolated to estimates of the *total jaw area* in the popular press, the numbers grew to nearly 200 tonnes for the total bite force of a 1.8–2.1 m (6–7 ft) Dusky Shark (*Carcharhinus obscurus*). This number is misleading because extrapolating from the force generated by the tip of a single tooth to an entire jaw assumes that *every* square millimeter of the jaw has the same force as that of the single, measured tooth and that all of these teeth come into contact with a prey item. This is not the case, and we know from lever mechanics that more force will be applied in the corners of the mouth than in the forward or anterior regions of the mouth. Anterior bite force is therefore

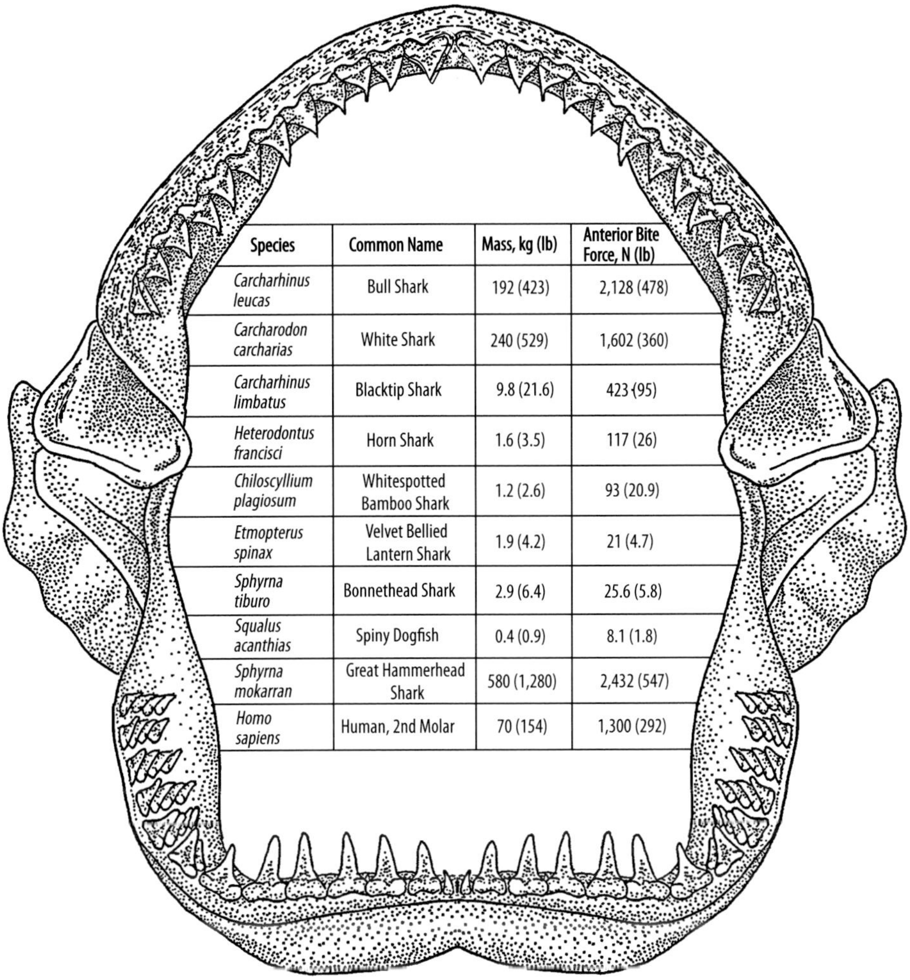

Species	Common Name	Mass, kg (lb)	Anterior Bite Force, N (lb)
Carcharhinus leucas	Bull Shark	192 (423)	2,128 (478)
Carcharodon carcharias	White Shark	240 (529)	1,602 (360)
Carcharhinus limbatus	Blacktip Shark	9.8 (21.6)	423 (95)
Heterodontus francisci	Horn Shark	1.6 (3.5)	117 (26)
Chiloscyllium plagiosum	Whitespotted Bamboo Shark	1.2 (2.6)	93 (20.9)
Etmopterus spinax	Velvet Bellied Lantern Shark	1.9 (4.2)	21 (4.7)
Sphyrna tiburo	Bonnethead Shark	2.9 (6.4)	25.6 (5.8)
Squalus acanthias	Spiny Dogfish	0.4 (0.9)	8.1 (1.8)
Sphyrna mokarran	Great Hammerhead Shark	580 (1,280)	2,432 (547)
Homo sapiens	Human, 2nd Molar	70 (154)	1,300 (292)

Figure 4.24. Bite forces in various sharks and humans (Adapted from: Habegger, M.L., Motta, P.J., Huber, D.R. and Dean, M.N., 2012. Feeding biomechanics and theoretical calculations of bite force in bull sharks (*Carcharhinus leucas*) during ontogeny. Zoology, 115(6), pp. 354–364).

a slightly more conservative estimate. Recent studies have provided estimates of anterior bite force for a number of species (fig. 4.24).

How do these forces compare to those of durophagous species, those that eat hard prey and typically possess hypertrophied jaw musculature (fig. 4.25), and teeth designed for grinding? Anterior bite force for a 1.6 kg (3.5 lb) Horn Shark (*Heterodontus francisci*) was only 117.2 N (26.3 lb), but maximum posterior bite force on the posterior molariform teeth was 338 N (76 lb). These seem low compared to measurements for Bull and White Sharks, but when they are considered relative to the mass of the Horn Shark, they represent one of the highest among sharks. One 2015 study[30] calculated that the Cownose Ray (*Rhinoptera bonasus*) would have a bite force of >500 N (112 lb).

Figure 4.25. Lateral view of a Horn Shark showing hypertrophied jaw musculature (Physiological Research Lab, Scripps Institution of Oceanography).

Bite Duration and Kinematics

Another metric of shark bite performance is the duration of the bite; that is, how long does it take for a shark to start in a resting position, open its jaws, protrude them, close them, and retract them to obtain food? This aspect of performance is crucial for any shark looking to catch its meal by surprise or consume it before a nearby competitor takes an interest.

For most of the 20th century, accurate measures of bite duration were limited to small sharks that were easily housed in tanks. High-speed cameras were set up at viewing windows, and sharks were fed and filmed. As technology improved and cameras could be taken into the field, a clearer picture of bite kinematics began to emerge across a wider variety of species.

Shark bites begin with an *expansive phase* in which Meckel's cartilage (the lower jaw) is lowered and the snout is elevated. After the gape has reached its maximum, the palatoquadrate begins to protrude outward and downward as Meckel's cartilage is elevated. This marks the *compressive phase*. Both phases are characterized by the actions of the jaws and are separated by the moment of maximum gape when the jaws are at their furthest distance apart (fig. 4.26). The relative timing and duration of those actions can differ dramatically from species to species.

Shark bites can happen in the blink of an eye. In 1985, biologist Timothy Tricas used footage of Australian White Sharks to calculate that total bite duration (expansive plus compressive phases) lasted between 0.264 and 0.740 seconds.[31] Surprisingly, this is not incredibly fast by shark bite standards. For example, juvenile Blacktip Sharks (*Carcharhinus limbatus*) less than a meter in length have been recorded biting as quickly as 0.117 seconds,[32] and several two-meter Sand Tiger Sharks (*Carcharias taurus*) showed an average bite duration of 0.14 seconds when their bites were studied with frame-by-frame analysis. The fastest bite recorded from these sharks lasted only 0.10 seconds![33]

There is still much to learn about how sharks modify their feeding performance under different conditions. For example, a 2023[34] study found that Blue Sharks (*Prionace glauca*) modulated the timing and duration of the expansive and compressive phases of their bites in response to food size (unsurprisingly, larger food

Figure 4.26. Bite sequence of a Blue Shark illustrating the expansive phase, maximum gape, and compressive phase of the bite.

required larger bites, which lasted longer). While collecting video footage for this study, researchers encountered a small Shortfin Mako Shark (*Isurus oxyrinchus*). When the footage was analyzed later in the lab, the researchers found that Blue Sharks bit more quickly when the Mako was close by. The conclusion to take from this case study is that among biting sharks, hasty generalizations about performance may obscure some interesting ecological considerations. Sometimes, variability in performance, which may include slower or less powerful bites, is just as interesting as the hardest, fastest bites. This is why we advocate responsible studies of sharks in their natural habitat as a complement to lab-based studies.

Gills

We conclude this chapter on functional anatomy with a discussion of gills to demonstrate that form-function relationships are just as crucial in the understanding of physiological processes as they are in locomotion or predation.

The organs of respiration in sharks are gills, which also play a role in ion exchange, pH balance, and excretion of nitrogenous waste products.[35] Here, we will focus on their respiratory function. Like respiratory structures from lungs of mammals to the gills of bony fishes, the gills of sharks have the following general characteristics:

- A short path for diffusion of respiratory gases (oxygen in, carbon dioxide out)
- A high density of capillaries (i.e., extensive vascularization)
- A large surface area over which the exchange of gases occurs
- Continuous wetness (a thin film of water in air-breathers)

The subclass Elasmobranchii translates to *plate gills*,[36] referring to the stacked architecture of the gills of this group, specifically the platelike *interbranchial septa* and *holobranchs* (gill arches; fig. 4.27). Here, the gill filaments are connected to these interbranchial septa, unlike the state in bony fishes, where a septum is absent, and the gill filaments are elongated and connected to the gill arch only at one end (fig. 4.27).

The four characteristics of respiratory structures are necessary but not sufficient to extract the required oxygen and eliminate the excess carbon dioxide. The last required ingredients are anatomical specializations within the gills that allow the water in the environment, which is presumably oxygen-rich or at least has a higher oxygen concentration than the blood, to flow over the gills in a direction exactly opposite to the flow of blood in capillaries in the gills.

These specializations, which are found in both chondrichthyans and bony fishes, are the secondary lamellae (also called *respiratory leaflets*), which are the only sites of respiratory gas exchange (fig. 4.27). The mechanism of blood flow through these secondary lamellae opposite water flow on the other side of a thin membrane is the same countercurrent flow principle (fig. 4.28) we see in muscles

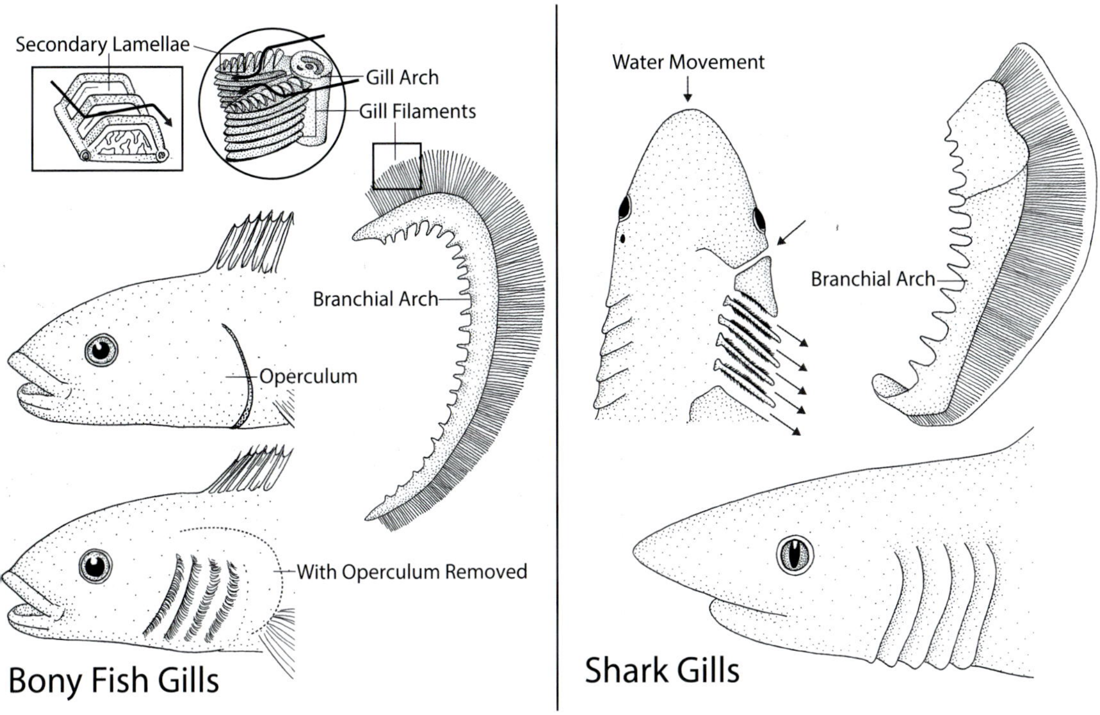

Figure 4.27. Gills of sharks and bony fishes. Gill filaments and secondary lamellae, shown for the bony fish, are also found in the shark gill (Redrawn from fig. 1, Wegner, N.C. *et al.* 2012. J. Exp. Biol. 215: 22–28).

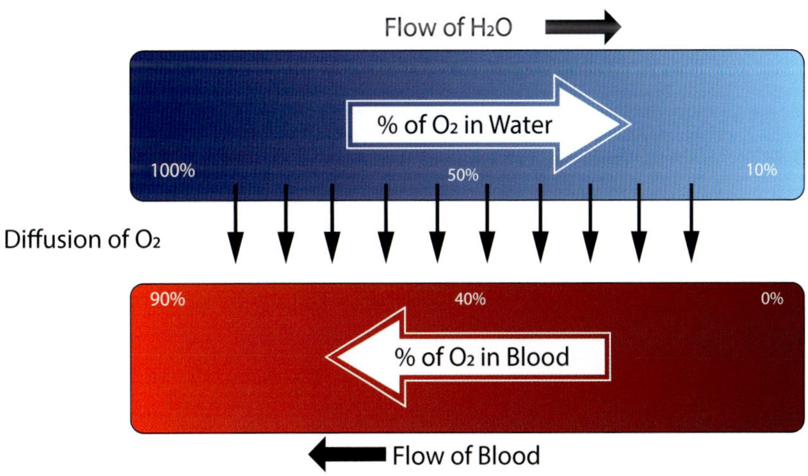

Figure 4.28. Diagram depicting the engineering principle *countercurrent exchange*, as it occurs in the gills of fishes. Values for oxygen percentages represent relative amounts.

of endothermic sharks (see chapter 7). Basically, water enters through the mouth (and spiracles in some) and flows over the secondary lamellae before exiting through the gill slits.

Most elasmobranchs have five bilateral, gill-bearing branchial arches and associated gill slits. This translates into nine *hemibranchs* (literally, *half-gills*, the gill filaments and lamellae on half of each gill arch, or *holobranch*), since the first gill arch (the *hyoid*) has filaments only on its posterior, and arches two through five have filaments on both the anterior and posterior sides. Elasmobranchs with six or seven gills have 11 or 13 bilateral hemibranchs, respectively.

Gill structure varies among elasmobranchs. In Collared Carpet Sharks (family Parascyllidae), there is no hemibranch on the posterior of the fifth gill arch.[37] Parascyllids occupy well-oxygenated waters, and have low oxygen requirements, so this hemibranch may not be missed.

The surface area of the gills for gas exchange also varies among elasmobranchs. The strongest correlation with gill surface area is with activity level/metabolic rate. The high-performance lamnid sharks thus have the highest gill surface area to body mass ratios, but the champion is the Bigeye Thresher (*Alopias superciliosus*),[38] perhaps because the latter is a diurnal vertical migrator that spends time in hypoxic waters of the oxygen minimum zone, where a bigger oxygen collector is needed. The gill surface area of Shortfin Makos, expressed as $cm^2/10$ kg body mass, is about 50,000 cm^2 (7750 in^2), which is close to 2.5 times that of Blue and Dusky Sharks and an entire order of magnitude higher than that of Atlantic Stingrays (*Hypanus sabina*) (2700 cm^2, or 418 in^2),[39] although it is only about half that of a similar-sized tuna.[40]

There are also differences in the dimensions of the water-blood among between more and less active elasmobranchs. The former (e.g., the Shortfin Mako) have a diffusion barrier to oxygen as small as one-tenth that of benthic species (e.g., the Small-spotted Catshark, *Scyliorhinus canicula*).

In addition to increased gill surface area and thinner barriers to diffusion, sharks capable of sustained high speeds, most notably lamnids (mackerel sharks) and some alopiids (threshers), have other modifications. First, their bodies have evolved to allow space in the branchial cavity to accommodate larger gills. For example, the pectoral fins of lamnid sharks are placed more posteriorly on the body, which allows the branchial chamber to be longer and thus more capacious. In alopiids, the gill chamber has expanded more cranially. Fast-swimming sharks also have larger gill slits, all of which are of similar size in each species.

In light of the high swimming speed of makos and their use of ram ventilation, their gills have modifications that stabilize them. They are stiffened by the interbranchial septum, and the filaments have one or two vascular sacs (fig. 4.29). Otherwise, the basic support structure is similar to that of other elasmobranchs.

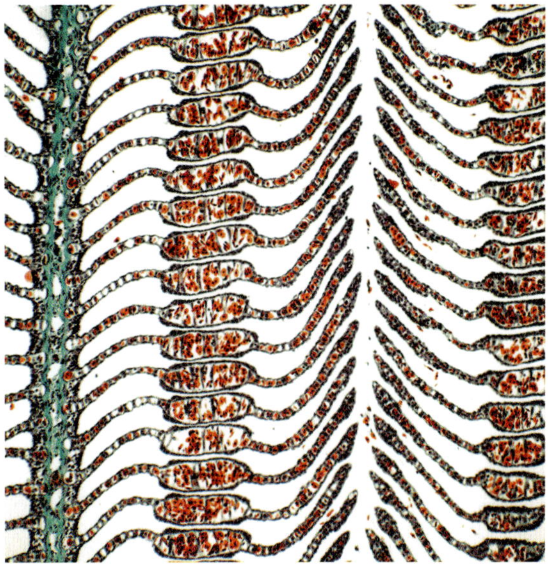

Figure 4.29. Stained section through gill filaments of a 9 kg (19.8 lb) Shortfin Mako showing red blood cells in vascular sacs, which may stabilize the filaments when water movement through the gills is increased during ram ventilation while swimming fast (Courtesy of Nick Wegner).

Concluding Comments

This chapter has highlighted only a limited sample of the anatomical adaptations of a small number of sharks, in part because more research is required if we are to make accurate generalizations that apply to more than 540 species of modern sharks. Less common species or those not amenable to captivity leave noticeable gaps in our body of knowledge. As these deficiencies are overcome, expect expansion and revision of what we know about the functional anatomy of sharks and rays and what it tells us about the ecology and evolution of elasmobranchs.

NOTES

1. Oeffner, J. and Lauder, G.V. 2012. J. Exp. Biol. 215: 785–795.
2. Bartholomew, G. A. 1986. BioSci. 36: 324–392.
3. Thomson K. S. and Simanek, D.E. 1977. Am. Zool. 17: 343–354.
4. Compagno, L.J. 1990. Environ. Biol. Fish. 28: 33–75.
5. Dolce, J.L. and Wilga, C.D. 2013. Bull. Mus. Comp. Zool. 161: 79–109.
6. White et al. 2023. In: Carrier et al. (eds.). Biology of sharks and their relatives, 3rd edition. CRC Marine Biology Series. CRC Press. 31–57 pp.
7. Sternes, P.C. and Shimada, K. 2020. Zool. 140: 125799.
8. Goto et al. 1999. Ichthyol. Res. 46: 281–287.
9. Emily R. Nelson, "Functional morphology and individual variation in movement performance of an apex marine predator" (Master's thesis, University of Miami, 2018), Open Access Theses (704). https://scholarship.miami.edu/esploro/outputs/graduate/Functional-Morphology-and-Individual-Variation-in/991031447423902976?institution=01UOML_INST. (Accessed 9/2/24).
10. Dean, B. and Bhushan, B. 2010. Philos. T. Roy. Soc. A 368: 4775–4806.
11. Rosenberger, L.J. 2001. J. Exp. Biol. 204: 379–394.
12. Porter et al. 2009 J. Morph. 270: 954–965.

13. Maia, A. and Wilga, C.A. 2016. J. Zool. 298: 139–149.

14. Lingham-Soliar, T. 2005. J. Morph. 263: 1–11.

15. This pesky, pedantic Latin phrase rears its ugly head again. Recall that it means *the essential element without which something is impossible.*

16. Oeffner, J. and Lauder, G.V. 2012. J. Exp. Biol. 215: 785–795.

17. Raschi, W.G. and Musick, J.A., *Hydrodynamic aspects of shark scales,* Special report in applied marine science and ocean engineering no. 272, a Report for NASA, (Virginia Institute of Marine Science) January 1984, 812 pp. https://scholarworks.wm.edu/reports/544/. (Accessed 9/2/24).

18. Raschi, W. and Elsom, J., 1986. "Comments on the structure and development of the drag reduction-type placoid scale," (Indo-Pacific Fish Biol. Proc. Second Int. Conf. on Indo-Pacific Fishes. Ichthy. So., Japan, 1986), 408–424 pp.

19. Motta, P. et al. 2012. J. Morph. 273: 1096–1110.

20. Kajiura, S.M. et al. 2003. Zool. 106: 19–28.

21. Payne et al. 2016. Nat Comm. 7: 12289. DOI: 10.1038/ncomms12289.

22. McComb, D.M. et al. 2009. J. Exp. Biol. 212: 4010–4018.

23. Strong, W.R. et al. 1990. Copeia 1990: 836–840.

24. Chapman, D.D. and Gruber, S.H. 2002. Bull. Mar. Sci. 70: 947–952.

25. Nakaya et al. 2016. Sci Rep. 6:27786. DOI: 10.1038/srep27786.

26. Kolmann, M.A. et al. 2014. J. Morph. 275: 862–881.

27. Moyer, J.K. and Bemis, W.E. 2017. Zool. 120: 101–109.

28. Luer et al. 1990. Copeia. 1990: 182–191.

29. P.W. Gilbert et al. (eds.). 1967. *Sharks, skates and rays.* Johns Hopkins U. Press: 331–337 pp.

30. Kolmann, M.A. et al. 2015. J. Anat. 227: 341–351.

31. Tricas, T.C. 1985. Mem. South. Calif. Acad. Sci. 9: 81–91.

32. Gardiner et al. 2017. PLoS One 9(4):e93036.

33. Moyer et al. 2019 J. Fish. Biol. 95: 881–892.

34. Moyer, J.K and Dodd, J. 2023. Environ. Biol. Fish. 106: 1207–1232.

35. An excellent comprehensive reference: Wegner, N.C. 2015. In: Elasmobranch gill structure. Fish Physiol. 34: 101–151. Academic Press.

36. This is sometimes interpreted as *strapped gills,* in reference to the tips of the gills not being free, rather constrained.

37. Goto, T. et al. 2013. Zool. Sci. 30: 461–468.

38. Wootton, T.P. et al. 2015. J. Morph. 276: 589–600.

39. Grim, J.M. et al. 2012. Fish Physiol. Biochem. 38: 1409–1417.

40. Wegner, N.C. et al. 2010. J. Morph. 271: 937–948.

5 / Sensory Biology

CHAPTER OUTLINE

Introduction: What Is Physiology?

Physiology is the study of biological function. One of the major themes in physiology is *homeostasis* (*homeo* = same; *stasis* = steady or staying), which refers to the roles of an organism's regulatory systems working in concert to maintain a relatively *constant internal environment* or *stable internal function,* even in the face of wide changes in the external environment. Physiology is a somewhat neglected topic, and one rarely encountered in books like this one. We begin with sensory physiology in this chapter.

Sensory Physiology

You have likely heard that sharks are *swimming noses* capable of detecting a drop of blood from a distance of several miles. This statement is not untrue, but it requires qualifying. Yes, sharks have an extremely well-developed sense of smell, but they are not the only bloodhounds of the ocean, since many bony fishes have similar olfactory sensitivities. For a shark to detect a drop of blood, specific molecules in that blood capable of stimulating the shark's olfactory receptors must physically come in contact with these receptors, and this requires time for the blood to reach the nasal organs of the shark.

Sharks have a suite of senses that allows them to interpret their surroundings and respond appropriately. This suite encompasses those that we humans and other vertebrates possess, and for good measure adds one totally foreign, the ability to detect vanishingly small electrical fields. These sensors specifically enable a shark to detect prey, predators, conspecifics (members of the same species), other organisms, and obstacles. They also allow a shark to orient itself.

A *sense* can be thought of as a group of specialized cells that work as part of a system to detect some form of physical energy or substance in the environment and transmit information (signals) about these to the central nervous system, which then may initiate a response.

The five traditional human senses are sight, hearing, touch, taste, and smell. Another way of describing these senses is by the form of energy they detect. Vision is a form of *photoreception*; that is, it detects energy in the form of light waves. Smell and taste represent types of *chemoreception*, responses to chemical stimuli. Touch and hearing represent *mechanoreception*, a means of detecting pressure or distortion. Sensory physiologists also recognize systems that detect other forms of energy or are specialized forms of the three types of reception discussed. *Thermoreception* detects temperature. *Proprioception*, also called *equilibrioception*, is the sense of one's own position and equilibrium. *Nociception* refers to the perception of pain. Discussion of these is beyond the scope of this book.

Sharks and rays possess photoreceptors, chemoreceptors, and mechanoreceptors. They also possess *electroreceptors*, sensors capable of detecting electrical fields, and possibly also *magnetoreceptors*, which detect, as you might suspect, magnetic fields.

These sensory receptors have several factors in common. First, they must be capable of detecting a physical or chemical stimulus. Each kind of receptor responds to its own *adequate stimulus*; that is, the type and intensity/concentration of stimulus to which the receptor is tuned.

Once the stimulus is detected, cells in the receptor *transduce* the stimulus, meaning that the detection is converted into a nerve impulse or impulses that travel to the brain or the spinal cord. There, the signal is interpreted, processed, and often integrated with information from other senses before an appropriate response is determined, typically nearly instantaneously.

Mechanoreception

In 2015, a Dutch Caribbean Coast Guard helicopter hovered over a shipwreck survivor in the water near Aruba, and was about to rescue him when, according to *The Guardian* newspaper,[1] a shark bit and killed him. Immediately, speculation surfaced that the low-frequency *whomp-whomp-whomp* of the helicopter's rotors played a role in attracting the sharks. Is that possible? What sounds attract sharks? We will answer these and other questions here.

Mechanoreception refers to a response to pressure, distortion, or displacement, including those associated with touch, sound, muscular contraction, and posture. The mechanoreceptors in sharks include the inner ear, lateral line, spiracular organ, touch receptors, proprioceptors, and hair cells. These all share similarities in the morphology of the receptor cells and have been collectively called the *acoustico-lateralis* (or *octavo-lateralis*) system.

Let us start with the ears. Yes, sharks have ears, though you would see them only upon dissection, the only external feature being small, paired *endolymphatic pores* (fig. 5.1A). The endolymphatic pores open into endolymphatic ducts that lead to the cartilaginous otic capsule, which houses the inner ears, each of which is typically described as a *membranous labyrinth.*[2] This labyrinth is structurally divided into upper and lower portions.

The upper part of the membranous labyrinth is composed of three *semicircular canals* arranged at right angles to each other. The lower portion has three sacs, or chambers (or organs), the *sacculus, utriculus,* and *lagena.* Each of these chambers has a sensory epithelium called a *macula.* A fourth macula, essentially a spot, called the *macula neglecta* (meaning *neglected spot*) is found at the intersection between the upper labyrinth (the semicircular canals) and the lower labyrinth. All parts are filled with a fluid called *endolymph* (fig. 5.1B).

The three lower chambers of sharks and rays contain *otoliths,* or *otoconia,* calcium carbonate structures the size of sand grains. The otoliths of sharks differ from those of bony fishes and are more specifically called *stataconia* to differentiate them. Stataconia do not lay down concentric rings and thus do not record age as otoliths do in bony fishes.

At the base of each of the upper labyrinth's semicircular canal is a swelling called an *ampulla.* There are bundled sensory hairs, the actual mechanoreceptors (in other words, the *functional unit)* on which the entire system is based, on a ridge called the *crista* inside the ampullae, as well as in the maculae of the three sacs in the lower chamber.

The tops of the hair cell bundles in the semicircular canals are embedded in a transparent structure called the *cupula,* which consists of mucilage, or thick mucopolysaccharide jelly (also called *gel*). Each bundle of hair cells includes a *kinocilium,* which projects the highest, and as many as 60 smaller *stereocilia.* These are connected at their base to sensory nerves, which will convey the information to the central nervous system when these cilia bend in response to a stimulus or stimuli (see pp. 202–203).

Let us discuss the upper labyrinth's semicircular canals first. These are organs in the *vestibular system*; that is, the primary structures concerned with equilibrium or balance (e.g., controlling pitch, roll, and yaw) rather than hearing. The semicircular canals detect angular acceleration; that is, the rate of change of the velocity of a rotating object, the object here being the shark. Thus, these organs sense what the

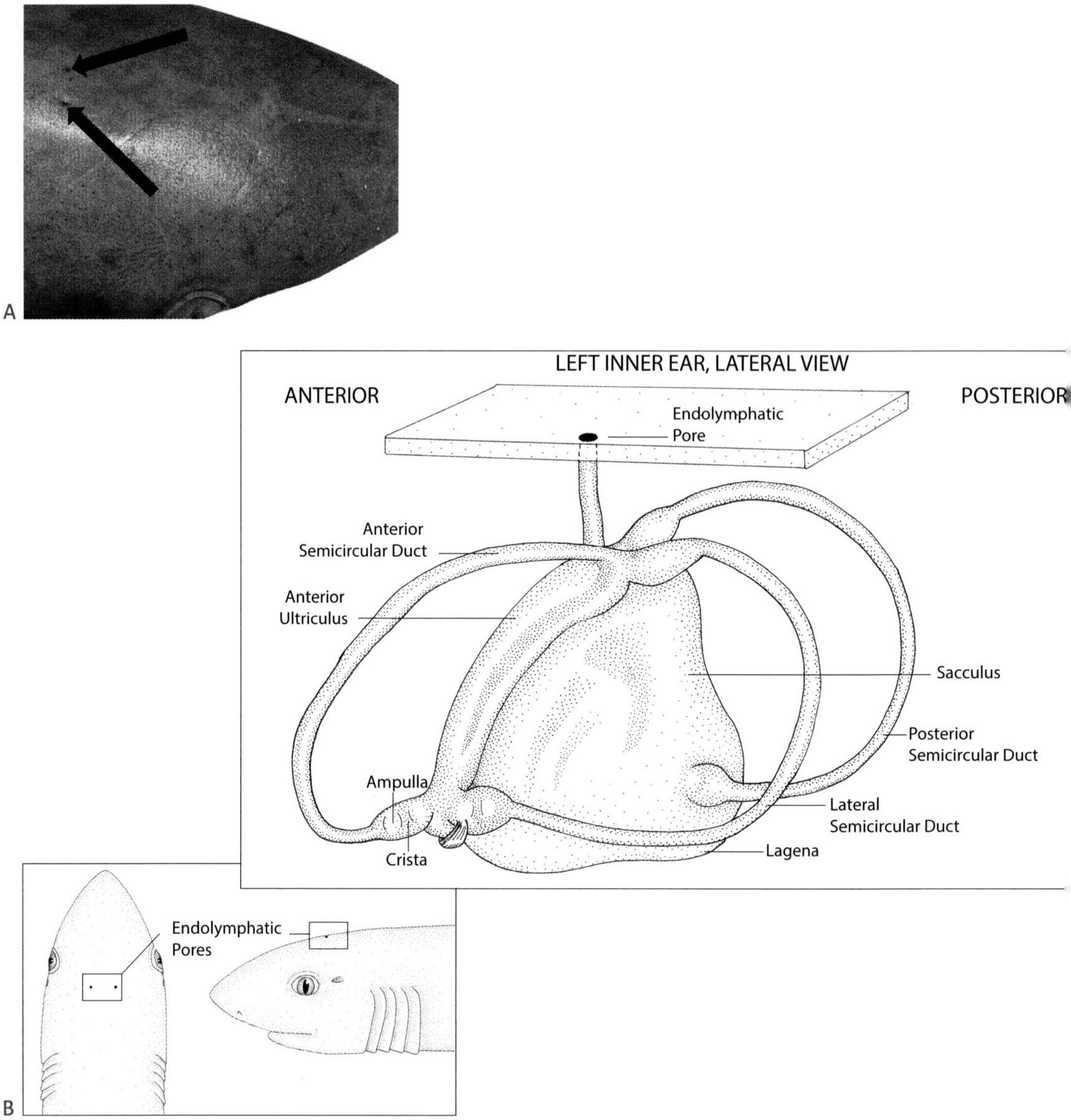

Figure 5.1. (A) Endolymphatic pores, external openings of the ears of sharks, from a Blacktip Shark. (B) Internal anatomy of the shark ear (Redrawn from Gilbert, S. 1973. *Pictorial Anatomy of the Dogfish*. University of Washington Press).

shark itself is doing and control the position of the shark in space, and they do not detect any events in the environment. For pelagic sharks in clear water, vision also plays a role in equilibrium, as does the sense of touch for benthic sharks.

During movement, there is an *inertial lag* of the gel compared to the mechanoreceptive hair cells. Inertial lag, in accordance with Newton's First Law, is a delay in the flow in response to a force, and it occurs in any fluids or matter of different densities. In sharks, this includes all of the extracellular fluids (e.g., endolymph [including gel], pericardial fluid, and blood), food in the digestive tract, etc. In this case, in response to movement, the hair cells will move at the same speed as the body (muscles, skeleton, organs, etc.) does, but the endolymph gel *lags behind*. The relatively faster movement of the hair cells embedded in the gel compared to the gel itself causes the hair cells to stick to the gel and to bend, which sends signals to the central nervous system on the rotation of the head such that equilibrium, or balance, is maintained.

The organs of hearing in elasmobranchs are exclusively the inner ear chambers (sacculus, utriculus, and lagena) plus the macula neglecta (bony fishes also have accessory structures, e.g., swim bladders, used for hearing).

Let us turn to the role of the sensory epithelium (the maculae) of the inner ears, which play the primary role in hearing but are also involved in balance. Like the semicircular canals, the maculae also possess a patch of hair cells. These hair cells are embedded in gel surrounding the stataconia that cover the maculae. They act as the inertial elements in the maculae, and function in a manner analogous to the endolymph in the semicircular canals. Mechanical deformation of these produces a signal that informs the central nervous system of straight-line acceleration as well as sound. The macula neglecta, which lacks stataconia but does have gelatinous cupulae, is also important to hearing.

At this point, it is necessary to describe the rudiments of the physics of sound transmission in water. The same physical principles apply to sound in water and air, but there are differences based on their physical properties. Sound travels faster in a solid material than in air. While both water and air are fluids, because water is much denser than air and is negligibly compressible, sound travels about 4.4 times faster in seawater than air at 35°C (95°F) and at any given frequency, the wavelength is about 4.8 times longer.

Sound starts with a vibration or disturbance that propagates within a medium and moves as the energy of one water particle transfers to the adjacent one. With some exceptions, sound in water does not rapidly attenuate because of water's relative incompressibility and thus propagates long distances. Low frequencies travel farther than high ones. Finally, the sources of sounds in water are difficult to localize.

Hearing

The first audiogram in sharks, in which they were presented with sounds of different frequencies, was conducted in 1961 on Bull Sharks (*Carcharhinus leucas*).[3] These sharks could hear frequencies between 100 and 1500 Hz (the range for humans is reported as 20–20,000 Hz) and were most sensitive to sound in the 400–600 Hz range. Audiograms on Lemon Sharks (*Negaprion brevirostris*) and Horn Sharks (*Heterodontus francisci*) extended the lower end of the range to 10 Hz.

In the 1980s, advanced neurological methods produced audiograms for a number of sharks and batoids slightly different from those already mentioned, with hearing in the range of 40–1500 Hz and highest sensitivity in the 200–400 Hz range.

These studies determined the range of *sensitivities* of the shark and batoid ear, but they did not paint a picture of the acoustic world in which sharks live; that is, the kinds of sounds sharks respond to in nature. The first study to understand the types of sounds that attracted sharks was conducted in 1963.[4] Noting that sharks were well-known to be attracted to the sounds of struggling fishes, researchers recorded the sounds from fish, most successfully from a Black Grouper, and acoustically characterized them. Then they played sounds like those of the struggling fish on coral reefs where no sharks were visible. The sounds that were successful in attracting sharks were low-frequency (<60 Hz) pulses, similar to those recorded from the speared, struggling fish. Using these sounds, the researchers attracted Bull Sharks, hammerheads (*Sphyrna* spp.), Lemon Sharks, a Tiger Shark (*Galeocerdo cuvieri*), and additional unidentified carcharhinids. Other sounds were not attractants. Subsequent studies confirmed these results.

From how far away can sharks detect sounds? Let us revisit the physics of sound in water. Sound has both particle motion and pressure components. The former produces sound that is rapidly attenuated as it moves from its source and is known as *acoustic near field* sound. Depending on the frequency, this is usually under 15 m (49 ft). Sound pressure extends up to hundreds of kilometers from the source and is thus *acoustic far field* sound.

The architecture of the inner ear of elasmobranchs is poised to hear only near field sounds; that is, it is sensitive to particle motion. Bony fishes with swim bladders are also capable of detecting sound in the far field.

Thus, we are left with a paradox—the contradiction between the verified observations that sharks are capable of responding to sounds in the kilometer range, and the structural limitations of their anatomy that apparently limit their range further. Some researchers have concluded that sharks orient to distant sound sources using a system that is more sensitive to particle motion than had been thought.

Irregular, low frequency sounds attract sharks, but do any sounds repel them? Yes, some sounds represent aversive stimuli. Lemon Sharks withdrew if sounds that might have been otherwise attractive were initiated when an individual was in sight and approaching. Similarly, Silky Sharks (*Carcharhinus falciformis*) and Oceanic Whitetip Sharks (*Carcharhinus longimanus*) withdrew if the intensity of the sound was changed when played back. Sounds simulating vocalizations of Orcas, natural predators of some sharks, likewise initiated withdrawals in Silky, Lemon, and other carcharhinid reef sharks, but not White Sharks. That naivete may have doomed the White Sharks (*Carcharodon carcharias*) killed by Orcas in South Africa recently.

Sharks thus are attracted to sounds of prey and some are repelled by sounds that may signal threats. How do they use sound during their daily activity when these sounds are absent? Put another way, do sharks also use hearing, perhaps in concert with other senses, to continually assess their environment? Sharks live in an environment that can be relatively noisy. Sounds under a frequency of 1 kHz in the marine environment that a shark might encounter and perceive include those associated with swimming fish schools, fish sounds, and waves.

For recent reviews of the shark auditory system and hearing, see the 2022 papers by Chapuis and Collin[5] and Mickle and Higgs.[6]

Back to the helicopter and the unfortunate swimmer. While the shark that killed the shipwreck victim could have been attracted to the swimmer using any of its suite of senses, yes, the sounds of the hovering helicopter unfortunately could have played a role.

Other Mechanosenses

You just learned that water conveys sensory information from mechanical disturbances (sounds and movements) at much greater distances than in air. Sharks and batoids, and bony fishes as well, are poised to sense this information via a series of external mechanoreceptors. In elasmobranchs, these include the lateral line, pit organs, spiracular organs, and vesicles of Savi (batoids only).

In all of these mechanoreceptors, the basic unit is the *neuromast*, which is a hair cell covered by a gelatinous cupula, similar to the auditory and vestibular hair cells in the inner ear. These mechanoreceptive neuromasts exist in an array of configurations that include canals with pores connecting to the environment, canals lacking pores, free (or superficial) neuromasts in grooves or pits, and subdermal or spiracular pouches. Together they allow the shark to detect predators, prey, other organisms, and conspecifics; to detect water flow and other physical characteristics of its surroundings; and for proprioception (perception of aspects of its own body). We discuss these other mechanoreceptors in sequence here. Note there is overlapping, integrated function among these sensory systems.

The lateral lines of sharks may either have pores, and thus be connected to the environment, or they may be non-pored (sealed and not open to the water). The pored canals run bilaterally along the flanks from the posterior of the head near the endolymphatic pores to tip of the tail (fig. 5.2). In batoids, the pored lateral line is on both dorsal and ventral surfaces but is concentrated on the former. Along the lateral line, the neuromasts are lined up in a water-filled tube, or canal, under the skin. The non-pored canals of sharks and other elasmobranchs are located only on the head region of sharks and the ventral surface of batoids.

Differences in function between the pored and non-pored canals are a direct result of the former's connection with the environment. Lateral lines with pores are said to sense *distant touch*, an experience terrestrial organisms lack. *Distant* in this case is a bit of exaggeration, since the lateral line is thought to detect only near-field signals; that is, those within one of two body lengths of the source. Hair cells of the neuromasts in the lateral line canal bend in response to a variety of stimuli, including low frequency sounds, vibrations, and water movement. This enables the shark to detect prey, predators, other organisms, and conspecifics, as well as to determine the velocity and direction of water flow around its body.

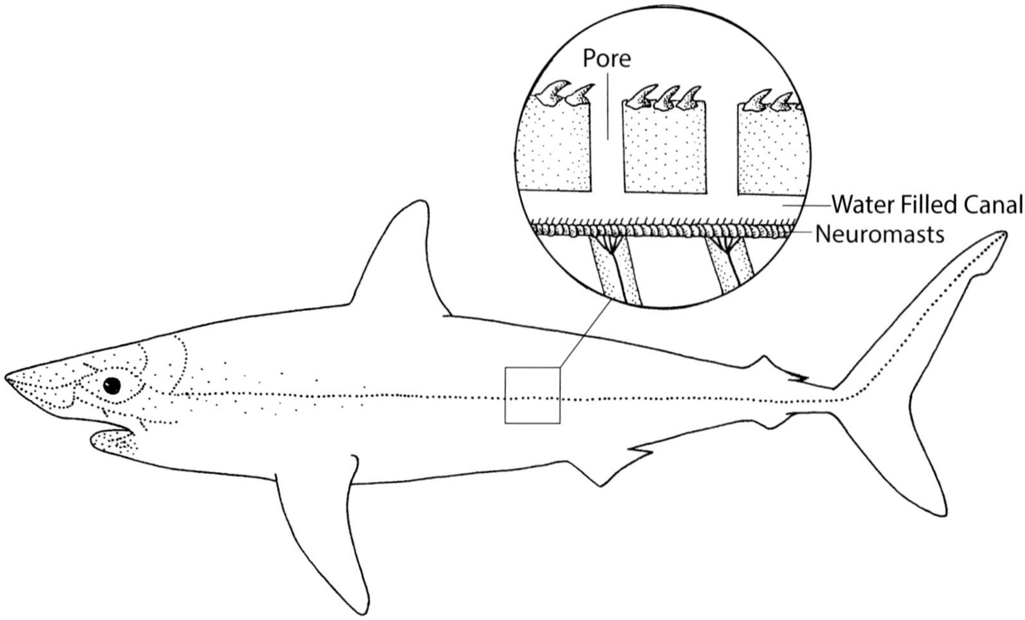

Figure 5.2. Diagram of the lateral line of a shark (Redrawn from Mojetta, A. 2004. *Sharks: White Star Guides Underwater World.* White Star).

Sensing water movement is what enables *rheotaxis* (the ability to orient with or against the current).

The non-pored components of the lateral line cannot react to external water movement. Instead, these neuromasts respond to internal fluid velocity, which requires a tactile external stimulus that deforms the skin, the kind that might occur on the head in sharks during mating, feeding, or when the body contacts the substrate in batoids.

Free, or unspecialized, neuromasts, which are also known as *pit organs*, are found on the tops and sides of head, trunk, and tail, as well as surrounding the spiracles and near the gills and ventral to the mouth (fig. 5.3). In batoids, pit organs are found in grooves on the skin whereas in sharks they are typically in shallow pits between modified scales. Pit organs are thought to detect external water movement. Active pelagic sharks (e.g., carcharhinids) have more pit organs than sluggish demersal species (e.g., heterodontids).

Spiracular organs, which are derived from lateral lines, are part of the first visceral pouch, or gill slit (also known as the *spiracular slit*), which in elasmobranchs is reduced to the spiracle. Their basic unit is the neuromast and their function is quite specialized; spiracular organs respond to movement of the cranial-hyomandibular joint. Although their function is enigmatic, they may serve as proprioceptors.

Vesicles of Savi are found on the ventral surface of three families of batoids (Torpedinidae, Narcinidae, and Dasyatidae). They are blind pouches (not

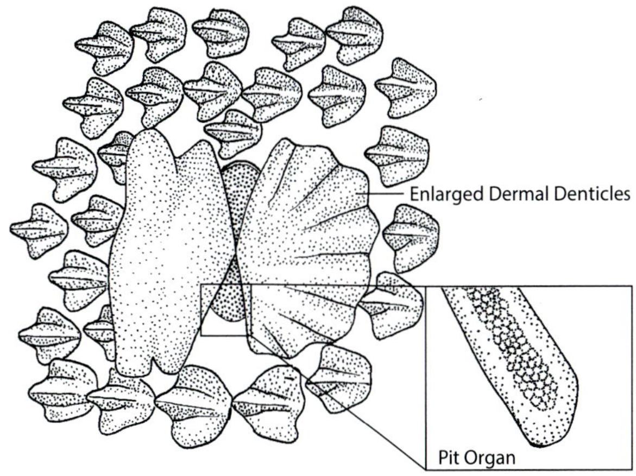

Figure 5.3. Pit organ, free neuromasts (sensory hair cells embedded in gelatinous cupula) in a small depression, or pit, between two enlarged dermal denticles. In some cases, the ridges of the dermal denticles direct water to the pit organ. Inset shows sensory surface of the pit organ (Adapted from fig. 2, Peach, M.B. and Marshall, N.J. 2009. J. Morphol. 270: 688–701).

connected to the environment) thought to be tactile receptors useful in prey detection or communication with conspecifics by sensing vibrations in the substrate or from direct contact with prey or conspecifics.

Chemoreception

In our annual *Biology of Sharks* course at the Bimini Biological Field Station in the Bahamas, one of the students' favorite activities is attracting and hand-feeding juvenile Lemon Sharks[7] in an isolated tidal lagoon. We spread out into a half-moon shape, and a *chum bag* (a mesh sack containing minced fish) is positioned on a stake in the sediment where the incoming tide slowly carries pieces of it deeper into the lagoon. After a period ranging from a few minutes to perhaps 30 minutes, we begin to see ripples representing the dorsal and caudal fins of the juvenile sharks breaking the water's surface as they wend their way toward us. After conditioning the sharks to take the squid or herring, the students each hand-feed one or two sharks.

This exercise is a microcosm of how a shark senses its environment. Since the lagoon is <100 m (330 ft) long, the sharks likely hear us, and initially move away. Vision, electroreception, and other forms of mechanoreception cannot come into play at such a distance. What remains is chemoreception, specifically olfaction (smelling), gustation (tasting), as well as a third form known as *common chemical sense.*

Chemoreception, which is the most ancient of the senses, having evolved over 500 mya, refers to detection of environmental chemicals. Chemoreception in water is quite different from chemoreception in air, since in the former the chemicals must be dissolved as opposed to in the gaseous state in the latter. Moreover, water is heavier and more resistant to flowing than is air. Thus, diffusion is slower in water, and the medium itself often moves more slowly than air. On the other hand, odors dissipate more quickly in air than in water, since water moves as parcels that are less likely to break apart than those in air.

Chemoreception is most important for feeding and reproduction. Chemical molecules sensed in aquatic environments are typically of small molecular mass, like amino acids and some steroids.

Back to the lagoon. The chemosensory system of these juvenile Lemon Sharks brings in the sharks. They then use their other senses in the final stages of locating the food source.

Let us start with the sense of smell. The organs of olfaction are paired *olfactory sacs* (also known as the *olfactory apparatuses*) on the ventral surface just anterior to the mouth, in cartilaginous nasal capsules. Water is typically channeled into the nares, the channels that lead into the olfactory sac (you might call these nostrils) by flaps on the lateral (outermost) incurrent, then out through the medial (innermost) excurrent openings (fig. 5.4). The small distance between incurrent and excurrent openings might mislead you to think that the water that enters travels only a short way before it exits. In reality, the water takes a circuitous path through a sensory epithelium consisting of lamellae analogous to those of the gills (fig. 5.4C). Called the *olfactory rosette,* this organ immensely increases the surface area for sensing environmental chemicals, and the slow transit time ensures that compounds in the water come into contact with the sensors in the rosette. Although the olfactory rosette contains cilia that may propel water and mucous through it, most sharks must be moving, or there must be a water current, in order for water to enter through the incurrent opening.[8] In sharks with nasoral grooves (e.g., Small-spotted Catshark), most of which frequently rest on the sea floor, the respiratory current (pumping water over the gills) also plays a role in water movement into the rosette. Olfaction in sharks is not linked to respiration and, unlike humans, they do not need to respire to smell, meaning that sharks smell their environment whenever water moves through their olfactory rosette, and thus the appellation *swimming noses.*

Molecules to which the shark is sensitive (e.g., those in mucus sloughed off of the skin of prey) stimulate receptors called *olfactory receptor neurons* (ORNs) on the epithelium of the olfactory lamellae. Cells comprising the ORN also add to the increased surface area through little projections called *microvilli.*

Stimulation of ORNs sends olfactory information via the short olfactory nerve to the *olfactory bulb,* an anterior extension of the brain that is located within the olfactory sac (fig. 5.5). Processing of the olfactory information takes place first in the olfactory bulb. Final processing of olfactory signals occurs in the olfactory area of the forebrain, or *telencephalon,* which functions in processing a broader range of sensory information than solely olfaction, as well as playing a role in complex behaviors.

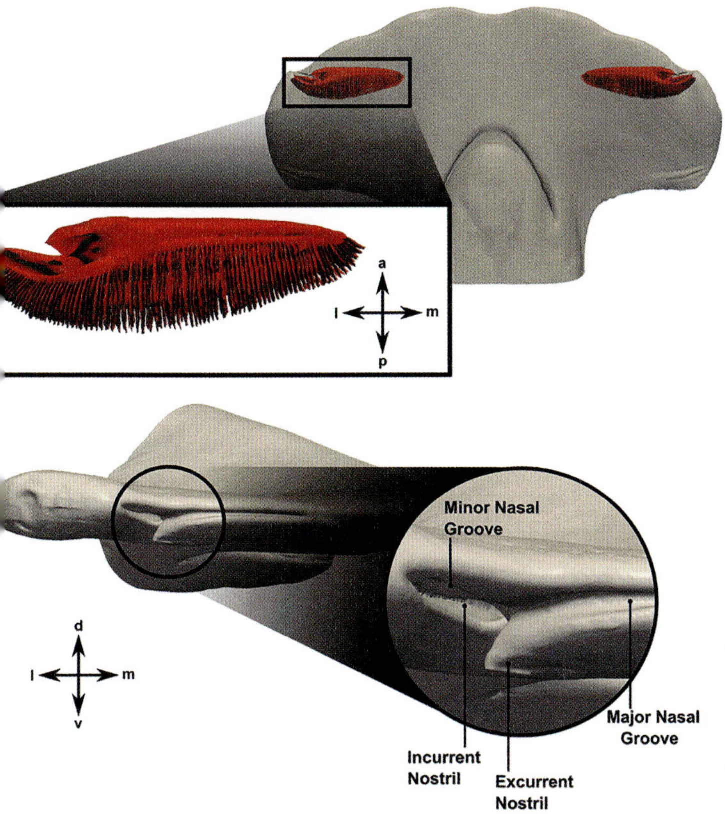

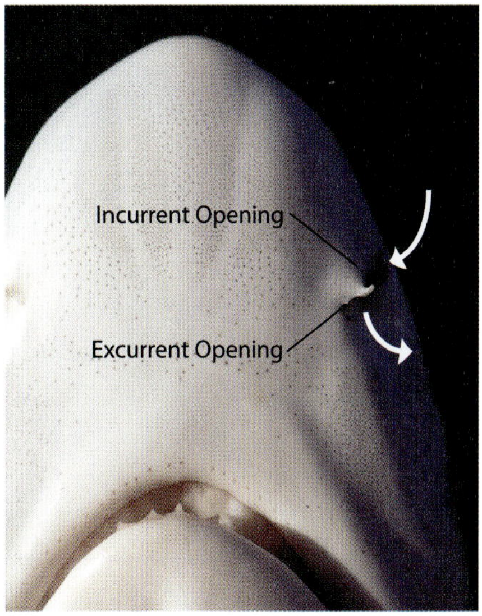

C

Figure 5.4. (A) and (B) Model of the olfactory rosette in a Hammerhead (From: Rygg, A.D. *et al.* 2013. PloS One, 8: p.e59783). (C) Incurrent and excurrent openings in olfactory sacs of a shark. White arrows depict direction of water flow.

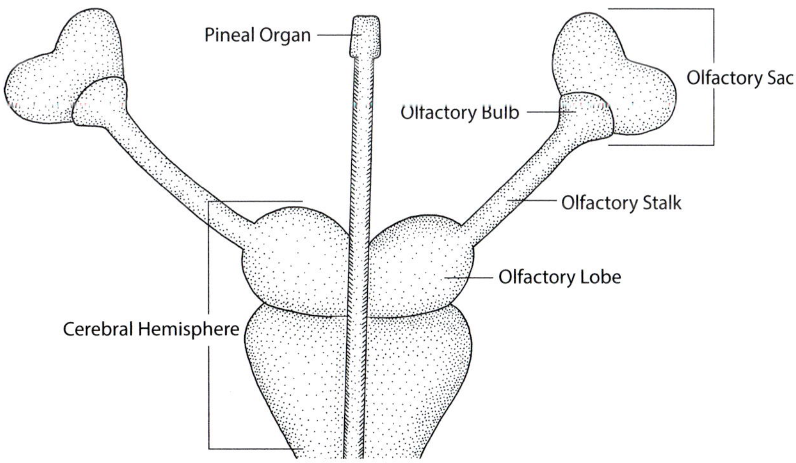

Figure 5.5. Olfactory nerve and bulb, anterior extensions of the brain, of a shark (Redrawn from Gilbert, S. 1973. *Pictorial Anatomy of the Dogfish*. University of Washington Press).

Olfactory rosette and olfactory bulb vary among elasmobranchs. As you might expect, this has been interpreted to correlate with the degree of reliance on olfaction by different species or by habitat. One study[9] that looked at 21 species of elasmobranchs concluded that lamellar number and surface area did not correlate with phylogeny but that benthopelagic (living near the bottom) species had more lamellae and higher surface area than exclusively benthic species. For example, the Port Jackson Shark (*Heterodontus portusjacksoni*) had the largest olfactory surface area for its body size (length and mass) as well as the largest rosette mass and deeply folded lamellae. Makes sense for a shark that spends much of its life sifting the substrate for invertebrates!

The size and shape of olfactory bulbs was also shown to vary.[10, 11] For example, the largest olfactory bulb known thus far, at 18% of total brain mass, belonged to the White Shark, while the smallest bulbs, at 3% of total brain mass, were found in species of *Carcharhinus*, as well as Blue Sharks (*Prionace glauca*) and Shortfin Makos (*Isurus oxyrinchus*).

The above disparities in relative size of the olfactory bulb raise questions about the correlation with reliance on olfaction, especially in the White Shark and Shortfin Mako, confamilials sharing numerous physiological and anatomical characteristics for their high-performance lifestyle. Researchers studying this have speculated that differences in the diets of these two predators may provide the explanation. White Sharks primarily consume marine mammals, which produce copious amounts of odorous defecation. Thus, the larger than average olfactory bulb of White Sharks may have evolved as an adaptation to detect marine mammal prey via their smelly emanations. This may also explain why humans gravitate toward fast food.

In general, trends in the size of the olfactory bulb are similar to those in lamellar surface area; that is, they parallel ecological niche more so than taxonomic relationship. The largest relative olfactory bulb sizes occurred in migratory pelagic sharks; deep-sea sharks also had large olfactory bulbs, implying the importance of smell for both groups. Reef-associated carcharhinids, hemiscyllids (longtail carpet sharks), and benthic batoids had the smallest olfactory bulbs.

The Role of Chemoreception

In an early 20th century study,[12] crushed crabs wrapped in cheesecloth (to minimize visual cues) fed to captive Dusky Smoothhounds (*Mustelus canis*) led to changes in their behavior, specifically a series of turns until the bait was located, seized, and removed, when the sharks moved to within a few feet of the crab packet.

When cotton plugs were inserted into both nares of the sharks, they ignored the crab packets. When cotton plugs were placed in only a single nostril, the sharks' turns were to the side opposite the plug. The study concluded that the presence of food elicited directed movement by the sharks, that odors in the food were the stimuli, and that the nasal organs detected the stimuli when they were in the odor stream.

This study also investigated responses to natural foods, exposing Tiger Sharks to decayed shark flesh and tuna extracts, and Blacktip Reef (*Carcharhinus melanopterus*) and Grey Reef Sharks (*Carcharhinus amblyrhynchos*) to a wide variety of potential prey, and concluded that all of the items tested served as attractants, but there was variation in responses to some items by some species. Extracts from oily-fleshed fish (e.g., tuna, grouper, and eel) elicited more activity in the sharks than extracts from snappers and other fish with drier flesh, a finding that shark fishers routinely exploit.

A variety of organic biomolecules serve as attractants to elasmobranchs, but amino acids (components of proteins), particularly natural ones, and nucleotides (the building blocks of the genetic material) seem to be most important.

Is the sense of smell in sharks as prodigious as commonly asserted? Studies have shown that elasmobranchs are capable of detecting concentrations of some odorants as low as one billionth of a mole per liter (10^{-9} mol L^{-1}), which represents an extremely high sensitivity. However, many bony fishes have similar sensitivities, so if sharks are swimming noses, then bony fishes are as well.

Before we move on to taste, let us consider hammerheads, which have widely spaced olfactory organs and deep narrow grooves running to these organs along the anterior margin of the cephalofoil (fig. 5.4). This degree of separation gives this group the potential to compare odor on one side to the other more effectively or at a greater distance than in other elasmobranchs. A 2009 study,[13] showed that overall Scalloped Hammerheads (*Sphyrna lewini*) were as or more sensitive to odorants than other sharks and bony fishes, although the authors cautioned that more comprehensive data were needed before conclusions could be drawn.

Gustation and Common Chemical Sense

Although there is debate about whether the sense of taste should be differentiated from olfaction in aquatic organisms, we consider it separately here. Gustation is assumed to have a role secondary to olfaction and is more specialized in its function, specifically to assess food quality; however, since taste is among the most poorly studied senses, its function is not well understood.

Taste receptors in elasmobranchs are located on taste buds found inside the mouth and on the gills. Modified placoid scales, known as *oral denticles,* are also present lining the mouth,[14] limiting the space for taste receptors. The taste buds sport papillae (fig. 5.6), which increase the surface area on which the receptors are located and thus increase sensitivity. There are also microvilli on the taste papillae, small projections that further increase the surface area for taste. Unlike in mammals, papillae are not concentrated on the tongue.

Studies[15, 16] in the Brown-banded Bamboo (*Chiloscyllium punctatum*) and Blue Shark showed that taste papillae were spread throughout the oropharyngeal cavity (gill arches and rakers, and dorsal and ventral surfaces of the mouth), but the highest densities of taste papillae were found on or near the oral valves, folds of tissue in the anterior of both jaws (maxillary and mandibular valves) behind the teeth. A 2011 study[17] concluded that taste papillae were more evenly distributed throughout the oropharyngeal cavity in benthic sharks than in pelagic species.

These different distributions of taste papillae in the oral cavity make good sense. Benthic species (e.g., Nurse Sharks [*Ginglymostoma cirratum*] and Brown-banded

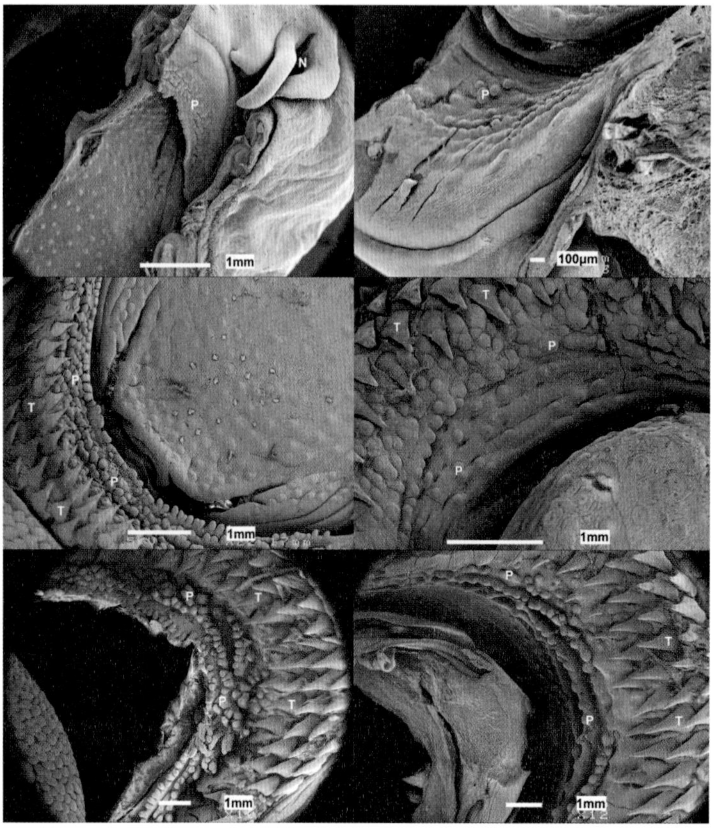

Figure 5.6. Scanning electron images of taste papillae (P) on taste buds of Brownbanded Bamboo Shark, *Chiloscyllium punctatum*. N = nostril (nare), T = tooth (from Atkinson, C.J. *et al.* 2016. Biology Open: bio-022327.)

Bamboo Sharks) often manipulate their prey in their mouths prior to swallowing, during which time assessing the palatability of the ingested item may occur.

In pelagic sharks, the location of the highest density of taste papillae in areas immediately adjacent to the teeth and the oral valves also appears adaptive. This area is the first point of tactile contact when an elasmobranch bites into its targeted food item. At this point, the predator must assess whether the item represents food on its menu and is worth continuing to consume, a process that may be instantaneous.

The concentration of oral papillae near the mouth also explains *bite-and-release* (also called *bite-and-spit*) shark bites. Along the SE US coast, during the summer months, there may be several reported shark bites that leave many of the victims with minor punctures that may not even require a trip to the emergency room. The culprits in these cases are thought to be 2 m (6.6 ft) long or smaller Blacktip Sharks (*Carcharhinus limbatus*), streamlined, swift-moving, bite-first, ask-questions-later fish-eaters. The waters in which these interactions occur are often quite murky, so much so that your hand disappears when immersed only inches. Add swirling schools of menhaden and mullet, so-called *baitfish*, nearshore to the mix of bathers and fast-moving feeding sharks, and you have a foolproof recipe for mistaken identity: Blacktip Shark bites bather.

Fortunately, what happens next most of the time, turns what might have been a fatal interaction into an *ouchie!* and a badge of courage. What likely occurs is what you might experience when you bite into spoiled food. Blech! Spit it out! Exactly what the Blacktip Shark does, and does so almost instantaneously, thanks to the evolution of taste receptors near the mouth and a rapid-fire nervous system that receives and processes the information and sends out a *stop!* signal to the jaws. While this may save victims of bites from Blacktip Sharks, if the perpetrator is a White Shark, and it mistakes a person for, say, a seal, then even if the same receptors indicated that, *oops!*, the shark made a mistake, the outcome will sadly be more than an ouchie.

Some elasmobranchs are capable of sampling water twice for olfactory, gustatory, and common chemical sensory information (fig. 5.7). Members of at least four families of sharks, Heterodontidae (horn sharks), Orectolobidae (carpet sharks), Parascyllidae (collared carpet sharks), Hemiscylliidae (longtail carpet sharks), and numerous batoids (e.g., Rajidae, Myliobatidae, Dasyatidae, Torpedinidae, and Urolophidae) possess paired *oronasal grooves,* channels that funnel water from the nares into the mouth, where taste papillae and common chemical sense receptors are free to sample it.

Lastly, we consider the mode of chemoreception called the *common chemical sense.* The receptors for this type of chemical sensitivity are not specialized cells or groups of cells, but are rather unspecialized, free nerve endings. Humans apparently use this sense when we feel that the air after a storm is crisp. The function of this sense in elasmobranchs is somewhat enigmatic, although most specialists feel that the common chemical sense serves to protect sharks from harmful chemicals. Chemical shark repellents and deterrents, both synthetic and natural (e.g., the skin of the Moses Sole) are thought to act on receptors of this system.

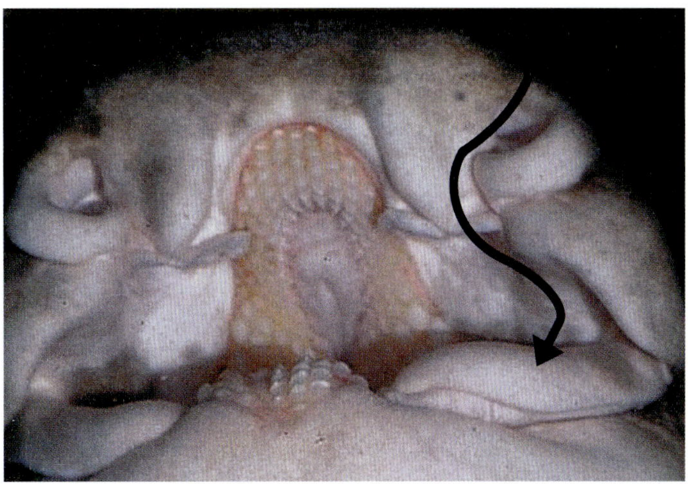

Figure 5.7. Oronasal groove in a Horn Shark. Arrow denotes movement of water through oronasal groove.

Orienting to Chemical Senses

We have now discussed the placement, morphology, and sensitivity of the three types of chemical senses in elasmobranchs. But how exactly do sharks zero in on the source of the chemicals (fig. 5.8)? Movements in response to a stimulus, in this case odor, are known as taxes (singular = taxis).

The first taxis, sampling the environment and turning toward the stimulus, is divided into two separate, distinct subcategories, *klinotaxis* (*klin* = to cause to lean) and *tropotaxis* (*tropo* = turn or response), which are differentiated by their temporal nature. Klinotaxis involves *successive* comparison of stimulus intensities between disparate spatial points (e.g., such as would occur by side-to-side head movements), whereas tropotaxis involves *simultaneous* comparison of stimulus intensities on two sides of the body (that is, at the same time).

Early work exploring this phenomenon pointed toward sharks being capable of simultaneously assessing odor intensities and turning toward the higher concentration; that is, tropotaxis. However, recent captive experiments with the Dusky Smoothhounds revealed for the first time that sharks steered using bilateral odor *arrival time* rather than concentration differences. And even with delayed pulses of higher concentration the sharks continued to turn toward the side stimulated first.

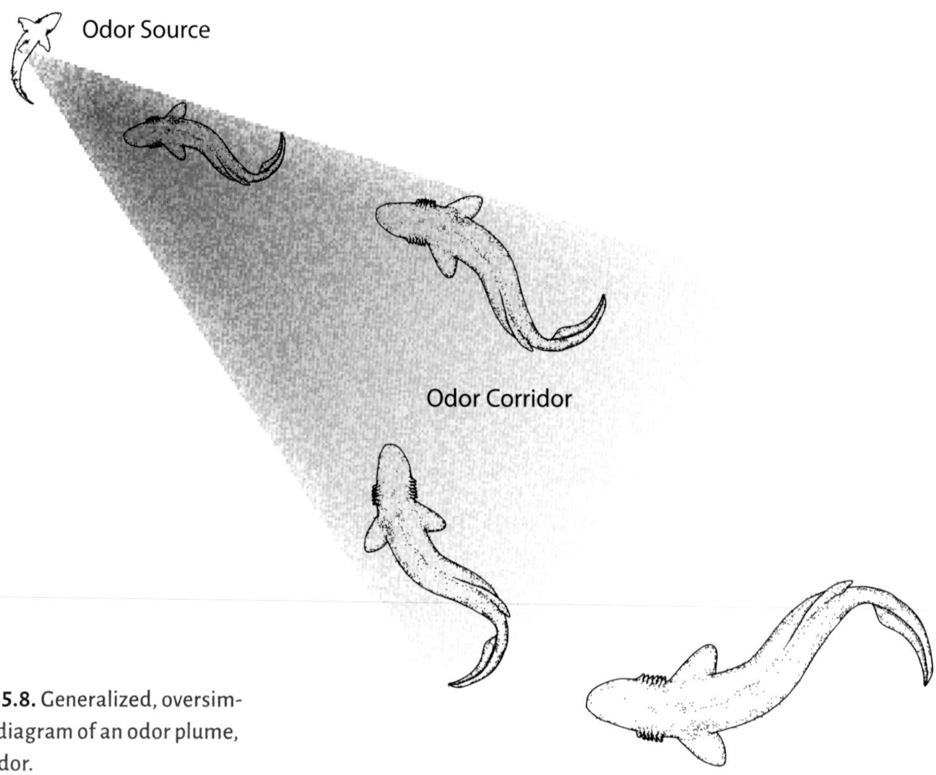

Figure 5.8. Generalized, oversimplified diagram of an odor plume, or corridor.

Another taxis employed by some elasmobranchs is *rheotaxis* (*rheos* = stream), in which an organism swims with or against a current. Rheotaxis is similar to tropotaxis, except instead of locating a chemical stimulus with search patterns, the individuals turn in the direction of the current and swim upstream. Lemon Sharks, for example, were shown to respond positively to chemical stimuli using rheotaxis and subsequently followed the current upstream.

Photoreception

The main organ of photoreception in sharks is, of course, the eye. Did you know that the visual system of sharks, as well as that of most other lower vertebrates, also includes another organ of photoreception, called the *pineal organ* (fig. 5.9)? All vertebrate groups except the hagfish possess the structure, which is more widely known as the pineal *gland*, because its primary function is secreting the hormone *melatonin*, which regulates both an organism's daily cycles (its *circadian rhythms*) and its reproductive hormones.

In lower vertebrates (cartilaginous, bony fishes, and amphibians), the pineal organ/gland is located near the top of the brain, where it receives light that enters through specialized areas on the top of the head, which explains a name the structure is no longer called—the *third eye*. That the morphology of the shark cranium is specialized to transmit more light to the pineal gland than surrounding tissues (fig. 5.9), up to seven times more, was discovered in 1975 by the shark biologist Sonny Gruber and associates.

The shark eye has been the focus of studies for 250 or more years. For most of that time, the subject of these studies was its anatomy. Sharks have a typical vertebrate, image-forming eye (fig. 5.10). The eyeball sits in a socket called the *orbit*, which is composed of the cartilaginous *sclera* posteriorly and the clear *cornea* anteriorly. The cornea has the same refractive index (ability to bend light) as seawater

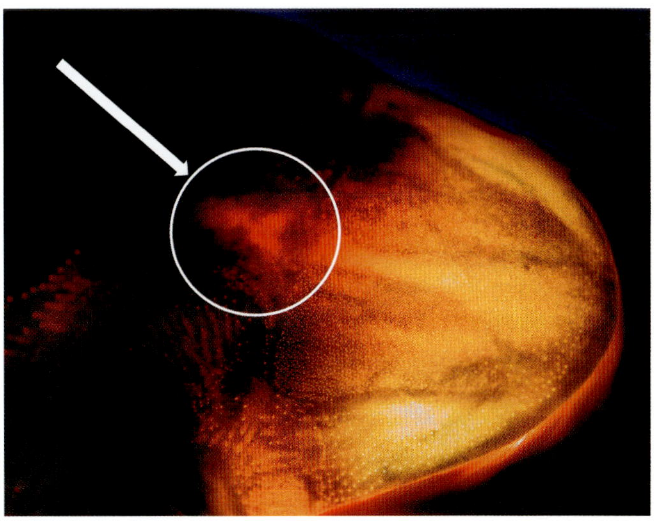

Figure 5.9. Pineal window (circle and arrow) of an Atlantic Sharpnose Shark illuminated by shining a bright light through its open mouth, on right.

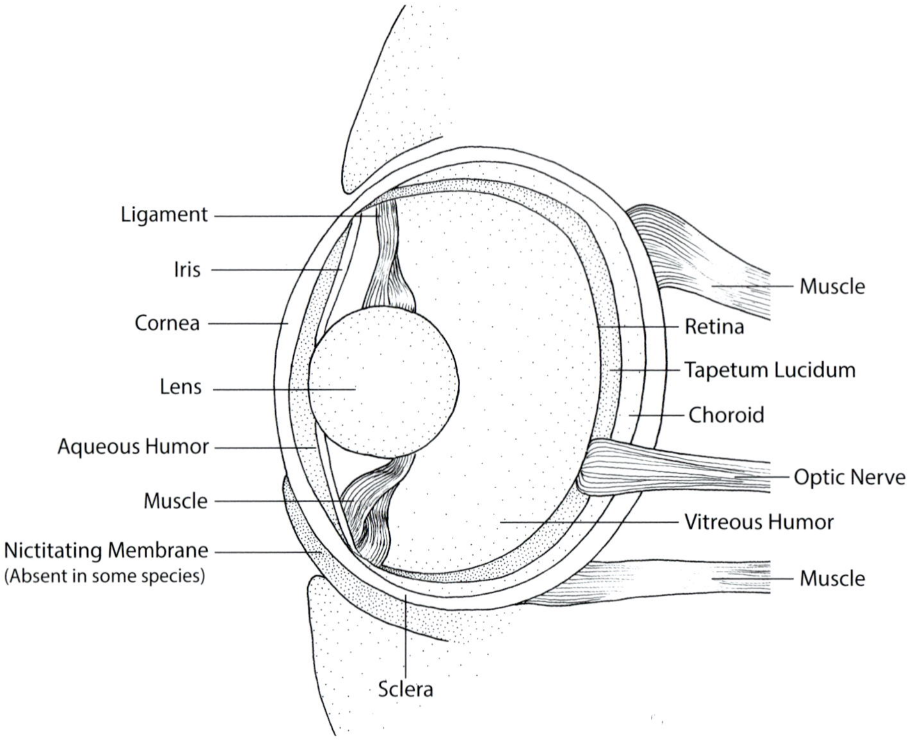

Figure 5.10. Anatomy of the eye of sharks (Redrawn from fig. 13.3, Hart, N.S. *et al.* 2006. Com. in Fish. 2: 337–392).

and thus serves only to protect the eye and not to focus the incoming light as it does in humans and other terrestrial vertebrates, where as much as 75% of the total refraction occurs. The tough, protective sclera surrounds and protects the eye.

Three additional specializations found in some sharks serve to protect the eye when faced with mechanical damage (e.g., during interactions with prey or predators). In lamnid sharks, the entire eye rotates 180° and the tough, fibrous sclera is positioned where the cornea had been. Hexanchid sharks use a different strategy. They use extraocular muscles to pull the entire eye back into the head several centimeters (as seen in fig. 3.7B).

Sharks also possess upper and lower eyelids, though they are fixed in most species and moveable in only a few, including Nurse and Swell Sharks (*Cephaloscyllium ventriosum*). In sharks of the order Carcharhiniformes, a fold of the lower eyelid develops into a tough, opaque, thin, third eyelid called the *nictitating membrane* (see fig. 1.12), whose external surface is covered with placoid scales, and which moves upward to protect the eye in a manner analogous to the rotation of the sclera in lamnids.

There are six muscles attached to the outside surface of the back of the orbit (and are thus called *extraocular* muscles), which is typical of vertebrates. These serve to allow coordinated movements of the eyes.

Lining the sclera is a layer called the *uvea tract*, which functions to deliver oxygen and nutrients to the tissues of the eye. The uvea tract includes a darkly pigmented layer called the *choroid* and parts of the *ciliary body*. The ciliary body and other ciliary structures perform a variety of functions, including supporting the lens, controlling the amount of light that enters the eye, and secreting the *aqueous humor* in the anterior portion of the eye between the cornea and the lens.

Part of the choroid layer lying behind the retina is the *tapetum lucidum*, a tissue composed of guanine crystals, which reduces internal glare and scattering of light, and improves vision in low-light conditions. This is the same tissue that is responsible for the *deer in the headlights* reflection from the eyes of cats, racoons, etc. The tapetum lucidum of a shark is twice as efficient as that of a cat.

The pupil, which regulates the amount of light that will strike the retina, comes in a wide variety of shapes in elasmobranchs (fig. 5.11). These include the typical vertebrate round shape, as well as horizontal, vertical, and oblique slits. Oddly shaped pupils are found in some benthic sharks and batoids, although any specialized functions of these pupils are not always clear.

The pupil in most elasmobranchs is capable of dilation and contraction. This is accomplished by means of a *dynamic* iris, the muscular structure controlling the diameter of the pupil. In some deep-sea species (e.g., the Ghost Catshark [*Apristurus*]), the pupil is fixed in the fully dilated state, which makes sense in light of its dark surroundings. In species with mobile pupils, the size of the pupillary opening correlates with light levels. In high light, the pupil contracts which, in addition

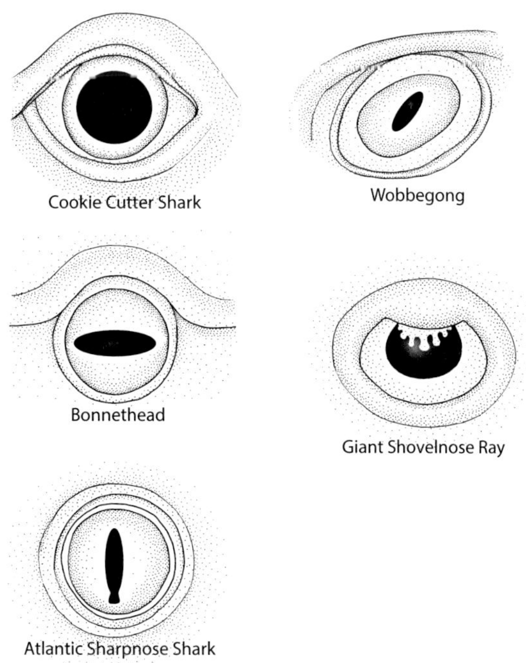

Figure 5.11. Diversity of pupillary shapes in sharks.

to reducing the light entering the eye, also increases the visual depth of field, just like in a camera.

The lens (fig. 5.10) in elasmobranchs is a transparent structure responsible for all of the refraction (bending of light so that it can be focused on the retina). It is composed of water and structural proteins. The lens of elasmobranchs also varies. Shapes include spherical, near spherical, and lenticular (ellipsoidal). The lens in most species is large and powerful. Unlike the lens of terrestrial vertebrates, accommodation, or maintaining focus at various distances, is achieved by moving the lens backward (for far vision) or forward (for near vision) to focus light on the retina rather than changing shape. This movement is accomplished by the six extraorbital (outside of the eye socket) eye muscles.

The retina is the eye's photoreceptor is the tissue responsible for converting the image that the lens focuses on into nerve impulses that it sends to the brain via a thick, white, ophthalmic nerve passing through the rear of the orbit. Sharks and rays have duplex retina; that is, the retina houses both rods and cones, the photoreceptive cells that are responsible for vision in low-light (*scotopic*) and bright light (*photopic*) environmental conditions. The presence of cones containing multiple visual pigments implies a basis for color vision, which has been experimentally verified in several species of sharks.

The size of eyes varies as well, most notably with depth distribution and habitat.[18] Epipelagic and upper mesopelagic sharks (e.g., Bigeye Thresher [*Alopias superciliosus*]) have the largest eyes, along with benthic and benthopelagic species (e.g., Portuguese Dogfish [*Centroscymnus coelolepis*]) that live in the mesopelagic zone. From an ecological perspective, benthopelagic and pelagic species whose prey are mobile and active also have larger eyes than less active species. The smallest eyes are found in coastal pelagic sharks and especially sluggish benthic sharks and batoids.

The visual field, defined as the space visible from the stationary eyes, also varies with species, specifically with differences in head morphology, and eye position and mobility. The extent of the visual field is important for prey detection, predator avoidance, and mating. In most batoids, and in some benthic sharks (e.g., angel sharks [*Squatina* spp.], wobbegongs *(Eucrossorhinus, Orectolobus,* and *Sutorectus),* and horn sharks [*Heterodontus* spp.]), the eyes are located dorsally. In most sharks and a few batoids, the eyes are laterally placed.

Recent studies[19, 20] measured aspects of the visual fields of several sharks and rays. They concluded that visual field was a function of phylogenetic age, habitat type, and locomotory patterns. The fastest-swimming and most evolutionarily derived batoid in the study, the Cownose Ray (*Rhinoptera bonasus*), had the largest binocular overlap and vertical visual field, which is consistent with the need for enhanced visual information in the rapidly changing landscape it encounters during swimming, while schooling, and while foraging on bottom invertebrates, primarily mollusks.

The three demersal (bottom-associated) batoids studied all bury themselves and had good overhead visual fields. Even though these species consume benthic infauna, they lacked good ventral vertical visual fields and instead relied on sensitive tactile (touch) receptors and electrosenses.

Among sharks, binocular overlap (an index of depth perception) in hammerhead species increased with the expansion of the head, and in the Winghead Shark (*Eusphyra blochii*) was nearly four times larger than in two carcharhinids. Anterior horizontal binocular overlap in the Bonnethead (*Sphyrna tiburo*) was smallest, a product not only of its smaller cephalofoil, but also because of the median placement of the eye in this species compared to the anterior position in the other two hammerheads in this study (see illustrations of these sharks accompanying species descriptions in chapter 3). All five hammerhead species tested had a 360° vertical visual field around the head.

Electrical and Magnetic Senses

Sharks and batoids are renowned for having a 'sixth sense,' the ability to detect very weak electric fields (1nV/cm–0.1μV/cm, the strength of electrical field within about 0.5 m or 1.6 ft of the source in salt water) of other organisms and even inanimate objects, using specialized sensory organs called *ampullae of Lorenzini* (fig. 5.12). These are connected to a network of canals that diverge from the ampullae and terminate on the skin as pore, which are concentrated on the head of sharks and batoids. Each ampulla consists of a bulb under the skin, which opens to the environment and connects to other ampulla via a canal. The ampullae are lined with a sensory epithelium (hundreds to thousands of hair cell receptors), and the canal and the bulb are filled with a jelly-like substance (glycoprotein) capable of conducting electricity, like seawater. The tissues underlying the ampullae are insulated from the current by structures called *tight junctions* between adjacent cells. In short, once an electrical signal is received, the *ampullae of Lorenzini* detects small differences in electrical potential between their two ends and transmits it, via neurons to the central nervous system (CNS) for processing and interpretation (see the section about shark brains and nerves at the end of this chapter).

The pioneering studies that illuminated that sharks possessed an electrosensory system were conducted by Adrianus Kalmijn, based in part on observations in 1935 by the Dutch scientist Sven Dijkgraaf that the Small-spotted Catshark (*Scyliorhinus canicula*) was very sensitive to metallic objects (a rusty steel wire). In the 1960s, Kalmijn demonstrated that electric fields with voltage gradients as low as 0.01 μV/cm evoked a reflex contraction of the eyelids ("winking of the eyes"). He and others showed that the organs responsible for electroception were the *ampullae of Lorenzini*, which were also sensitive to temperature and mechanical stimuli. Next, Kalmijn was interested in exploring how sharks and batoids might use this extraordinary electrosense; he did this through designing a series

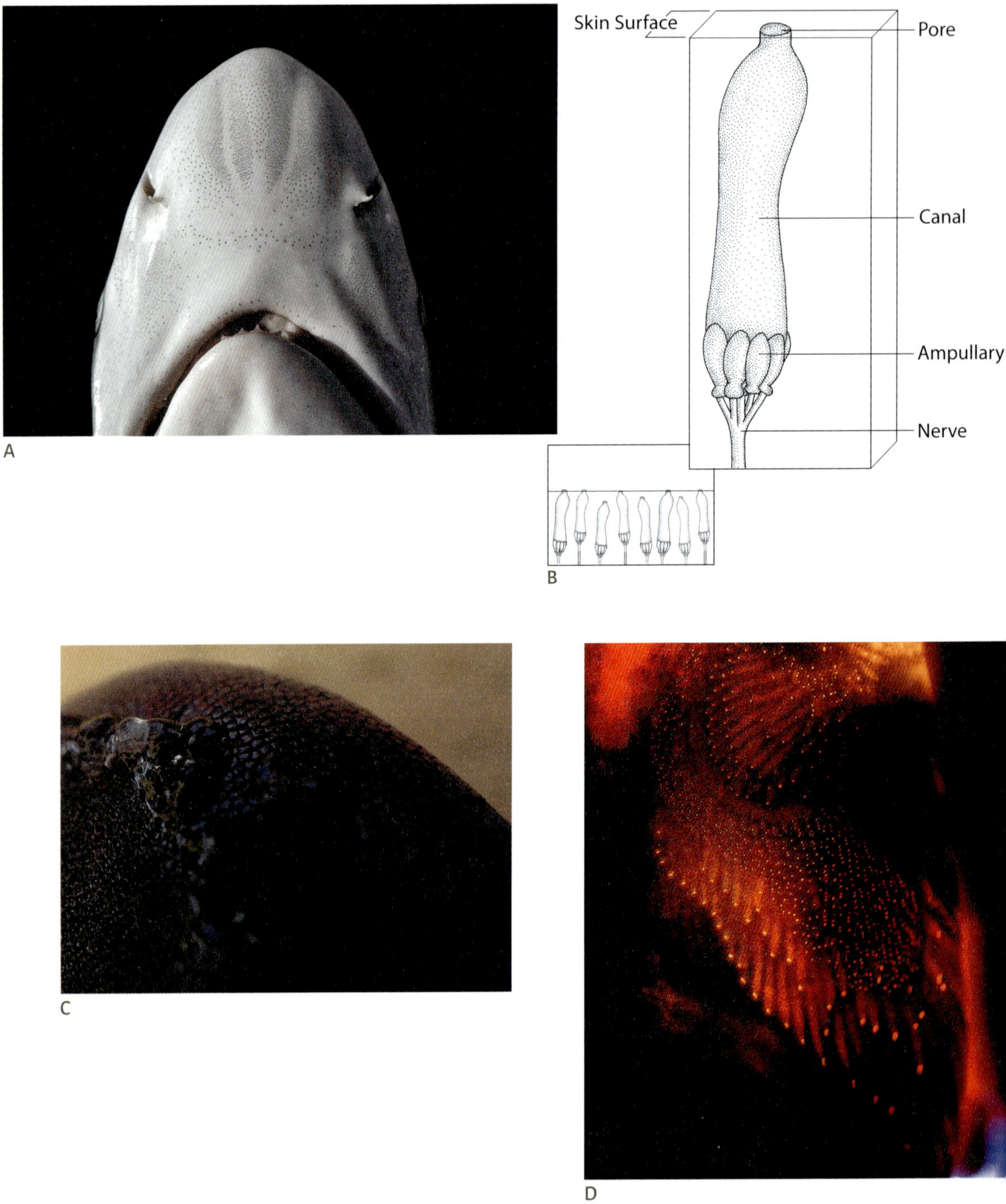

Figure 5.12. (A) Photos of ampullae of Lorenzini (Courtesy of Caroline Collatos). (B) Diagram of a single ampulla (Adapted from fig. 13.2, Hart, N.S. *et al.* 2006. Com. in Fish. 2: 337–392). (C) Gel squeezed from ampullae on snout of shark (Courtesy of Caroline Collatos). (D) Ampullary canals illuminated on the right half of the head of an Atlantic Sharpnose Shark by shining a light through its mouth (Courtesy of Robert Johnson).

of captive behavioral experiments examining the feeding responses of catsharks and thornback rays. His test animals were presented with a live flat fish buried in the sand (Fig 5.13) and incredibly were able to find hidden prey despite the absence of visual, mechanical, or chemical cues. However, when the bioelectric cues (i.e., those created by the release of charged ions during osmoregulation) were

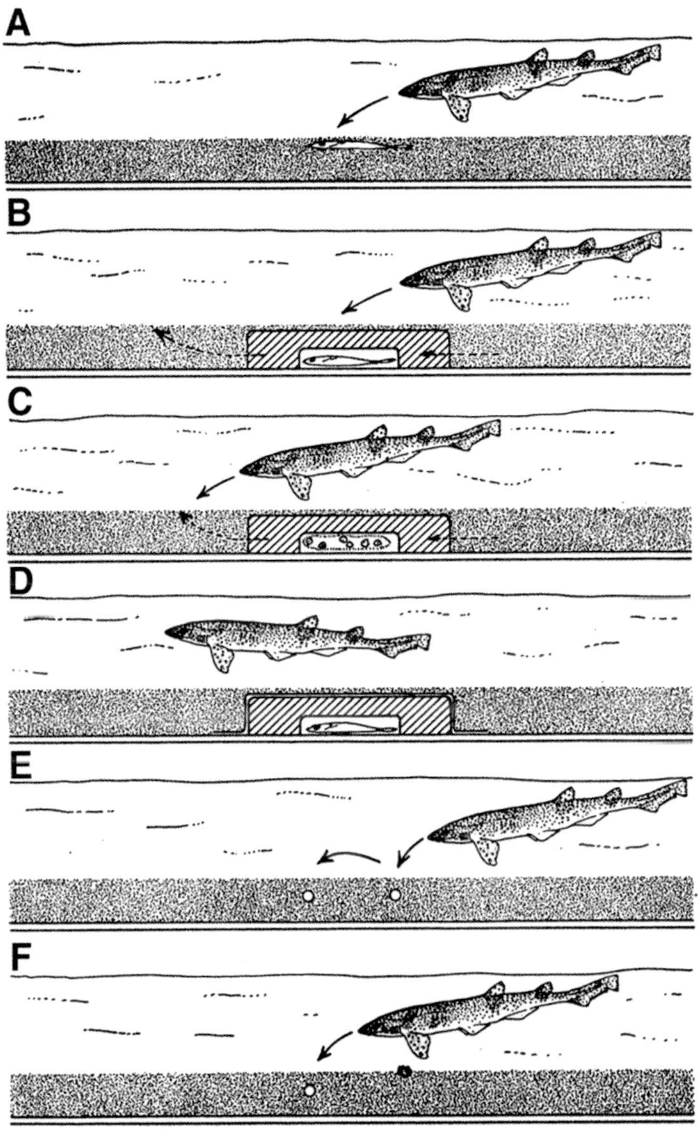

Figure 5.13. Kalmijn's classic experiment on the role of electroreception. The figure depicts the feeding responses of the shark *Scyliorhinus canicula* to (A) Plaice under sand, (B) Plaice in agar chamber, (C) pieces of Whiting in agar chamber, (D) Plaice in agar chamber covered with plastic film, (E) electrodes producing electric dipole field, and (F) piece of Whiting and electrodes (only one shown). Solid arrows: responses of shark; dashed arrows: flow of sea water through agar chamber (Reproduced with permission from Kalmijn, A.J., 1971. J. Exp. Bio. 55(2): 371–383).

concealed, by using agar with a plastic film to hide the flat fish, the sharks and rays were unable to detect the prey. Finally, to truly prove their electroreceptive ability, Kalmijn showed that the sharks and rays responded to buried electrodes that simulated the prey's electric stimuli.

Since these groundbreaking experiments, researchers have been fascinated by this alien-like sixth sense. I mean, wouldn't you? So many questions to explore about sensitivity, form, and function. Indeed, as we have learnt earlier in this chapter, there is considerable variation across species in their sensory capabilities and anatomy, think the small beady, almost colorless eyes of a horn shark (*Heterodontus* Spp.) compared with the huge, green eyes of a gulper shark (*Centrophorus* Spp.). These differences can be determined by phylogeny but are also linked to behavior (e.g., foraging strategies) and ecology (e.g., habitat use).

Let us compare the ampullae pore count and distribution of two unique, little-known sharks, the Megamouth, *Megachasma pelagios* and the Daggernose Shark, *Carcharhinus oxyrhynchus.* The former has relatively few pores (225) with most (75%) distributed dorsally around the anterior margin of the mouth, whereas the latter has up to 25 times as many pores (3943 ± 1663.81) with a high density in the ventral region of the snout. These differences highlight that the Daggernose Shark is likely an electrosensory specialist, with a highly evolved, high-resolution system, whereas the Megamouth, although likely using passive electroreception to maximize feeding efficiency, relies on a less derived system with low electrosensory resolution. Delving into these two species' behavior and ecology provides further explanation. The Daggernose Shark has a geographically restricted range and inhabits turbid waters of the Amazon plume. It has very small eyes with an elongated snout which maximizes the coverage of the search area by the electrosensory system, and it feeds on fast moving fish. The Megamouth, on the other hand, is a broadly distributed, open-ocean species that filter-feeds on plankton and bait fish from the side and below. Interestingly, the Daggernose Shark pore count increases dramatically as it reaches adulthood, with three times as many pores compared to its juvenile life stage. Ontogenetic changes in diet and habitat use are common for sharks (see chapter 9) and accompanied sensory system shifts likely provide important enhancements critical for survival at later life stages.

Since Kalmijn's experimental research on electroreception, various sharks including Bonnetheads (*Sphyrna tiburo*), Port Jackson Sharks (*Heterodontus portusjacksoni*), and Blacktip Reef Sharks (*Carcharhinus melanopterus*) have been behaviorally tested, confirming their preferences for active electrodes that emitted prey-simulating stimuli over control electrodes. Sharks and batoids were sensitive to stimuli ranging from 5–107 nV/cm at distances of 22–44 cm (8.5–17.5 in).[21] Although, the very latest behavioral experiments found that adult Sandbar Sharks (*Carcharhinus plumbeus*) displayed greater sensitivity to dipole electric fields than juveniles, and adults had the lowest sensitivity threshold (0.002 nV/cm) of any elasmobranch examined to date.[22] This finding suggests that adults can detect electric fields at a greater distance than juveniles, highlighting the importance of behavioral electrosensitivity testing across life stages, as well as accompanying anatomical work. Combining captive tests with CT scanning (see chapters 4 and 13 for latest developments and uses) that can offer more detailed, accurate pore morphometric data will be an important next step to deepen our understanding of electroreception function and sensitivity.

Sharks as prey? Surely not! Even the iconic White Shark is not safe if there are Orcas about, but do sharks use their electrosense to avoid and detect predators? As we learned in chapter 6, many sharks lay eggs that are deposited on the sea floor, often between rocks, coral, or attached to seaweed. The embryos typically develop for several months in their protective cases with no parental care or defense leaving them exposed and at risk from predators. So how do they survive? It turns out these embryos have a neat trick! Lab experiments monitoring the development of Bamboo Sharks (*Chiloscyllium punctatum*) and Small-spotted Catsharks (*Scyliorhinus canicula*) have demonstrated that they exhibit an innate response to predator-simulating electric stimuli; they stop ventilating, a type of freeze response.[23] Despite being trapped in a small space, the embryonic sharks were able to recognize dangerous stimuli and react accordingly, ceasing their own bioelectric signal to avoid detection by a predator. Surprisingly, Small-spotted Catshark embryos also reacted with a freeze response when the electrical field was turned off. This indicates that the change in the electrical field, rather than its presence elicited the response.

Given the importance of the electrosense for finding hidden prey, it seems intuitive that sharks and batoids would also use this sense to find each other. A series of intricate experiments were conducted on Round Stingrays (*Urolophus halleri*), finding that both male and females in the wild use their electrosense to detect and locate each other during the mating season. Indeed, the males used their sense to locate prospective females for possible mating opportunities, whereas the females used their electrosense to locate other females to refuge with, perhaps confusing the searching males with their overlapping bioelectric fields.[24] It seems likely that benthic shark species such as angel (*Squatina* Spp.) or catsharks (*Scyliorhinus* Spp.), which hide and aggregate in rock crevices or under the substrate, may use their electrosense in a similar way.

Lastly, it has been hypothesized that sharks use their electroreceptors in navigation and orientation. Indeed, the earth's geomagnetic field (GMF) is an extremely reliable, omnipresent sensory cue that exists at all locations on the planet, from the depths of the ocean to the uppermost part of the atmosphere. It is caused by an interaction between rotation in the planet's magnetic core and electrical currents. Several geomagnetic field elements, e.g., *inclination angle* and *total intensity*, vary predictably across the surface of the earth, and these can be used to maintain a consistent heading in a particular direction (that is, as a *compass*) or to determine position (as a *map*). Many animals are known to use the GMF, for example, crustaceans, mollusks, turtles, and fish. For sharks, enter our *Physics Wizard* Adrianus Kalmijn again, who first theorized that sharks indirectly detected the GMF using the mechanism of electromagnetic induction, and thus would not need a true 'magnetoreceptor cell.' He proposed that as a shark swims through the GMF, its electroreceptors detect the voltage drop of the induced current that flows through the seawater around it (fig. 5.14). Working with Leopard Sharks in captivity, Kalmijn noted that in the mornings, they rested in a part of the tank that corresponded to magnetic north. In subsequent behavioral trials, he used a magnetic coil to null (block) the effect of the GMF, which led to the sharks taking up random positions within the tank, suggesting that they orient to magnetic cues.

Since these early trials, scientists have continued to explore this theory, revealing that Scalloped Hammerheads (*Sphyrna lewini*) and Sandbar Sharks (*Carcharhinus plumbeus*) can be trained (via reward conditioning) to respond to geomagnetic polarity and intensity. Tracking studies of wild sharks, conducted in the 1980s off Baja California, also revealed remarkable relationships between swimming trajectory and the local magnetic extremes that extended from seamounts to feeding grounds. Considering the extraordinary migrations that many sharks undergo, traversing the open ocean and huge deepwater trenches, as well as the precise and highly directed nature of these movements—they must have a *trick up their fin*! Perhaps their electrosense is just that?

Interestingly, despite the aforementioned studies, it was only very recently that sharks were definitively shown to use map-like information from the GMF as a navigational aid (see chapter 10 for further description). A reoccurring critique of

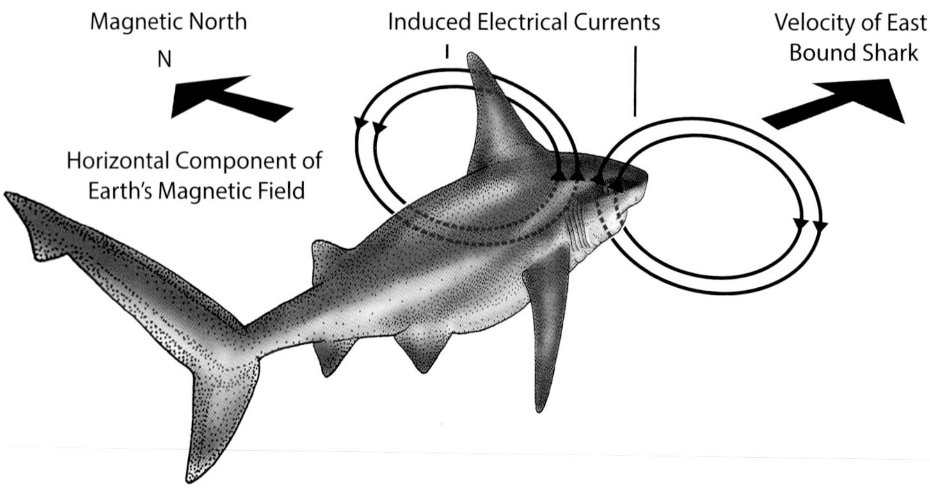

Figure 5.14. Active mode induction of weak electric field at the head as the shark swims through the earth's geomagnetic field. This can provide a physical basis for an electromagnetic compass sense (From Kalmijn, A.J. 1974. In: *Handbook of Sensory Physiology*. III/3: 147–200).

many previous studies had been that changing the magnetic field generates a transient electrical field, so sharks were simply responding to the electric field and not the magnetic one. Thus, experimental procedures are complicated as they must use measures to remove this possible source of uncertainty. Indeed, with this in mind, scientists working with juvenile Sandbar Sharks were able to quantify associated transient electrical artifacts, showing that any behavioral responses by the sharks were highly unlikely to be attributed to such artifacts. They went on to conclude "that the perception of magnetic fields . . . might involve the electrosensory system and putative magnetoreceptive structures located in the shark's naso-olfactory capsules."[25]

Clearly, there is much to learn about the mechanisms underpinning shark electroreception, and we point the reader to a summary of the late Adrianus Kalmijn's life's work for further discussion, *Theory of Electric and Magnetic Orientation in Sharks and Rays Revisited*.[26]

Hierarchy of Senses

Although the entire suite of senses often integrates with each other to paint a complete sensory picture of the environment, different senses in elasmobranchs come into play at various distances and for specific functions (fig. 5.15).

Brains and Cranial Nerves

The Shark Brain: The Basic Layout

Now that we have surveyed the sensory systems and structures of elasmobranchs, a new question of functional significance arises: when sensory structures such as

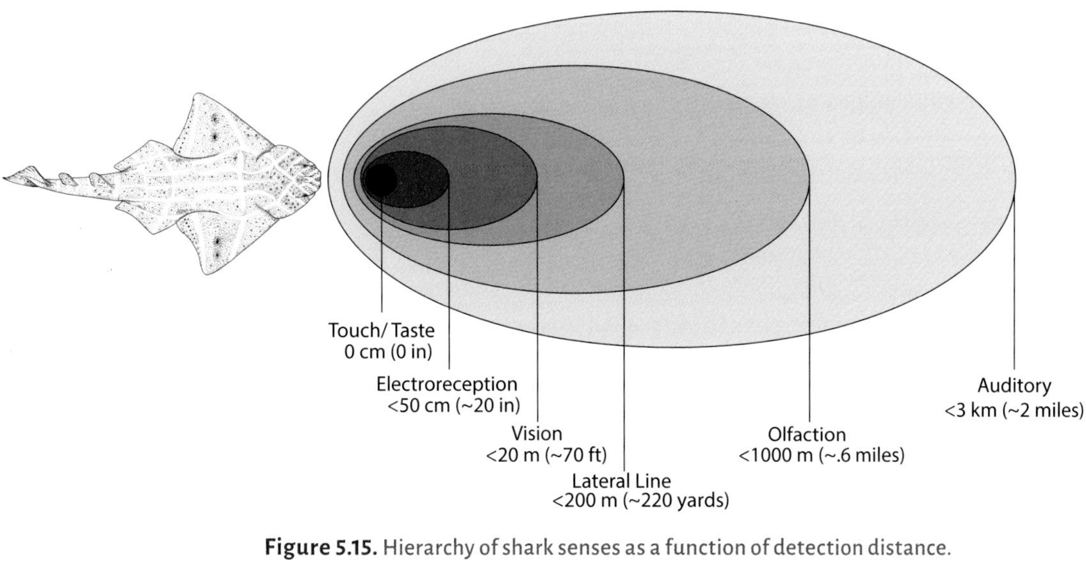

Figure 5.15. Hierarchy of shark senses as a function of detection distance.

the photoreceptors of the eye, the chemoreceptors of tastebuds, or the electroreceptors of ampullary organs are stimulated, how is that information relayed and processed to affect a response? This is a critical question because it links sensory biology with ecology and behavior (or, in terms of this volume, it links this chapter with chapters 9 and 10).[27]

To answer the question, it helps understand the anatomical layout of the elasmobranch brain. Not only is the organization of shark brains relevant because sensory impulses are received and processed in the brain, but for many students destined for medical or veterinary school, shark brains are often the first stop on a survey of vertebrate neurobiology. This is due in large part to sharks' basal position on the gnathostome tree of life. Consequently, a generalized shark brain could be thought of as representing a prototype for the vertebrate brain[28] and serve as a useful reference for understanding the evolutionary origins of our own brains.

A shark's brain sits in its chondrocranium, and, like a human brain, it is divided into five main regions plus the olfactory bulbs.[29] From anterior to posterior (front to back), those regions are the olfactory bulbs and telencephalon, diencephalon, mesencephalon, cerebellum, and medulla oblongata. Figure 5.16 shows the general layout of a shark brain and its regions. At this point, our students usually pose a functional question, "What does each region of the brain do?" It is important to remember that many complex behaviors and physiological feedback loops involve more than one part of the brain. Here, we will provide a broad overview of shark brain functionality and leave more extensive treatments of cell types, nerve pathways, and developmental neuroanatomy to comparative neurobiologists.

The telencephalon of elasmobranchs and other fishes has historically been interpreted as the region of the brain dedicated to smell. This is not inaccurate—it is critical for olfactory processing, and the olfactory bulbs, which are directly connected to the olfactory capsule where the sensory epithelium is, are outgrowths of the telencephalon. However, more recent studies tracing the paths of nerves to and from the telencephalon suggest that this region of the brain is also involved in sensory integration. In addition to olfaction, a shark's telencephalon functions in behavioral modulation as well as spatial awareness and learning.[30, 31] For the comparative anatomists among us, it is worth noting that the telencephalons of modern chondrichthyans show an interesting evolutionary trend. The telencephalons of holocephalans and squalomorph sharks have two bilaterally symmetrical hemispheres in which neurons do not move very far from where they started in the developing embryonic brain. In more derived taxa of galeomorph sharks and batoids, migration of neurons in the developing telencephalon are more complex, as is the region's overall structural complexity.[32] The functional significance of this will become more apparent when we consider ecological trends in each of these groups (see the section on neuroecology).

Behind the telencephalon we find the diencephalon. The diencephalon of sharks is closely associated with relaying sensory input, especially photosensory input, and maintaining internal body conditions (homeostasis). Therefore, it should

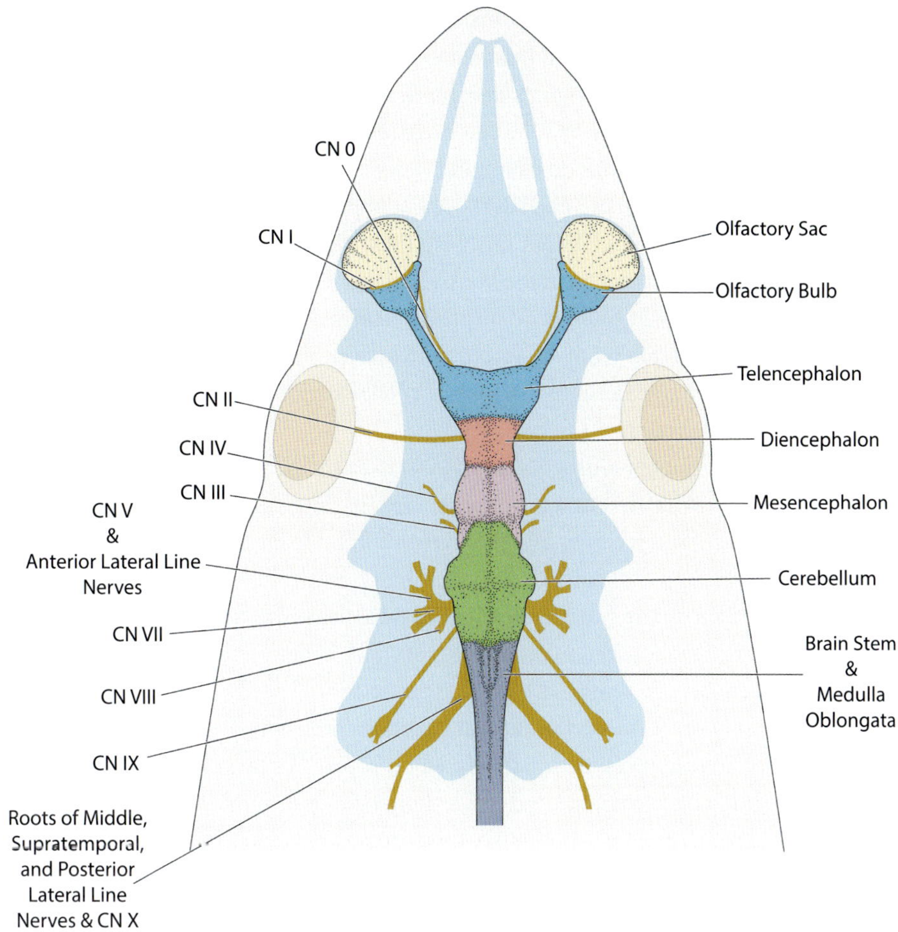

Figure 5.16. Diagram of a generalized elasmobranch brain showing the orientation of five brain regions, which are distinguished by color, and the relative arrangement of cranial nerves as seen from above.

come as no surprise that many structures associated with hormonal regulation such as the pineal gland, hypothalamus, and pituitary gland are part of the diencephalon. Additionally, regions of the elasmobranch diencephalon receive information from more posterior parts of the brain where electrosensory and mechanosensory stimuli are received from sensory structures. In short, the diencephalon may be regarded as a relay station of sorts within the brain for sensory information.[33]

Here it is worth pausing to acknowledge the difference between *receiving* a message from a sensory structure and *processing* that message into something that results in a coordinated physiological or behavioral response. Our next brain region, the mesencephalon, is associated with sharks' ability to process visual stimuli.[34] A particular part of the mesencephalon, the *optic tectum,* allows the shark to process the majority of visual stimuli in relation to its field of view. In other words, when signals from photoreceptors of the retina reach a shark's brain, it is the optic tectum that assigns that stimulus to a region of space relative to the shark.[35] But it is not just visual stimuli that reach the mesencephalon. This region of the brain also receives nervous signals from more posterior regions of the brain, responsible for relaying auditory and mechanosensory stimuli. Other regions of the mesencephalon act as a midbrain command generator, responsible for initiating some coordinated movements, such as those of the paired fins. Other regions still contain dopaminergic nerve fibers, meaning they rely on dopamine, which is a neurotransmitter that is critical in the brain's internal reward system.[36] Given its multifaceted role, it should come as no surprise that elasmobranch mesencephala (plural of mesencephalon) contain varied networks of *efferent* neurons that carry signals away from the mesencephalon and *afferent* neurons carrying information to the mesencephalon.[37]

In elasmobranchs, the remaining brain regions, the cerebellum and medulla oblongata constitute what is known as the *hindbrain*. The cerebellum helps coordinate movement and receives a high degree of input from *proprioceptors*, specialized nerve endings that detect self-movement and action.[38] Although experimental research on the functionality of chondrichthyan cerebella (plural of cerebellum) is scarce compared to other fields of shark biology, most researchers agree that elasmobranch cerebella are key in motor control pathways governing body movements based on environmental factors. In other words, this region of the brain helps control physical actions based on copies of sensory impulses that it gets from other brain regions. In effect, a shark takes information from the environment and modifies its physical movements accordingly, thanks to its cerebellum. A good example of a task in which this ability would be relevant is target tracking during predatory pursuit.

The posterior-most region of a chondrichthyan brain is the medulla oblongata, which narrows into the spinal cord. This region of the brain contains nerves associated with cutaneous (within the skin) sensory structures (e.g., mechanoreceptors that respond to movement, nociceptors that detect pain from tissue damage, and gustatory receptors). Nerves conveying impulses from electroreceptors, the lateral line, and inner ear (which includes vestibular sensory structures that detect equilibrium and acceleration) also connect to the medulla oblongata. Additionally, this region contains motor neurons that help coordinate contraction of facial and branchial muscles (on a human, we would informally call these face and neck muscles).

Another crucial role of the medulla oblongata is filtering out self-generated bioelectrical and mechanical "noise."[39] To put it another way, if a shark can feel the displacement of water made by a prey item or its bioelectrical field, then the shark could also sense its own movement or bioelectric field. This would be problematic when trying to pinpoint that prey item unless the shark could neurologically "filter out" its own sensory signature. This task falls to a shark's cerebellar-like structures.[40] These structures get their name from a morphological similarity to regions of the cerebellum, but make no mistake, they are part of the medulla oblongata. One way to understand the importance of these structures in sharks is to recognize that in many cases, they are a relay station for nervous signals between the spinal cord and the rest of the brain.

Before we conclude our brief tour of shark brains, let us compare shark brains to something a little nearer and dearer to us, the human brain. If you looked at the diagram of a generalized shark brain in Figure 5.16 and thought, "that looks nothing like the bulbous, walnut-shaped organ in my cranial cavity," you are not alone. Many of our students offer similar remarks. Nevertheless, human brains (in fact, all gnathostome brains) can be divided into the same regions as shark brains. Obvious differences in appearance are due in large measure to differences in the relative sizes of the brain regions (our telencephalon is dominated by a large mammalian neocortex responsible for many executive functions, such as cognition and reasoning) and the *flexures* that develop as human embryos grow into fetuses and regions of the brain develop and fold over other regions. These flexures are not observed in sharks or other fishes.

Cranial Nerves

Having a brain capable of receiving and processing sensory input in the form of nervous impulses is important, and there is no question that shark brains represent an evolutionary innovation of early gnathostomes. But without a way to convey information from a sensory structure to the brain, much of the utility would be lost. Likewise, jaws and branchial muscles are also evolutionary innovations, but the brain still needs to coordinate their movements by sending nervous impulses to the corresponding muscles. For vertebrates, including sharks and humans, a particular subset of nerves stemming from the brain facilitate these tasks. These are the cranial nerves.

Each cranial nerve has a name, and most get a number indicated by a Roman numeral. We say "most" because cranial nerves were first numbered in humans. As you know by now, humans lack a lateral line, but sharks do not. That means that the six cranial nerves responsible for relaying mechanosensory information from a shark's lateral line to its brain are identified by abbreviations rather than numerals.[41] For example, the optic nerve is cranial nerve two and is abbreviated CN II, but the anterodorsal lateral line nerve is abbreviated ADLLN. This can be a bit confusing at first, but not to worry—we identified each cranial nerve by name, number or abbreviation, and function in a shark in fig. 5.17 and Table 5.1.[42] In total, chondrichthyans have 17 cranial nerves.

In sharks, as in all vertebrates, some cranial nerves innervate sensory structures (e.g., the olfactory nerve, CN I, carries impulses from the olfactory receptors to the olfactory bulbs that arise from the telencephalon). Other cranial nerves innervate muscles, making them essential in motor pathways of the head and branchial regions (e.g., the oculomotor nerve, CN III, branches out from the mesencephalon and innervates several muscles that control the movement of the eyeball). Still other cranial nerves contain the long, thin axons of both sensory and

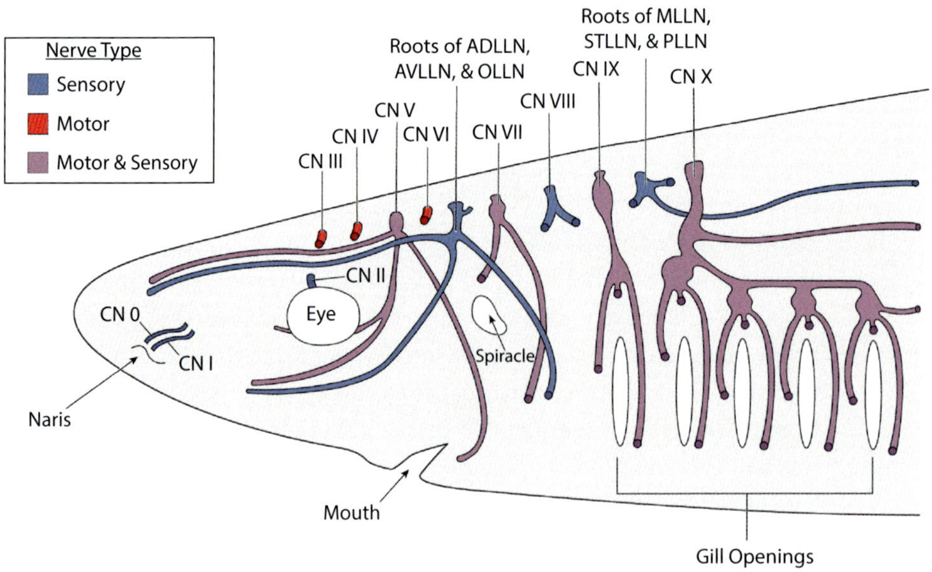

Figure 5.17. Generalized schematic of cranial nerve arrangement in the head of an elasmobranch. Note that this is not illustrated to scale.

motor neurons (e.g., the trigeminal nerve, CN V, emanates from the medulla oblongata and carries tactile sensory information from what we would informally refer to as the shark's face to the brain while also innervating jaw muscles).

Sharks and humans share ten cranial nerves, and although the shape and length of their branches may differ, the functions of those nerves are conserved. In other words, you would not expect to find that a cranial nerve with a sensory function in sharks has a motor function in humans or *vice versa*. As you might expect, the cranial nerves that sharks have and humans lack innervate sensory structures in sharks that we also lack, namely the ampullary electroreceptors and the lateral line. These two sensory structures are very different with respect to their stimuli and the order of use by a hunting shark. But to neuroanatomists tracing the pathways of cranial nerves, the lateral line and electroreceptors are undeniably linked in sharks.[43]

Six cranial nerves innervate sharks' lateral lines and emanate from the medulla of the hindbrain. The posterior-most of these nerves is aptly named the posterior lateral line nerve (PLLN) and receives sensory input from the lateral line mechanoreceptors in the canals running the length of a shark's body, posterior to the brain. In a shark's head, an anterior dorsal lateral line nerve (ADLLN), anterior ventral lateral line nerve (AVLLN), and otic lateral line nerve (OLLN) receive input from the lateral line mechanoreceptors on the top of the snout, bottom of the snout, and around the eyes respectively (fig. 5.17). Much to the relief of anatomy students everywhere, these names are relatively descriptive and informative. But why is there no anterior electroreceptor cranial nerve?

Table 5.1 Cranial nerves found in elasmobranchs.

Nerve	Sensory or Motor	Role
Terminal (0)	Sensory	Transmits signals in chemoreception, likely pheromones
Olfactory (I)	Sensory	Transmit signals from olfactory epithelium
Optic (II)	Sensory	Transmit signals from retina
Oculomotor (III)	Motor	Innervates some extrinsic eye muscles and ciliary muscles of the eye
Trochlear (IV)	Motor	Innervates extrinsic eye muscles, specifically the dorsal oblique
Anterodorsal lateral line nerve (ADLLN)	Sensory	Innervate neuromast mechanoreceptors of the lateral line canals above the eye, beneath the eye, and on the snout and sides of the head
Anteroventral lateral line nerve (AVLLN)	Sensory	
Otic lateral line nerve (OLLN)	Sensory	
Trigeminal (V)	Both	Four branches (ophthalmic, deep, maxillary, mandibular regions) involved in sensation and muscular innervation
Abducens (VI)	Motor	Innervates extrinsic eye muscles, specifically the lateral rectus
Facial (VII)	Both	Innervates facial and jaw regions and hyoid muscles; also transmits signals from taste receptors
Vestibulocochlear (VIII)	Sensory	Transmits signals from inner ear
Middle lateral line nerve (MLLN)	Sensory	Innervate neuromast mechanoreceptors of the lateral canals in the posterior regions of the head and along the trunk of the shark
Supratemporal lateral line nerve (STLLN)	Sensory	
Posterior lateral line nerve (PLLN)	Sensory	
Glossopharyngeal (IX)	Both	Transmits sensory signals and innervates muscles of the pharynx and anterior branchial region
Vagus (X)	Both	Innervates remaining branchial arches and multiple internal organ groups (heart, stomach, intestine, etc.)

As we know from our review of elasmobranch electrosensory organs, each ampullary canal has sensory cells at the base of the canal. When those cells are stimulated by an electrical field, afferent neurons register the detection of an electrical field by the sensory cells and carry the information to the brain. Rather than evolving a completely new cranial nerve dedicated to electroreception, the afferent neurons at the base of sharks' ampullary organs connect to branches of the anterior dorsal lateral line nerves (ADLLN).

As a thought exercise, we encourage readers to take what they have learned about shark sensory biology and trace the sequence of events from detection to neurological transmission and then to processing and integration in one or more regions of the brain. For example, a shark being filmed by a researcher using a camera on a pole may detect the camera's electrical field. Sensory cells in the ampullary organ send a signal to neurons that branch off an anterior lateral line nerve. That cranial nerve sends impulses to the medulla of the brain, which in turn relays the signal to the mesencephalon for further processing and integration with other sensory input, often leading to a behavioral response. In a very real sense, neurobiology is the bridge between sensory biology and behavior.

Neuroecology of Sharks

Before leaving the topic of shark neurobiology and while the sensory biology of sharks is still fresh in our minds, we should consider a field of shark biology that has seen increased attention in recent years: *neuroecology*. Neuroecology examines relationships between the physical structure of the brain and ecological factors, such as habitat and diet.

There are different ways to describe the physical structure of a vertebrate brain, including the number and density of neurons within a given brain region. In general, this sort of data is lacking for elasmobranchs, although strides are being made to fill some of the gaps.[44] Historically, a more common approach to identifying neuroecological correlations in sharks has been to examine the relative size of each region of the brain with respect to mass or surface area. If all the regions of the brain add up to 100% of a shark brain, then what percentage of the brain is accounted for by any one region?

In 1996, researchers Leo Demski and Glenn Northcutt published their examination of the brain and cranial nerves of a White Shark (*Carcharodon carcharias*)[45] and reported that in the specimen they examined, the olfactory bulbs alone accounted for 18% of the overall brain mass, and the remainder of the telencephalon accounted for an additional 28%. Was that exceptional by shark standards? To find out, they compared the relative masses of those regions to other sharks including dogfish in the genus *Squalus* and *Mustelus* as well as hammerheads in the genus *Sphyrna* and requiem sharks in the genus *Carcharhinus*. Across all these sampled taxa, the olfactory bulbs of the White Shark accounted for the highest percentage of brain mass. What might this tell you about the importance of olfaction in the ecology of White Sharks?

Demski and Northcutt took their neuroecological reasoning a step further and noted that the telencephalon of carcharhinids and sphyrnids (requiem sharks and hammerheads) showed a proportionally more developed telencephalon than the White Shark, particularly in the region associated with sensory integration and control of complex behaviors. They inferred that this was due to the more elaborate social and territorial systems of these sharks.

Subsequent neuroecological studies tend to support Demski and Northcutt's inferences and have revealed even more nuances. For example, sharks and rays that live predominantly in coastal reef habitats have more developed telencephala (plural of telencephalon) than those of the deep-sea.[46] This stands to reason given the

BOX 5.1

Research, Conservation, and Education Focus

Advances in our Understanding of the Chondrichthyan Brain

Kara E. Yopak

University of North Carolina Wilmington
Wilmington, NC, US

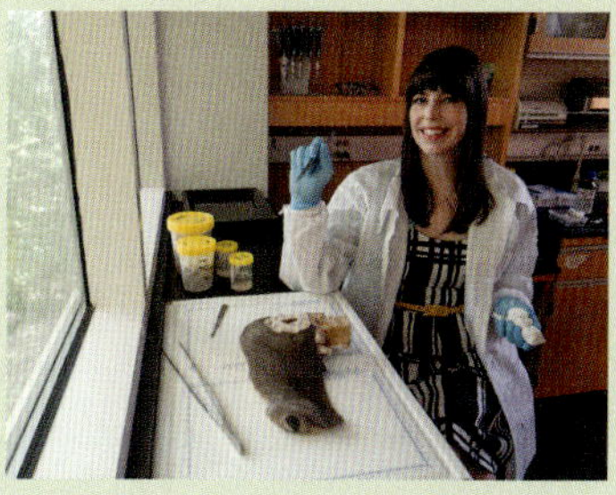

The cartilaginous fishes (Class: Chondrichthyes) represent the appearance of the first complete vertebrate brain "blueprint," comprised of the major brain regions shared across jawed vertebrates.[1] This common brain plan includes the forebrain (olfactory bulb(s), telencephalon, and diencephalon), midbrain (optic tectum and tegmentum), and hindbrain (cerebellum and medulla oblongata).[2,3,4] An increased interest in characterizing inter- (between species) and intraspecific (within a single species) variation in these brain regions (figure a–f) and understanding learning, memory, and cognition, coupled with advanced phylogenetic statistics, has greatly enhanced our ability to map evolutionary trends in this group.

The last two decades have also seen a surge of technological advances, ranging from explorations of *in situ* hybridization and gene expression, brain lesion studies, and applying high-resolution bioimaging to the study of the shark brain. Bioimaging techniques, including computed tomography (CT) (figure h) and magnetic resonance imaging (MRI) (figure i–j), are unique in their ability to acquire high-resolution, digital, 3D data, while leaving the specimen intact. These techniques allow for highly accurate and efficient visualization and "digital dissection" of internal anatomy that have been applied to the brain of both extinct[5] and extant chondrichthyans.[6]

Within the conserved brain plan, there is a high degree of interspecific variation in encephalization (a larger than expected brain for an animal's body size) and brain organization (the relative size and complexity of major brain regions) across cartilaginous fishes (see figure). The most encephalized species, such as Sphyrnids, Carcharhinids, and Myliobatiform rays, are chiefly found in reef-associated or oceanic habitats and show a higher degree of sociality, which may reflect greater cognitive demands.[7,8] There are also correlations between maternal investment and encephalization, where matrotrophic species (where there is additional provisioning to the embryo beyond the yolk sac) have brains that are 20–70% larger than lecithotrophic (yolk-only) species, which suggests a cognitive advantage at birth.[9,10] Although the degree to which encephalization directly confers cognitive capabilities in chondrichthyans is unclear, some studies have shown an empirical link between larger brains and learning, navigation, flexibility, and mate discrimination in bony fishes.[11]

In 1973, Jerison proposed the "Principle of Proper Mass," which asserts that the relative size of a brain region should reflect the relative importance of its function.[12] Accordingly, the relative size of major brain regions is often used as a neuroanatomical proxy for functional specialization. In cartilaginous fishes, correlations among brain organization and phylogenetic, ecological, and life history parameters similarly suggest that interspecific variation in the brain may confer functional capacity and/or the relative importance of different sensory systems.[13] For example, reef-associated cartilaginous fishes have patterns of brain organization that reflect selection for the visual and cognitive demands of spatially complex reef habitats, with relatively large telencephalons; large, highly foliated cerebellums;[14,15] relatively large optic tecta;[16] and relatively small olfactory bulbs.[17] In contrast, deep-sea chondrichthyan brains reflect the requirements for living in deep, low-light habitats with a scarcity of prey and conspecifics, and are characterized by relatively small telencephalons;[18] optic tecta;[19] small, smooth cerebellums;[20] but relatively large olfactory bulbs;[21] and a clear hypertrophy of the termination sites for primary projections from the octavolateralis senses.[22]

Comparatively far less is known about intraspecific variation in the brain in this group. Importantly, cartilaginous fishes (as well as jawless and bony fishes, some amphibians and reptiles) experience indeterminate growth, where the brain (and other organs) continue to grow throughout life. Of the few species studied to date, the brain continues to increase in size as the animal matures, with the steepest period of growth in earlier juvenile stages, which tapers at sexual maturity.[23] Patterns of brain organization also show ontogenetic shifts, with significant differences in the relative size of brain regions, the complexity of key structures (e.g., the cerebellum), and variation in the rate of brain region growth throughout life, potentially reflecting changes in metabolic requirements, locomotor performance, and the relative importance of different senses as they move to new

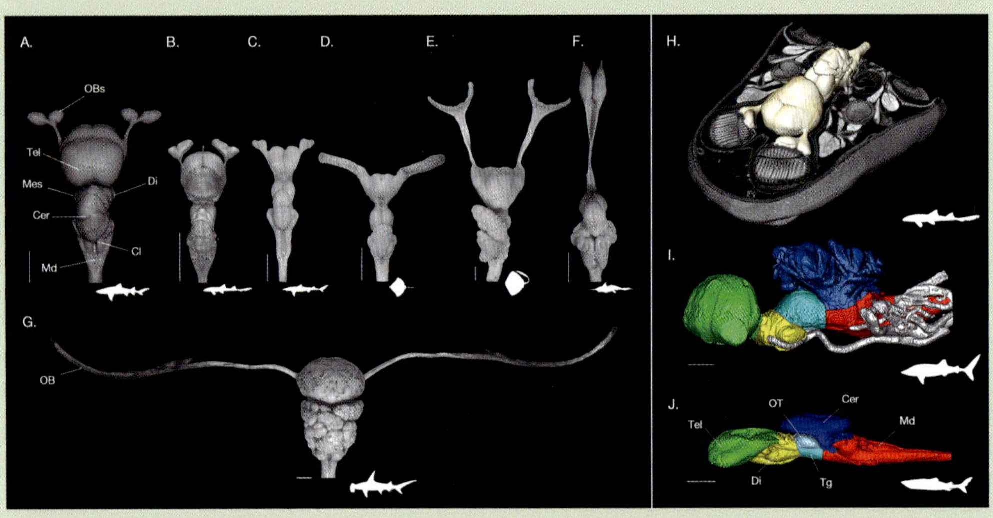

Brains from representative species of cartilaginous fishes to illustrate the diversity in the morphology and size of the major brain regions found in this clade. Dorsal view of the brain in a (A) Lemon Shark (*Negaprion brevirostris*); (B) Brownbanded Bamboo Shark (*Chiloscyllium punctatum*); (C) Spiny Dogfish (*Squalus acanthias*); (D) Argus Skate (*Dipturus polyommata*); (E) Cowtail Stingray (*Pastinachus atrus*); (F) Narrownose Chimaera (*Harriotta raleighana*); and (G) Great Hammerhead (*Sphyrna mokarran*). (Brain images A and C are adapted from Collin *et al.* 2015. Image B is adapted from Camilieri-Asch *et al.* 2020a. Images D, E, and G are adapted from Yopak *et al.* 2015. Image F is adapted from Yopak 2012). (H–J) demonstrate the application of bioimaging to chondrichthyan neurobiology. (H) Image of the brain of the *C. punctatum in situ*, created from x-ray μCT images of the head (Camilieri-Asch *et al.* 2019, 2020a). (I) Lateral digital segmentation of the brain of the Whale Shark (*Rhincodon typus*) and (J) Greenland Shark (*Somniosus microcephalus*), segmented from MR data (I is adapted from Yopak and Frank 2009, and J is adapted from Yopak *et al.* 2019). OBs, olfactory bulbs; Tel, telencephalon (green); Di, diencephalon (yellow); mesencephalon (Mes), comprised of the OT, optic tectum (light blue) and Tg, tegmentum (turquoise); Cer, cerebellum (royal blue); Md, medulla oblongata (red). Note that the Mes and Di are partially obscured from a dorsal perspective, and OBs are absent in F, I, and J. Scale bar = 1 cm. Figures adapted from Yopak, 2022.

ecological niches.[24] Studies have also explored variation in the brain between male and female conspecifics,[25] with some evidence for sexual dimorphism in the brain. More recently, questions have arisen as to the extent to which anthropogenic disturbance may impact normal patterns of brain growth. Environmental changes, including ocean acidification, stress, and elevated water temperature, can potentially lead to changes in learning, brain growth, aerobic potential, and neuro oxidative stress in cartilaginous fishes;[26] however, much more work is required to understand the neural impacts of a changing ocean.

---

1. Striedter, G. F. 2005. *Principles of Brain Evolution*. Sinauer Associates, Inc.

2. Northcutt, R. G. 1978. In: Hodgson, E.S. & Mathewson, R.F. (eds.). Sensory Biology of Sharks, Skates, and Rays. Office of Naval Research. 117–194 pp.

3. Yopak, K. 2022. In: Carrier, J.C. *et al.* (eds.). Biology of sharks and their relatives, 3rd. ed. CRC Press. 105–142 pp.

4. Yopak, K. E. *et al.* 2007. Brain Behav. Evolut. 69: 280–300.

5. Pradel, A. *et al.* 2009. PNAS 106, 5224–5228.

6. E.g., Yopak, K. *et al.* 2019. Sci Rep 9L: 10022. And Peele, E. *et al.* 2023. Fishes 8: 611–631.

7. Yopak, K. 2022. In: Carrier, J.C. *et al.* (eds.). Biology of sharks and their relatives, 3rd. ed. CRC Press. 105–142 pp.

8. Yopak, K. E. 2012. J. Fish Biol. 80: 1968–2023. doi:10.1111/j.1095-8649.2012.03254.x.

9. Mull, C. *et al.* 2020. Am. Nat. 195: 1056–1069, doi:https://doi.org/10.1086/708531 (Accessed 10/2/24).

10. Mull, C. *et al.* 2011. Mar. Fresh. Res, 62: 567–575.

11. E.g., Kotrschal, A. *et al.* 2015. Behav. Ecol, 26: 527–532. And Corral-Lopez, A. *et al*, 2017. Sci. Adv. 3: e1601990.

12. Jerison, H. J. 1973. *Evolution of the Brain and Intelligence*. Academic Press.

13. E.g., Collin, S. *et al.* 2015. In: Shadwick R.E. *et al.* Fish Physiology: Physiology of Elasmobranch Fishes Vol. 34A. Elsevier. 19–99 pp.

14. Yopak, K. E. *et al.* 2007. Brain Behav. Evolut. 69: 280–300.

15. Lisney, T. J. *et al.* 2008. Brain Behav. Evolut. 72: 262–282.

16. Yopak, K. E. and Lisney, T. J. 2012. Brain Behav. Evolut. 80: 108–126.

17. Yopak, K. E. *et al.* 2015. Brain Struc. Func. 220: 1127–1143 doi:10.1007/s00429-014-0705-0.

18. E.g., Yopak, K. E. *et al.* 2007. Brain Behav. Evolut. 69: 280–300.

19. Yopak, K. E. and Lisney, T. J. 2012. Brain Behav. Evolut. 80: 108–126.

20. Yopak, K. E. and Montgomery, J. C. 2008. Brain Behav. Evolut. 71: 287–304.

21. E.g., Yopak, K. *et al.* 2019. Sci. Rep. 9: 10022.

22. Yopak, K. E. and Montgomery, J. C. 2008. Brain Behav. Evolut. 71: 287–304.

23. E.g., Wong, S. *et al.* 2021. J. Fish. Biol. 99: 990–998.

24. E.g., Lisney, T. J. *et al.* 2007. Raffles B. Zool. 14: 7–15.

25. E.g., Puzdrowski, R. L. and Leonard, R. B. 1992. Neurosci. Let. 135: 196–200.

26. E.g., Vila Pouca, C. *et al.* 2019. Behav. Ecol. Sociobiol. 73: 93.

telencephalon's role in spatial learning and the complexity of coral reefs. Coastal pelagic sharks, which includes the White Shark, have larger olfactory bulbs, which are all the better for detecting and following a scent trail where prey may be hiding or just around the bend or beyond the breakers. Open ocean sharks, like the Blue Shark (*Prionace glauca*) and the mako sharks (*Isurus* spp.) tend to have larger than average optic tecta of the mesencephalon. The same is true for many requiem sharks, such as the Grey Reef Shark (*Carcharhinus amblyrhynchos*).[47] In both cases, the well-developed optic tecta suggest that vision plays an important role in how these species perceive their surroundings and interact with other animals. Interested readers can learn more about the neuroecology of sharks by reading *Research, Conservation, and Education Focus* Box 5.1, written by Kara Yopak.

Concluding Comments

The sensory world in which sharks live is rich in information, but at the same time can be confusing. What remains after the organism detects a stimulus is more complicated and interesting: how does it process to effect a response that is appropriate? As you have seen in this chapter, sharks possess an exquisite set of sensors, and they collect and integrate sensory information to make sense of this sensory information, which in large part has contributed to their success!

NOTES

1. Staff and agencies, "Shark kills man during rescue attempt by coastguard in Caribbean," *Guardian*, December 22, 2015, https://www.theguardian.com/environment/2015/dec/22/man-killed-by-shark-in-caribbean-while-being-rescued-by-coast-guard. (Accessed 10/1/24).

2. *Labyrinth* is an anatomical term that refers to structures with communicating cavities or canals.

3. Kritzler, H. and Wood, L. 1961. Science 133: 1480–1482.

4. Nelson, D.R. and Gruber, S.H. 1963. Science 142: 975–977.

5. Chapuis, L. and Collin, S.P. 2022. Rev. Fish. Biol. Fisher. 32(2): 521–554.

6. S Mickle, M.F. and Higgs, D.M. 2022. Mar. Biol.: 169(1), 12.

7. We recognize that hand-feeding wild animals is a controversial activity, and there are good reasons to oppose the activity. We consider this a valid educational activity, and it is done infrequently such that the likelihood of long-term impact is low.

8. Theisen, B. et al. 1986. Acta Zool. 67: 73–86.

9. Schluessel, V. et al. 2008. J. Morphol. 269: 1365–1386.

10. Northcutt, R.G. 1978. In: Hodgson, E.S. and Mathewson, R.F. (eds.). Sensory biology of sharks, skates, and rays. U.S. Office of Naval Research. 117–193 pp.

11. Demski, L.S. and Northcutt, R.G. 1996. In: Klimley, A.P. and Ainley, D.G. (eds.). Great white sharks. Academic Press. 121–130 pp.

12. Parker, G.H. 1914. Bull. U.S. Bur. Fish. 33: 61–68.

13. Tricas, T.C. et al. 2009. J. Comp. Physiol. A 195: 947–954.

14. Atkinson, C.J. and Collin, S.P. 2012. Biol. Bull. 222: 26–34.

15. Atkinson, C.J. et al. 2016. Biol. Open, bio-022327.

16. Collin, S.P. et al. 2015. In: Fish Physiology 34. Academic Press. 19–99 pp.

17. Atkinson, C., "The gustatory system of elasmobranchs: morphology, distribution and development of oral papillae and oral denticles," PhD diss., (University of Queensland, 2011).

18. Lisney, T.J. and Collin, S.P. 2007. Brain Behav. Evolut. 69: 266–279.

19. McComb, D.M. and Kajiura, S.M. 2008. J. Exp. Biol. 211: 482–490.

20. McComb, D.M. et al. 2009. J. Exp. Biol. 212: 4010–4018.

21. Newton, K. C. et al.2019). J. Fish Biol. 95(1): 135–154.

22. Crawford, L. M. et al. 2024. Environ. Biol. Fish. 107(3): 257–273.

23. Ripley, D. M. et al. 2021. Conserv. Physiol. 9, coab045: https://doi.org/10.1093/conphys/coab045. (Accessed 09/07/2024).

24. Tricas, T.C. et al. 1995. Neurosci. Let. 202(1–2): 129–132.

25. Crawford, L. M. et al. 2024. Environ. Biol. Fish. 107(3): 257–273.

26. Kalmijn, A.J. 2024. *Theory of electric and magnetic orientation in sharks and rays revisited; physical principles, biological evidence, and rebuttal of misconceptions.* Springer.

27. In a practically minded if slightly dated chapter, a neurobiological approach to the question "How will sharks react to the presumably adverse conditions created by the user of a protective device?" exemplifies another benefit to understanding elasmobranch brains and sensory structures. Aronson, L. 1963. In: Gilbert, P.W. Sharks and survival. D. C Heath and Company. 165–241 pp.

28. A phrasing respectfully borrowed from Liem et al. 2001. *Functional anatomy of the vertebrates: An evolutionary perspective,* 3rd Edition. Harcourt College Publishers.

29. Note that many texts regard the olfactory bulbs as the anteriormost region of the telencephalon because they develop as outgrowths of that brain region (e.g., Striedter, G.F. and Northcutt, R.G. 2020. *Brains through time: A natural history of vertebrates.* Oxford University Press.) Others recognize the highly specialized nature and position of olfactory bulbs and regard them as their own brain region (e.g., Yopak, K. 2022. In: Carrier et al. Biology of sharks and their Relatives, 3rd Edition. CRC Press. 105–142 pp.).

30. Fuss et al. 2014. J. Comp. Physiol. 200: 37–52.

31. Ari, C. 2011. Open Anat J 3:1–13.

32. Northcutt, R. G. In: Hodgson, E.S. and Mathewson, R.F. Sensory biology of sharks, skates, and rays. Office of Naval Research, Department of the Navy. 117–194 pp.

33. Yopak, K. 2022. In: Carrier et al. Biology of sharks and their relatives, 3rd Edition. CRC Press. 105–142 pp.

34. Notice that here we deliberately reference vision and not photoreception. Vision, as enabled by the eyes, is a type of photoreception, but so is the light-dark distinction done by the pineal gland. The difference is that the pineal gland does not form an image the way the eyes do.

35. Yopak, K.E. and Lisney, T.J. 2012. Brain Behav. Evolut. 80 :108–126.

36. Quintana-Urzainqui et al. 2012. Brain Behav. Evolut. 80:127–141.

37. A helpful rule of thumb for neurobiology: the central nervous system (brain and spinal cord) is often the point of reference. Efferent and afferent are usually used in relation to the CNS. *E*fferent nerves *e*xit the CNS, and *a*fferent nerves *a*rrive at the CNS. One carries impulses away from the brain, and the other carries impulses to the brain.

38. An interesting sidenote: among extant, or living, vertebrates, chondrichthyans represent the oldest lineage of gnathostomes (jawed vertebrates), and the presence of a fully formed cerebellum appears to have occurred around the same time that jaws and more robust paired fins evolved. This makes sense because as gnathostomes evolved to occupy higher trophic levels and actively hunt, pursue, and consume prey, coordination of body movements would have been essential.

39. Montgomery et al. 2012. Brain Behav. Evolut. 80:152–165.

40. Striedter G. F. and Northcutt, R. G. 2020. *Brains through time.* Oxford University Press.

41. While sharks and other fishes retain lateral line nerves that we secondarily lost as mammals, amniotes (represented by reptiles and mammals) have two cranial nerves that do not develop in sharks: the eleventh cranial nerve (the accessory nerve) and the twelfth cranial nerve (the hypoglossal nerve). In mammals, these nerves innervate muscles of the neck and a muscular tongue respectively, two structures that sharks lack.

42. Notice that cranial nerve numbering starts at zero, and zero is not a Roman numeral. The reason for this numerical oddity is that cranial nerves are numbered based on the order in which they emanate from the brain. Working in an anterior to posterior direction, the olfactory nerve was designated CN I. The problem is that after the olfactory nerve was identified, a small cranial nerve, the terminal nerve, was found to be widespread in vertebrates and slightly anterior to the olfactory nerve. Thus, the terminal nerve got the number zero to reflect that it is anterior to the olfactory nerve.

43. Northcutt, R.G. 1990. Brain Behav. Evolut. 36:116–140.

44. Yopak et al. 2014. Brain Behav. Evolut. 84:74.

45. Demski, L.S. and Northcutt, R.G. 1996. In: Klimley, A.P. and Ainley, D.G. Great white sharks: the biology of *Carcharodon carcharias.* Academic Press. 121–130 pp.

46. Collin et al. 2015. In: Shadwick et al. Fish physiology: physiology of elasmobranch fishes, Vol 34A. Elsevier. 19–99 pp.

47. Yopak, K.E. and Lisney, T.J. 2012. Brain Behav. Evolut. 80: 108–126.

6 / Reproduction

CHAPTER OUTLINE

Introduction: The Unparalleled Diversity of Chondrichthyan Reproduction[1]

Shark sex. It does not get any better than that, especially if the amazing array of reproductive styles and strategies piques your biological interest.

As a whole, the diversity of specializations for reproduction in this group is nothing short of astonishing, especially in the ways in which developing embryos are nourished. Some sharks lay eggs, while others are born live, but the myriad ways in which these are accomplished will amaze you. Sharks vary from most bony fishes in that all fertilization is internal. Instead of producing thousands to millions of eggs like many bony fishes, few of which have a chance to hatch and survive until adulthood, sharks produce a small number of larger eggs, and the resulting offspring are born or hatched at a larger size that gives them a greater likelihood of surviving. Fecundity, the number of offspring per female pregnancy, ranges from 1 to >300 in this group. And, unlike some species of bony fishes, there is no parental care after birth.

Sexual Dimorphism

Sexual dimorphism denotes observable differences between males and females of a species. *Secondary sexual characteristics* refer to those beyond external sex organs and may include size, color, morphological adaptations, or even behavior.

Sexual dimorphism usually comes into play in courtship and reproduction. Charles Darwin recognized this phenomenon and called it *sexual selection.* If natural selection can be summarized as *survival of the fittest,* then sexual selection can be thought of as *survival of the sexiest.* This means that in some species one sex, typically the females, selects the male with whom she will mate based on some sexually dimorphic characteristic that stands out to her.

Among chondrichthyans, the most obvious difference between males and females is the presence of claspers (see fig. 1.5) on the males, which is a *primary* sexual characteristic. Claspers, also called *mixopterygia,* are the intromittent organs, which is the general term describing the structures responsible for sperm transfer for organisms using internal fertilization.

Claspers are the rearward tubular extensions of the inner margin of the pelvic fins. Claspers of immature elasmobranchs are pliable and soft, but they calcify and harden during maturity. A single clasper is inserted into the cloaca (the common urogenital and anal opening) of the female (fig. 6.1), followed by insemination.

Once the clasper is inserted into the female, its tip splays out and, depending on the species, features a series of structures (spines, hooks, flaps, and/or spurs),

Facing page

A

Figure 6.1. (A) Photo of clasper inserted into cloaca of a female Nurse Shark (Courtesy of Jeff Carrier). (B) Graphic illustrating the sequence of events (from top to bottom) occurring during courtship and mating in Angel Sharks, culminating in the insertion of a clasper and insemination. Lowermost image depicts a pregnant female (Marc Dando/Save Our Seas Magazine).

Figure 6.2. Sexual dimorphism in Green Lantern Shark. Which one do you think is the female?

which hold the clasper in place as mating occurs. See the section on Copulation in this chapter for how the claspers transfer sperm.

In sharks and batoids, the second most obvious example of sexual dimorphism is a difference in size. In viviparous species, those that carry their developing embryos inside their body and give birth to live young (about half of all chondrichthyan species), as opposed to laying eggs, the females are larger (fig. 6.2). This makes sense. During pregnancy, irrespective of whether a female gives birth to a single neonate per uterus, as in the case of the False Catshark (*Pseudotriakis microdon*), or 40–60 young, as occurs in the Tiger Shark (*Galeocerdo cuvier*), the embryos take up an increasing amount of space as they grow. In some species, there is considerable maternal input of calories and nutrients into the developing embryos (matrotrophy), but even in species with no post-fertilization maternal input, to accommodate the developing embryos, the females of live-bearing species are substantially larger than the males.

Among egg-laying chondrichthyans, there is no size dimorphism, since eggs are typically deposited early in the development of the embryos and thus do not require the addition of space or maternal resources beyond those to the point of deposition.

Another sexual dimorphism, though it is not externally visible, is the evolution in females of thicker skin. This is a strictly defensive adaptation, evolved in response to bites from males during courtship and copulation in some species (fig. 6.3). During mating season, females of numerous species can be found with mating scars on their pectoral fins, head, and body, ranging from minor abrasions to severe wounding. Some of these bites would be capable of inflicting serious damage were it not for this thicker skin, which is often just deeper than the size of the teeth of the male, thus protecting the internal organs from potentially lethal puncture wounds or exposure.

Male rajiforms (skates), and some myliobatiforms (stingrays), exhibit a sexually dimorphic characteristic also related to biting females (fig. 6.4). In these species,

Figure 6.3. Mating scars (male courtship bites) on the flanks of a Large Bull Shark (Courtesy of Joel Blessing).

A B

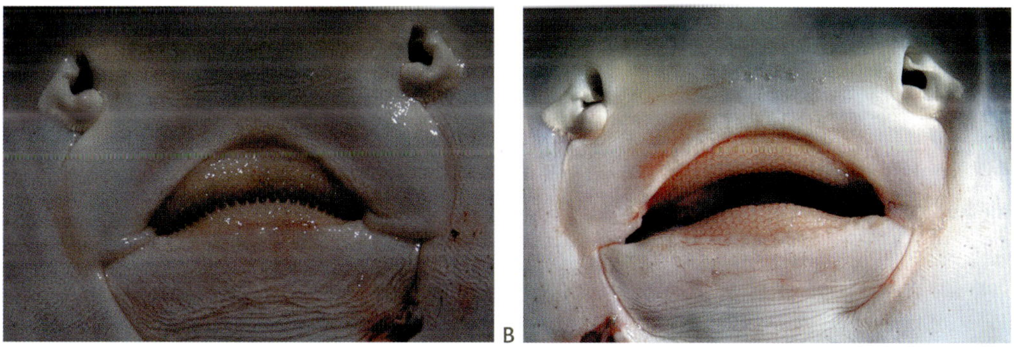

Figure 6.4. Examples of sexual dimorphism in the Clearnose Skate. The teeth of the female (A) are specialized for feeding, whereas those of the male (B) are for both feeding and grasping the female.

female teeth are designed exclusively for feeding, where male teeth are strongly cusped and are used for grasping onto the female in addition to feeding. These teeth are lost after mating season and are replaced with non-cusped teeth.

In female rajiforms, there are rows of large, pointed, sharp spines, or thorns, on the dorsal and lateral surfaces of the tail. These spines, which are enlarged dermal denticles, are also present in males, but are much smaller. Females use them to discourage mating attempts by some males, since mating events may entail pursuit by several males at a time.

Not to be outdone, male rajiforms also have a sexually dimorphic characteristic, a series of modified dermal denticles called *malar* and *alar* spines on the cheeks and pectoral fins, respectively. These spines are used to anchor the male to the female during mating. Alar and malar spines are deciduous and drop off following mating season.

Internal Fertilization

All chondrichthyans use internal fertilization, whereas <2% of bony fish species do. Internal fertilization results in offspring that are not larva or postlarva, that is, born at a very small, often microscopic size, like bony fishes, rather are basically miniature versions of adults, poised to fend for themselves from birth. Having a higher survival probability is central to the chondrichthyan way of life, since there is no parental care.

The evolutionary success of chondrichthyans does not come without costs. This mode of reproduction involves a high investment of resources by the female. Because of these resources, which include nourishment for the developing embryos and physical space for them to develop, fecundity (the number of offspring) is lower than in bony fishes.

Copulation

Males of all chondrichthyan fishes have two claspers. Only one of the claspers, however, is inserted into the female at any time. Why, then, are there two? The answer lies in how courtship proceeds, how the clasper is inserted into the female, and the competition among males (fig. 6.5). Before a clasper is inserted, flexor muscles cause the clasper to flex and rotate 90° or more across the body. The clasper on the shark's left side is thus only useful if the male is on the left of the female, and vice versa. Thus, if multiple males are attempting to copulate with a single female, or if the female only allows access from one side, the male has access.

During copulation, the orientation of the male and female varies with the species. In many kinds of sharks, the male or males will swim alongside the female and attempt to insert one of the two claspers while biting onto the pectoral fin or flank. In the family Scyliorhinidae (catsharks), the male grabs the flank or even the first dorsal fin of the female and envelops her entire body, surrounding her like a donut. Some batoids, like Cownose Rays (*Rhinoptera bonasus*), often copulate belly-to-belly while in others the male is situated on the female's back. Southern Stingrays (*Hypanus americanus*) employ both methods of copulation.

Transfer of sperm from the male to the female in chondrichthyans occurs via a groove in the claspers. When the clasper bends and folds across the body, the tip splays open and anchors itself into the cloaca, a method so effective that females must often shake vigorously to disgorge the clasper.

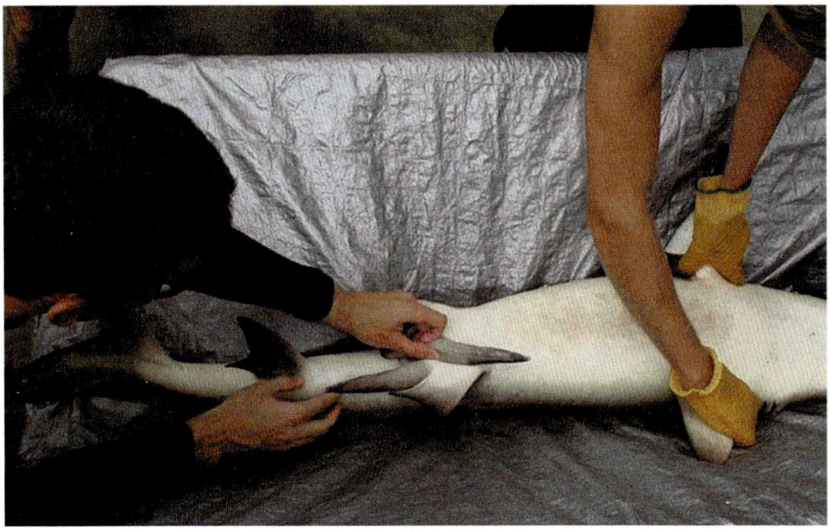

Figure 6.5. The right clasper of a Whitetip Reef Shark flexed as if it were being inserted into a female on its left (Courtesy of Nick Whitney).

The shape and size of claspers and the mechanism of anchoring vary by species. In squaliform sharks, the tips of the splayed claspers are very sharp hooks used for anchoring. In skates, there are no well-developed hooks but rather the splayed open clasper opens like a webbed hand, anchoring the clasper inside the cloaca. The cow sharks, a primitive group, have a simple clasper with two cartilages that are inserted with almost no protrusions for anchoring. These sharks are also unusual in that the clasper sits in a sheath analogous to the foreskin of the human male.

Reproductive Anatomy

What follows is an overview of what we have called the *reproductive plumbing*, the organs of reproduction, their location in the body, and their function. Like other systems in sharks and their relatives, there is some variation. The reproductive tracts of egg-layers, for example, differ from that of live-bearers, as you might have already concluded. But some aspects of the process are conserved irrespective of the species. Ova produced by the ovaries in the process of oogenesis are fertilized by sperm made in the testes by the process of spermatogenesis. The developing embryo receives its initial nutrition from yolk deposited in the egg. After the yolk is absorbed, the variation (see pp. 255–270), begins.

All chondrichthyans have two entire reproductive tracts (i.e., two male or two female). In some species (e.g., sharks in the genus *Carcharhinus*), both tracts are simultaneously functional, and a single copulation event, or sequential fertilization by two males, can fertilize ova in one or both tracts. In other species, only one of the two ovaries is functional, and it supplies ova to both reproductive tracts (i.e., to two uteri). In the Cownose Ray (*Rhinoptera bonasus*) and the Little Gulper Shark (*Centrophorus uyato*), species with a fecundity of one, only a single uterus is functional.

Anatomy of the Male Reproductive Tract

The male reproductive system of elasmobranchs consists of testes, a series of genital ducts where the sperm mature as they move through the system, a convoluted ductus deferens, a urogenital papilla, siphon sacs, and claspers (fig. 6.6).

The organ of sperm production is the testis, and males have paired testes in the anterior portion of the body cavity attached to the dorsal body wall at the anterior end by a thin tissue layer called the *mesorchium* (fig. 6.6). The testes of

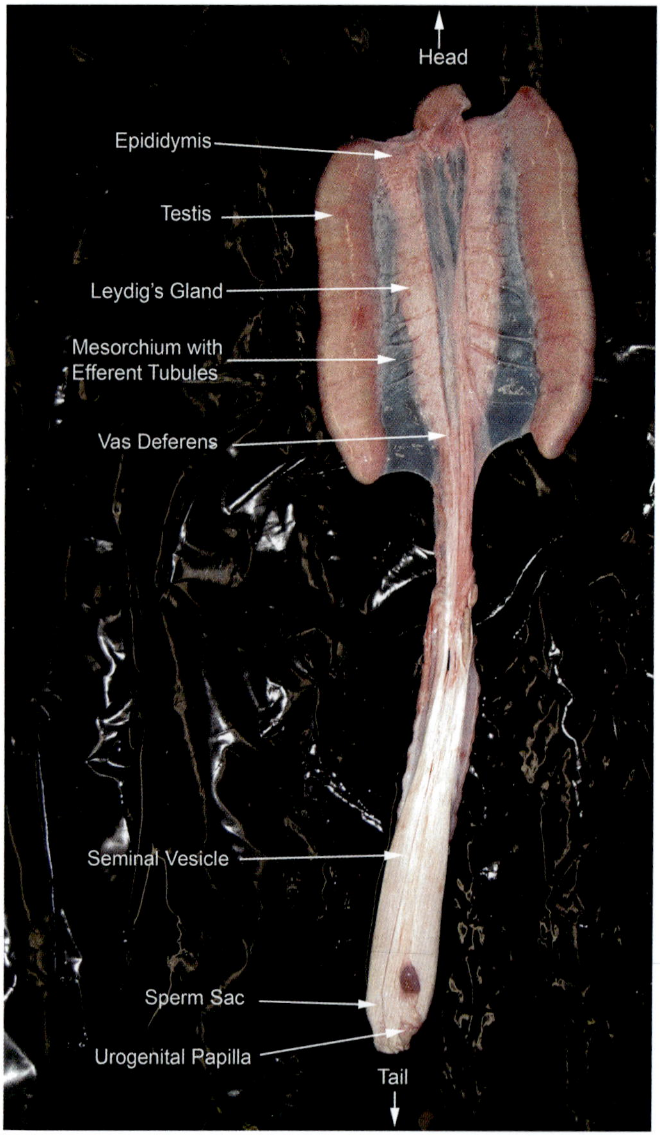

Figure 6.6. Labeled photo of the male reproductive tract from the Atlantic Sixgill Shark.

elasmobranchs, but not chimaeras, are fully or partially encased in the *epigonal organ*, a structure unique to sharks and rays which is responsible for hematopoiesis (production of both red blood cells and those of the immune response system).

Testes in chondrichthyan fishes are relatively simple and are similar to those of the primitive bony fishes like sturgeon and gar. Chondrichthyans have three basic types of testes,[2] which differ in where the sperm cells form and mature. The most common form of testes in elasmobranchs are *diametric* (located at the diameter). In diametric testes, which are found in all squalomorph and galeomorph sharks (excluding lamniforms), developing sperm are located along a strip of the testes, and these sperm go through two phases to mature as they migrate across each testis. The other types are *radial* (in lamniform sharks) and *compound* (in many batoids).

Within each testis are microscopic, spherical structures known as *spermatocysts*. Within each spermatocyst are immature little hairlike germ cells called *spermatids*, which are destined to become sperm or, more accurately, *spermatozoa*. The spermatocysts also contain clumps of Sertoli cells, which produce hormones that help in sperm maturation. Once mature, chondrichthyan sperm (fig. 6.7) look like wiggly hairs (due to their helical head), have a flagellated tail, are long (> 30 μm), and lack the expanded head characteristic of mammalian sperm.

Once sperm have neared maturity in the testes, they swim through a series of efferent ducts (also called *ductus efferens*; *effere* = to carry out) in the thin mesorchium into the next genital duct, the paired *epididymi* (fig. 6.6), highly convoluted structures just ventral to the anterior portion of the kidneys, which are flattened, ribbon-like, and embedded in the dorsal body wall. Each epididymis is lined with secretory cells whose discharges promote sperm maturation and vitality.

Posteriorly, the epididymis leads to another sperm storage organ, the *ductus* (or *vas*) *deferens*. At this stage in mature male sharks, the anterior portion of the kidney has been modified into an organ known as Leydig's Gland, which produces

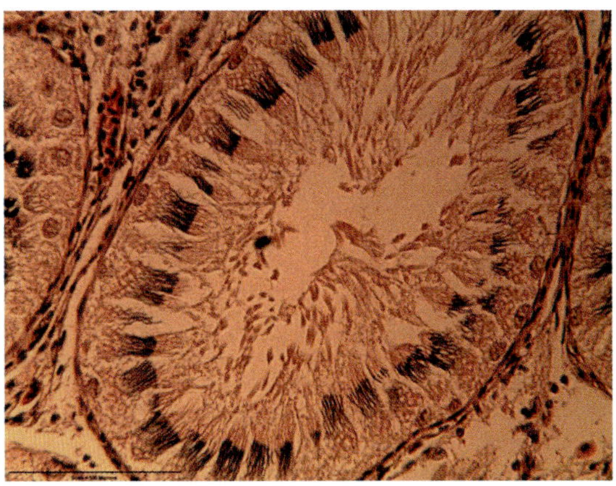

Figure 6.7. Mature sperm (dark lines) from Finetooth Shark (Courtesy of Amanda Brown).

hormones that it secretes into the ductus deferens to nourish the sperm. The ductus deferens also has specialized secretory cells as well as ciliated cells. Their job is to continue to promote sperm viability and to create currents that move the sperm along.

Sperm make their way down the ductus deferens to a *seminal vesicle,* where they are stored until mating begins. At that point, a sphincter muscle relaxes, and the sperm move into the *sperm sac,* which is also known as the *alkaline gland.* This gland is packed with secretory cells that create an alkaline (high pH, from 8.9 to 9.2) environment inside the sac by releasing carbon dioxide and hydroxide ions. What is the reason that the sperm are bathed in a non-neutral solution? The answer is twofold. First, the next structure in the system is a single urogenital papilla, into which the sperm sac empties as does the urinary system. Thus, the sperm are exposed to the acid characteristic of urine, which could damage or destroy them were they not protected by the alkaline contents of the sperm sac buffering the urine's acid. Secondly, research has shown that the alkaline gland contains proteins that function at high pH and increase sperm motility, which becomes critical to their success at fertilizing ova.

The next step in the copulation process is to get the sperm from the sperm sac into the clasper, which is not connected to it. This is where the rotation of the claspers comes into play. Each clasper has a groove, and this groove on the clasper contacts the urogenital papilla in the cloaca.

There still is one major hurdle to overcome: how to flush the sperm from the sperm sac to the female, since there is no mechanism for ejaculation within the male reproductive tract. To solve this problem, sharks and batoids (but not chimaeras) have another trick, use of a structure called the *siphon sac* (fig. 6.8). The siphon sac is a dead space on the ventral surface of male sharks running almost the entire length of the abdominal space, which translates into about one third of the body length, between the skin and the underlying muscles. This sub-

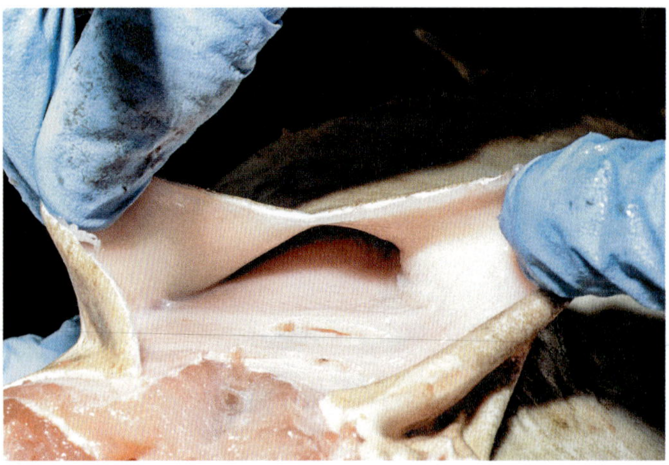

Figure 6.8. View from the rear of exposed siphon sac from a male Blacknose Shark on its back (Photo by Joshua C. Bruni).

cutaneous space has only a single opening, a pore located near the urogenital papilla. Prior to copulation, using repeated clasper flexing as a pump, seawater is sucked into the space.

In some of the species whose siphon sac has been examined, there are cells that produce and secrete the chemical 5-hydroxytryptamine (5-HT). This chemical has been shown to stimulate the uteri to contract repeatedly, which facilitates transport of the sperm to the ova. Then during copulation, the male bends, compressing the siphon sacs and forcing the contents through the pore, through the cloaca, and into an opening on the now-erect clasper (now rotated across the body) called the *apopyle.* Simultaneously, sperm pass through the urogenital papilla, and into the apopyle, where they are entrained by the seawater from the siphon sac and flushed into the female reproductive tract. This method of intromission is unique among vertebrates.

Anatomy of the Female Reproductive Tract

The female reproductive system consists of ovaries, genital ducts called *oviducts* (which include oviducal glands and uteri),[3] a vagina, and cloaca (fig. 6.9).

The organs that produce the ova, or egg cells, in females in the process of oogenesis, as well as hormones that control their development, are the ovaries (fig. 6.9). Females have paired ovaries in the anterior of the body cavity, although in some galeomorph genera (e.g., *Prionace, Scyliorhinus, Pristiophorus, Carcharhinus, Galeus, Mustelus,* and *Sphyrna)* the left ovary atrophies and only the right ovary is functional. In some myliobatiform batoids (e.g., *Urolophus*), the right ovary becomes vestigial, and the left ovary is functional. In some species (e.g., *Prionace glauca*), the ovaries are embedded in the epigonal organ. In immature species, the ovaries are reduced to a small strip, whereas in mature females they are large and globular, with bright yellow follicles containing the ova, on the surface. Like the testes, ovaries in elasmobranchs are similar to those in primitive modern bony fishes.

There are two types of ovaries. Lamniform sharks have a single, *internal* ovary (the right one). *Internal* refers to the fact that this ovary lies within the epigonal organ. In non-lamniform sharks, the ovary (either the single functional one, or both) is *external* (sometimes referred to as *naked,* or *gymnovarium* type), either on the surface of the epigonal organ or attached to the body wall by a mesovarium, a thin mesentery that connects the ovaries to the dorsal wall of the body cavity. The size and number of ova vary between these two types, with the former producing many small ova (3–5 mm or 0.1–0.2 in diameter) whereas external ovaries produce fewer but larger ova (often over 25 mm or 1 in but as big as 100 mm or 4 in diameter). In both types of ovaries, the developing ova (known as oocytes) are *vitellogenic* (full of yolk). The differences in size and number of ova of each ovary type are important to the mode of embryonic nutrition employed by each group (see pp. 255–270).

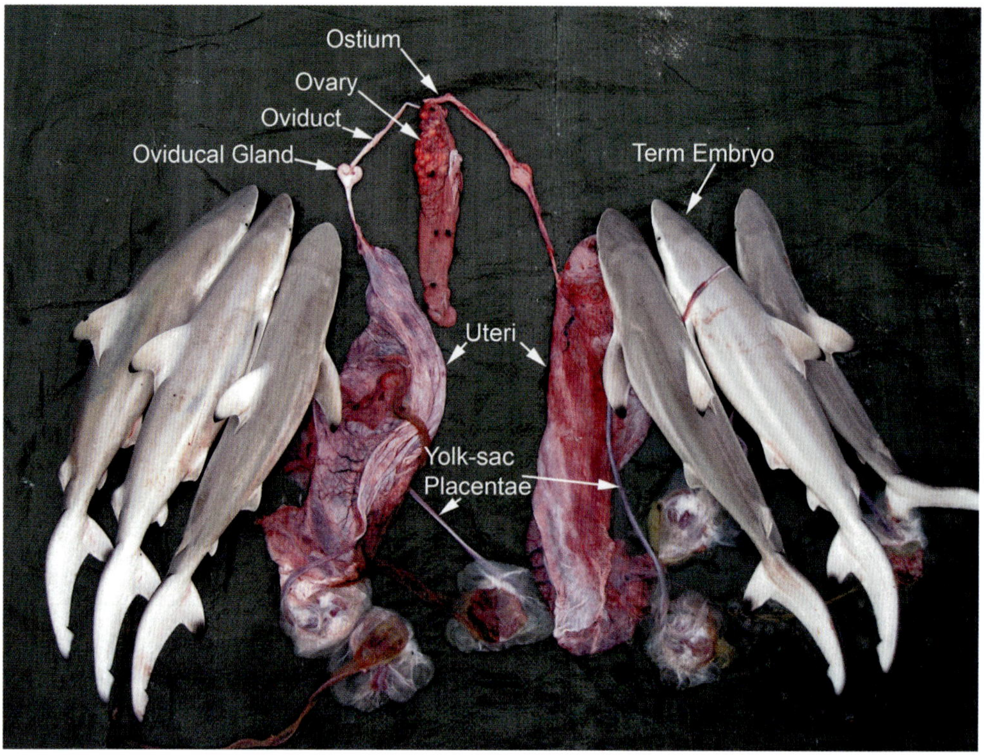

A

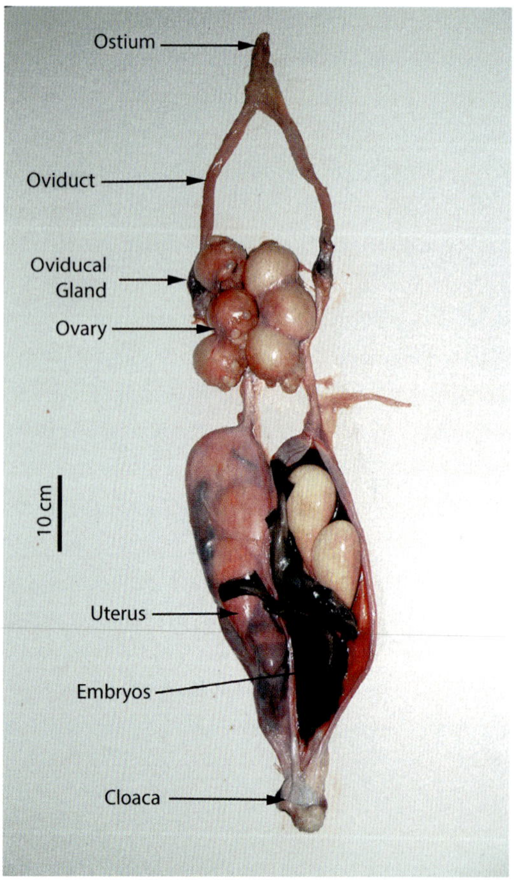

B

Figure 6.9. Labeled photos of the female reproductive tracts from (A) Blacktip and (B) Gulper Sharks.

The ova produced by each ovary next move into the *oviduct* (also called a *Müllerian tube*), through the *mesovarium*, a two-layered tissue resembling plastic wrap that suspends the ovaries in the body cavity, similar to the male mesorchium.

The oviduct varies among species, but in all cases is essentially a long tube with some differentiation along its length beginning with a funnel called the *ostium* (or *falciform ligament*), an oviducal gland (also known as a *nidamental* or *shell gland*), a connecting segment (the *isthmus*), and the *uterus*. These components function to collect and transport the ova, receive and store sperm, promote fertilization, form a membrane or egg case (also known as a *mermaid's purse*) around fertilized ova, and protect and nourish, to varying degrees, the developing embryo. Primitive bony fishes like sturgeon have a similar architecture, whereas in advanced bony fishes, the ova are shed straight into the environment.

Single ova are shed from the ovary through the ostium into the oviduct and into the oviducal gland. In egg-laying (*oviparous*) species, the oviducal gland is well-developed, and it is in this location that the eggshell is developed around the fertilized ovum. In those species that bear live young, the oviducal gland is relatively small, and typically deposits only a membrane. The oviducal gland is also the site of sperm storage and fertilization. Sperm swim up the uterus and oviduct into the oviducal gland where they are stored and maintained viable for periods that have been documented to be seven years or longer. The phenomenon of sperm storage in elasmobranchs is not fully understood. It has been observed in numerous species, but apparently not in lamniforms. Biologist John Morrissey provided fertilized ova (viable embryos) to researchers for over seven years from female Chain Catsharks (*Scyliorhinus retifer*) that were caught in the wild and maintained in captivity, but which had never been in captivity with males. Multiple parentage, litters sired by more than one father, is known from many elasmobranchs as well, and this may also involve sperm storage (see p. 247) In some species (e.g., Lemon Sharks, Blacknose Sharks, Small-spotted Catsharks, Thornback Skates), all or nearly all litters are the result of eggs fertilized by multiple fathers. The record is from a Bluntnose Sixgill Shark in which a single litter of 71 pups was sired by up to nine fathers! In contrast, most litters of some other species (e.g., Bonnethead, Gummy Sharks, Spiny Dogfish) are sired by a single father.

The fertilized egg, now called an embryo, moves into the isthmus of the oviduct and then to the uterus, primarily through contractions of smooth muscle. In oviparous (egg-laying) species, the uterus is not well-developed and serves mainly to convey the egg to the cloaca and then to the environment. In contrast, the uterus of viviparous (live-bearing) species is expanded and houses the embryo while it develops. Additional roles of the uterus in the nutrition of developing embryos are discussed beginning on p. 260.

Figure 6.10 shows a reproductive tract of the Atlantic Sharpnose Shark (*Rhizoprionodon terraenovae*) after fertilization.

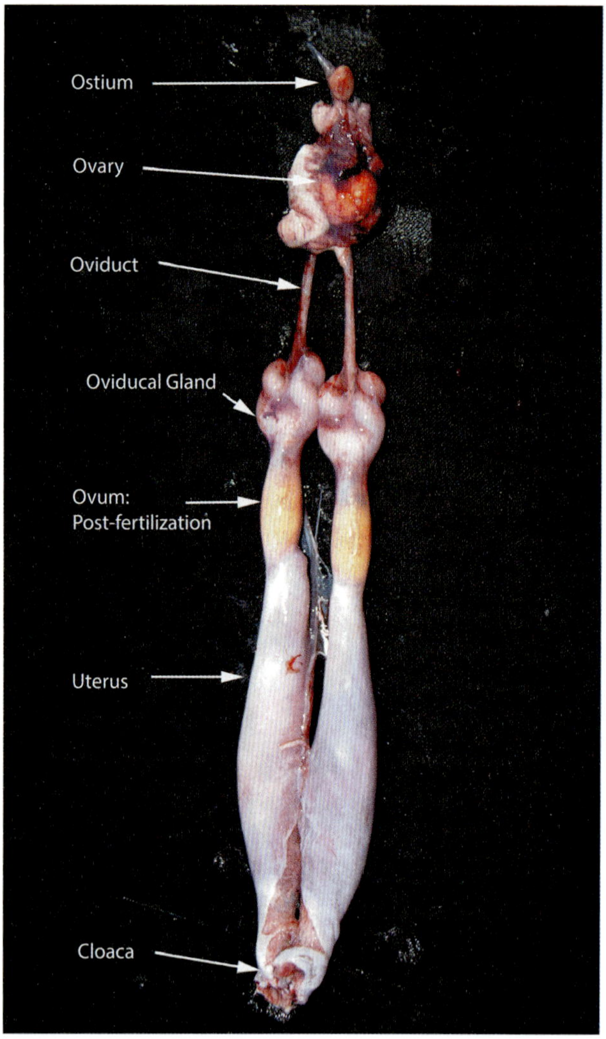

Figure 6.10. Labeled photo of the reproductive tract of a female Atlantic Sharpnose Shark showing ova exiting the oviducal glands after being fertilized.

Reproductive Cycles

Sharks and their relatives exhibit three main reproductive cycles: *continuous, seasonal,* or *punctuated* breeding.[4]

Continuous breeding involves year-round reproduction in which females are pregnant for up to a year or more, with vitellogenesis (development of ova) occurring during pregnancy in preparation for mating and subsequent fertilization. Females typically mate soon after parturition (birth), with no resting phase. Continuous reproduction is characteristic of many live-bearing sharks where there is limited to no matrotrophy (maternal input); for example, deep-sea squaliforms that are continuously pregnant, and some batoids.

In *seasonal* breeders, reproduction is restricted to certain times of the year (typically spring and summer) with shorter pregnancies and annual reproduction frequency. Seasonal breeding occurs in most myliobatiform rays, egg-laying skates, and some live-bearing sharks.

The final category is *punctuated* breeding, in which females are pregnant often for a year or more, then are dormant for an intervening period before becoming pregnant again. In other words, they experience a two-year or even three-year reproductive cycle. Dusky Sharks, for example, have a three-year cycle that includes a 20-month gestation period, followed by a 16-month gap. Punctuated breeding is found in placental live-bearing sharks where maternal input to offspring is high.

Endocrine Control of Reproduction

The production of sex cells and preparation for pregnancy are under the control of the endocrine system, the complex series of glands that secrete hormones that regulate physiological function in organs and organ systems throughout the body.

Endocrine glands like the adrenals, pituitary, thyroid, and ovaries and testes are highly conserved among vertebrates over evolutionary history; that is, they are present in the entire extant lineage. You, a trout, and a shark all possess the same glands, although trout have some unique glands, and the chemicals produced in different taxa are also conserved, though they are not identical.

The protein or steroid hormones produced by endocrine glands and secreted into the bloodstream are chemical messengers that go to other glands and signal them to start or stop producing other hormones, or they go to organs and signal them to produce an enzyme or other chemical. They bind to receptors in the cell membranes of target tissue, and this binding often triggers biochemical response within the cell (e.g., production of another hormone or maturation of a cell). Regardless of the system, there is a hormone cascade associated with development. The timing of the release of hormones is influenced by environmental cues, the most important being temperature and day length.

Reproduction in all vertebrates, including sharks, is regulated by what is known as the HPG (*hypothalamus-pituitary-gonadal*; *gonadal* refers to testes and ovaries) axis. These three separate glands function in unison in regulating the reproductive process.

How to Assess Reproductive State in a Shark

In order to understand important aspects of the life histories of sharks, it is vital to be able to tell if a shark is mature or immature. If the shark is a mature female, it is even more useful to know if she is pregnant or postpartum. Fortunately, there are ways of making these determinations, though some require chemical analysis or specialized instrumentation like ultrasound machines. For many commonly encountered sharks, researchers have used some of these techniques to publish sizes of maturity for males and females. These published values, however, should be used only as rough estimates, since populations from different areas mature at different ages and sizes, or the timing of reproduction varies. The following methods can be used to determine the stage of maturity and reproduction in an individual, or for an entire population.

In males, the first method, examining the claspers, is a non-invasive one; that is, it does not require an incision or removal of muscle tissue or blood. In immature males, the claspers are short and flexible, the latter due to the lack of calcification, although claspers lengthen and become less flexible as a shark approaches maturity. Other ways to assess male maturity include measuring the size and weight of the testes (requiring lethal sampling) or measuring levels of reproductive hormones in the bloodstream. In males, testosterone levels in the blood begin to increase during spermatogenesis and remain elevated until after mating. If you examine a large number of male sharks or batoids over a broad range of sizes, you can assign them to two categories, immature or mature, based on the above criteria. Then, you can use the vertebral centra (as described in chapter 9) to determine the age at which the shark is sexually mature.

In females, unless they are visibly pregnant, there is no external method of determining maturity comparable to that used with claspers. The main nonchemical method of determining maturity of female sharks is to sacrifice them or use animals captured in fisheries and examine their reproductive tracts. The size of the largest ova and widths of the oviducal gland and uteri indicate whether a shark is mature or not, whether she has mated before, and whether she is or was pregnant.

Alternatively, levels of circulating reproductive hormones can also be measured. Levels of hormones such as estradiol and progesterone begin to elevate during the vitellogenesis stage of ova production, when the eggs begin to become yolkier. Then these levels may decrease during copulation, since by that time they are fully developed, and no additional hormonal signals are required.

Finally, a method increasingly being employed to assess pregnancy in sharks is high-resolution ultrasound (see fig. 6.15).

Sex Determination, Hermaphroditism, and Virgin Births

Chromosomes in the cell's nucleus carry the genes that build an organism. Biological sex in most animals is determined genetically by sex chromosomes. You may have learned that humans have 23 pairs of chromosomes and the last pair determines sex, where females usually have two X chromosomes (XX) and males usually have a single X and a single Y chromosome (XY). The XX/XY system is most familiar because it is the sex determination system for humans, but many other systems exist in the animal kingdom (platypus have 10 sex determining chromosomes!). The karyotypes (complete set of chromosomes) for sharks and rays have only recently been determined. The number of chromosomes varies from 28 (14 pairs) in many stingrays to 106 (53 pairs) in the Zebra Shark. Like humans, sharks employ the XX/XY sex determination system which evolutionarily traces back hundreds of millions of years.

Hermaphroditism (the presence of both male and female reproductive organs in an individual) occurs but is uncommon in the animal kingdom, except in bony fish, where it is the norm in more than 20 families (e.g., wrasses, groupers, and sea basses). Among bony fish, the hermaphroditism is sequential (one sex transforms into the other) and not simultaneous, except in the mangrove rivulus, *Kryptolebias marmoratus,* in which self-fertilization occurs and the offspring become clones of the parent.

Sequential hermaphroditism is not possible among the chondrichthyans due to developmental architecture of the reproductive tracts, as the male tract terminates at the urogenital papilla, but the female tract opens to the cloaca. Intersex individuals, possessing a mix of primary and/or secondary sexual characteristics of both males and females, have been found in dozens of species of sharks and rays. Incidence rates of intersexuality is usually very low. However, in the lantern shark (*Etmopterus unicolor*) researchers[5] examined 70 specimens and found about 23% intersex individuals, 30% gonochoristic males, and 47% females. The authors speculated that the cause was likely *environmental contamination.*

Normal hermaphrodism is rare among chondrichthyans. The Longhead Catshark, *Apristurus longicephalus,* was shown to be a normal rudimentary hermaphroditic species, where all females possess uncalcified claspers and all males possess ovotestes (gonads with both ovarian and testicular tissue). As more research is conducted, it is likely other species of sharks and rays will be found to exhibit normal rudimentary hermaphroditism.

Another reproductive phenomenon is *parthenogenesis,* or virgin birth, where an embryo develops from an ovum with no male contribution. In many groups (e.g., in some plants, insects, and bony fish), parthenogenesis occurs as a natural form of asexual reproduction. Parthenogenesis was unknown in sharks, however, until it was reported for a captive Bonnethead (*Sphyrna tiburo*)[6] and has now been described in at least six species of captive galeomorph sharks and four species of batoids.[7] Captive female Zebra Sharks (*Stegostoma tigrinum*) have been shown to switch repeatedly from sexual reproduction to parthenogenesis when a male is excluded.[8] One study also reported parthenogenesis occurring in the wild and leading to viable offspring in the Smalltooth Sawfish (*Pristis pectinata*).

Modes of Embryonic Development

You may have learned that animals either lay eggs or give birth to live young, which is true as far as it goes. Chickens are *oviparous,* as are most bony fishes. And mammals are *viviparous,* except egg laying monotremes like platypus and echidnas. But these terms are named only for the point at which either the egg or live young leave the mother. If you look backward from the point to the time the ovum is fertilized (except in organisms like bony fishes where fertilization is external) until either the eggs are deposited or young are born, there is much diversity in the developmental pathway among vertebrates, especially in how the developing embryo is nourished. And no vertebrate group shows greater variety in the modes of embryonic nutrition than sharks and their relatives. The term *lecithotrophic* (*lecitho*=yolk; *trophic*=feeding) denotes that embryos receive their nutrition exclusively from yolk, a fat-, protein-, and often vitamin-rich substance produced by the mother during oogenesis. In contrast, in *matrotrophic* (mother feeding) species, most of the nutrition throughout gestation comes from the mother. Among the chondrichthyans, modes of embryonic nutrition exist on a continuum between these extremes but fall into four major categories and here we divide these into 10 subcategories, but it is important to understand that substantial overlap occurs among the categories.

The modes of embryonic nutrition in elasmobranchs are:

- **Oviparity** (extended oviparity, lecithotrophic oviparity)
  - Single (immediate)
  - Retained (delayed)
- **Yolk-sac Viviparity** (previously called *ovoviviparity*)
  - Lecithotrophic viviparity
  - Limited (mucoid) histotrophy
  - Embryotrophy
- **Oophagy**
  - Oophagy
  - Embryophagy
  - Oophagy with lipid histotrophy
- **Histotrophy**
  - Trophonematic viviparity (Lipid Histotrophy)
  - Placental viviparity (Yolk-sac Placental Viviparity or Placentotrophy)

A Few Words About Oviparity

Oviparity refers to egg-laying. Here, the source of nutrition throughout the developmental period is the yolk deposited by the mother during vitellogenesis. Thus, once the yolk is deposited, there is no additional maternal contribution to embryonic nutrition.

An important consideration in embryonic development, one that has implications for both the mother and the newborn, is whether the embryo gains or loses weight from either the point the eggshell is deposited in oviparous species or from when the egg is fertilized to hatching in viviparous species, to when the neonate is born. Researchers determine the weights of embryos and neonates by drying them in an oven at 60°C (140°F) until a constant weight, called the *dry weight*, is reached. At this point, all of the water has been evaporated and all that remains is dry organic matter.

In oviparous species, since all of the nutrition that the developing embryo can receive is limited to what is contained in the yolk in the sealed egg, the end-weight cannot be greater than the original. It cannot be the same either, since the transfer of energy from yolk to body mass is not perfect.

Why is this? It is thermodynamics 101, the same reason why you do not ultimately gain a half-pound when you eat a half-pound of food, since the conversion of the energy contained in that half-pound of food to your body weight involves loss of over 50% of the energy as heat, as well as use of some of that energy to fuel your body's routine functions and the elimination of waste products.

Similarly, the developing embryos are using energy for respiration and metabolism, and they also defecate and urinate. So, thermodynamically speaking, there must be a decrease in the dry weight of the neonate prior to birth; this decrease is typically in the 20% to 25% range.

Here are some facts about both categories of oviparous sharks and their relatives:

- Young are smaller than in viviparous species.
- 40% of sharks and rays are oviparous, including the largest family of sharks, the Pentanchidae (deepwater catsharks), and all skates (i.e., half of batoids).
- All chimaeras are oviparous.
- Ova are large. After being fertilized, they are enclosed in an egg case while in the oviducal gland, are deposited, then hatch after periods of 12 weeks to 15 months, during which time all development is external of the mother.
- Eggs are laid on the substrate or attached to bottom structures.
- The embryo is nourished solely by the yolk sac.
- Oviparous sharks are primarily small and benthic.
- Body functions of the embryo (e.g., urination and defecation) occur in the egg.
- Nearly halfway through development, a seam opens in the egg case allowing the flushing of waste and inflow of oxygenated water into the protective shell.

Single Oviparity (Immediate)

In the case of single oviparity, *single* does not refer to the number of eggs. It means that, in contrast to retained (or delayed) oviparity, described further in this chapter, the eggs are not retained within the mother during the entire developmental period but are deposited soon after being fertilized.

Single oviparity is characteristic in the order Rajiformes (skates), which has over 290 species; the shark families Scyliorhinidae (catsharks) and Pentanchidae (deepwater catsharks), the most speciose family of sharks with > 120 species; all members of the order Heterodontiformes; many species in the order Orectolobiformes; and all members of the order Chimaeriformes,. In most oviparous

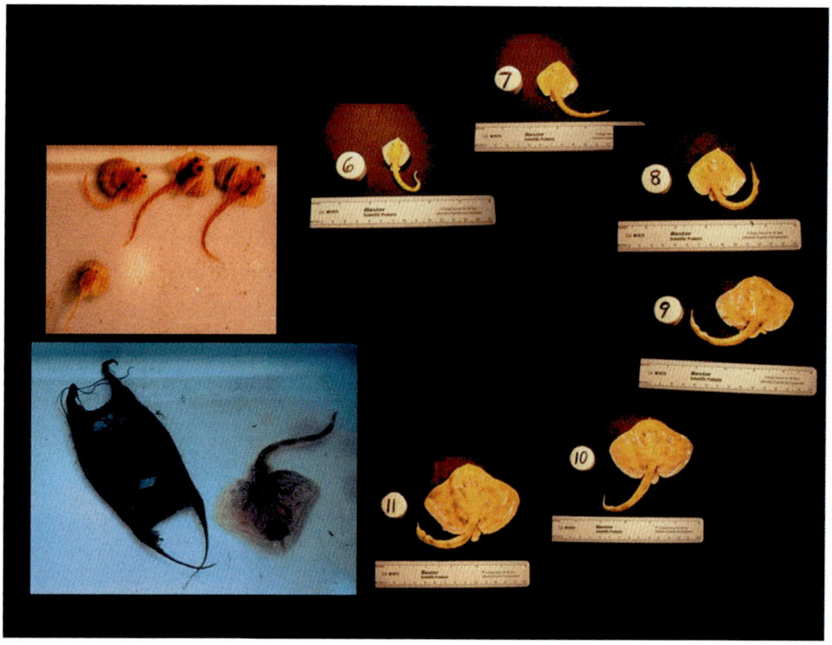

Figure 6.11. Development in the Clearnose Skate, an example of single oviparity.

elasmobranchs, the egg case surrounds a single fertilized ovum. However, in some skate species, up to four embryos are encased in a single egg case.

Let us look at development in the Clearnose Skate (*Rostroraja eglanteria*; fig. 6.11), which takes about 12 weeks from fertilization. The egg case is sealed off from the environment when it is deposited, but the accumulation of waste products and the decrease in dissolved oxygen would eventually make for a toxic, lethal brew if it were not removed.

At between five and six weeks, a seam along the horns of the egg case dissolves and allows for the influx of seawater and the efflux of the gradually toxifying contents. Replacement of the fluid in the egg case is facilitated by the presence of a very long tail, whose motion sets up a current that flushes out the tainted water. This means that at 5–6 weeks of development, the seawater environment is identical inside and outside the egg case.

Horn sharks (heterodontids), which typically inhabit rocky coastal substrates, lay individual corkscrew-shaped eggs (fig. 6.12), which are adapted to become lodged between rocks so they are protected and do not get washed onto the beach. Horn shark mothers have been observed carrying the eggs in their mouths and positioning them among the rocks.

Retained (or Delayed) Oviparity

Here, the shell is put around the eggs in the oviducal gland, but the eggs are retained in the mother's uteri, so egg-laying is delayed. Similar to the situation in

Figure 6.12. Corkscrew egg of horn shark, and near-term embryo.

single oviparity, since there is no additional maternal nutrient input, there is a 20–25% weight loss experienced by the time of hatching or birth.

In several orectolobiform sharks, the fertilized eggs are retained in the oviducts of the mother for a portion of development and are then deposited on the seafloor prior to hatching. In some cases, the eggs are not laid but instead are incubated within the mother until hatching occurs inside her uteri and the offspring are born live. A classic example is from the Nurse Shark (*Ginglymostoma cirratum*), described in a beautifully illustrated paper.[9] As the embryos reach about the halfway point of development, they hatch from their eggs within the uteri and the now-empty egg case remnants are shunted out of the mother. The remaining development occurs in the mother. The external yolk sac has already been absorbed before that point, but it has not been used up. Instead, there is an internal yolk stomach that contains the yolk, which is still available as a source of nutrition until birth.

Whale Sharks (*Rhincodon typus*) were once considered as exhibiting single oviparity based on the discovery of an egg case containing an embryo; however, closer examination revealed an egg case that likely would not hold if deposited on the substrate, suggesting that it may have been aborted. Subsequently, Taiwanese researchers cut open a pregnant Whale Shark that was being sold for meat and found about 300 embryos in egg cases at various stages of hatching, similar to the reproductive mode described in the Nurse Shark.

After reading the next section, you may wonder why we categorized Nurse Sharks and Whale Sharks as oviparous and not viviparous, given that young are born alive. This illustrates that the modes of embryonic nourishment we describe are not always discrete but exist on a continuum. Nurse Sharks and Whale Sharks are often described as exhibiting lecithotrophic live birth. However, other live-bearing species do not have a thick egg case around the embryo during early development, and hence the reason for the inclusion of Whale and Nurse Sharks among species exhibiting retained oviparity here.

A Few Words About Viviparity

Viviparity refers to live birth, where the developing embryos are protected within the mother while they develop and grow, and embryos are typically born at a size that increases their potential to survive. In your readings, you will sometimes find shark viviparity divided into *aplacental* and *placental* modes. We avoid the term aplacental because it is uninformative since it describes what it is not (i.e., not placental) and it masks the diversity in developmental modes, including some that are equal in matrotrophic inputs to that in placental live bearers. We broadly divide viviparity in elasmobranchs into three groups based on the major source of nutrition—a yolk sac, the mother's unfertilized eggs, or histotroph secreted from the mother's uterus (uterine milk), though in all cases, the developing embryo initially gains its nourishment from the yolk sac.

Yolk-sac Viviparity (Ovoviviparity)

The most widely employed type of live birth in the elasmobranchs is yolk-sac viviparity, where the developing pup receives the bulk of its nourishment from its yolk sac and is thus heavily lecithotrophic. In yolk-sac viviparous species, the oviducal gland is reduced and unspecialized, since no true shell is deposited, and sperm storage is likely limited. Instead of a shell, a membrane is deposited around the developing eggs. In some cases, a single membrane is deposited around all eggs fertilized in the oviducal gland, forming a structure referred to as a *candle* (see fig. 6.13C). The candles move into the two uteri and serve to maintain intrauterine conditions required for the embryos to develop. The embryos break free of this membrane about halfway through development and are then free in the uterus. Overall, about 40% of extant species of elasmobranchs practice yolk-sac viviparity. This includes all 180+ species of sharks of the superorder Squalomorphii (i.e., angel, frilled, saw, dogfish, gulper, sleeper, lantern, and cow sharks), plus a few of the Carcharhiniformes (e.g., some smoothhounds) and Orectolobiformes (wobbegongs). No members of the orders Heterodontiformes or Lamniformes utilize yolk-sac viviparity. Among batoids, the Torpediniformes (torpedo rays) and Rhinopristiformes (guitarfish and sawfish) are yolk-sac live-bearers.

Lecithotrophic Viviparity and Limited (Mucoid) Histotrophy

Most yolk-sac viviparous species are strictly lecithotrophic, meaning embryos receive all nourishment from the yolk sac, similar to what occurs in oviparity, but there is no eggshell. Over the course of their development, the embryos exhibit a

20–25% weight drop, due to the absence of additional maternal nutrition. In species in which there is no additional maternal input, females do not require a resting phase (discussed further in this chapter) and will physiologically prepare for their next pregnancy while already pregnant. Thus, it is not surprising that most mature female squalomorphs are almost always pregnant. Interestingly, the squalomorph sharks employing yolk-sac viviparity have among the longest gestation periods of any vertebrates, two years or more, and some, like the Little Gulper Shark (*Centrophorus uyato*) produce only a single pup following this long gestation.

Moreover, since most of the sharks of the superorder Squalomorphii are deep-sea, where seasonal cues for the timing of reproduction are absent or muted, it is not unusual to find sharks in all stages of development at any time (fig. 6.13A and B.).

Lecithotrophic viviparity is best-known in the genus *Squalus* (fig. 6.13). In *Squalus*, all of the embryos in each reproductive tract, typically three or four, are placed in a single candle. They continue to absorb their yolk while in the candle, and they break out when they are one-third to halfway developed and continue to absorb their yolk.

In the Gulper Shark (*Centrophorus granulosus*), the mean dry weight loss between egg and neonate is 19.5%, but in the Great Lantern Shark (*Etmopterus princeps*), the decrease is only 8%. This suggests that the Gulper Shark embryos get all nourishment from the yolk sac (i.e., are lecithotrophic), and the Great Lantern Shark was receiving some maternal contribution, likely through mucoid secretions from the mother's uterine wall (termed *limited histotrophy*). The maternal contribution may be *limited* compared to those employing uterine milk, but in representatives of some families (e.g., Triakidae) there is weight gain up to 350%, which implies this additional maternal contribution of nutrients can be significant. In these triakids (and some other species) there are frilly extensions of the uterine wall that secrete a mucous that has nutritive ability and may also aid in respiration by providing additional oxygen.

Some batoids, such as the electric rays, guitarfishes, and sawfishes, are yolk-sac live-bearers; however, at least one species of electric ray employs limited histotrophy. It is unknown if others in these groups are strictly lecithotrophic or limited histotrophs. Speaking of sawfish, ever wonder how female sawfish avoid getting wounded by the toothed rostra of embryos *in utero*, or when they give birth to young? Here is how they do it: the toothed rostrum of each embryo and newborn is covered by a tough skin-like sheath that sloughs off over the course of days after birth (fig. 6.14).

Embryotrophy

Embryotrophy, described by Castro and others,[10] occurs only in Tiger Sharks and is a mode of yolk-sac viviparity in which embryos are sequestered in sacs containing a clear fluid. The fluid, termed *embryotroph*, is not the protein- and fat-rich *uterine*

A
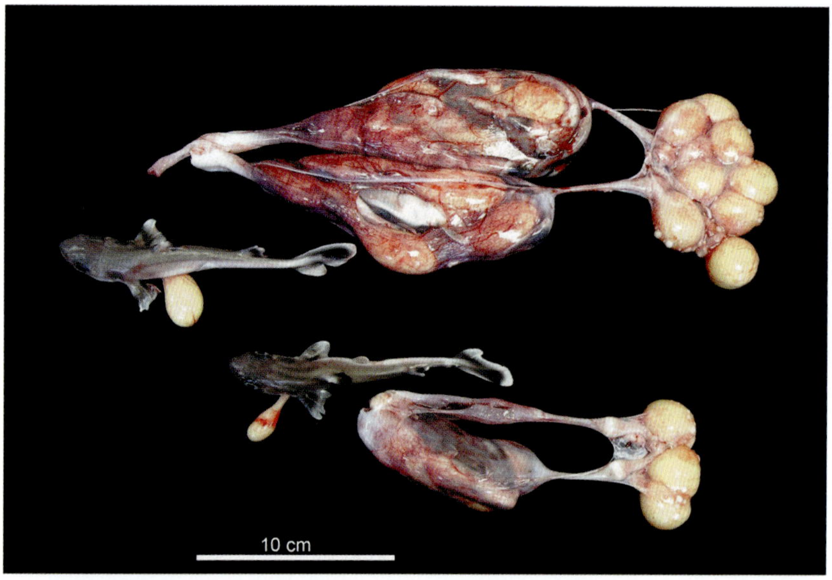

B
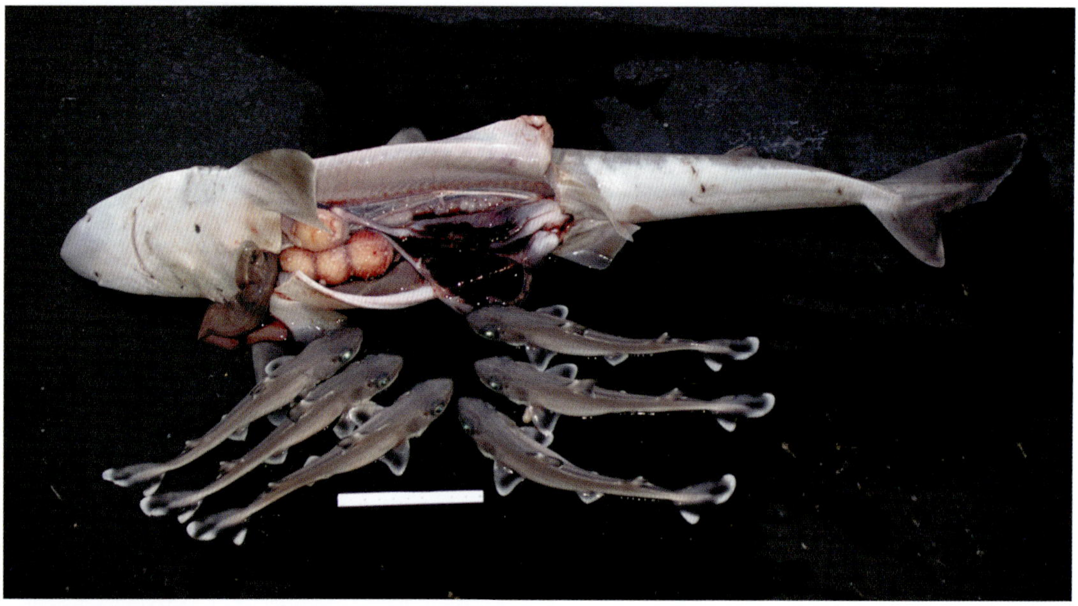

C
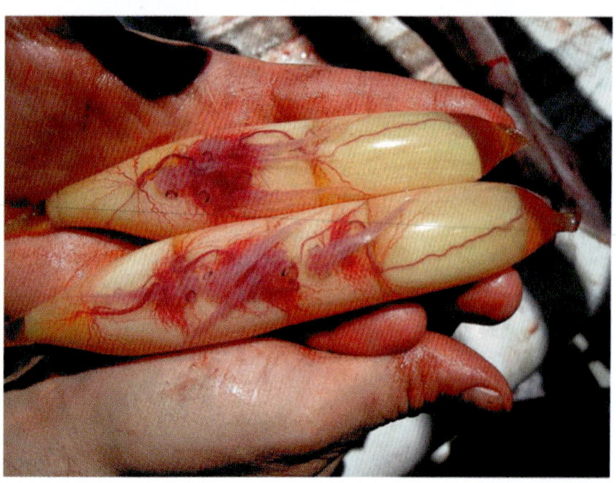

Figure 6.13. Lecithotrophic viviparity with mature embryos and ova in two members of the genus *Squalus*: (A) Genie's Dogfish and (B) Hawaiian Shortspine Spurdog. Note the variation in fecundity between two closely related species. (C) Two candles from Genie's Dogfish, showing seven embryos.

Figure 6.14. Sheath around rostrum of neonate Smalltooth Sawfish.

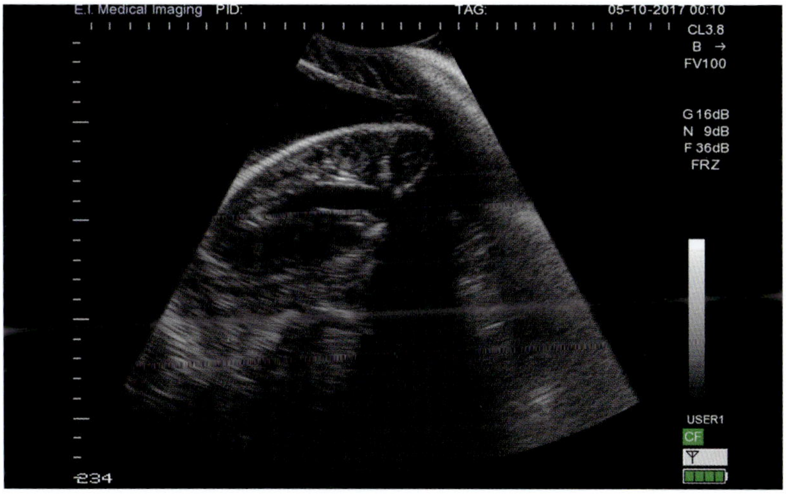

Figure 6.15. Ultrasound image of a Tiger Shark embryo *in utero* (Courtesy of Matt Smukall).

milk we discuss later, but it has nutritive value, leading to dry weight gains of as much as 1092% in this species. In contrast to lecithotrophy, in which nutrition is provided by the yolk, embryotrophy is a form of *matrotrophy* (*matro* = mother); that is, much of the nutrition comes from the mother.

Embryotrophy represents a novel form of nutrition among sharks, and a deviation from that of closely related carcharhinid and sphyrnid sharks, which all use placental viviparity. This is one of the reasons Tiger Sharks are in their own monotypic family Galeocerdidae. Embryotrophy has contributed to the reproductive success of this species. Tiger Sharks are known for having among the largest litters, typically 45–60 pups (fig. 6.15), of all sharks.

Oophagy

Oophagy means egg-eating, and it occurs in all lamniform sharks (fig. 6.16) in at least three species in the deep sea carcharhiniform family Pseudotriakidae (false catsharks) and, oddly enough, in at least one population of the orectolobiform tawny nurse shark, *Nebrius ferrugineus*. Here, like all other elasmobranchs, embryos begin development using the stored energy of their yolk sac. However, as yolk is depleted, the embryos eat unfertilized eggs provided by the mother. This matrotrophic means of embryonic nutrition results in dry weight gains of > 1000%.

Lamniform Oophagy

Female lamniform sharks initially only ovulate eggs to be fertilized, the resulting embryos initially using the yolk-sac for nutrition. At about the same time that most of the yolk has been absorbed by the embryos, the mother begins ovulating many unfertilized ova. The embryos swim within the uteri and consume the new ova and develop large egg stomachs. All lamniform sharks utilize oophagy for embryonic nourishment and the high matrotrophic contribution through unfertilized eggs typically results in few offspring. For example, oophagy in the deep-sea

A

B

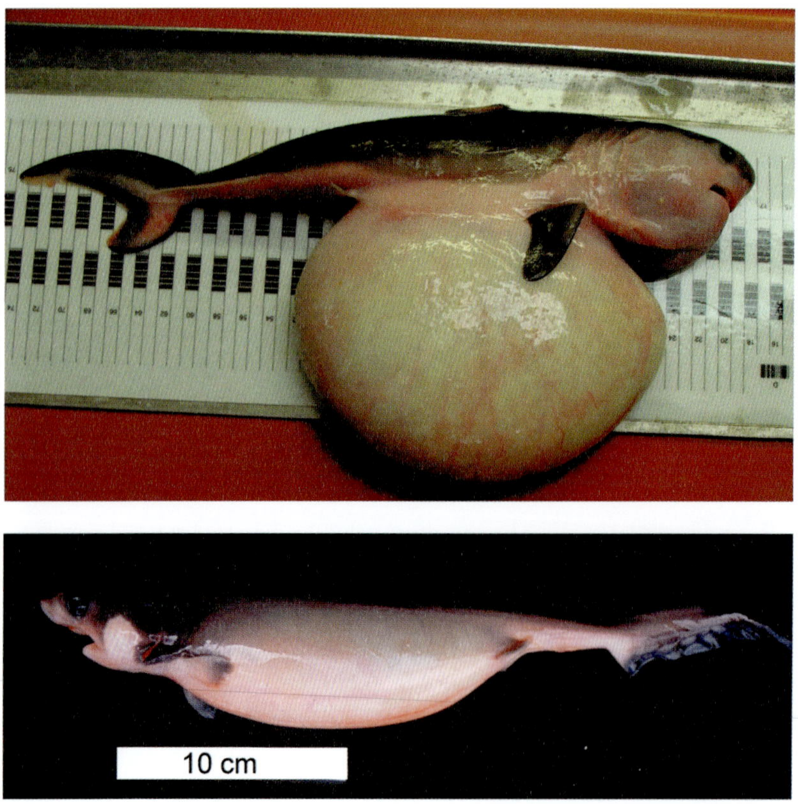

Figure 6.16. (A) Embryo of Salmon Shark exhibiting its yolk stomach filled with ova it has consumed in utero (Courtesy of Ken Goldman). (B) Crocodile Shark Embryo.

Crocodile Shark (*Pseudocarcharias kamoharai*) (fig. 6.16B) typically results in only four young (two per uterus) and the Pelagic Thresher produces only two offspring.

Embryophagy (Adelphophagy)

The lamniform Sand Tiger (*Carcharias taurus*) takes oophagy to the next level, in that the fastest-growing embryo in each uterus eats its siblings, a form of nutrition called *embryophagy* (embryo-eating), *adelphophagy* (sibling-eating), or *intrauterine cannibalism*. In embryos of most shark species, teeth are among the last major anatomical structures to develop. In most cases they remain internal and do not break through the gums until slightly before birth. Sand Tiger embryos, however, exhibit precocious tooth development, an adaptation for intrauterine predation. When the largest embryo with precocious tooth development reaches about 12 cm (4.7 in), it starts eating as many as a dozen other embryos in its uterus, as well as unfertilized ova, and stores the energy in its yolk stomach. The result is that females give birth to two large young (one in each uterus), each > 1 m (3.3 ft) and thus formidable predators with a high survival probability at birth. Talk about sibling rivalry!

Carcharhiniform Oophagy

The False Catshark, *Pseudotriakis microdon,* was the first non-lamniform shark reported to use oophagy, employing a different mechanism. In oophagous pseudotriakids, a single egg is fertilized and deposited in each uterus along with thousands of tiny unfertilized ova, contained in the same egg envelope. Unlike in the lamniforms, where the female ovulates eggs during gestation, pseudotriakid embryo are limited to those initially deposited ova. Also unlike in the lamniforms, which store the ingested eggs in their yolk-stomach after the yolk-sac is depleted, the pseudotriakid embryos reportedly can transfer ingested eggs into the external yolk-sac, thus replenishing it.

Lipid Histotrophy with Oophagy

The reproductive mode of White Sharks has been reported as oophagous based on late-term embryos observed having large yolk stomachs like other lamnid sharks. However, recent studies have shown that during early embryonic development, thick lamellae on the uterine wall produce energy-rich lipids for nourishment. This is accomplished using a novel mechanism where the outer cell layer (epithelium) of the uterus is shed releasing lipids contained in the cells. This process is likely repeated until the female begins ovulating unfertilized eggs for the embryos to then feed on. It may be the other, if not most, lamniform sharks also use lipid histotrophy through uterine secretions relatively early in development. Prior to this discovery, lipid histotrophy in elasmobranchs was only known from the myliobatiform stingrays using the trophonemata we will discuss next. This is another example of how the categories we use to classify modes of embryonic nourishment are not always discrete.

Histotrophy

Histotrophy, uterine feeding, is a very strongly matrotrophic form of nourishment, meaning the embryos rely primarily on maternal energy sources. This differs from the limited or mucoid histotrophy we discussed earlier. As in all sharks and rays, initial nourishment to the developing embryo is supplied from the yolk sac. After yolk is depleted, the mother secretes copious nutrients, proteins, or lipid-rich histotroph or uterine milk. There are two primary types of histotrophy based on the mode of maternal transfer of this nourishment through uterine villi called *trophonemata* or through a yolk-sac placenta.

Trophonematic Viviparity

Trophonematic (*tropho* = feeding; *nema* = thread) nutrition, which occurs only in a single order, the Myliobatiformes (about 234 species), refers to the presence of long villi, or filaments, in the uterus called *trophonemata* (fig. 6.17). These are referred to as placental analogues because they mimic the function of placenta by secreting uterine milk, although there is no direct connection between the uterine wall and the embryo. Myliobatiform ray embryos begin with a yolk sac, but the mother develops trophonemata, from the epithelium of the uterus. These trophonemata secrete a lipid-rich (fat- and oil-filled) nutritive histotroph (uterine milk). The trophonemata are bright red before producing milk, then turn pink when milk is produced.

Trophonematic embryonic nutrition may be the most advanced type of development strategy in the group as embryos swim in this rich nutritional medium that bathes every orifice and exchange surface, thus they cannot *avoid* nourishment. Wet weight gain by developing stingray embryos have been reported up to 5000%, with dry weight gains predicted to be comparable to oophagy.

Myliobatiformes produce relatively small litters, as few as one (e.g., Cownose Rays, Devil Rays), but they are very large, and their survival probability is high. Also, Cownose Rays have knobby extensions on the yolk stalk that may aid in respiration, though this structure and its function have not been fully studied. The single pup of Cownose Rays is born after a 12-month gestation period. There is a huge increase in weight for the developing embryo.

Myliobatiform rays that possess barbs are born with a fleshy tip on their barb, thus avoiding a sticky situation at birth. The tip disintegrates quickly, with a fully calcified spine ready to go, if needed.

Placental Viviparity (Yolk-sac Placental Viviparity or Placentotrophy)

As the name implies, *placental viviparity* refers to a form of live birth in which the developing embryo receives nutrition from a structure (called a placenta in mammals) that connects the embryo to the mother through the uterine wall. However, the placenta in sharks is analogous (similar in function) rather than

A

B

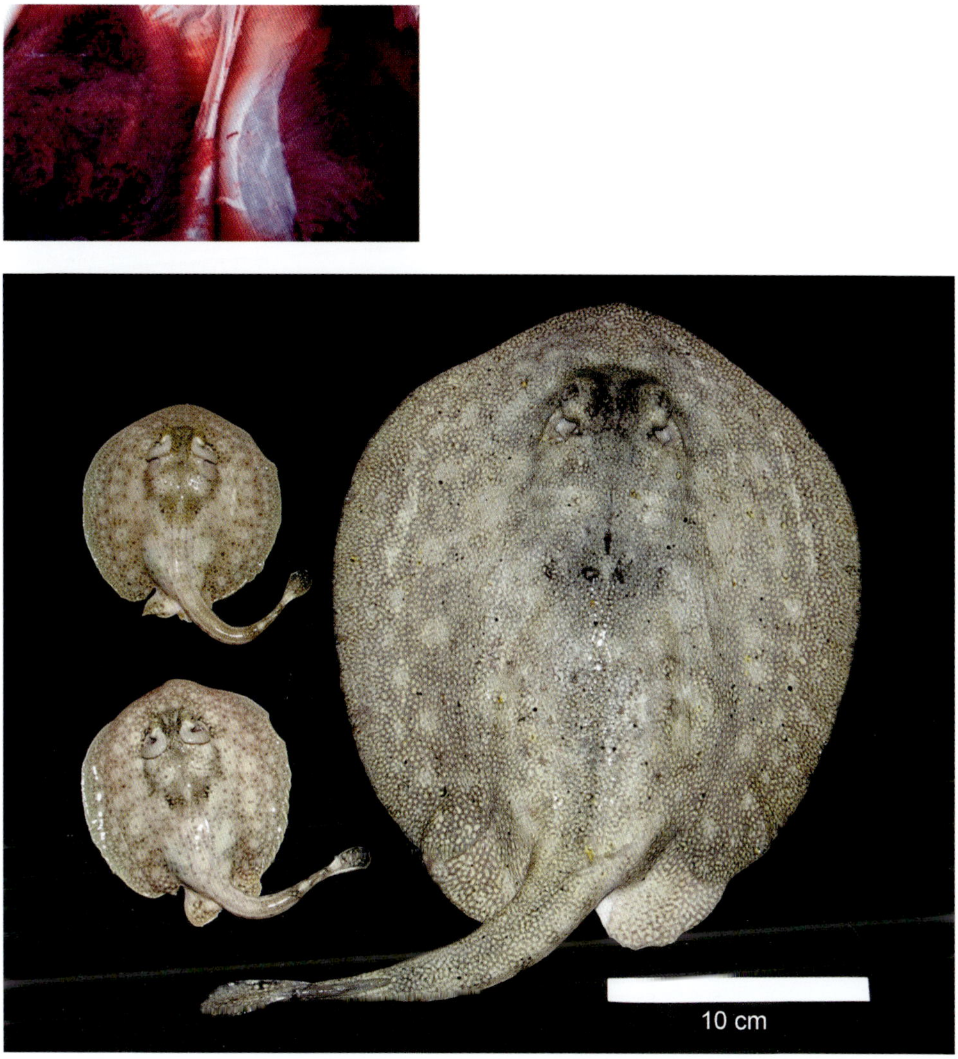

Figure 6.17. (A) Trophonemata in the myliobatiform Smooth Butterfly Ray in late term of pregnancy. (B) Mother and term embryos in the Yellow Stingray, also a trophonematic species.

homologous (similar in derivation or origin) to the mammalian placenta. It thus might more appropriately be given the name *yolk-sac* placenta, although this might create confusion with the lecithotrophic form of embryonic nutrition which, you will recall, is also known as *yolk-sac* viviparity. We will stick with the name placental viviparity for this category, although you may see this referred to as *yolk-sac placental viviparity* or *placentotrophy* in the literature. No matter the name, placental viviparity represents the opposite extreme to oviparity.

In sharks with placental viviparity, after the ova are fertilized, they are encased in a membrane in the oviducal gland, and then delivered into the uterus, where

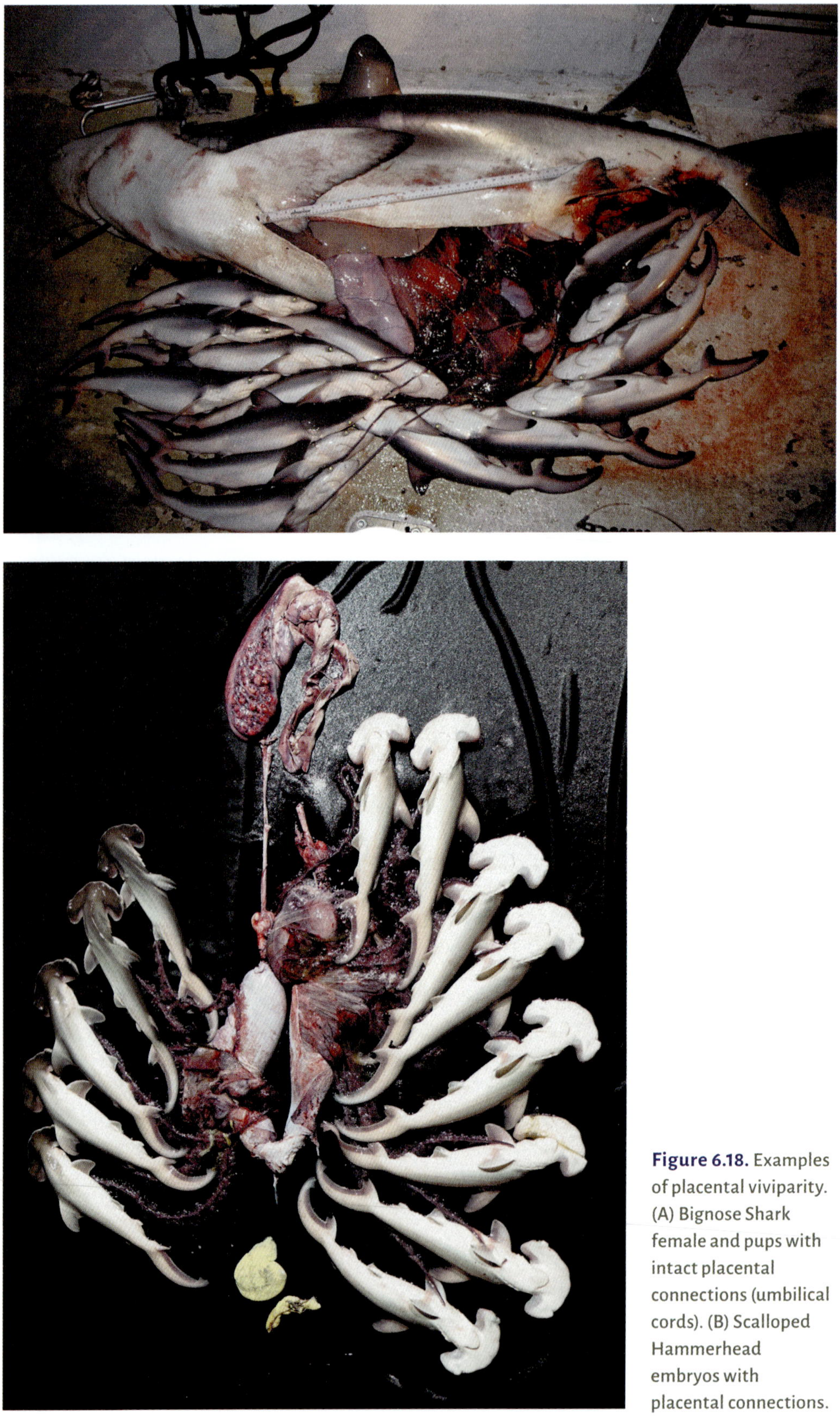

A

B

Figure 6.18. Examples of placental viviparity. (A) Bignose Shark female and pups with intact placental connections (umbilical cords). (B) Scalloped Hammerhead embryos with placental connections.

the membrane-encased sharks are stacked (fig. 6.18). Once in place, in many species the uteri compartmentalize into crypts, thus separating the developing embryos. During early development, as is the case in all of the forms of embryonic nutrition we have described, the embryos rely on the yolk sac for nutrition. After the yolk is exhausted, the empty yolk sac interdigitates with the heavily vascularized uterine wall, forming the placental connection. The mother then feeds the embryos directly through uterine secretions, rather than through the blood stream as in placental mammals. In some species (most sharpnose sharks, about half the hammerhead species), yolk-sac "umbilicus" also develops fleshy protrusions called appendiculae, structures believed to facilitate gas exchange and aid in respiration for the embryos. It is unclear why some species possess appendiculae, and some do not, even within the same genus of sharks. The placental connection separates from the embryo during birth, and neonates typically feature what could be called a belly button but is more appropriately called a chest button or *umbilical scar.* Immediately after birth, the scar is in fact not entirely healed, a process that can take a month or more, and which can be used (along with other characteristics, e.g., size) to estimate the age of the pup.

Placental viviparity occurs in about 18% of shark species. All of these are in the order Carcharhiniformes, but only in the families Carcharhinidae (requiem sharks), Triakidae (houndsharks), Hemigalidae (weasel sharks), Leptochariidae (barbeled houndshark), and Sphyrnidae (hammerheads). Interestingly, placental viviparity can co-occur with forms of yolk-sac viviparity in the same family and, in some cases, the same genus (e.g., *Mustelus*).

Placental viviparity is energetically costly to pregnant females. Little energy is put into vitellogenesis, from which recovery of the female is relatively quick (like in lecithotrophic sharks; see p. 261). Instead, after placentation, the developing embryos require constant nutrition, and the pregnant female must expend considerable energy swimming around carrying the embryos (fig. 6.19). Thus, many sharks in this category (e.g., Sandbar and Blacktip Sharks [*Carcharhinus plumbeus* and *C. limbatus*, respectively]) take a year off after giving birth, and the Dusky Shark (*C. obscurus*) has a three-year cycle. The reproductive cycle in some species is also condition-dependent. In years when environmental conditions are more favorable (i.e., food is abundant), individuals may not need to wait a year or more before becoming pregnant.

Placental viviparity also entails a potential problem—the potential for umbilical cords to become tangled or disconnected (fig. 6.20A). The likelihood of entanglement is minimized (but not eliminated) by sequestering the embryos in crypts (fig. 6.20B). This may protect the embryos by compartmentalizing them, but this method also comes with a potential danger—the lack of a way to abort or eject an embryo that may be injured or dead, respectively. However, if an embryo dies within a crypt, only other embryos in the same crypt might be affected, not the entire assembly in that uterus.

Figure 6.19. (A) Pregnant and (B) postpartum Sandbar Sharks.

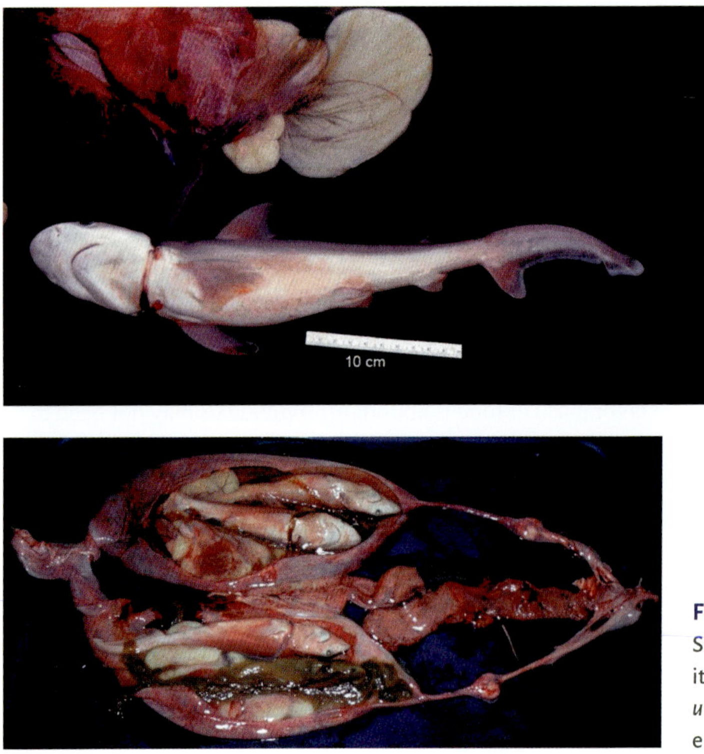

Figure 6.20. (A) Sandbar Shark pup strangled by its umbilical cord *in utero*. (B) Blacktip Shark embryos in crypts.

Concluding Comments

We didn't mislead you at the outset of this chapter when we marveled at the diversity of reproductive modes of elasmobranchs. Who knew that being predatory extended into the womb for some species? The diversity is even more remarkable when you consider the relatively low number of species, about 1250, many of which (e.g., deep-sea species), are still poorly known. Stay-tuned for more variations, possibly![11]

NOTES

1. Still one of the most authoritative references on the subject: Hamlett, W.C. ed., 2011. *Reproductive biology and phylogeny of chondrichthyes: sharks, batoids, and chimaeras, 3.* CRC Marine Biology Series. CRC Press.

2. Pratt Jr., H.L. 1988. Copeia 1988: 719–729.

3. The terminology becomes confusing because some authors define the oviduct as the genital tube leading from the ovaries to the nidamental (oviducal, or shell gland) and the nidamental glands and uteri as components separate from the oviducts.

4. Maruska, K.P. and Gelsleichter, J. 2011. In: Norris, D.O. and Lopez, K.H. (eds.). Hormones and reproduction of vertebrates. Fishes 1: 209–237. Academic Press.

5. Yano, K. and Tanaka, S. 1989. Jap. J. Ichthyol. 36: 338–345.

6. Chapman, D.D. et al. 2007. Biol. Let. 3(4): 425–427.

7. Feldheim, K.A. et al. 2022. Front. Mar. Sci. 9: p.963542.

8. Dudgeon, C.L. et al. 2017. Sci. Rep. 7(1): 1–8.

9. Castro, J.I. 2000. Env. Biol. Fishes 58: 1–22.

10. Castro, J.I. et al. 2016. Mar. Biol. Res. 12: 200–205.

11. Blackburn, D.G., Hughes, D.F., 2024. Biological Reviews, 99:1314–1356.

7 / Circulation, Respiration, and Metabolism

Introduction

In the early 1980s, the following question appeared on a TV game show: *According to marine biologists, who has a bigger heart, a human or a Great White Shark?* The marine biologist in this case was one of this book's authors.

If you were the contestant, how would you have answered the question? The answer that was accepted as correct was the Great White Shark (*Carcharodon carcharias*; also known as the White Shark). However, to be accurate, before answering the question, we first must qualify that we mean a human and White Shark *of the same size.* A White Shark weighing 900 kg (2000 lb) would have a substantially larger heart than even an extremely large person. But an average adult human would have a heart weighing nearly 2.4 times more than that of the White Shark of the same weight as the human.

What if the question was *Who has a bigger heart, a human or a Horn Shark?* The human would, regardless of the size, since Horn Sharks (*Heterodontus francisci*) are substantially smaller. If the human and Horn Shark were of equal weight (one *scary* Horn Shark!), the human heart would weigh nearly six times more than that of the Horn Shark.

Even though the game show writer thought he was asking a trivial question, what he was really asking was more meaningful: *What is the heart adapted for and how does this differ in taxonomically and ecologically distinct organisms like humans and White Sharks?* In this chapter, we will answer these and other questions about shark physiology, starting with the circulatory system.

Circulation in Sharks

Anatomy of the Heart and Circulatory System

Vertebrates, including elasmobranchs, possess a closed circulatory system, one in which blood is contained in blood vessels. A typical closed circulatory system consists of a pump (the heart), a high-pressure distributive system (the arterial system), an exchange system (the capillaries), and a low-pressure reservoir (the venous system). The blood in the circulatory system transports oxygen, carbon dioxide, nutrients, waste products, hormones, and other substances.

Unlike mammals, fishes (both osteichthyans and chondrichthyans) have a single circuit system; that is, one main loop and some secondary loops in which the blood goes from the heart to the gills to the body (head and trunk), then back to the heart (fig. 7.1).

Blood vessels downstream of the heart make up the arterial system and these vessels get progressively smaller in diameter and more numerous until they reach their destinations as capillaries, where exchange of some of the transport products (listed previously) occurs.

In fishes, even though there is only a single major loop, the arterial circulation is given different designations. First is the *branchial,* or gill circulation

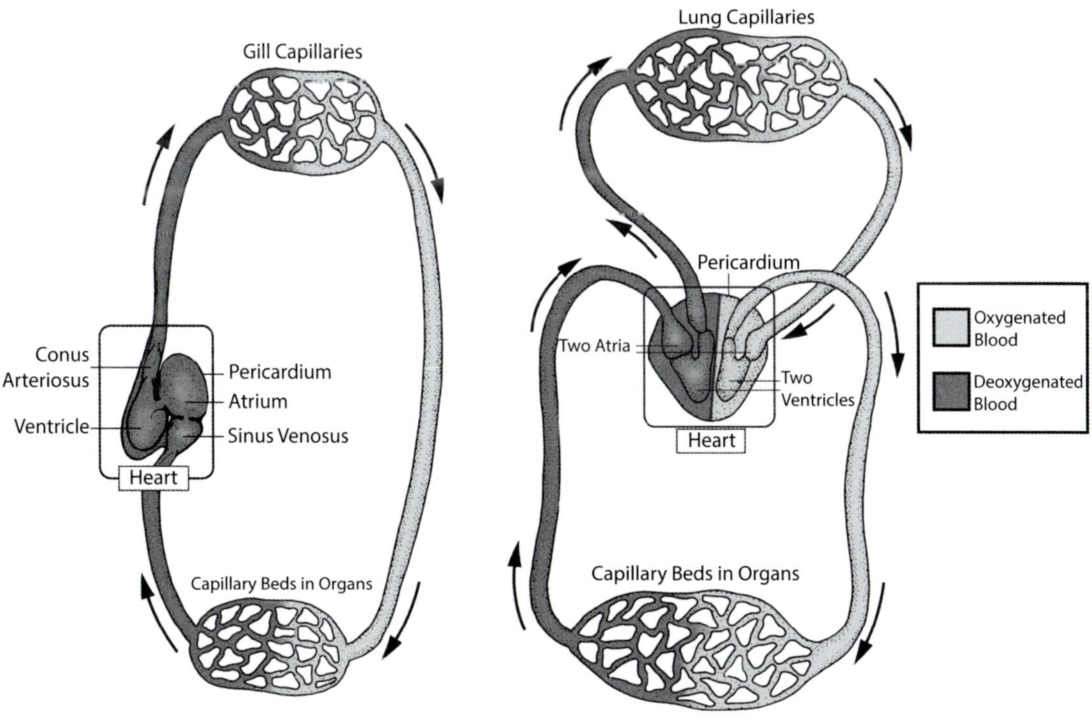

Figure 7.1. Generalized diagram of the closed circulatory system of chondrichthyan fishes (A) and mammals (B). Degree of oxygenation of the blood is also shown (Redrawn from fig. 9, Furst, B. 2015. J. Cardiothor. Vasc. Anesth. 29: 1688–1701).

(figs. 7.2 and 7.4), the part of the circulatory system from the heart through the gills. In all fishes, the artery leaving the heart is the *ventral aorta*. Once it exits the heart, it almost immediately branches into pairs of *afferent*[1] *branchial arteries*, which carry the blood to the gills, where these subdivide ultimately into capillaries, and gas and ion exchange occur. After exiting the gills, these vessels coalesce into *efferent*[2] *branchial arteries* before the now-oxygenated blood is sent to the head and trunk (muscles and viscera).

A subdivision of the branchial circulation is the *coronary* circulation, which arose first evolutionarily in chondrichthyans. The coronary circulation of chondrichthyans branches off the gills and supplies the entire myocardium (heart muscle) with oxygen, nutrients, etc. (fig. 7.3).

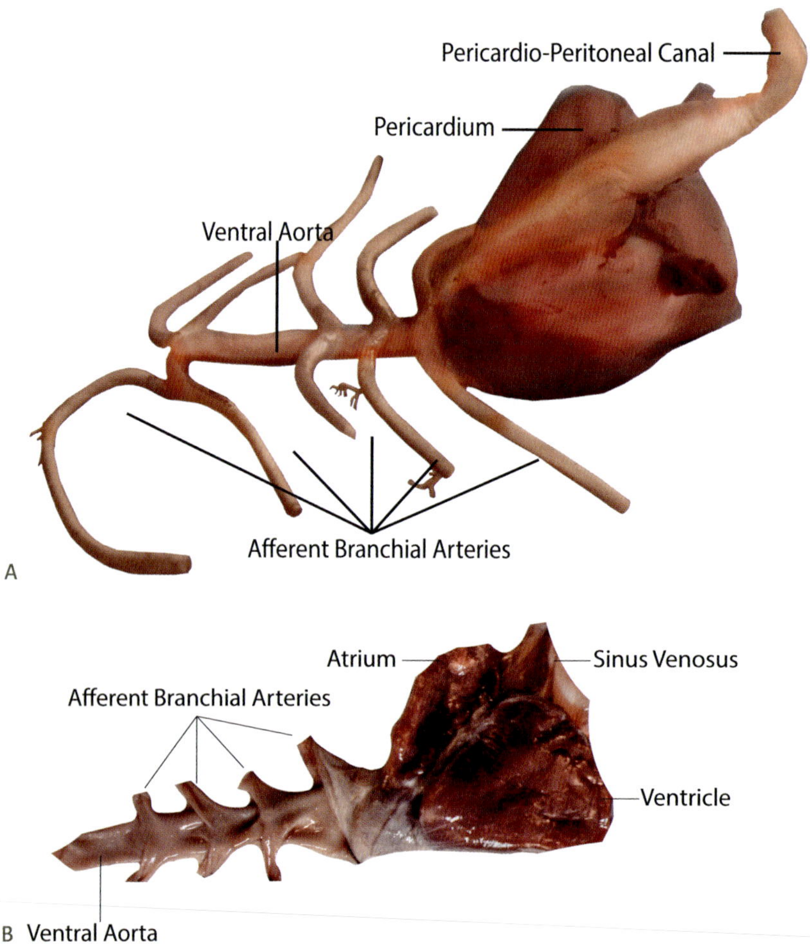

Figure 7.2. (A) Silicone rubber cast of the pericardium, ventral aorta, pericardio-peritoneal canal, and afferent branchial arteries from a 2 m (6.6 ft) Blue Shark. The heart is contained within the pericardial space and thus is not shown. (B) Heart, ventral aorta, and afferent branchial arteries removed from a Blacknose Shark.

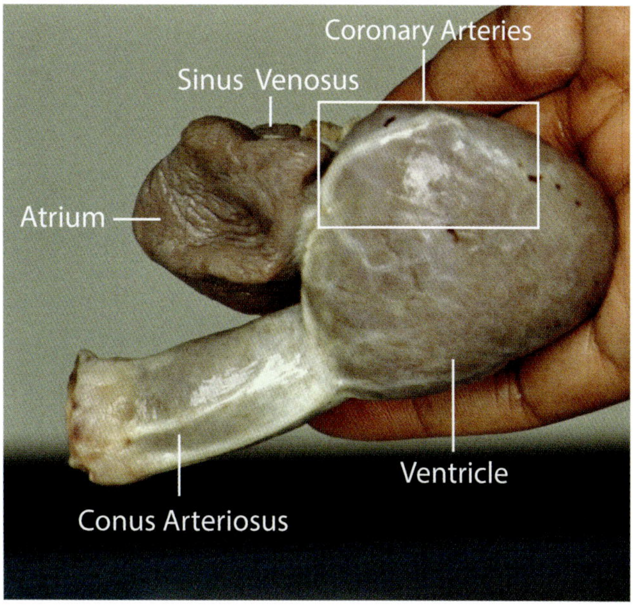

Figure 7.3. Labeled photo of the heart of a Blue Shark showing coronary vasculature.

The other major subdivision of the arterial system is the *systemic* circulation, in which blood goes to the tissues in the head and trunk (fig. 7.4). The efferent branchial arteries of the gills deliver blood to a *dorsal aorta,* the largest blood vessel and the main distributive channel in all fishes. The dorsal aorta sends blood into either the *cranial* circulation (to the head) or *caudal* circulation (posteriorly, to the trunk). Finally, the caudal circulation divides into *visceral* and *somatic* branches. The visceral branches supply the internal organs whereas the somatic branches supply some muscles, the kidneys, and body wall.

The venous system (fig. 7.4) serves as a low-pressure reservoir of blood. The venous system of chondrichthyans differs from that of other vertebrates in that the blood is not always in discrete blood vessels but prior to entering the heart is contained in large sinuses[3] (fig. 7.5). A sinus is a cavity, in contrast to the tubular architecture of blood vessels. Sinuses can accommodate more blood and thus are a ready reservoir when metabolic demands require that the heart pump more blood (e.g., when in pursuit of prey or avoiding predators).

In chondrichthyans, there are three venous sinuses, the *anterior cardinal, posterior cardinal,* and *hepatic.* The anterior cardinal sinus is located in the head, and the other two are at the terminus of the venous system immediately upstream from where blood enters the *sinus venosus* of the heart (see pp. 277–279).

The anterior cardinal sinus drains the head before leading to the heart. The posterior cardinal sinus drains the posterior body, including the kidneys, through posterior cardinal veins, and the hepatic sinus accommodates blood from the liver before it goes into the heart.

During dissection of recently deceased sharks and batoids, we have observed greatly expanded hepatic sinuses on the verge of rupturing, like overfilled balloons.

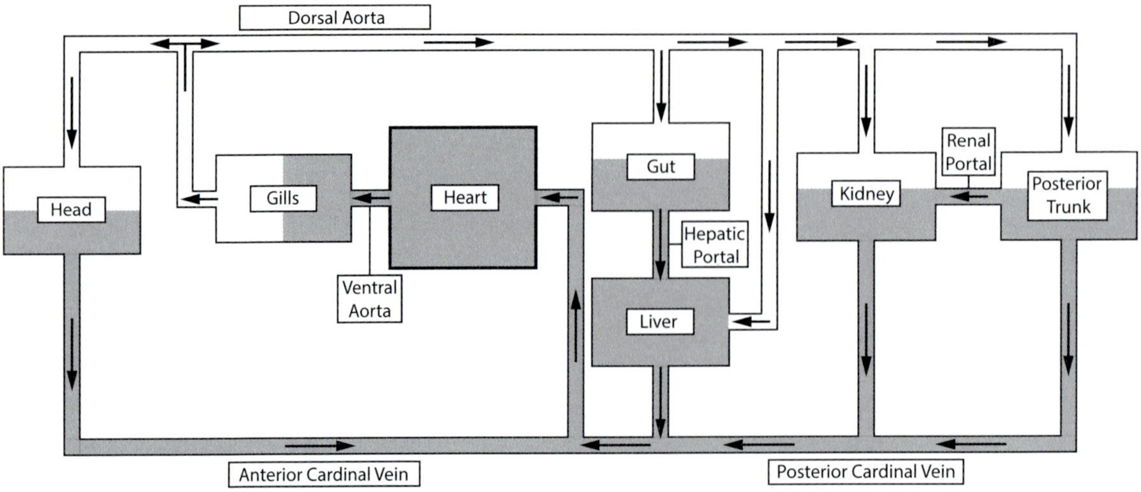

Figure 7.4. Systemic circulation of shark. Darkened parts represent the venous system. Portal systems carry blood from the capillary system of one organ to another before returning to the heart. Portal systems allow some additional degree of processing of blood or delivering blood components in the second capillary bed (Redrawn from fig. 6.3, Gordon, M.S. 1982. *Animal Function: Principles and Adaptations*. MacMillan).

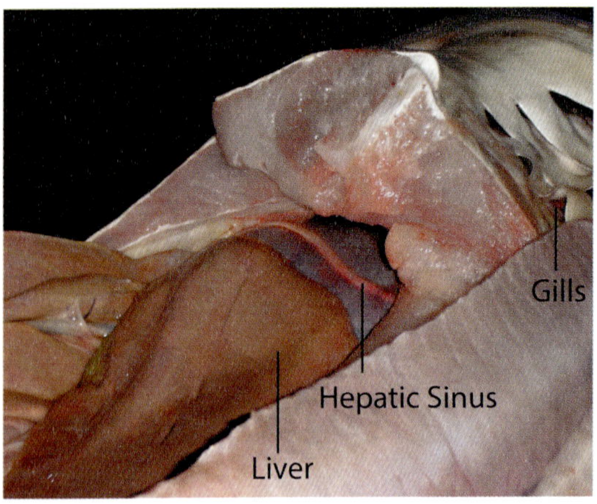

Figure 7.5. Large hepatic blood sinus of the venous system on the liver of a Lemon Shark. The hepatic sinus serves as a reservoir of blood immediately upstream of the heart that could be mobilized when needed. The shark is on its side with its head at upper right (note the gills).

Explanations for this observation are unclear, and include diminished cardiac performance associated with capture stress and/or pooling caused by the effects of gravity and unusual postures when the animal was struggling or lying on the deck without the buoying support of the seawater. Regardless of the direct cause, it is possible that rupture of one or more of the venous sinuses could contribute to stress-related or post-release mortality in some species.

Circulatory specializations in warm-bodied (i.e., *endothermic*) sharks are discussed in chapter 8.

The Heart of Sharks

Do you know how many chambers the chondrichthyan heart has? The answer is four, although unfortunately many textbooks of biology frequently mistakenly state that chondrichthyans have two-chambered hearts.

The explanation for this widespread, monstrous mistake is *mammalian egocentrism*. Mammals have four contractile chambers, two *atria* and two *ventricles*, that constitute their heart. Chondrichthyans have a single atrium and a single ventricle. Even though they possess two other chambers composed of contractile, cardiac muscle (the *sinus venosus* and *conus arteriosus*), our human bias resulted in degrading lower vertebrates like sharks and their relatives by not counting their additional chambers as part of the heart, which indeed they are. Thus, the heart of chondrichthyan fishes is a *bona fide* four-chambered heart (fig. 7.6). Unlike the mammalian heart, in which there are two separate atrium-ventricle pairs *in parallel* with each other, and hence two separate circulations, the chondrichthyan heart consists of four chambers *in series* and only a single circulation, as we described earlier (see fig. 7.1).

In chondrichthyans, all venous blood returns to the first chamber of the heart, the *sinus venosus*, then it moves to the atrium, ventricle, and the *conus arteriosus* (Figs. 7.4 and 7.6) before being sent to the gills. The chondrichthyan heart thus pumps relatively deoxygenated blood.

How does a muscle that depends on aerobic metabolism (.i.e., requiring a continuous source of oxygen) ceaselessly pump if the blood it is pumping has lost most of its oxygen before returning to the heart? First, the blood within the heart, called the *luminal* blood (named because it is found in the lumen, or space, of a tubular structure), is mostly but not completely deoxygenated. Thus, the small amount of oxygen remaining in the luminal blood supplies the myocardial tissue, albeit minimally. This method of supplying oxygen to the heart likely constitutes the primitive condition among all chordates, with jawless fishes being the only extant chordates utilizing luminal blood as the only source of oxygen for the myocardium.

Second, the chondrichthyan heart has *coronary vasculature* (heart arteries) that carry blood to the heart muscles from arteries that peel off from the circulation after the blood has been oxygenated in the gills (fig. 7.3). This constitutes the major source of oxygen for the elasmobranch heart. A recent study[4] showed 3–4% of the entire *cardiac output* (the volume of blood pumped by the heart per unit time) of Sandbar Sharks (*Carcharhinus plumbeus*) constituted coronary circulation.

All four cardiac chambers of sharks are contractile and thus contain cardiac muscle, but each chamber has a different function and architecture. The chambers are separated from each other by valves, which are closed, preventing backflow, until pressure develops during contraction of each chamber and opens them. Refer to Figures 7.4 and 7.6 as we describe the chambers.

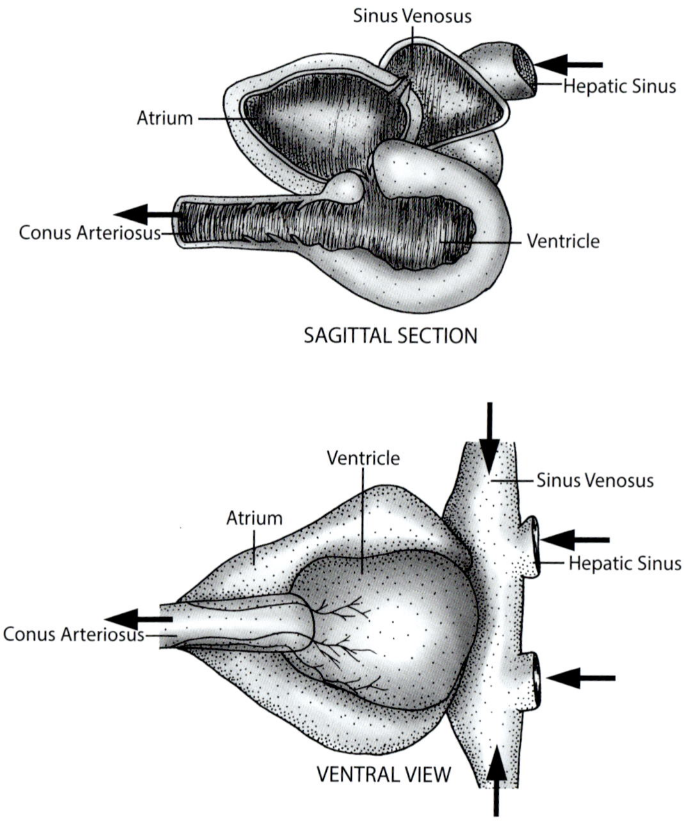

A

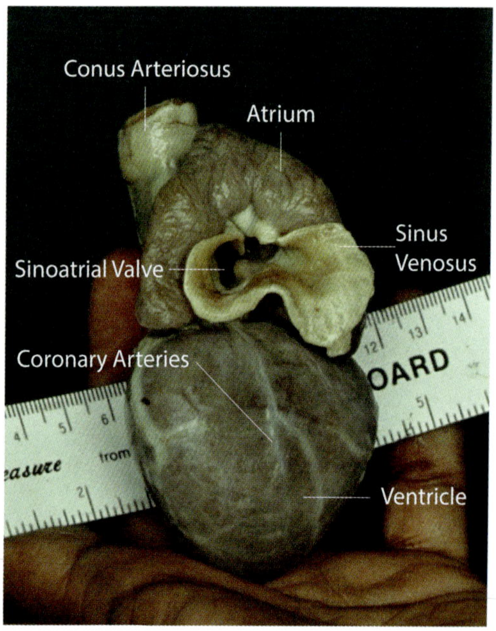

B

Figure 7.6. (A) Anatomy of the shark heart. Top: Side view. Bottom: Ventral view. Arrows depict direction of blood flow (Redrawn from Gilbert, S., 1973. *Pictorial Anatomy of the Dogfish*. University of Washington Press). (B) The heart of a Blue Shark.

The first (most upstream) part of the heart, the *sinus venosus,* is a small, very thin-walled chamber that receives the blood from the venous circulation. Do you have a sinus venosus? Not *exactly.* Chordates, of which you are one, possess a sinus venosus at some point in their embryonic development. It persists to adulthood in hagfish, bony fishes, chondrichthyans, and amphibians, but is found in only embryos in other chordates. As the mammalian embryo develops, the sinus venosus is modified and no longer exists as a chamber, but rather as specialized assemblages of cells, including the *sino-atrial node,* the pacemaker of the mammalian heart that initiates the heartbeat. Next time you see a fish, acknowledge its role in initiating your heartbeat!

Blood entering the sinus venosus moves into the atrium through a *sino-atrial* valve. In contrast to the small size of the sinus venosus, the chondrichthyan atrium is capacious—essentially a large sac—though it is still thin-walled.

Blood flows from the atrium through an *atrio-ventricular* valve into the ventricle. The ventricle is the workhorse of the chondrichthyan heart. It is the heaviest of the chambers, and it contracts with the greatest force. The outer layer of the ventricle consists of *compact myocardium* overlying a thicker later of *spongy myocardium.* The lumen of the chondrichthyan ventricle is small relative to the size of the ventricle. Most of the blood in the ventricle resides in its spongy myocardium, which is said to be *trabeculated;* that is, containing supporting beams of cardiac tissue that create a meshwork, much like a sponge.

Finally, blood passes from the muscular ventricle through a *semilunar* valve into the *conus arteriosus,* another contractile chamber, but more tubular and with a variable number of its own semilunar valves. Unlike the *bulbus arteriosus* in bony fishes, the elasmobranch conus arteriosus is a contractile chamber with myocardial muscle in its walls. Contraction of the conus arteriosus does not greatly increase pressure but rather maintains the pressure developed by contraction of the ventricle for a longer period as blood moves into the ventral aorta, thus protecting the gills from large pressure swings. A conus arteriosus is found only in dipnoans (lungfish), holosteans (bowfin and gars), and chondrichthyans.

A recent paper[5] documented the presence of bulbus arteriosus tissue at the distal end of the elasmobranch heart between the conus arteriosus and the ventral aorta. The function and phylogenetic significance of the bulbous arteriosus tissue in elasmobranchs is unknown.

The chondrichthyan heart's electrical activity can be seen in an EKG, or electrocardiogram, although it differs from that of a human in that contraction of the sinus venosus and conus arteriosus are evident in the former. A very dangerous EKG pattern in humans is known as *Shark Fin Electrocardiogram* because of the resemblance of this abnormal wave form to a shark's dorsal fin.

The heart of all vertebrates is surrounded by a *pericardium,* also called a pericardial sac (figs. 7.2, 7.7, and 7.8).

Three features of the elasmobranch pericardium distinguish it from that of other vertebrates. First, the pericardial walls largely adhere to surrounding tissue,

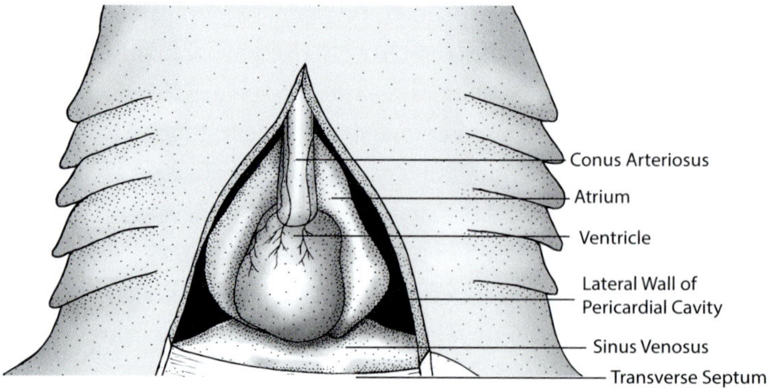

Figure 7.7. Diagram showing a ventral view of the heart of a shark in the pericardial space, which is extremely spacious and whose walls are connected to surrounding muscle and cartilage (with the exception of the transverse septum). Note that the atrium partially envelops the ventricle. The lightest-colored chamber is the conus arteriosus.

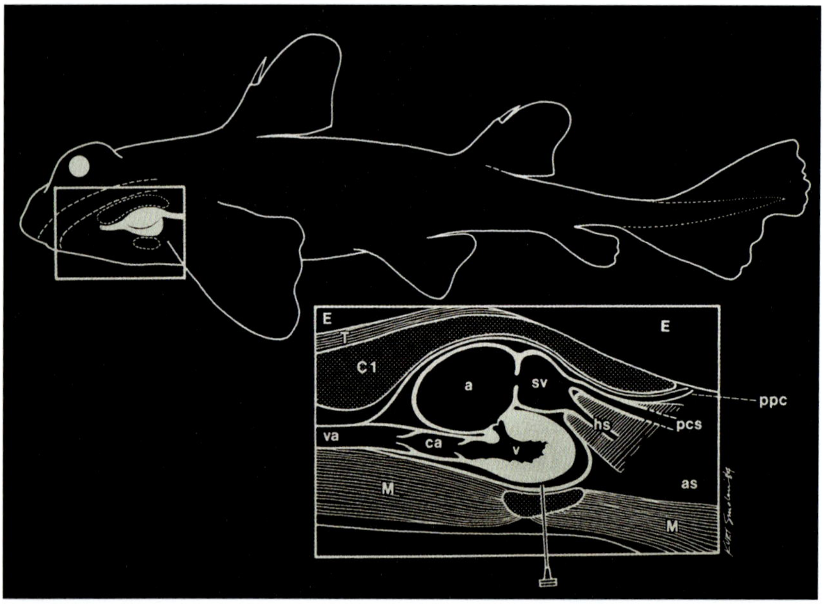

Figure 7.8. Diagram of the pericardium and pericardio-peritoneal canal (labeled *ppc*). Other labels: a = atrium; as = abdominal space; C1 = basibranchial plate (cartilage); C2 = coracoid bar (cartilage); ca = conus arteriosus; E = esophagus; hs = hepatic sinus; M = muscle; pcs = postcardinal sinus; sv = sinus venosus; v = ventricle; va = ventral aorta; T = tongue (Courtesy of Kurt Smolen).

mostly muscle. Because of this, the elasmobranch pericardium is frequently called *semi-rigid* or relatively *non-complaint*. Only the transverse septum (fig. 7.7), a fibrous sheet that separates the pericardial cavity from the abdominal (peritoneal) cavity and is attached only along its edges to surrounding tissue, is free to move. Second, the pericardial space of elasmobranchs is capacious. Under physiological conditions (normal activity level), it contains around 2.0 ml/kg (2 ml in a 1 kg shark) in the Horn Shark compared to about 0.3–0.8 ml/kg in humans. The shape of the pericardial space differs between batoids and dorso-ventrally flattened sharks (e.g., angel sharks and carpet sharks) on the one hand, where it is expanded more laterally, and fusiform sharks and batoids (e.g., sawfish) on the other, where it is expanded vertically. Correspondingly, heart shape also varies as a result of these different pericardia architecture.[6] Third, elasmobranchs have a duct that connects the pericardial and peritoneal spaces called the *pericardio-peritoneal canal* (PPC; fig. 7.8), which we included in chapter 1 as an unrecognized distinguishing feature of this group. A connection between these two spaces is a feature in the early embryological development of all vertebrates, but persists only in elasmobranchs, some agnathans (jawless fish), and sturgeon. The significance of these three features to heart function is described here.

Let us see how these pieces work in unison to do what a cardiovascular system does—pump the appropriate amount of blood to where it is needed precisely when it is needed.

We begin with the heart, specifically with a description of a mechanism so simple and elegant that it is included in every textbook in which heart function in sharks is discussed. Unfortunately, it is only partially accurate.[7]

According to this model, when the muscular ventricle contracts within the semirigid pericardium, two actions occur. First, the blood in the ventricle is propelled to the conus arteriosus and into the branchial (gill) circulation. This much is in fact true, and physiologists call this *vis a tergo* (force from behind).

Second, since most of the pericardial walls are affixed to surrounding muscles, they behave as a semirigid box (think thin cardboard or plastic milk bottle, not concrete), and are only minimally drawn in as the volume in the pericardial space is reduced when the muscular ventricle contracts and pumps blood into the conus arteriosus. Only the transverse septum that separates the pericardial and abdominal spaces is moveable, and it may be pulled inward slightly. This part of the story is also true.

As a result, according to the model prevalent in physiology textbooks, pressure in the pericardial space obeys the laws of physics and declines and becomes very negative, or subambient, like the pressure in a straw the moment you begin to draw up the fluid (fig. 7.9). This diminished pressure pulls open the walls of the thin-walled atrium, immediately upstream of the ventricle, and the suction developed within the atrial space aspirates, or sucks, blood from the upstream sinus venosus and blood sinuses into the atrium. Physiologists refer to this phenomenon as *vis a fronte* (force from the front). This is the part that is not entirely accurate.

In the 1980s, a new conceptual model[8] emerged that showed that pericardial pressures were only *slightly* rather than *very* negative under routine, non-stressed conditions, and that this pressure played a role, in concert with muscular pumps in the sharks' trunk (similar to those in your legs), to return blood to the heart. There are one-way valves in some of the veins to prevent backflow of blood.

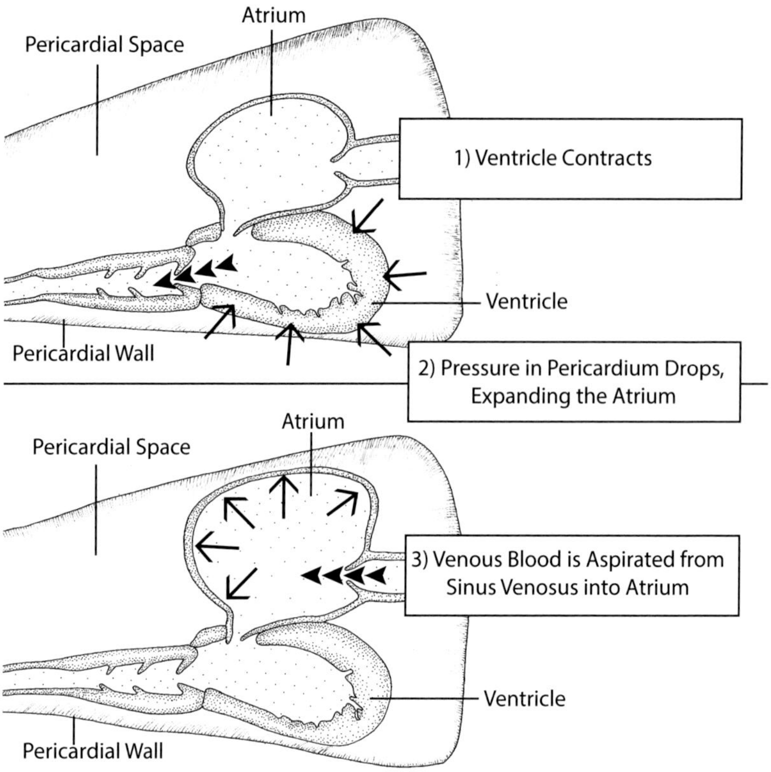

Figure 7.9. Series of drawings showing motions of heart within pericardium accompanied by pressure tracings (Redrawn from fig. 1, Burggren, W.W. 1988. Experientia, 44: 919–930).

However, as we explain in the subsequent section, under certain factors (e.g., when accelerating to catch prey or repositioning food in their mouths) after a transient increase in pericardial pressure, the pressure drops to the more negative levels assumed in the old model. We will see that the PPC plays a pivotal role in this process, and the new conceptual model is even more elegant than the one it replaced.

Responses of the Heart and Circulation to Activity

As in all vertebrates, the heart plays a major role in supporting increased activity levels and meeting energetic demands that may accompany predation, avoiding being preyed upon, etc. During these times, muscle and other tissues require additional oxygen and fuel and they must eliminate carbon dioxide and potentially toxic metabolites (end-products) like lactic acid. Thus, the *cardiac output* (the volume of blood pumped per unit time) must be increased under these circumstances. Cardiac output is numerically equal to the volume of blood pumped per beat (the *stroke volume*) multiplied by the number of beats per minute (the *heart rate*) and is typically reported as $ml \cdot min^{-1} \cdot kg^{-1}$ (kg represents body weight and is included to account for differences in body size).

There are only two pathways to achieving increased cardiac output in vertebrates—increasing stroke volume and/or heart rate. In sharks and batoids, faster swimming will push more blood back into the heart (recall the muscular pumps we alluded to), supplementing the suction we previously described. In response to the additional blood, the ventricle will expand and then contract with greater force and thus will increase stroke volume. This is an important physiological principle known as *Starling's Law of the Heart* (or the acronym SLOTH), an exquisite way of quickly reacting to exigent situations. You could think of this as, once again, the tail wagging the dogfish. Small increases in heart rate can occur in chondrichthyans, but these are much more limited in scope than in other vertebrates.

Heart rate varies among elasmobranchs. In the Horn Shark, average routine heart rate was about 40 beats per minute (bpm), and in the low- to mid-50s in Lemon (*Negaprion brevirostris*), Sandbar, and Shortfin Mako Sharks. Maximum heart rate found in sharks has not been shown to exceed 60–70 bpm. In contrast, maximum heart rate can be greater than 200 bpm in tuna.[9] Thus, increases in stroke volume constitute the largest contributor to cardiac output in sharks and batoids.

As this chapter's opening anecdote portrayed, sharks with high performance cardiovascular systems, like White Sharks and Shortfin Makos, have a larger heart, especially the ventricle, than a Horn Shark, and may have cardiac output several times higher than the Horn Shark's during activity, even if the heart rates are only slightly different. A recent study[10] in our lab documented smaller heart sizes in deep-sea sharks, which makes sense in light of the decreased activity of these sharks and is consistent with the finding that heart size correlates with activity level in elasmobranchs.

Role of the Pericardio-peritoneal Canal (PPC)

The importance of the PPC to the life of sharks cannot be overstated, improving heart function in elasmobranchs in two general ways. First, it protects the heart. Second, it facilitates enhanced cardiac performance. Moreover, these two functions occur simultaneously, a physiological *buy one, get one free* exquisite cardiovascular adaptation, as we explain here.

Experiments with Horn Sharks, Blue Sharks (*Prionace glauca*), Swell Sharks (*Cephaloscyllium ventriosum*), Shortfin Makos, Dusky Smoothhounds (*Mustelus canis*), Thornback Skates (*Platyrhinoidis triseriata*), and others showed that the PPC is normally closed. During feeding and acceleration or prolonger rapid movement, actions that caused pericardial pressure to increase, the PPC opened and ejected a variable amount of pericardial fluid from the pericardial space into the peritoneal space through the PPC,[11] which lowered the pericardial pressure. This decompression of the pericardium protected the heart from potentially dangerously high pressures (cardiologists call pathological compression of the heart *tamponade*) that might occur in the routine life of sharks, e.g., during biting and swallowing food, as well as during faster swimming as the animal bends and the skin tightens. Here's the bonus function: the decreased pericardial pressure resulting from the shifting of pericardial fluid from the pericardial space to the peritoneal space translated into an increased return of blood to the heart. Recall that this increased blood return can be important for faster swimming through the SLOTH mechanism we discussed. One structure, two (or more) functions. Evolution rocks!

Blood in Chondrichthyans

Information on characteristics of the blood of chondrichthyans is somewhat scarce. Moreover, blood characteristics vary among species.

Whole blood is composed of blood cells in a fluid medium known as plasma, which is mostly water and dissolved substances (gases, proteins, glucose, minerals, etc.). The percentage of red blood cells, the dominant cellular component of blood, in whole blood is known as *hematocrit*. The pigment in the red blood cells that binds to and transports oxygen in vertebrates is *hemoglobin*.

Most information on blood of chondrichthyans (fig. 7.10) comes from two applied areas, animal husbandry[12] and biomedical science. In combination with additional scientific studies,[13] these have provided the following information:

- Chondrichthyans possess the cells characteristic of vertebrate blood, namely erythrocytes (red blood cells, which function in gas exchange), thrombocytes (platelets, important in clotting), and leukocytes (white blood cells, part of the immune response system).
- Chondrichthyans lack bone marrow, the primary tissue responsible for hematopoiesis (blood formation) in mammals, as well as lymph nodes. In their place they have a thymus and spleen, secondary sites of hematopoiesis in mammals. Elasmobranchs also possess two hematopoietic organs unique to them,[14] the epigonal organ (in the gonads) and Leydig's organ (in the lining of the esophagus). In bony fishes, hematopoiesis occurs in a part of the kidney.
- Erythrocytes of elasmobranchs are nucleated and about 2.5 times larger than those of mammals, whose red blood cells lack nuclei and are 3–5 times larger than those of bony fishes.
- Elasmobranchs have fewer erythrocytes than bony fishes, although elasmobranch erythrocytes have more hemoglobin than those of bony fishes.
- Bony fishes have higher overall mean hematocrit, hemoglobin concentration, and erythrocyte counts than elasmobranchs, which translates in greater capacity to carry oxygen. Among elasmobranchs, sharks have higher values than batoids.
- Among sharks, hemoglobin and hematocrit levels are higher in endothermic (i.e., warm-bodied) species (e.g., White Shark and Shortfin Mako [*Isurus*

Figure 7.10. Blood being withdrawn from the caudal artery or vein on the ventral side of a Sandbar Shark (Courtesy of George Boneillo).

oxyrinchus]) than in ectothermic species (most elasmobranchs). Likewise, values are higher in active sharks (e.g., Blacktip Shark, *Carcharhinus limbatus*) than in more sedentary species (e.g., Port Jackson Shark, *Heterodontus portusjacksoni*).

- Elasmobranchs typically have higher blood volumes than bony fishes, although there is considerable overlap. Elasmobranch blood volume measurements range from 18 to 80 ml/kg body mass compared to 30 to 70 ml/kg for bony fishes. In both cases, higher volumes are associated with more active fish. Because of their large blood volume and fragility of the blood sinuses, dissections of sharks can be very bloody.

In summary, like that of all vertebrates, the closed circulatory system of sharks and batoids operates under a range of situations to meet the needs of the animal and at the same time participate in whole body homeostasis, which is one of the major goals of physiological systems.

Respiration and Metabolism

In 1982, Scripps Institution of Oceanography physiological ecologist Jeff Graham and colleagues designed and supervised the construction of one of the biggest treadmills on the planet, a behemoth nicknamed the *elasmotunatron* (fig. 7.11) that was built on a 3.3 × 5.3 m (10.8 × 17.4 ft) pallet.

Figure 7.11. Shortfin Mako swimming in the elasmotunatron (Courtesy of Chugey Sepulveda).

The elasmotunatron, also called a *water tunnel respirometer* for reasons that will become clear later, was not destined for a fish gym but the NOAA research vessel *David Starr Jordan*. You could not walk or run on it, because it was a *water* treadmill, also known as a *flume*. The rather large device was constructed with the goal of swimming a Shortfin Mako or an Albacore (tuna) to determine the amount of oxygen consumed as it swam against currents of different intensity.

The rate of oxygen consumption is a proxy for an animal's metabolic rate, or energy use per unit time. The branch of science concerned with metabolic rates is *energetics*. Understanding how much energy an animal uses is a key ingredient to understanding its *energy budget*, an accounting of all of the energy an animal requires and all of the ways this energy is expended.

Back to the elasmotunatron. Because of the logistical difficulties of capturing a Shortfin Mako, one of the most metabolically-charged sharks in the ocean, while minimizing the stresses associated with capturing such a beast, only a handful of Shortfin Makos have been successfully tested. Do you think marine biologists will ever swim a mature White Shark in a specially designed water treadmill? Not any time soon, we suspect (see p. 288).

Why It Is Difficult to Breathe in Water

Acquiring oxygen from water is far more challenging than doing so from air. The first huge advantage goes to air, which contains an almost invariable 20.95% oxygen, or about 30 times the average amount (0.7%) in water. Some aquatic ecosystems are even devoid, or almost devoid, of any oxygen. The second, third, and fourth big advantages also go to air; it is lighter, flows more easily, and oxygen diffuses more readily in it.

The upshot is: getting access to 1 liter oxygen takes a volume of air weighing 0.0062 kg (0.01 lb)—an almost negligible amount—but getting the same volume of oxygen in water takes 143 kg (315 lb) of the medium! Thus, breathing water requires not only a greater expenditure of energy, but also the evolution of means to efficiently extract the limited oxygen. And these together impose limitations on the upper metabolic rates of bony fishes, sharks, and rays compared to marine mammals, which breathe oxygen-rich, lighter, more flowable air (although, to be fair, mammals are unable to breathe underwater, which imposes its own problems).

Metabolism and Energetics

Given that conservation of energy is central to the struggle for existence in animals, how an organism obtains and uses energy underpins all aspects of that organism's life, including its physiology, ecology, behavior, and evolution. Metabolism has been called a *central pillar of ecology*[15] because the acquisition and use of energy is fundamental to ecosystem structure and function. One practical application of knowing a fish's energy use involves fishery management, for which scientists and policymakers should know growth rates for all of the targeted species and bycatch in the fishery, which depends in part on energy consumption.

Central to understanding the energetics of any organism is knowing its daily energy budget. An energy budget accounts for how an animal uses, wastes, and eliminates the energy it requires for all of its energy-consuming processes (e.g., pumping the Na^+ and Cl^- ions during osmoregulation; see chapter 8).

An energy budget for an organism can be represented mathematically as:

$$C = G + M + E$$

Where,

C = consumption of energy (i.e., calories obtained from food)
G = growth in mass (i.e., energy stored in tissue)
M = metabolism (energy expended in daily life)
E = excretion (i.e., energy lost in urine and feces; these are sometimes separated into two terms, U = urinary excretion and F = egestion)

Sonny Gruber of the University of Miami and his students[16] calculated one of the first shark energy budgets, for a 1 kg (2.2 lb) juvenile Lemon Shark, as:

$$100\%\ C = 22.4\%\ G + 49.4\%\ M + 28.2\%\ E$$

Metabolism, which constitutes a central component of an organism's energy budget, is apportioned among a number of different uses, including the energy costs of locomotion, ventilation, growth and maintenance, reproduction, and specific dynamic action (i.e., the additional energy expended during assimilation of food, e.g., digestion and protein synthesis).

Measuring an organism's metabolism (its *metabolic rate*) may seem easy in theory but in practice it is very difficult, especially for sharks. Think about why this may be. First, you need a shark that is not stressed, a condition that is extremely difficult to achieve in recently captured or other captive animals. Second, there are several levels of metabolism, and ideally a biologist wants to know all of them, from the minimum amount of energy expended during rest to that associated with the maximum level of activity, as well as points in between. Measuring all of these requires a combination of expensive instrumentation, cooperative sharks, and good luck. Third, metabolism varies with a number of factors, e.g., water temperature, oxygen concentration, and condition of the shark. Finally, seemingly insurmountable logistical problems exist for measuring metabolic rates in rarely encountered sharks (e.g., Goblin Sharks, *Mitsukurina owstoni*), sharks living in extreme environments (the deep-sea, like Gulper Sharks, *Centrophorus granulosus*), big sharks (e.g., Whale Sharks, *Rhincodon typus*), or highly active sharks (e.g., Porbeagles, *Lamna nasus*).

Given these limitations and difficulties, you might be surprised to learn that metabolic rates have been measured or estimated for dozens of species. Before we tell you how these studies have informed scientists about metabolic rates of sharks, we will provide more background.

How is metabolic rate measured? Actually, the true rate at which an organism uses energy is not directly measured, since there is no easy way to measure the conversion of energy; that is, the number of ATP molecules (the biocompounds, frequently called a cell's *energy currency*, converted to the lower energy ADP molecules *in cells in every organ in the body*. Instead, *proxies* (another measurement that is proportional to or is a close estimate of metabolic rate) for metabolic rate are used.

The method most routinely employed involves measuring the amount of oxygen used, a process called *respirometry*. Since most of metabolism is aerobic (oxygen-using), oxygen consumption is one of the most valid estimates of metabolism. It is measured using sharks contained in respirometers, in which sharks either swim in circles (annular/circular respirometers) or against a current (swim tunnels/water treadmills).

Other methods include biotelemetry sensors, which, once implanted, can collect data on heart rate, tail-beat frequency, muscular contractions (electromyograms, or EMGs), muscle temperature (for endotherms only), production of metabolic water, and what is called overall dynamic body acceleration (ODBA).

The last of these, accelerometry, is being used more frequently since the advent of inexpensive accelerometers (found in, e.g., smartphones, drones, automotive airbags, and smart watches). Acceleration refers to the rate of change of velocity. As sharks swim, they expend energy for contracting the swimming muscles, and as this happens, there is acceleration of the body and its parts in different planes, and studies have demonstrated that acceleration is proportional to metabolic rate.

Accelerometers must first be calibrated using conventional techniques (e.g., swimming a shark with accelerometers in a water treadmill at various speeds).

One final but rarely used method involves measuring the activity of key regulatory enzymes involved in metabolic processes. These include citrate synthase, lactate dehydrogenase, and gatekeeping enzymes important in cellular respiration.

Early investigations of the metabolic rates of sharks were conducted on two smaller species, which were most easily captured and maintained in captivity, the Small-spotted Catshark (*Scyliorhinus canicula*) and the Spiny Dogfish (*Squalus acanthias*). These results suggested that sharks were more metabolically constrained than ecologically similar bony fishes. In other words, bony fishes were thought to have higher metabolic rates than elasmobranchs. However, the two shark species studied are no more representative of typical sharks than a couch potato or Olympic athlete are of typical humans.

Since the metabolic rates of the Small-spotted Catshark and Spiny Dogfish were measured, similar studies have been conducted on numerous other elasmobranchs. What these studies have shown is that standard metabolic rate (SMR), the minimum energy required for essential biological functions during rest, standardized for body size, was not significantly lower in elasmobranchs than in bony fishes, and that it correlated positively with environmental temperature and activity. Similarly, maximum metabolic rate was highest in more active species. The most active sharks (e.g., Shortfin Makos) are obligate ram ventilators and, as you would expect, have the highest mass specific metabolic rate. Recently, researchers employed accelerometers and speed sensors to estimate field metabolic rates (FMRs) of free-swimming Shortfin Makos. Although only a small number of comparative studies on sharks exist, the estimated FMRs for Shortfin Makos[17] in this study was higher than that of the Greenland Shark (*Somniosus microcephalus*)[18] and other large high-latitude sharks, as well as the Basking Shark (*Cetorhinus maximus*).[19]

The Shortfin Mako and other sharks with high metabolism also have higher blood oxygen capacities than less active sharks, and these are achieved through a combination of adaptations that include larger blood volumes, increased hematocrit levels, and higher hemoglobin concentrations.

A 1993 study[20] was among the first to hypothesize that sharks and rays have reduced metabolic rates compared to ecologically similar bony fishes. In this study, researchers did not measure oxygen consumption, but the speed of chemical reactions of the three enzymes considered biochemical indices of aerobic and anaerobic capacity in tunas and lamnid sharks: *citrate synthase, pyruvate kinase,* and *lactate dehydrogenase.* They found that tunas appear to have more and better biochemical specializations, in addition to anatomical and physiological, for high performance than lamnid sharks, even as the two endothermic groups have converged. They also concluded that the metabolic rates of non-lamnid sharks compared favorably to those of bony fishes with similar ecological roles.

Studying metabolism and energetics of free-ranging sharks, including larger animals and those inhabiting the deep sea, is important for managing fisheries and understanding ecological energetics (the sum of all energy flows in a marine ecosystem). Moreover, it is becoming critical to know the energetics of organisms in marine ecosystems as global climate change potentially impacts food supplies, activity levels, and other physiological processes.

For the comprehensive review of energetics of elasmobranchs that is beyond the scope of this book, we refer you to Gleiss et al.[21]

Concluding Comments

Sharks are not heartless, though their hearts are smaller than those of similarly sized mammals, and they function differently as well. Remarkably, these relatively small hearts work efficiently to meet the needs of Dwarf Lanterns Sharks, Whale Sharks, and all elasmobranchs in between, size-wise. And despite the difficulty of breathing in water, elasmobranchs have evolved a respiratory system that works in concert with the circulatory system to extract sufficient oxygen and deliver it and other important products to muscles and other tissues as needed and remove wastes. Elucidating how these functions are accomplished continues to enthrall and amaze us and, we hope, you too!

NOTES

1. In physiology, the term *afferent* is always used for something that is directed toward a tissue, organ, or organ system. In this case, that something is the gills.
2. Similarly, *efferent* refers to something directed away from a tissue, organ, or organ system.
3. Bony fishes and other vertebrates have greatly reduced sinuses embedded in the liver.
4. Cox, G.K. et al. 2017. J. Comp. Physiol. B 187: 315–327.
5. Durán, A.C. et al. 2008. J. Anat. 213: 597–606.
6. Männer, J., 2024. Preprints 2024, 2024070983. https://doi.org/10.20944/preprints202407.0983.v1. (Accessed 7/21/24).
7. See, Abel, D.C. et al. 1986. Fish Physiol. Biochem. 1: 75–83 and Lai, N.C. et al. 1996. Am. J. Physiol.Heart and Circ. Physiol. 270: H1766 - H1771.
8. Abel, D.C. et al. 1986. Fish Physiol. Biochem. 1: 75–83.
9. As cited in Brill, R.W. and Lai, N.C. 2015. Fish Physiol. 34: 1–82.
10. Larsen, M.E. et al. 2020. Mar. Biol. 167: 169.
11. Abel, D.C. et al. 1994. Fish Physiol. Biochem. 13: 263–274.
12. Walsh, C.J. and Luer, C.A. 2004. In: Smith, M. et al. (eds.). The Elasmobranch husbandry manual: captive care of sharks, rays and their relatives. 307–323 pp. https://serviettegroup.com/img/publications/collecting-elasmobranchs.pdf. (Accessed 10/1/24).
13. Haines, A.N. and Arnold, J.E. 2014. In: Smith, S.L. et al. Immunobiology of the shark. CRC Marine Biology Series. CRC Press.
14. Holocephalans apparently lack both organs, and use the thymus, spleen, and lining of the orbit for hematopoiesis.
15. Gleiss, A.C. et al. 2022. Biology of sharks and their relatives, 289–321 pp.

16. Gruber, S.H. 1984. Ann. Proc. Am. Ass. Zool. Parks and Aq. (AAZPA) 1984: 341–373.

17. Waller, M.J. et al. 2023. J. Fish. Biol. 103: 864–883.

18. Ste-Marie, E. J. 2022. Exp. Biol. 225: jeb242994.

19. Sims, D.W., 2000. Mar. Ecol. Prog. Ser. 200: 289–296.

20. Dickson, K.A. et al. 1993. Mar. Biol. 117: 185–193.

21. Gleiss, A.C. et al. 2022. Biology of sharks and their relatives, 289–321 pp.

8 / Thermal Physiology, Osmoregulation, and Digestion

CHAPTER OUTLINE

Introduction

Let's say you're a juvenile, meter-long (3.3 ft) Sandbar Shark (*Carcharhinus plumbeus*) swimming into Winyah Bay in NE South Carolina in late spring, having migrated there from overwintering in Florida as part of an annual ritual. You would not be unlike the people, some call them *snowbirds*, who travel from the NE United States to Florida to escape the frigid cold and return when the conditions become more clement. Similarly, you will return to your Florida digs when the water cools in late fall.

Being an estuary, Winyah Bay is connected to the ocean on one end, and to several rivers on the other. Thus, the salinity typically ranges from full-strength seawater at the mouth of the bay to full freshwater in the bay's upper reaches. You settle for an area where the salinity is intermediate, about mid-bay, where the ocean and river waters mix.

Your migration and selection of habitat have been driven by your physiological tolerances to and preferences for certain temperatures and salinities, as well as the advantages (e.g., abundant food, protection from predators) that your seasonal habitats may afford you. In this chapter, we'll examine closely the effects of temperature and salinity on elasmobranchs (including juvenile Sandbar Sharks) and their often-amazing physiological responses. We will also examine the process of digestion. (You can now return to being a person again).

Temperature

Introduction

Temperature is one of the most important environmental factors for all living organisms and is becoming more so as climate change impacts materialize. It affects not only their diversity, abundance, and distribution, but also their behavior, activity, growth, development, metabolism, heart rate, digestive rate, blood characteristics, and other aspects of their physiology at all levels from the molecular to organ systems.

Physiologically, these impacts typically influence the rates of chemical reactions, a phenomenon described by the *temperature coefficient,* or *Q10,* of a reaction or process. Q10 numerically equals the factor by which the reaction rate increases for every 10°C (18°F) increase in temperature. Q10s typically are around 2, that is, the reaction rate *doubles,* or in some cases, triples (i.e., Q10 = 3), for every 10°C temperature increase. For example, the Q10 for metabolic rate for the Horn Shark (*Heterodontus francisci*) is 2.01.

Organisms have both temperature *optima* and *preferences.* Exposure to temperatures outside of this range (beyond an organism's upper and lower thermal limits), or *thermal tolerance scope,* can lead to thermal stress, which may be directly lethal or have sublethal effects that will reduce an organism's fitness (its ability to forage, avoid being eaten, compete, mate, etc.).

The highest temperature an organism can tolerate before it begins a suite of irreversible actions leading to its death is called its *incipient lethal temperature* (ILT) or *critical thermal maximum* (CTmax). Temperatures in this range lead to a cascade of deadly physiological problems, including disrupting cellular function by destroying membranes and denaturing (disrupting the physical structure of) proteins.

The lowest temperature an organism is capable of tolerating is its *critical thermal minimum* (CTmin). Like heat-related death, cold death involves a cascade of physiological impacts; for example, central nervous system and cardiac malfunction, as well as metabolic, blood, and water balance changes.[1]

Both CTmax and CTmin can vary by species, season, life-stage, geography, and other factors. Given the range of factors that affect an organism's CTmax and CTmin, as well as the logistical difficulties inherent to studying sharks and batoids, it is not surprising that few of these have been determined for elasmobranchs, and few are based on observations of animals killed by temperature extremes in their environments more so than in controlled laboratory experiments.

The first CTmax determined for a shark, published in 1924 for the Spiny Dogfish (*Squalus acanthias*), was 29.1°C (84.4°F).[2] In laboratory studies, the CTmax for Atlantic Stingrays (*Hypanus sabina*) from St. Joseph's Bay, Florida, maintained in aquaria at 10.8°, 20.5°, or 35.1°C (51.4°, 68.9°, and 95.2°F) were 35.7°, 39.3°, or 43.2°C (96.3°, 102.7°, and 109.8°F), respectively. In other words, sharks and rays living in warmer waters had higher tolerances than those in cold, and the higher the temperature of acclimatization in captivity, the higher the CTmax and the tolerance to high temperatures.

CTmin for Atlantic Stingrays acclimatized as mentioned earlier were 0.8°, 4.8°, or 10.8°C (33.4°, 40.6°, 51.4°F), respectively.[3] Thus, the lower the temperature of acclimatization, the lower the CTmin and the higher the tolerance for lower temperature.

Caution should be exercised, however, in extrapolating these findings to conclude that elasmobranchs will similarly increase or decrease their thermal tolerances as their habitats warm or cool as a result of climate change. Evolution of physiological tolerances is a complex phenomenon that may require longer durations than the accelerated pace of climate change-associated temperature changes.

Within the normal range of temperatures in which organisms are found, other variables being equal (e.g., prey availability), there are advantages to living nearer to the higher temperatures. Physiological processes (e.g., rates of digestion, processing of sensory information, muscle contractions, and rates of chemical reactions) will occur more quickly, along with development *in utero* and growth rate.

For aquatic organisms, however, it is hard to be warm in anything but tropical water because of the thermal properties of water, including its high heat capacity (3000 times that of air) and its thermal conductivity (25 times that of air). These translate into rapidly losing heat in water, about 75,000 times more heat than is lost in air. Some animals (e.g., marine mammals, some fishes) have evolved ways to keep warmer than their surroundings, but most aquatic organisms have a body temperature the same as that of the environment in which they reside.

We call animals whose body temperature is determined primarily by the temperature of the environment *ectotherms* (from *ekto* = out; *therm* = temperature).[4] All invertebrates and most bony fishes and chondrichthyans are ectotherms.

Consider the effect of temperature on just one factor—metabolic rate (how fast an organism uses energy). The metabolic rate of ectotherms decreases in colder temperatures and increases in warmer ones. This translates into being less active with less food required for energy in the former and more active with more food in the latter. Deep-sea sharks, specifically those that do not migrate vertically daily, live in an environment whose temperature is about 36–41°F (2–5°C) day and night, across all seasons. These animals have a lower metabolic rate than, say, a Blue Shark (*Prionace glauca*) at the surface. Deep-sea sharks thus expend less energy and require less energy from prey than shallow-water counterparts, which is convenient since the deep-sea is a food-poor environment. This deep-sea lifestyle often leads to relatively delayed life-history characteristics, such as age at maturity or gestation.

While ectothermic sharks do not possess physiological specializations that allow them to maintain body temperatures above those of their environment like the *endothermic* sharks described here later, many migrate seasonally to warmer waters and some select higher water temperatures on shorter time scales, presumably as an adaptation that confers upon them one or more of the advantages of the higher temperatures.

In California, female Leopard Sharks (*Triakis semifasciata*) congregate in warm shallow coastal waters during the summer, in the process elevating their core body temperature by as much as 3°C (5.4°F), and in doing so increase their metabolic rate by about 17%.[5] Moreover, the sharks in the warm shallow waters darkened their skin, further increasing their heat gain. These behavioral and morphological adaptations apparently make these sharks better predators, although observational verification is lacking.

These Leopard Sharks leave these same areas in late afternoon as these shallow waters cool faster than deeper ones. Movement to deeper and cooler waters is thought to slow the rates of digestion and defecation, in the process improving the efficiency of extracting nutrients from the food being digested. In other species that feed more frequently, however, the opposite is thought to occur (feeding while cool and more quickly digesting and evacuating while warm).

Similarly, Blacktip Reef Sharks (*Carcharhinus melanopterus*) aggregate in warm shallow waters during daylight and disperse at night.[6] These consistently exhibited core body temperatures of about 1.3°C (2.3°F) above that of their environment. This temperature increase persists for a variable time (depending on the shark's body size and activity as well as the water temperature) even after leaving the warm shallows, a phenomenon called *thermal inertia*.

Bluntnose Sixgill Sharks (*Hexanchus griseus*) migrate vertically in the water column diurnally. In a study conducted in the NW Atlantic, Gulf of Mexico, Central Pacific, and Bahamas, individuals spent days at deeper depths than in evenings, although the depth disparity varied from < 50 m (160 ft) in the NW Atlantic to > 300 m (990 ft) in other locations. It would be tempting to attribute this exclusively to following prey as they migrate vertically, but closer examination showed that these differences in depth in all four locations reflect temperature preference for 5°C (41°F) in the day and 16–17°C (61–63°F) in the evening.

Sandbar Sharks also migrate, at least in part based on water temperature. In Chesapeake Bay,[7] Sandbar Sharks were caught on experimental longlines from late-May until mid-October, with most immigration in mid-June when bottom temperature was above 20.0°C (68°F). Emigration, however, correlated more with day length (photoperiod) than temperature, although the study caught no Sandbar Sharks when bottom temperature was below 15.9°C (60.6°F).

Endothermic Sharks

Only a handful of bony fishes and elasmobranchs have diverged from the mainstream and have evolved to be *endotherms*; that is, their body temperature is determined to a greater extent by the metabolic heat they produce—their metabolic furnace—than in ectothermic sharks and bony fishes. Recall that chemical reactions that involve energy conversions are inherently inefficient, about 40% or so in living systems, meaning that as much as about 60% of the energy is lost as heat.

The major problem an aquatic organism faces in evolving endothermy is preventing heat loss. For bony fishes and sharks, most of the heat is lost across the skin and gills. Those same architectural traits of gills that enable them to obtain oxygen and eliminate carbon dioxide, specifically their large surface area and thinness, are huge liabilities for retaining heat since the conductivity of heat is so much higher than that of respiratory gases. As much as 80–90% of the heat of blood is lost in a single passage through the gills, which represents between 20–40% of the total heat loss.

Given that the gills of fish have adapted as gas exchange organs with large surface areas, there have been no specializations in evolution to conserve heat there. Instead, the focus in evolution has been on preventing heat loss from the body wall. Marine mammals and seabirds (e.g., penguins) have become masters of heat conservation by using thermal mass, blubber, fur, and/or feathers to trap heat inside. It helps that all marine mammals and birds breathe air, and thus do not experience heat loss through the gills (since they lack them), avoiding the 20–40% heat loss of fish.

Members of only seven fish families, Scombridae (14 species of tunas), Lampridae (one species of Opah), Istiophoridae (marlin and other billfishes), Xiphiidae (swordfish), Lamnidae (five mackerel sharks), Alopiidae (three species of threshers), and Mobulidae (at least two species of mantas) have evolved ways of trapping heat and elevating their body (or specific parts of their body) temperatures above that of the environment, a phenomenon known as *regional endothermy*[8] (so named because there are still temperature gradients, i.e., body regions that are warmer than others by a degree or more). Moreover, anatomical characters of at least two other sharks in the order Lamniformes, the Smalltooth Sand Tiger (*Odontaspis ferox*) and the Basking Shark, suggest that these two species may also exhibit some degree of regional endothermy.[9] Since scombrids, lamprids, istiophorids, and xiphiids on the one hand, and lamnids, alopiids, and mobulids on the other, are also in two different taxonomic classes that diverged as long ago as 450 million years, their regional endothermy is an example of *convergent evolution,* in which distantly related or unrelated organisms evolve similar anatomical, physiological, or behavioral traits (see chapter 2). We will focus on lamnid sharks, but the general principles *mostly* apply to scombrid bony fishes as well, although there are anatomical differences.

The ability to retain heat in lamnids is due to the evolution of structures in the circulatory system that allow the engineering principle called *countercurrent exchange* to take place. In countercurrent exchange, there will be virtually complete transfer of some substance or property, in this case heat (but in gills, oxygen as well as heat), from one fluid to another across a permeable membrane if the separated fluids are juxtaposed and running in opposite directions for a distance long enough for the transfer to occur (see fig. 4.28). In co-current (or concurrent) exchange (i.e., when the fluids run in the same direction), only 50% of the heat at best could be transferred.

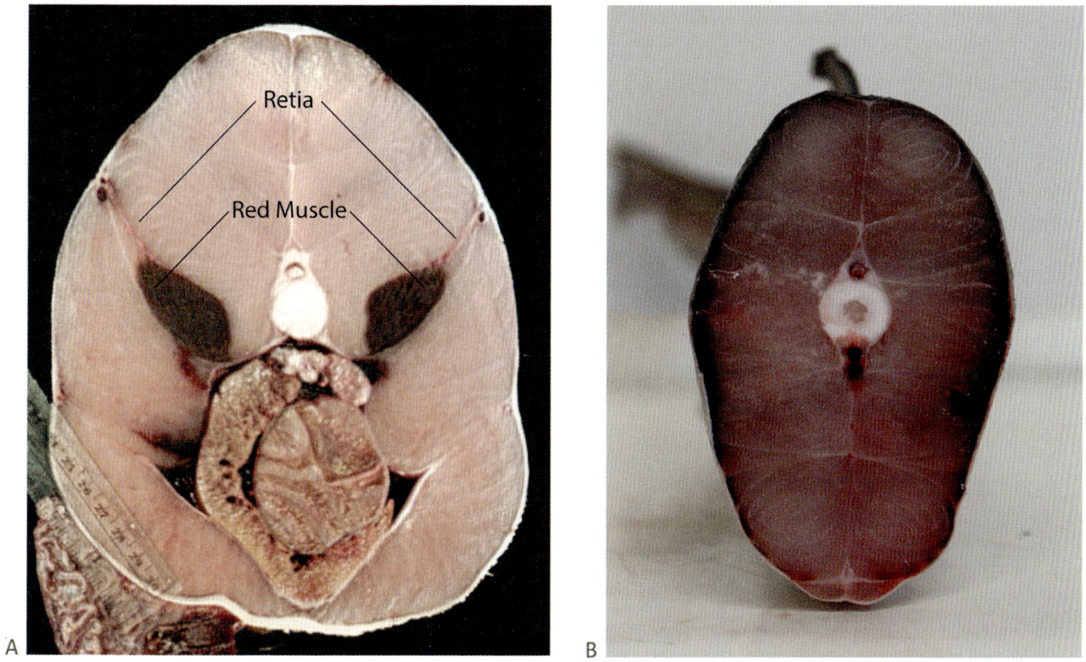

Figure 8.1. (A) Cross-section through an endothermic shark, a 125 cm (4.1 ft) Shortfin Mako, showing its retia mirabilia, centralized red muscle, and white muscle (Courtesy of Jeff Graham Lab, Physiological Research Laboratory, Scripps Institution of Oceanography). (B) Section through an ectothermic shark, here, a Bonnethead, showing the narrow strip of red muscle confined to the periphery (Courtesy of Bryan Keller).

These countercurrent exchangers are known as *retia mirabilia* (singular: *rete mirabile* = wonderful net) and consist of intermingled blood vessels, one set (venules, or small veins) carrying blood warmed near the shark's core outward, surrounded by another set (arterioles, or small arteries) carrying cooler blood inward from the periphery (near the skin; fig. 8.1). By the time the warmer blood reaches the periphery, the heat has been transferred to the cooler blood moving toward the core, and thus the core retains the heat.

Lamnid sharks (and tunas) are capable of using countercurrent exchange because they have a special arrangement of their main blood distribution system. In most sharks and bony fishes, there is a central blood distribution system, with blood from the gills going to a large dorsal aorta, then to the tissues (muscles and organs), after which it returns through the post-cardinal vein (fig. 8.2).

The major vessel distributing blood from the gills in lamnids sharks and tunas is not a single dorsal aorta, but, depending on the species, one to four artery-vein pairs just under the skin (fig. 8.2). The lateral artery or arteries carry cooled blood from the gills, and the lateral vein or veins are the main route via which blood returns to the heart. The retia are comprised of the arterioles that branch off of the lateral arteries running next to venules that branch off of veins nearer to the backbone (or body core).

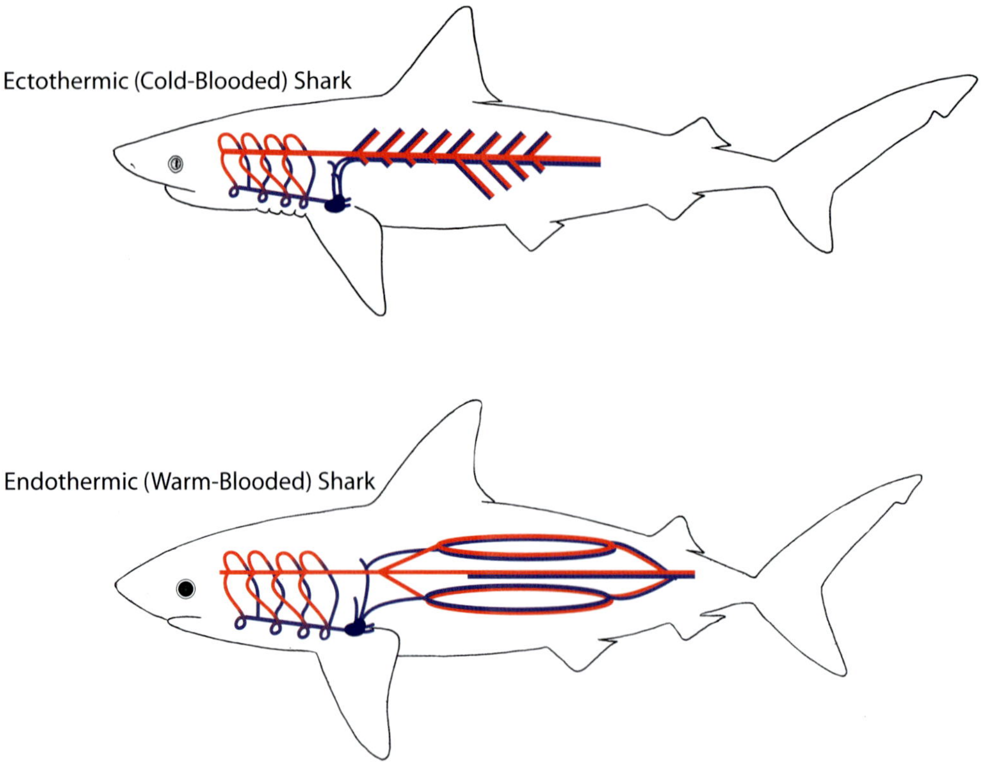

Figure 8.2. Diagram of differences in arterial blood distribution in ectothermic and endothermic sharks (Redrawn from Carey, F.G. 1974. In: Wessells, N.K. (Ed.). *Vertebrate Structures and Function*. Freeman. pp. 236–244).

In lamnid sharks, the retia are located not only in the locomotory muscles (the *lateral cutaneous* retia) but also in the cranium near the eyes (*orbital* retia) and in the viscera (*suprahepatic* retia). These produce varying levels of red muscle endothermy, cranial endothermy, and visceral (i.e., organ) endothermy, respectively.

The main source of heat for the lateral cutaneous retia is *red muscle* (fig. 8.1)—muscle that is red in color due to high vascularization and myoglobin (a relative of hemoglobin that stores oxygen in muscle) content. Red muscle is used primarily for cruising or slow-speed swimming and is characterized by being slow-twitch, aerobic, and having small-diameter fibers. Interestingly, the amount of red muscle in endothermic and ectothermic sharks is similar, about 2% of body mass. However, the location of the red muscle bundles differs between endothermic and ectothermic sharks. In the former, the red muscle is near the backbone, whereas in the latter it is found near the skin. Thus, in endothermic sharks, contractions of the deeper red muscle generate the heat that the countercurrent exchangers conserve.

The countercurrent exchangers allow lamnid sharks to maintain temperature differentials from 8°C (14°F) over ambient in the Shortfin Mako (*Isurus oxyrinchus*) to 14.3°C (26°F) in the White Shark (*Carcharodon carcharias*) to as high as 21.2°C (38°F) in the Salmon Shark (*Lamna ditropis*).[10]

Two hypotheses have been offered to explain the major advantages of red muscle endothermy.[11] The first is the *thermal niche expansion* hypothesis, which asserts that endothermy allows fish to move independently of temperature—that is, to tolerate a broader range of temperatures both latitudinally and vertically within the water column—than their ectothermic counterparts.

The second hypothesis is that endothermy enables elevated cruising speeds. All endothermic fishes are fast-swimming, highly mobile predators. Thus, advantages related to their high-performance lifestyle include more frequent prey encounters, more efficient prey capture, enhanced sensory capacity, and more efficient digestion.

Both explanations seem plausible. A 2015 paper[12] supported the cruising speed hypothesis, using modern tagging techniques on free-swimming animals. Their results, that cruising speeds and maximum annual migration ranges of tunas and sharks with red muscle endothermy are 2–3 times greater than ectothermic relatives, also validated the niche expansion hypothesis. Two Shortfin Makos fitted with bio-loggers in a different study[13] had a sustained cruising speed of 0.90 m per s (2 mph), and a 2 m female Shortfin Mako swam for bursts of 5.02 m per s (11.2 mph). Both measured speeds were higher than those of ectothermic sharks. Another study[14] showed that cruising speeds of four regionally endothermic fishes, including two sharks, were about 1.6 times faster than 12 ectothermic fishes (10 of which were sharks). This same report reviewed data from 89 published studies of the range of temperatures encountered by endothermic and ecothermic bony fishes and elasmobranchs and rejected the thermal niche hypothesis. The contradictory conclusions of these two studies suggest that the niche expansion hypothesis requires further examination.

Irrespective of these two hypotheses, there are other physiological advantages of regional endothermy. Warming the brain specifically (cranial endothermy) increases sensory acuity and processing, since temperature increases rates of chemical reactions. Thus, endothermic animals likely are superior prey finders and predator avoiders as well.

Theoretically, the main cost of being endothermic is that the metabolic furnace generating the heat must be continually fed. An estimate of the food required by an organism of a particular life stage is its *daily ration*, which is typically reported as the mean percentage of an organism's total weight that is consumed over a 24-hour period. Surprisingly, however, so far these do not clearly demonstrate that endothermic sharks and tunas necessarily consume more food daily on average than their ectothermic relatives.

For ectothermic sharks, daily ration estimates cover a wide range, including 0.2% for adult Broadnose Sevengill Sharks (*Notorhynchus cepedianus*), 2.1% for juvenile Lemon Sharks (*Negaprion brevirostris*), and > 4% for adult and juvenile Bonnetheads (*Sphyrna tiburo*).[15]

For endothermic sharks, daily ration has been studied only in adult Shortfin Makos, with estimates ranging from 2.2–3.0% in one study[16] to 4.1% in another.[17] The latter calculation was based on data not available to the former and is likely more accurate.

The lower-than-expected daily ration calculations for endothermic sharks and bony fishes could be real and explained by higher digestive efficiencies or more calorically dense prey. Alternatively, they could be artifacts of using inexact data (e.g., metabolic rates, growth rates) in models and a general lack of studies on the subject in a wide variety of sharks and tunas.

Two final points. First, no endothermic sharks are small as adults. The explanation for this is basically the same as why there are no small marine mammals, and it invokes a little geometry and physics. Basically, a larger animal stores more heat in its thermal mass than a smaller one, and it also gains and loses heat more slowly. Smaller sharks are not capable of keeping the metabolic heat that is produced, or the heat they gain from behavioral thermoregulation, for very long. Thus, it is easier for large-bodied sharks, with their smaller surface area to volume ratio compared to small-bodied sharks, to maintain a stable body temperature when they enter cooler waters below the thermocline. This phenomenon was recently shown in Whale Sharks,[18] which do not possess countercurrent heat exchangers.

Finally, a study[19] that looked at isotope levels in the teeth of extant and fossil sharks concluded that *Otodus megalodon*, aka Megalodon, was very likely a regional endotherm.

Osmoregulation

Osmoregulation[20] refers to the phenomenon of actively maintaining water and ion balance in an organism. Among vertebrates, for proper internal function, both the volume and the solute composition (dissolved components) of an organism's fluid compartments must be regulated so that homeostasis (the constancy of the structure and function of the internal environment) is maintained.

In what ways does the fluid environment affect biological function? Here are just a few:

- The structure of proteins, one of the most important classes of biological molecules, is directly affected by the ionic composition of the internal fluid environment. The function of proteins as enzymes, hormones, and energy, in transportation and as structure, are all highly dependent on their complex 3-dimensional structure, which can be disrupted by changes in the fluid environment.
- Ionic composition influences pH, which affects protein structure and function.
- Cells rely on electrical gradients across their membranes for transportation of atoms and molecules and for the development and conduction of nerve impulses. These gradients rely on the proper ionic environment.

The problem that virtually all aquatic vertebrates must face, in fresh or salt water, is that their internal environments *differ* from their external environments in terms of both *ionic* and *osmotic* characteristics; that is, in terms of the concentration and type of atoms (ions) and molecules as well as the percentage of water. These differences result in *gradients* (differences in the concentration over a distance or across a barrier) of the ions and water inside and outside of the organism. More specifically, these are called *ionic diffusion gradients* and *osmotic gradients*, respectively.

According to Fick's Laws of Diffusion, ions and water will each diffuse from where they are in higher concentration to where they are in lower concentration. We do not often think of the concentration of water in a solution, but the concentration of the dissolved salts or ions. But, for our purposes, we can assume that if solution 1 is more concentrated in terms of its ionic composition than solution 2, then the opposite will apply to their water concentrations (i.e., solution 2 has a higher water concentration than solution 1).

An important point here is that to achieve homeostasis (i.e., to regulate their internal environments), aquatic vertebrates can regulate their *water balance*, their *salt balance*, or *both*.

Osmoregulation in Bony Fishes

Let us first consider osmoregulation in saltwater bony fishes; for example, a tuna. If you measure the internal concentration of all of the solutes in the blood and other extracellular fluids of a tuna (thus, ignoring the fluid environment *inside* the cells, which is regulated differently), you will find that they are not nearly as concentrated as the salt water in which the tuna resides.

At this point it is necessary to introduce some terminology. We caution you not to be intimidated, because the concepts underlying osmoregulation in both bony fishes and sharks and their relatives are fairly straightforward and rely on only simple physical principles, even if the terminology may take some time to learn.

The salt concentration of seawater is referred to as its *salinity*, and historically this was measured in *parts per thousand* (ppt); that is, the weight of the salts (expressed in grams) per 1000 g (1 kg or 2.2 lb) of seawater. The average salinity of seawater, expressed as ppt, is about 35 (written as 35‰, or 3.5%). Oceanographers continue to use units of ppt, but salinity is typically measured using conductivity rather than density and is thus reported as dimensionless units (e.g., 35 rather than 35 ppt).

Physiologists do not refer to the internal *salinity* of an organism. Instead, they refer to the *osmolarity* or *osmolality* of the internal and external environments of living organisms, but the concepts are similar. Osmolarity and osmolality are based on the concentration of osmotically active particles (atoms, ions, molecules) in a solution. *Osmolality* is measured as the solute concentration per unit *mass* and is thus for our purposes the same entity as salinity (although in different units).

Osmolarity refers to the solute concentration per unit *volume* instead of mass. Numerical values of these are close but not identical for the systems we will discuss. We will primarily use osmolality, whose units are milli-osmoles per kg (expressed as $mOsm \cdot kg^{-1}$).

If the environment on one side or the other (does not matter which) is *more concentrated* (i.e., has more solutes, or osmotically active particles) than the opposite side, the more concentrated side is said to be *hyperosmotic* to the less concentrated side. If the environment on one side or the other (again, does not matter which) is *less concentrated* (i.e., has fewer solutes, or osmotically active particles) than the opposite side, the less concentrated side is said to be *hypoosmotic*[21] to the more concentrated side. Unless the osmotic concentrations are equal (i.e., *isosmotic*), then the more concentrated side will be hyperosmotic to the less concentrated side, which is hypoosmotic to the more concentrated side. Confounding terminology perhaps, but a simple concept.

Unless the animal is totally impervious to the solute (typically salt) and water, which none are, the salt will move (diffuse) from the hyperosmotic (more concentrated) side to the hypoosmotic (less concentrated) side. Water will move in the opposite direction, from where it is more concentrated to where it is less concentrated.

Now, let us return to that saltwater fish, the tuna. The solutes in the tuna's blood and other extracellular fluid, mainly sodium (Na^+) and chloride (Cl^-) ions, are less concentrated than that of the external environment. Thus, the internal environment is hypoosmotic to the external environment. The solutes in seawater (again mainly but not exclusively Na^+ and Cl^-) will diffuse into the tuna, and water will leave. The latter process is referred to as *osmosis* (diffusion of water).

How quickly and how much movement of salts and water occurs depend on several factors (e.g., temperature, permeability of the gills and skin, etc.), but the most important is the magnitude of the gradient (how much difference there is in the salt and water concentrations *inside* and *outside* the tuna) in the seawater. Saltwater bony fishes like our tuna maintain an internal osmotic concentration of between 300–500 $mOsm \cdot kg^{-1}$. Seawater has an osmotic concentration of around 1000 $mOsm \cdot kg^{-1}$. Thus, there is a concentration gradient of around 500–700 $mOsm \cdot kg^{-1}$, with the internal environment hypoosmotic to the external environment (i.e., less concentrated than).

Saltwater fish therefore have a *water conservation problem* because water moves from the fish, primarily across the gills, which have a large surface area and are only a thin barrier to seawater. To compensate for this water loss, tuna and other marine bony fishes drink saltwater (3 to 10 times as much as a freshwater fish drinks) and water is absorbed across the intestine. Urine flow is low, which also conserves water.

At the same time, Na^+ and Cl^- ions diffuse from the seawater into the bloodstream of the tuna across the gills, and also enter into the body when the tuna eats and drinks. *Monovalent* ions (those with only a single electrical charge), here sodium (Na^+) and chloride (Cl^-), are eliminated by specialized, mitochondria-rich *chloride cells* at the gills. The process of eliminating the monovalent ions is energy-intensive and involves active transport, which explains the high concentration of mitochondria in the chloride cells. *Divalent* ions (those with two electrical charges), specifically magnesium (Mg^{++}) and sulfate (SO_4^{--}), are excreted via the kidneys. The combination of regulating both the ions and water results in osmotic and ionic homeostasis.

This homeostasis is not free—the energetic cost of osmoregulation in marine bony fishes ranges from 8–17% of their resting energy expenditures.

Osmoregulation in Sharks and Rays

Sharks and rays have taken a different evolutionary path to osmoregulation in seawater. Instead of having an internal solute concentration significantly lower than that of the seawater in which they reside, the case in bony fishes, elasmobranchs maintain their internal osmolarity at about the same or slightly higher osmotic concentrations than that of their environment. We call animals in which this occurs *osmoconformers*, because their internal water concentration conforms to that of the environment and thus water levels do not require regulation. For example, we found the mean osmolality of the blood of 12 Blue Sharks to be 1021, with seawater they resided in at 999 $mOsm \cdot kg^{-1}$.

Although the absolute value of the osmolality of the internal fluid of sharks and other chondrichthyans is at or, more typically, slightly above that of the sea water in which they live, this does not mean that the two different fluid compartments (i.e., the interior and exterior environments) are composed of the same constituents. There are differences, and the ions responsible for these differences are regulated.

Seawater is mostly Na^+ and Cl^- ions. About 50% of the internal fluids of sharks is Na^+ and Cl^- ions, a slightly higher level than is characteristic of bony fishes. Why not just elevate the internal Na^+ and Cl^- concentrations to that of seawater? Among vertebrates, only the hagfish is isosmotic (same osmolality) to seawater with similar principal ionic constituents (Na^+ and Cl^- ions about 500 $mOsm \cdot kg^{-1}$ each). However, this is the exception, and it occurs in a group whose evolutionary success is dubious (perhaps, at least in part, explained by physiological limitations due to being isosmotic with seawater). If elasmobranchs had levels of Na^+ and Cl^- of the magnitude in hagfish, it would create the kinds of problems we began this section describing (i.e., causing unfolding of proteins, or changing the electrical gradients across cell membranes required for normal function).

Instead of Na^+ and Cl^-, the remaining components that elevate the internal osmotic concentration of elasmobranchs to at or above that of seawater are two organic compounds that sharks and rays synthesize from metabolic waste products, *urea* and *trimethyl amine oxide* (TMAO; fig. 8.3). The use of these two

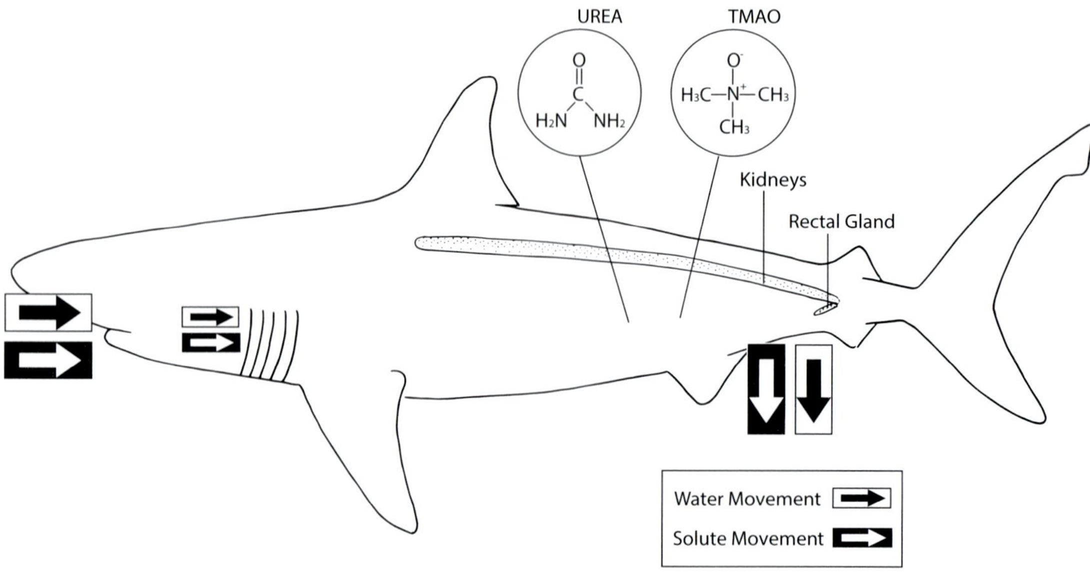

Figure 8.3. Relationships between distribution of solutes and water between a shark and its environment.

osmotically active particles (also called *osmolytes*) was discovered in the late 1920s and early 1930s by renowned physiologist Homer Smith.[22]

If we ended the story here, we would be negligent by omitting the monumental problems that had to be overcome to allow urea-based osmoregulation. Consider these four major drawbacks:

1 / *Urea is a potent metabolic poison that destabilizes or denatures proteins.*

Proteins are arguably the most diverse biomolecules in organisms. Central to their functioning effectively is that each protein's unique three-dimensional structure is not altered. Urea severely disrupts the structure of a protein, such that the protein either is unable to fulfill its functions, or it does so less effectively.

Uremia, or urea in the blood, is a horrifying, often terminal, human disease in which urea accumulates because of kidney failure. Symptoms are far-ranging and include fatigue, weakness, nausea, and a long list of other neurological, ocular, respiratory, cardiovascular, and other effects. Urea poisoning sometimes occurs in cows in which there is too much urea in the feed (used as a source of nitrogen) and frequently leads to death.

2 / *Urea is energetically costly to synthesize.*

Recall that the energy currency of the cell is ATP. To synthesize a single urea molecule requires about five molecules of ATP, a significant expenditure of energy.

3 / *Urea is easily lost to the environment because of its high diffusibility.*

Urea and TMAO, a nitrogen-containing waste product in a class of chemicals called *methylamines,* are small, uncharged, highly soluble molecules that are easily lost to the environment through diffusion and excretion by the kidneys.

4 / *According to Fick's Laws of Diffusion, individual particles (ions, molecules) will diffuse independently of each other.*

Even if the internal and external concentrations of ions are close, each ion will tend to move along *its own* concentration gradient. Thus, there is a large diffusion gradient for Na^+ and Cl^- from outside to inside in sharks and rays, and thus both ions diffuse from seawater into the body. Also, salts enter the shark's body through the food that it eats. And urea and TMAO will tend to diffuse out.

How Do You Solve a Problem Like Urea?[23]

How have these problems been solved such that urea can be retained in the body of sharks and other chondrichthyans?

The solution to the first problem with urea, its toxicity, is the role that TMAO plays in protecting proteins from urea's pernicious effects. Although the mechanism of its action is not entirely clear, the disruptive effects of urea on protein structure and function are prevented in the presence of TMAO. A 1980 study[24] discovered that the ratio of urea to TMAO and related methylamines was 2:1, and that methylamines evolved as osmolytes due to their stabilizing effects on urea-sensitive enzymes. Because of its protective role, TMAO has been called a *chemical chaperone*.

The solution to the second problem, how costly urea is to synthesize, is best explained by the wisdom of evolution. It would seem that two processes being equal, the one that maximizes conservation of energy should be evolutionarily favored. The process that requires greater expenditures of energy, here, synthesis of urea to achieve a higher internal osmotic concentration, should in some way disproportionately enhance survival to be favored or, alternatively, be a part of a *larger* process that is favored by natural selection. Being hyperosmotic or isosmotic to seawater represents a successful, alternative way of achieving osmotic homeostasis to that of marine bony fish, so the energetic costs may not matter that much. Moreover, becoming superior predators with metabolic systems capable of providing enough ATP early in their evolutionary history may have allowed the use of urea even if it is an energy-hogging osmolyte.

The loss of urea through urine and the gills, the third problem, was solved by reabsorbing 99% of the urea in the kidneys before it was excreted and blocking urea loss by having tissue impervious to the compound in the gills.

Solving the final problem involves compensating for the influx of Na^+ and Cl^- along their concentration gradients in accordance with Fick's Laws of Diffusion. Marine bony fishes have a similar problem with these and other ions, which they solve with specialized, mitochondria-packed *chloride cells* that expend energy to pump the Na^+ and Cl^- ions back into the environment.

Just as in marine bony fishes, the kidneys of sharks are not able to excrete the excess Na^+ and Cl^-, in part because doing so would require upsetting the animal's water balance by increasing the flow of urine required to eliminate these ions.

What about using chloride cells to excrete the excess Na^+ and Cl^-, like bony fishes do? Sharks and batoids appear to have cells (called *chloride-type* cells) in their gills similar to the chloride cells of bony fishes, but these cells lack the complexity of the bony fishes' chloride cell and thus the ability to pump out ions. A study[25] on the Atlantic Stingray (*Hypanus sabina*) suggested that this species, which is *euryhaline* (capable of tolerating a wide range of salinities), has a mechanism for pumping Na^+ and Cl^- ions *in* when in fresh water but not *out* while in seawater. We will discuss elasmobranchs in fresh water later in this chapter.

The Rectal Gland

The solution to the influx of Na^+ and Cl^- ions is a finger-shaped, bean-shaped, or globular tube called the *rectal gland* (also known as the *caecal* or *digitiform* gland; fig. 8.4), which is also found in the coelacanth. It is also a mitochondria-packed structure (i.e., one capable of expending lots of energy), which selectively accumulates Na^+ and Cl^- in its lumen (fig. 8.4) to twice the levels found in the blood and secretes these ions into the *cloaca*, from where they are dumped into the environment. Thus, the continuous inward diffusion of Na^+ and Cl^- ions from seawater is compensated for by mechanisms primarily in the rectal gland, with limited excretion by the kidneys, whose main responsibility is to eliminate divalent ions like Mg^{2+} and SO_4^{2-}, just as in marine bony fishes.

Sharks and Batoids in Fresh Water

We know that in salt water, sharks remain osmotically neutral by elevating their internal osmolality to approximately that of seawater, and they eliminate the Na^+ and Cl^- ions that diffuse in or enter through their food primarily using the rectal gland, with the kidneys assisting mainly with the divalent ions.

But what about euryhaline and freshwater elasmobranchs? Let us again start by considering bony fishes, freshwater ones this time, which face an entirely opposite suite of problems from those of the saltwater bony fishes we discussed previously. For one, body fluids of a freshwater fish are hyperosmotic to (more concentrated than) the environment, since fresh water has few dissolved salts. Thus, the tendency is for the fish to lose ions, primarily through the gills but also through the urine, because a freshwater fish continually gains water through osmosis from the external environment and frequently urinates to get rid of this excess water. The freshwater bony fish solves the first problem, loss of ions, by using its chloride cells to pump Na^+ and Cl^- back into the bloodstream of the fish from the environment, as well as through eating.

Among elasmobranch species, about 5% (mostly batoids) enter fresh water and 3–4% (exclusively batoids) are obligatorily freshwater (i.e., they spend their entire

A

B

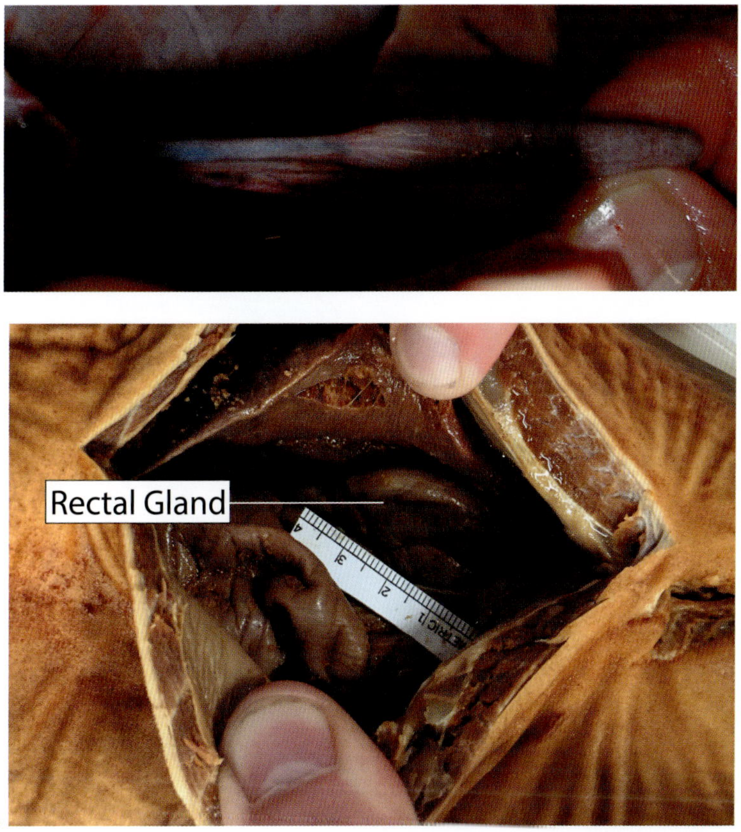

C

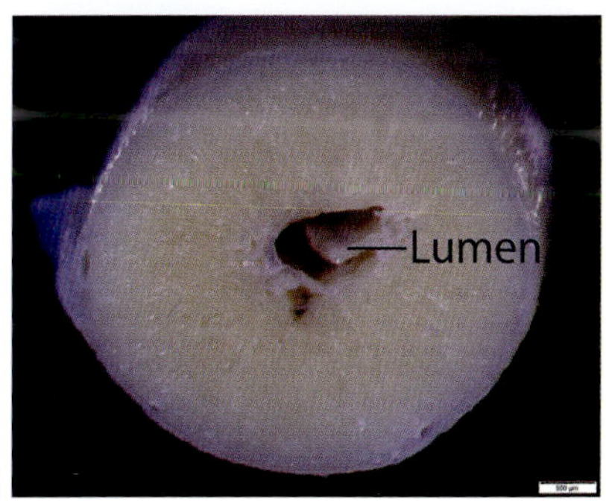

Figure 8.4. Rectal glands. (A) Digitiform rectal gland from a Blacktip Shark. (B) Globose rectal gland from a Broadnose Sevengill Shark. (C) Cross-section of a digitiform shark rectal gland showing the central lumen, where the Na^+ and Cl^- ions will accumulate before being excreted (Courtesy of Matthew Larsen).

lives in fresh water). A 2010 review paper[26] subdivided freshwater tolerant elasmobranch into three groups based on the stage of freshwater colonization. Note that this classification does not include elasmobranchs that live full-time in brackish (intermediate salinity) waters, or move between high salinity seawater and brackish estuaries, a list that includes adults of more than 50 species of sharks, as well as juveniles of some species such as Sandbar Sharks (*Carcharhinus plumbeus*; fig 8.5). These juveniles may use habitats of lower salinity than adults, and in doing so, may find refuge from predation from conspecifics and other larger sharks, at least temporarily. Preliminary work that we have done on Sandbar Sharks has

Figure 8.5. Author Abel with tagged juvenile Sandbar Shark caught in low salinity waters in Winyah Bay, SC, immediately prior to release (Courtesy of George Boneillo).

shown that these juveniles inhabit lower salinities during low tide (likely for protection from larger sharks and for food), and that physiologically they lower their solute concentrations (Na^+, Cl^-, urea) just like Bull Sharks (see pp. 309–311).

The first group is euryhaline elasmobranchs capable of routinely entering and leaving fresh water, primarily the Bull Shark (*Carcharhinus leucas*), three species of river sharks (genus *Glyphis*), Atlantic Stingray (*Hypanus sabina*), and Largetooth Sawfish (*Pristis pristis*). No members of this group live and reproduce entirely in freshwater systems. A recent publication[27] recounts the saga of several Bull Sharks that had occupied a landlocked, largely freshwater golf course lake near Brisbane, Australia, from 1996 to 2013—17 years! Apparently, flooding of their nearby river residence allowed the juvenile Bull Sharks to penetrate the adjacent pond, where they became trapped once the floodwaters receded. Thereafter, the sharks feasted on the "heap of fish species" in the pond, augmented, as you might expect, by meat supplied by the golf club's staff. Apparently, most of these Bull Sharks left the pond during subsequent flooding events; only a single shark was observed dead.

The second group consists of species that live permanently in fresh water but can tolerate more saline waters. This group includes one population of Atlantic

Stingrays (fig. 8.6) that lives full time and reproduces in fresh water in Florida's St. Johns River system, the Smooth Freshwater Stingray (*Dasyatis* [= *Fontitrygon*] *garouaensis*) from Western Africa, the White-edge Freshwater Whipray (*Fluvitrygon signifer*) from Southeast Asia, and the giant Mekong Freshwater Stingray (*Hemitrygon laosensis*), which is found only in the Mekong and Chao Phraya Rivers in Laos and Thailand (fig. 8.6).

The third and final group is found exclusively in fresh water and has lost the ability to tolerate even moderate salinities. These are the rays of the family Potamotrygonidae (fig. 8.7), a mostly landlocked, obligate freshwater assemblage found only in South America.

Stop for a moment and think what problems these elasmobranchs might face in dilute salt water or fresh water and how they might have adapted to these conditions. First, let us consider a Bull Shark, a species known from temperate and tropical shallow marine, estuarine, riverine, and lake ecosystems circumglobally.

If you just caught this shark in the Atlantic Ocean near, say, Recife, Brazil, and *instantaneously transported* it 4000 km (2500 mi) up the Amazon River, where the species has been documented, there would be enormous differences in the osmolality of the fluids inside the Bull Shark, which would measure about 1050 $mOsm \cdot kg^{-1}$, and that of the river, whose osmolality would be not much higher than zero. Thus, there would be huge gradients for both water inflow and salt and urea loss.

Of course, it would take a lot longer for a Bull Shark to move from a marine environment to a freshwater one, and thus there would be more time for physiological adjustments to occur, especially since these are mediated principally by hormones, whose action is not immediate. The first order of business, physiologically, is to drop the total osmolality of the internal fluid environment as much as possible so that the gradients for salt and water movement are reduced. There is in fact about a 40% drop. The levels of Na^+ and Cl^- are reduced by about 30% each, but urea levels are reduced by about 50%. The reduction in blood urea levels in freshwater-acclimated Bull Sharks appears to be due to decreases in urea synthesis as well as decreased reabsorption in the kidneys more so than changes in gill permeability to urea.

Even with lowered levels of Na^+ and Cl^-, and thus a lower concentration gradient for diffusion out, these ions will still be lost to the environment. We know that freshwater bony fishes compensate for this loss (i.e., gain salt) by using chloride cells and eating. But what about our Bull Shark? It would make a nice story to report that the rectal gland simply reverses its function from pumping Na^+ and Cl^- *out* when in full-strength seawater to pumping them *in* while in fresh water. However, this is not the case, since the rectal gland is not in direct contact with the environment like the chloride cells of the gills. All that happens in the rectal gland is that its Na^+/K^+-ATPase activity in fresh water is reduced to near zero (i.e., Na^+ and Cl^- are not removed from the blood and are thus not lost). Some research has shown a reduction in size of the rectal glands of freshwater-acclimated Bull

A

B

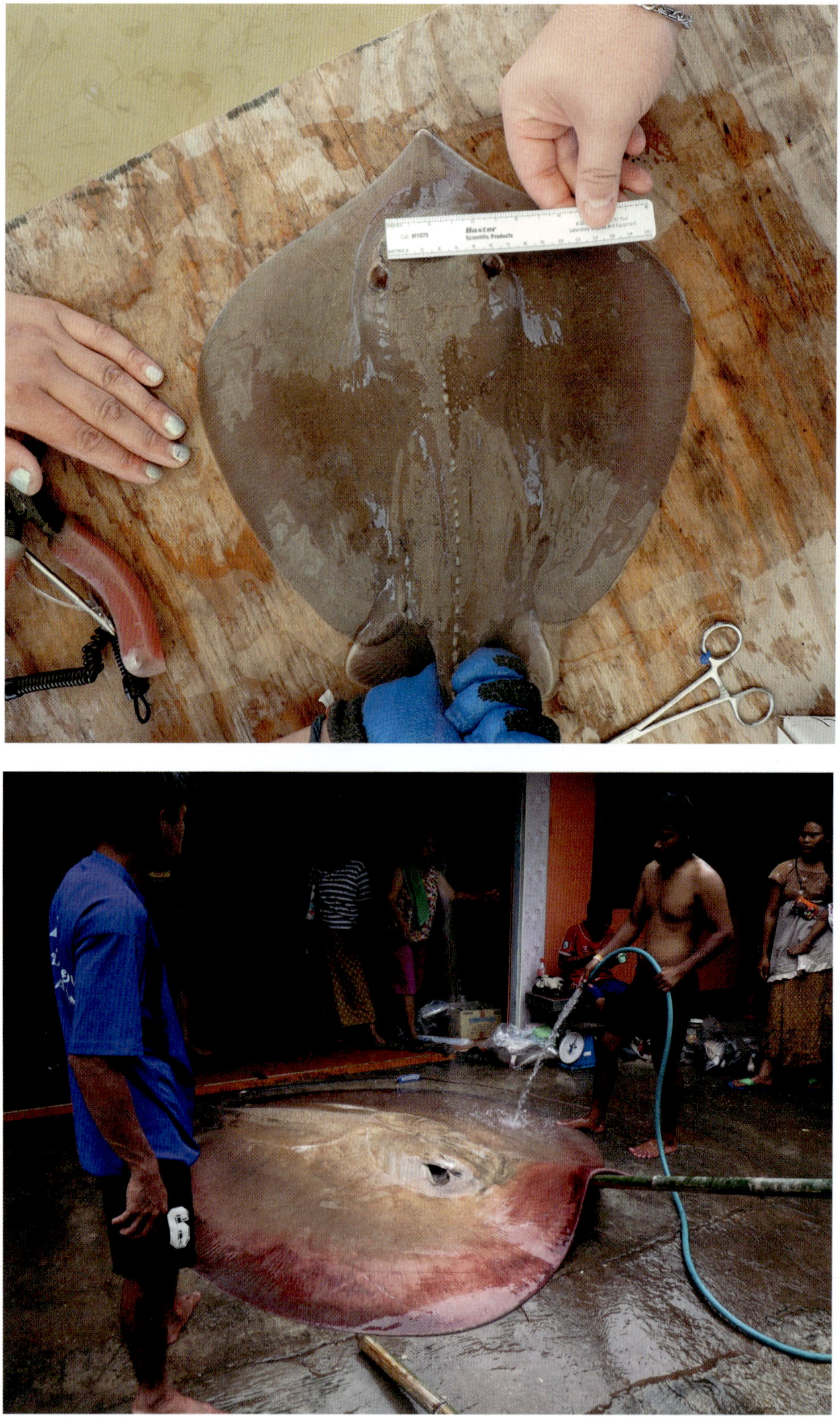

Figure 8.6. Batoids that live in fresh water but can tolerate salt water. (A) Atlantic Stingray. (B) Mekong Freshwater Stingray (PhumjaiFcNightsky/Shutterstock.com).

Figure 8.7. Potamotrygonid rays in a public aquarium.

Sharks, which makes sense since the organ is superfluous in fresh water. Loss of ions in other tissues is compensated for by eating, increased uptake at the gills through specialized chloride-type cells, and reabsorption from the kidneys.

Why not drop urea levels to zero by excreting all of it, especially since urea is so energetically costly to synthesize? The answer is the Bull Shark and other elasmobranchs in this category only secondarily moved into fresh water. Their entire physiology evolved in the marine environment, including the evolution of enzymes and other proteins in the presence of urea and TMAO. Thus, in spite of the toxic nature of urea and the need for TMAO to protect biomolecules, it turns out that enzyme systems of Bull Sharks have become so accustomed to the presence of urea and TMAO over evolutionary time that now they will not work in their absence. We call this *Osmoregulatory Stockholm Syndrome.*[28] In the same manner as when hostages become emotionally attached to their captors, enzymes will not work without the presence of the potential toxic compound urea, even though it is not required for osmoregulation under these conditions.

What about elasmobranchs that spend most of their time in fresh water? These sharks and batoids have adaptations similar to those of the Bull Shark, although some differences have been seen. For example, rectal glands of rays in the St. Johns River system were 80% smaller than those of fully marine conspecifics.

Finally, consider the potamotrygonid rays that are exclusively freshwater. This group has effectively become as close to osmoregulating like freshwater bony fishes as its constraints as an elasmobranch allow. First, the rectal gland is nonfunctional and vestigial. They reduce their Na^+ and Cl^- levels, and they lower their urea to about 1 $mOsm \cdot kg^{-1}$. To allow for the loss of urea, their gills are 20 to 50 times leakier than those of the Spiny Dogfish.[29] Additionally, their enzymes and other proteins have lost their urea sensitivity. Moreover, there is an energetic advantage (as much as 14%), since no ATPs must be diverted to synthesize urea.

One final question concerning freshwater elasmobranchs: Why are there so few of them? Notwithstanding the success of most potamotrygonids in freshwater, overcoming the osmotic, sensory, and reproductive constraints of being a freshwater elasmobranch, including the main obstacle of no longer retaining urea, would be extremely difficult to replicate.[30] Additionally, abandoning the retention of urea would need to co-occur with the evolution of urea-insensitive proteins. Also, the decreased electrical conductivity of fresh water may impose constraints on the electrosensory ability of the ampullae of Lorenzini. In potamotrygonids, the ampullae are much smaller than those of marine elasmobranchs and have been called *microampullae*. However, Paddlefish (*Polyodon spathula*) and Sturgeon (*Acipencer*) have ampullae of Lorenzini that they use in fresh water, so this may not be an insurmountable problem. Other problems would include overcoming the reduced viability of sperm in fresh water (recall that male elasmobranchs flush sperm into the female with seawater) and living in an environment that might be less stable thermally. Sharks may also be somewhat disadvantaged energetically in freshwater, since the absence of salts in the water translates into less buoyancy, and thus a faster swimming speed is required to prevent the shark from sinking. Here again the advantage goes to the potamotrygonids and their benthic existence. Furthermore, freshwater environments may have limited niche spaces for additional shark species, in other words, *No Vacancy*. In summary, the physiological, anatomical, and ecological specializations required for sharks to inhabit freshwater ecosystems are so extreme that evolution did not favor them except in a few cases.

Role of the Kidney

The elasmobranch kidneys are paired, elongate, flattened structures that run lengthwise along the dorsal wall of the abdominal space (fig. 8.8). The major roles of the elasmobranch kidney are to eliminate divalent ions (e.g., magnesium and sulfate), to produce dilute urine, and to reabsorb urea—as much as 90% of the urea that is filtered from the blood—from the urine before it is excreted.[31] The molec-

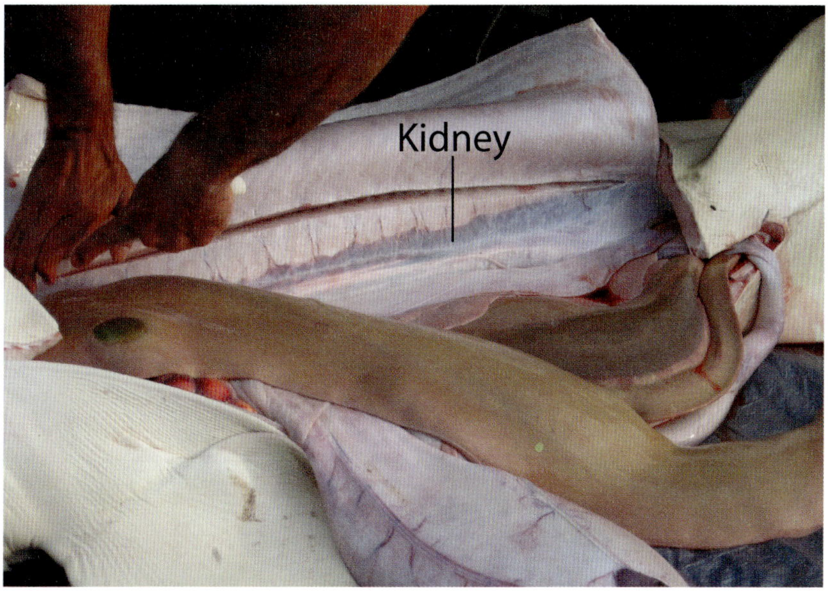

Figure 8.8. The shark kidney, here in a Lemon Shark.

ular mechanisms underlying urea reabsorption are not entirely clear, but likely include the usual transporters used by the kidneys, namely Na^+/K^+-ATPase, Na^+-K^+-Cl^- cotransporters, aquaporins (proteins that form water channels) as well as other urea transporters.[32]

Digestion

When you look at a shark, do you ever wonder what it has eaten?

Previously we have discussed the anatomy of the jaw and the roles of the teeth in feeding. Now, we move to the anatomy and physiology of the remaining processing of food. Refer to Figure 8.9 as we describe the anatomy of the shark digestive tract.

The protein- and fat-rich food of sharks and rays is either ingested whole or is sheared into pieces of varying sizes in the orobranchial chamber. The digestion of these items consists of mechanical, chemical, and microbial processing followed by absorption of the nutrients.

The gastrointestinal, or alimentary, tract consists of a series of organs (fig. 8.9) lined by a *mucosa* containing cells that secrete mucous and digestive enzymes, underlain by supporting connective tissue and, typically, longitudinal and circular smooth muscle.

Food moves from the mouth through the pharynx and into a short but distensible esophagus, which lies dorsal to the moveable cartilage called the basibranchial plate (see fig. 7.8). The esophagus secretes mucous and has striated muscles that facilitate the movement of food into the stomach.

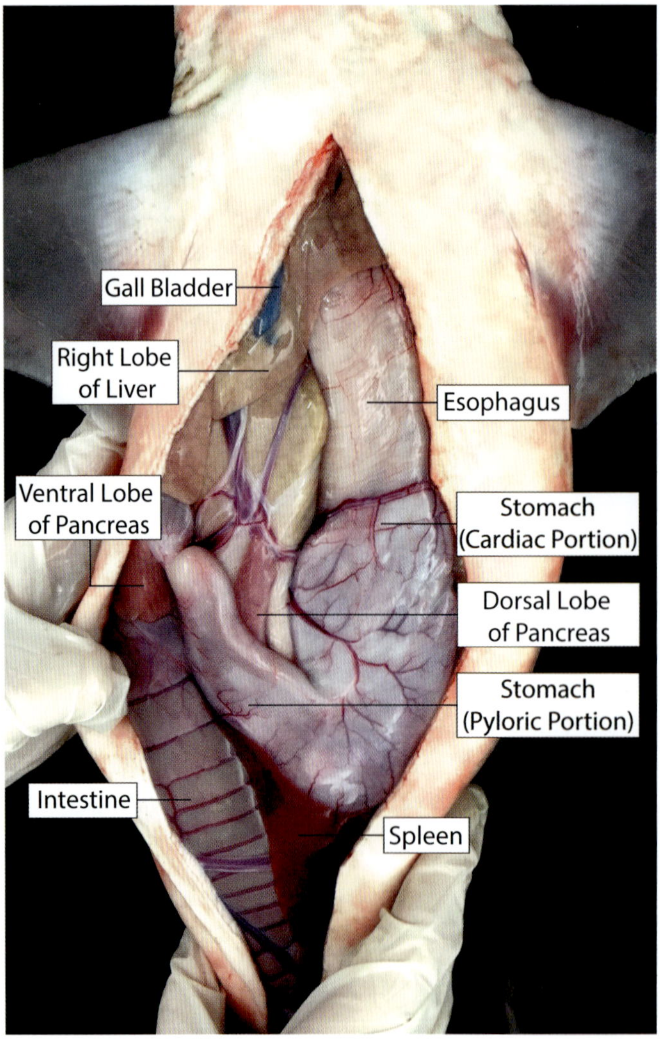

Figure 8.9. Gastrointestinal tract of a female Spiny Dogfish.

The stomach of sharks varies in form, with the most common being J-shaped (also called *siphonal*). It is divided into an anterior *cardiac stomach* and posterior *pyloric stomach.* It terminates at the *pyloric sphincter.* The stomach stores food and begins digestion of proteins, which is facilitated mechanically by contraction of the stomach muscles. Secretions include acidic (i.e., low pH) gastric juices, mucous that protects the stomach lining, and a precursor that is converted into the proteolytic (protein-digesting) enzyme *pepsin.* The low pH of the stomach also starts to emulsify lipids and helps begin the digestion of carbohydrates and proteins.

An important function of the stomach is to ensure that the partially digested food is not emptied prematurely. The hormone cholecystokinin (CCK), which has been found in Spiny Dogfish (*Squalus acanthias*), Porbeagle (*Lamna nasus*), and Shortfin Mako, is released by the gastric mucosa in the presence of proteins, and it serves to slow the rate at which the stomach empties. It may also curb the shark's appetite if there is food that is still being digested.[33]

Figure 8.10. Stomach of shark showing rugae (Photo by Joshua C. Bruni).

It is generally true in biological systems that surfaces specializing in absorption or exchange of substances (e.g., gills) often have elaborations that increase their surface area, and the stomach is no exception. The stomach contains longitudinal ridges formed by infoldings of the stomach wall known as *rugae* (fig. 8.10). These unfold and allow the stomach to distend to accommodate food that has moved into it. This increased volume occurs without any increase in gastric pressure, a phenomenon known as *receptive relaxation*. The stomach thus stores this food until it begins digesting it.

The pH within the stomach is either continuously low or it varies, depending on the species. A 2005 study[34] showed that during periods of fasting, the stomach pH in Nurse Sharks (*Ginglymostoma cirratum*) rose to > 8, compared to between 2 and 3 when digesting food. Other studies[35] demonstrated persistent pH of 1–2 in Leopard Sharks (*Triakis semifasciata*), Blacktip Reef Sharks (*Carcharhinus melanopterus*), and Spiny Dogfish, regardless of whether food was present.

The result of digestion in the stomach is conversion of whole or partial prey into a semi-liquid material called *chyme*. This chyme slurry moves into the small intestine through the pyloric sphincter. The intestines in sharks are not strictly analogous to those of other vertebrates, and a 2017 review[36] suggested that the terms *proximal*, *spiral*, and *distal* intestines be used.

The proximal intestine, sometimes referred to as the duodenum, receives bile from the gallbladder and digestive enzymes, some in an inactive form, from the pancreas. Bile serves to emulsify fats. Pancreatic digestive enzymes may include *proteases* (to digest proteins), *amylases* (to break down large carbohydrates), and *lipases* (to digest fats).

We have already discussed the varied morphologies of the section called the *spiral* intestine. Similar to the stomach rugae, the spirals (from 2 to 50), scrolls, and rings serve to increase the surface area for absorption, and thus slow the movement of the chyme (now called *digesta*), although experimental verification is sparse. The intestine also conserves space (it does not receptively relax like the stomach), which is a limited resource given the crowding by a large liver and possibly embryos.[37]

The intestine leads into the more muscular distal intestine, where feces accumulate and dries until released from the rectum, the lower part of the distal intestine, through the anus into the cloaca, and from there, the environment.

A spate of publications has recently reported the presence of micro- and macroplastics in the digestive tracts of numerous neonate, juvenile, and adult sharks and rays that are meso- and apex predators. In one case[38], a microplastic fragment translocated out of the digestive tract into the abdominal cavity in an Atlantic Sharpnose Shark. The impacts of plastics on elasmobranch digestion and overall health are not well-known.

Concluding Comments

At this point, we have discussed the physiology of the senses, reproduction, circulation, respiration, metabolism, temperature, osmoregulation, and digestion. These are all fascinating in their own right, and there are still many unanswered questions about the internal function of sharks and their relatives. Recall, however, that all physiological processes arise through natural selection and occur in service to ensuring survival and reproduction. Thus, our coverage of physiology appropriately segues into the relationship between the animal and its living and non-living environment, that is, ecology, in the following chapter.

NOTES

1. Iwama, G.K. et al. 1999. Am. Zool. 39: 901–909.
2. Huntsman, A.G. and Sparks, M.I. 1924. Contr. Can. Biol. Fish. 2: 95–114.
3. Fangue, N.A. and Bennett W.A. 2003. Copeia 2003: 315–325.
4. Ectotherms have also been called *cold-blooded* and *poikilotherms*. The former is incorrect because some ectotherms live in warm or hot environments and these ectotherms are not cold at all. Poikilotherms refers to having a variable body temperature. Many ectotherms are poikilotherms, but not all. If you are a fish living on a coral reef, in the tropics, then you are ectothermic but functionally *homeothermic* (with a body temperature that stays the same).
5. Hight, B.V. and Lowe, C.G. 2007. J. Exp. Mar. Biol. Ecol. 352: 114–128.
6. Speed, C. et al. 2012. Ecol. Prog. Ser. 463: 231–244.
7. Grubbs, R.D. et al. 2007. In: McCandless, C.T., Kohler, N.E. and Pratt Jr., H.L. (eds.). American Fisheries Society Symposium. Am. Fish. Soc. 50: 87–108.
8. Some authors differentiate between *systemic* and *regional* endothermy as follows: reserving the former for fishes whose whole body is warmed (those families listed in the text) and the latter for elevated temperatures in certain organs (e.g., the brains or the eyes).
9. Dolton, H.R. et al. 2023. Biol. Letters 19: 20230331 and Dolton, H.R. et al. 2023. Endan. Sp. Res. 51: 227–232.
10. Dickson, K.A. and Graham, J.B. 2004. Physiol. Biochem. Zool. 77: 998–1018.
11. Dickson, K.A. and Graham, J.B. 2004. Physiol. Biochem. Zool. 77: 998–1018.
12. Watanabe, Y.Y. et al. 2015. Proc. Natl. Acad. Sci. 112: 6104–6109.

13. Waller, M.J. et al. 2023. J. Fish Biol. 103: 864–883.

14. Harding, L. et al. 2021. Funct. Ecol. 35: 1951–1959.

15. As cited in Table 8.1, p. 250 of Wetherbee, B. et al. 2012. In: Carrier, J.C. et al. (eds.). Biology of sharks and their relatives. CRC Marine Biology Series. CRC Press. 3–31 pp.

16. Stillwell, C.E. and Kohler, N.E. 1982. Can. J. Fish. Aq. Sci. 39: 407–414.

17. Wood, A.D. et al. 2009. Fish. Bull. 107: 1–15.

18. Nakamura, I. et al. 2020, J. Exper. Biol. 223: jeb210286. doi:10.1242/jeb.210286.

19. Griffiths, M.L. et al. 2023. Proc. Nat. Acad. Sci. 120: e2218153120.

20. Strictly, *osmoregulation* refers to regulating *osmotic pressure* in an organism, which is achieved by regulating water and/or solute balance.

21. *Hypoosmotic* is pronounced hy-po-osmotic. If you think the word should have a hyphen (i.e., hypo-osmotic), we agree.

22. Smith, H.W. 1936. Biol. Rev. 11: 49–82.

23. Dad joke, courtesy of co-author Abel, who thought readers of the first edition would universally recognize this as a pun referencing the song *How do you solve a problem like Maria?* from the 1965 musical *The Sound of Music.* Turns out anyone younger than 60 had literally no idea that this was supposed to be clever. We'd say it went over like a lead balloon, but *that* phrase dates back to 1924. Let's just call it an epic fail, and move on.

24. Yancey, P H. and Somero, G.N. 1980. J. Exper. Zool. 212: 205–213.

25. Piermarini, P.M. and Evans, D.H. 2000. J. Exp. Biol. 203: 2957–2966.

26. Ballantyne, J.S. and Robinson, J.W. 2010. J. Comp. Physiol. B 180: 475–493.

27. Gausmann, P., 2024. Mar. & Fish. Sci. (MAFIS). 37: 5–25.

28. You will not get any hits in Googling the term, as no *sensible* physiologist has ever used it. But it makes sense to us.

29. Goldstein, L. and Forster, R.P. 1971. Comp. Biochem. Physiol. 39B: 415–421.

30. Ballantyne, J.S. and Robinson, J.W. 2010. J. Comp. Physiol. B 180: 475–493.

31. Hyodo, S. et al. 2014. Am. J. Physiol. Reg., Integr., and Comp. Physiol., 307: R1381-R1395 and references therein.

32. Hyodo, S. et al. 2014. Am. J. Physiol. Reg., Integr., and Comp. Physiol., 307: R1381-R1395 and references therein.

33. Leigh, S.C. et al. 2017. Rev. Fish Biol. Fisher. 27: 561–585.

34. Papastamatiou, Y. and Lowe, C. 2005. Comp. Biochem. Physiol. A 141: 210–214.

35. Leigh, S.C. et al. 2017. Rev. Fish Biol. Fisher. 27: 561–585.

36. Leigh, S.C. et al. 2017. Rev. Fish Biol. Fisher. 27: 561–585.

37. Wetherbee, B.M. et al. 2004. In: Carrier, J.C. et al. (eds.). Biology of sharks and their relatives. CRC Marine Biology Series. CRC Press. 225–246 pp.

38. Pullen, E., "Microplastics in the Digestive System of the Atlantic Sharpnose Shark (Rhizoprionodon terraenovae) in Winyah Bay, SC" (M.Sc thesis, Coastal Carolina University, 2019).

PART III
Ecology and Behavior

9 / Ecology

CHAPTER OUTLINE

Introduction

Ecology is the study of the relationships between organisms and their living and non-living environment. Its subdivisions include:

- Ecophysiology (or physiological ecology), which examines physiological adaptations of organisms to their environment, for example osmoregulation of Bull Sharks (*Carcharhinus leucas*) in rivers. We have considered physiology previously and will discuss behavioral ecology in chapter 10.
- Population ecology (or autoecology), which examines the dynamics of populations of individual species. This entails *stock assessments*, which are critical to management of elasmobranchs, e.g., the Sandbar Shark (*Carcharhinus plumbeus*) as the species recovers from overfishing along the US Atlantic coast.
- Community ecology (or synecology), which studies the interactions between species within an ecological community, e.g., relative abundances of sharks in an estuary or on a coral reef, and predator-prey interactions.

- Ecosystem ecology, which looks at energy and matter flows through ecosystems, e.g., the trophic, or feeding, structure of an ecosystem.

We begin our discussion of the ecology of sharks with an overview of the important questions that shark ecologists ask.

Big Ecological Questions Related to Sharks and Rays

Studies into the ecology of sharks (i.e., how sharks live their lives and interact with other species and the environment) revolve around searching for answers to a basic suite of questions.

1) / Where do sharks live and what habitats do they require? The answers to these questions vary by species and by life stage. Juveniles often use different habitats than adults, and males and females may segregate for much of their lives.

2) / Where do sharks move and what drives these movements? Some species of sharks live in very small home ranges (smaller than a Piggly Wiggly[1]) whereas others migrate across entire ocean basins.

3) / What and how often do sharks eat, and what eats them? Some shark species eat a huge variety of prey whereas others specialize in feeding on a few prey types. Some may eat one big meal that lasts for weeks, whereas others must feed daily. Also, we typically think of sharks only as predators, but it is important to note that most sharks are also potential prey, often to other sharks. In fact, all sharks are subject to being eaten, at least as juveniles.

4) / What life-history characteristics do specific shark species possess? Fundamentally, it is critical to know how long it takes a shark to become sexually mature and how many offspring it can produce over its life, in order to manage and conserve populations. Shark life histories vary widely and are directly a function of the energy available to them and how they use it and thus depends on where they live (e.g., tropics versus the deep sea), where they fall in the food web (top versus meso-predators), how much energy they spend moving (sedentary versus highly migratory), and which reproductive modes they employ (ranging from laying scores of eggs to live birth of one pup).

5) / What roles do sharks play in the ecosystem? This is perhaps the most difficult ecological question, and its solutions rely on the answers to all of the previous questions. Simplistically, sharks are often characterized as apex predators that maintain ecosystem health by controlling prey populations and eating the weak and diseased. In reality, empirical evidence to support this notion has been hard to find and, while some sharks are indeed apex predators, most sharks are not, but instead fill roles in the mid-to-upper tiers of the food web. Sharks play many roles in marine ecosystems that are often complex and not always obvious. In the world of shark research, we are just scratching the surface in answering this fundamental question.

Challenges of Studying Sharks

Unraveling the ecological mysteries of sharks is more complicated with different challenges than studying terrestrial predators like wolves or lions. For one, sharks live underwater, obviously, in a concealing environment rendering observation by air-breathing scientists difficult. Depending on the questions being asked, researchers may need to capture sharks or simply observe them. In some cases, both approaches can be relatively inexpensive. Your authors capture and release thousands of sharks every year in the coastal zone for a variety of ecological studies using small vessels. In clear water, sharks can be observed relatively inexpensively using SCUBA or baited cameras. However, the expense of ships and equipment increases rapidly when studying deep-sea and pelagic species, except in areas such as oceanic islands where the deep-sea slope is very close to shore. Deep-sea research often requires the use of large ships for many days and specialized equipment to capture (e.g., using trawls, baited hooks, or traps) or even to observe (e.g., using submersibles or remotely operated vehicles) the subjects (fig. 9.1).

Figure 9.1. Author Grubbs studied deep-sea sharks from a submersible in the Bahamas as part of an OceanX expedition, during which he photographed this Bluntnose Sixgill Shark. Spearguns used to tag sharks from the submersible are shown.

BOX 9.1

Research, Conservation, and Education Focus

Challenges and Successes in Efforts to Promote Shark and Ray Conservation in Sudan

Igbal Elhassan

Fisheries Department
University of Bahri
Khartoum, Sudan

The Sudanese Red Sea is 750 km (466 mi) long, a significant geographic region in Sudan. It is known for its warm waters and diverse marine life, particularly its thriving coral reefs. This area provides a home for numerous sea species.

The economic crisis, years of economic sanctions, and the recent war have had a significant impact on Sudan's academic infrastructure. Although there are many institutions in the country and qualified staff, the main issue is the lack of adequate research facilities and adequate funding. In the past, Sudan has partnered with countries like Norway and the UK, which have provided technical support and worked together with Sudanese institutions. Since 2013, Norway has restarted its collaboration with Sudanese institutions in the Red Sea State through the United Nations Industrial Development Organization (UNIDO) project.

The Marine Fisheries Law of 2008 banned fishing for sharks. Yemen poaching boats and some local fishermen targeted sharks illegally and intensively after the ban of fishing. Pups, juveniles, and some adult sharks and rays were bycatch in the trawl fishery. In 2016, the Marine Fisheries Administration amended the law to include the trade of any parts of sharks.

My master's thesis in 2002 was the first study on sharks in Sudanese waters and the surrounding countries. The study focused on various biological aspects of sharks and their fisheries. It also investigated the socioeconomics of the shark fishery and trade. Additionally, the study determined the geographical distribution and mating and pupping seasons of these species, and identified nursery areas for the main species targeted by fisheries.

In addition to collecting extensive data on sharks, I also aimed to raise awareness among fishermen, fish processors, and fish traders about the importance of shark conservation. I have incredible stories to share about it.

Since beginning my master's study to date, I have successfully identified 26 different species of sharks in the Sudanese Red Sea. The majority of these identifications were confirmed through genetic analysis. I also identified 14 batoid fish, which were confirmed through genetic identification conducted by Dr. Naylor Gavin, recently.

Obtaining data for studying sharks in Sudan posed challenges. The fishermen processed sharks at sea and brought them to the fish market as trunks, with only the second dorsal, pelvic, and anal fins attached. I trained myself to identify different shark species based on these fins and other distinguishing characteristics on their trunks. Unfortunately, there was a lack of literature on shark biology and fisheries in institutional laboratories in Sudan. By collecting data on the catch from the fish market and the fishing sites, I also accompanied a foreign long-line vessel in Sudan to collect data on migratory shark species. I also collected data on bycatch from trawling and tons of sharks by coaching boats.

Through the funds I received from the Save Our Seas Foundation in 2013–2014, I was able to make three trips to Yemen, including Al Hudaydah, Aden, Hatramoot, Almukalla, and Gusaer in the Arabian Sea, to collect data for my PhD study. During my time there, I experienced some intense moments. Two instances were a significant explosion in Sanaa and Aden, prompting an evacuation alarm at the Marine Fisheries Corporation in Al Mukalla, where I had gone to seek approval for collecting samples. Despite these incidents, I will always remember the warm welcome and helpfulness of the authorities and staff at Al Hudaydah Marine Fisheries Corporation and Al Mukalla Marine Fisheries Corporation and the cooperation of fishermen and fish processors at the fishing sites I visited.

A Green Sawfish (*Pristis zijsron*) is measured.

The funds also allowed me to receive training in genetics and statistical genetics at the SOSF shark laboratory, equipping me with the necessary skills to effectively analyze my data under their supervision.

In 2012, I conducted individual research on the Green Sawfish (*Pristis zijsron*) with partial funding from the Mohammed Bin Zayed Conservation Fund. This research allowed me to gather valuable biological data on the species. In 2019, I formed a team with the Marine Fisheries Administration and applied for a small grant from the Shark Conservation Funds. Our collaboration led us to survey the green sawfish and confirm the existence of nursery areas in Agig Bay on the southern coast and Ras Kasar on the south coast of Sudan. The participation of Marine Fisheries Administration officers in this project significantly increased their awareness of the importance of conserving this endangered species. Furthermore, the local fishermen played a vital role in preserving the green sawfish. They released live pups and juveniles back into the sea when they encountered them and reported these sightings to our team.

To effectively protect the green sawfish habitats in the Sudanese Red Sea and the surrounding region, it is crucial to collaborate with neighboring countries, especially Eritrea. Joint research efforts can help us track the movement of this fish and ensure its conservation.

Most readers of this book will find that these challenges are serious impediments to studying sharks. Now, envision adding to this the economic and other impacts of a recent war. For *Research, Conservation, and Education Focus* Box 9.1, we invited indomitable conservation biologist Igbal Elhassan of Sudan to describe the important work that she and her colleagues are doing in, understatement acknowledged, less than ideal conditions.

If we are fortunate enough to overcome these roadblocks and capture sharks we wish to study, handling them often presents a new suite of problems. Sharks do not want to be handled and many of them have large teeth and will bite. We know many colleagues that have suffered serious injuries from shark bites and even small sharks can be dangerous to the researcher—author Grubbs, having tagged thousands of juvenile Sandbar Sharks in Virginia waters, had his thumb nearly removed by a 3-foot-long Sandbar Shark in 1996, and in the summer of 2019 author Abel was gnawed by a smaller beast (fig. 9.2). But Grubbs found that studying sawfish can get even more treacherous, as the animals swing a huge tooth-bearing rostrum at the researchers with incredible speed and force. In many cases, researchers seek to release the animals unharmed after all data and sample collection are completed. Many sharks are much more fragile than their reputations would suggest, so researchers try to protect the sharks from damage and release them as quickly as possible while trying to protect themselves from being bitten by a shark, slapped by its tail, or bludgeoned by a sawfish rostrum.

A final challenge to studying the ecology of sharks lies in achieving the meaningful sample sizes. In ecological research, we seek to answer questions or describe

Figure 9.2. Author Abel during an unsettling two-minute period waiting for the small Sandbar Shark to release his gloved finger. The interaction left shallow puncture marks resembling paper cuts (Courtesy of Jim Luken).

patterns within our data that apply beyond our samples or the animals from which they came. In other words, the results allow us to make inferences that apply to whole populations, species, or ecosystems. The sample size needed to answer these questions confidently depends in part on the variability in the measured trait within the population and how well that variability is described in our samples. As an amusing example, years ago we had the opportunity to examine the gut contents of three Blue Sharks (*Prionace glauca*) that were captured together. The only items in their stomachs were green beans—yes, the kind you eat with your family during the holidays. Most likely these were scarfed up after being tossed from a ship. But if we attempted to describe the diet of Blue Sharks based on our available sample size of three, we would conclude that they are herbivores that frequent Piggly Wiggly in search of green beans. In reality, Blue Sharks feed on a wide variety of pelagic fishes and squid and hundreds of Blue Sharks are required to be studied to describe their diet quantitatively. This is why for biological questions that are amenable to experimentation, model organisms such as fruit flies, zebra fish, and white mice that can be produced in large numbers are used.

And herein lies the challenge in studying sharks. Energy flow in any ecosystem dictates that top predators like large sharks naturally occur in low densities. If we visualize the pelagic food web as the classic trophic pyramid (fig. 9.3), the

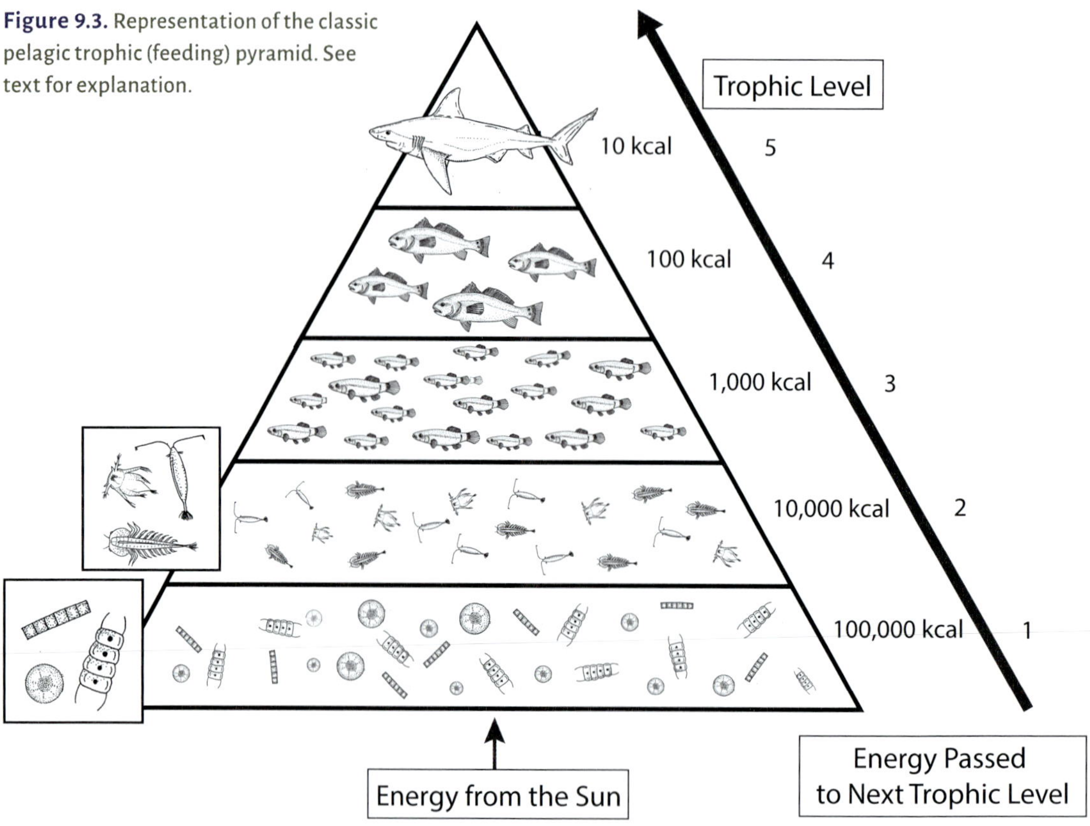

Figure 9.3. Representation of the classic pelagic trophic (feeding) pyramid. See text for explanation.

primary producers (phytoplankton), are the food source for the primary consumers (zooplankton). Zooplankton are eaten by small fishes which are in turn eaten by larger fishes like tunas and sharks. As a rule of thumb, only about 10% of the energy in one trophic level is available to create biomass at the next highest trophic level. This is the *trophic transfer efficiency*. So, with our Blue Shark example, 10,000 kilocalories (kcal, a measure of energy more commonly referred to as *calories* in labels on human food) of phytoplankton would support 1000 kcal of copepods, which could support 100 kcal of herring, leading to 10 kcal of Blue Shark tissue. Shark muscle is close to 1 kcal per g, but liver and kidney tissues are much higher. If we assume the overall energy density of a Blue Shark is 2 kcal per g, and that of phytoplankton is roughly 0.6 kcal per g, then an average size Blue Shark weighing 50 kg (110 lb) would require primary production of 167,000 kg (370,000 lb) of phytoplankton (or green beans for that matter). That is roughly 12 semi-tractor trailer loads of green beans to produce one adult Blue Shark! It should be obvious now why achieving appropriate sample sizes to answer ecological questions about shark populations is a challenge. This is, of course, compounded when studying species that are truly apex predators or those that are rare, endangered, or live in the deep sea.

Where Do Sharks Live and How Many Are There?

Sharks inhabit nearly all marine environments from tropical coral reefs to cold arctic waters and from estuaries and coasts to the deep sea and open ocean. There are even sharks and rays, though relatively few in number, which live in fresh water. Why does any animal live where it does? The distribution of animals and the habitats they use reflect a combination of congruent and opposing forces.

First, an animal must be physiologically capable of living in a given environment, which requires adaptations to abiotic (non-biological) factors such as temperature, salinity, dissolved oxygen, pH, pressure (depth), and light, to name a few. Temperature tolerances are among the most important factors limiting shark distributions, but also recall the special adaptations required for sharks to live in fresh water (see Chapter 8).

Next, there are a series of ecological constraints that not only influence where sharks live, but also the density (how many) of sharks living there. All animals must eat—that is one of their defining characteristics. Thus, the distribution of suitable resources (prey) is a major driver of where sharks live. However, there is often intense competition for those prey from members of the same species or others, so the ability to compete for prey also determines where sharks live. Of course, just as one species of shark knows where to find its prey, its predators (often larger sharks) also know where to find them. This balance between competition for resources and predation risk plays a large role in determining where animals, including sharks, live. Ultimately, an animal's evolutionary fitness is tied to its ability to pass its genes to the next generation (i.e., to reproduce). The probability of finding a mate also influences where a shark will spend a portion of its time.

It may surprise you to learn that over half (~53%) of all species of sharks live their lives in the deep sea (defined as deeper than 200 m, or 656 ft). Most deepwater species of sharks are small (< 100 cm or 3.3 ft) and this high diversity is likely driven by the relatively low competition for resources and the lower predation risk than in shallower environments. But sharks do not live in the *very* deep sea. The average depth of the world's oceans is 4000 m (13,100 ft), but sharks rarely occur deeper than 3000 m (9800 ft). Explanations for these depth limits are debated, but likely involved both ecological (food limitations for predators) and physiological (e.g., constraints on producing liver oil and synthesizing TMAO under high pressure) factors. About 40% of shark species live in coastal tropical to warm temperate environments, from estuaries to continental shelves. In these regions, there is high food biomass and diversity along with very high habitat diversity. Together, these drive niche specialization, which leads to the rise of new species and thus high diversity among coastal sharks. There are very few truly pelagic species of sharks (< 3%), primarily due to limited prey resources in the oligotrophic (low nutrient) open ocean and the lack of barriers to separate them long enough to drive speciation. Thus, most pelagic shark species are found circumglobally. The remaining sharks (~7% of the diversity) are composed of Arctic and freshwater species.

How Do We Know Where Sharks Live? Catching Sharks

Fishery-independent Surveys

In the arsenal of methods used to study the ecology of sharks, tag-and-release fishing has a special place among both the public and scientists. To answer many questions about the biology and ecology of sharks and their relatives, scientists need to handle the specimens so that they can identify the species, make accurate measurements, take tissue samples, and apply tags.

Methods used by shark biologists rely on the same techniques used by commercial and recreational fishers (e.g., longlining, gillnetting, trawling, and hook-and-line fishing; see Chapter 11). Some aspects of shark ecology are learned from data and samples collected directly from commercial and recreational fisheries, which is referred to as fishery-dependent data. However, much of our knowledge comes from *fishery-independent* sources, primarily from scientific surveys. Why is this important? Fishery management as well as understanding fish ecology relies on obtaining estimates of how abundances of sharks vary over space and time. Catch rates (CPUE = catch per unit effort), such as the number of sharks caught on 100 hooks in an hour, number caught in an hour gillnet soak, or number caught per hour trawling, are used as the estimate of relative abundance. If we are interested in the distribution of a population of sharks and what habitats they use, we not only must sample where sharks are but also where they are not, regardless of cost or effort required, in order to define the edges of the distribution (i.e., where CPUE goes to zero). To accomplish this, stations are chosen using a random design (e.g., at different depths or geographic locations; scientists would say *stratified by depth* or *spatially balanced*, respectively). If we are interested in changes in the populations over time, we must sample in the same way across the same part of the population repeatedly using identical methods. These surveys may use a randomized design or fixed stations that are sampled repeatedly for many years. It would be foolish for commercial and recreational fishers to fish in these ways. Fishers weigh the costs and benefits of fishing different areas and target the highest density of fish that is legal and economical.

Fishery-independent shark surveys are often critical to stock assessments for fishery management. Fisheries often use methods that only catch specific portions of the populations of sharks (e.g., longline fisheries tend to catch intermediate-sized sharks because hooks are too large to catch small sharks and the gear is designed to allow large sharks to break free), whereas fishery-independent surveys can be designed to catch sharks that represent the population (e.g., using multiple hook sizes). This provides managers with a more complete picture of the population structure and can highlight critical changes in the stock, such as juvenescence (decreases in the mean age and size), which is a prime indicator of overfishing (fig. 9.4).

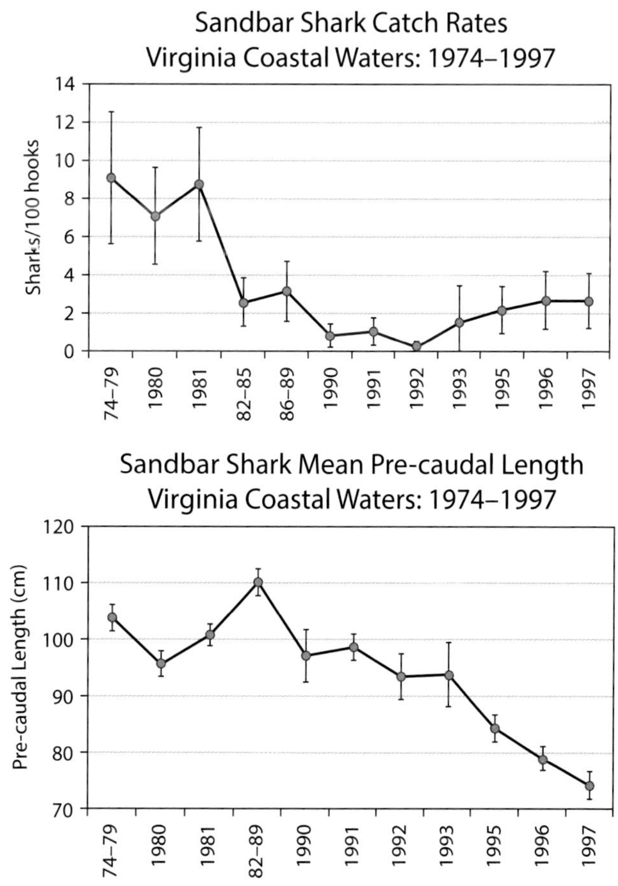

Figure 9.4. Examples of fishery-independent data important to fishery managers. Top: Sandbar Shark *Catch Per Unit Effort* (CPUE, see text), that is, sharks per hundred hooks, caught on experimental longlines in Virginia coastal waters from 1973–1997. Bottom: Mean lengths of these sharks. Note the period beginning in 1993. Though catch abundance (top graph) began to increase after the fishery management plan went into effect in 1993, the mean size of these sharks (bottom graph) continued to decline, a classic example of juvenescence from overfishing (Grubbs and Musick, unpublished data).

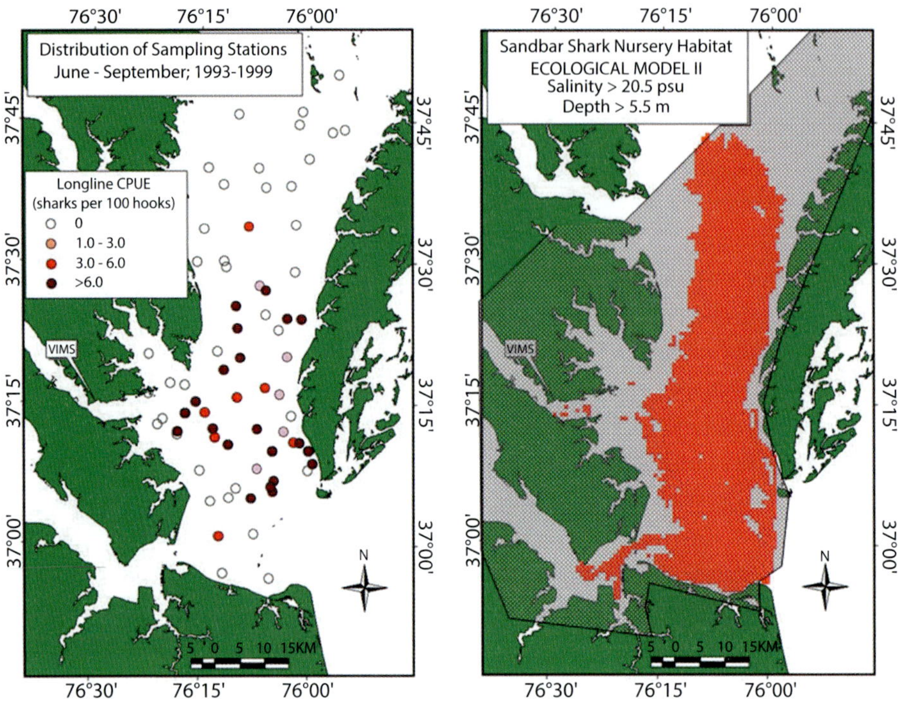

Figure 9.5. Map showing Catch Per Unit Effort (CPUE, sharks per hundred hooks) for Sandbar Sharks on a randomized longline survey (left) used to define primary nursery habitat in lower Chesapeake Bay based on depth, salinity, and catch data (right). These catch data informed fishery managers, which led to the designation of the area as a Habitat Area of Particular Concern (polygon in figure on right) in the fishery management plan (Modified from Grubbs and Musick 2007).

Surveys can be used to delineate critical areas for overfished or imperiled species that may be closed to fishing. A survey for juvenile Sandbar Sharks in Chesapeake Bay was used to delineate primary nursery habitat for this species, which led to the federal designation of a *Habitat Area of Particular Concern* (HAPC) for that species in the management plan (fig. 9.5).[2,3]

Fishery-independent surveys can also be used to determine how fishery-induced mortality may change if fisheries changed their methods. For example, the Cuban Dogfish (*Squalus cubensis*) is one of the most common bycatch species in deep snapper and grouper fisheries in the Gulf of Mexico, and 97% are reported as released alive. Fishery-independent surveys were used in conjunction with stress physiology and post-release monitoring to determine that even when handled carefully, about 50% of those released alive die within 24 hours[4] (fig. 9.6). Co-author Grubbs and

Facing page

Figure 9.6. Graphic showing experimental methods used in a study on the post-release mortality of Cuban Dogfish (Courtesy of Read Frost and Brendan Talwar).

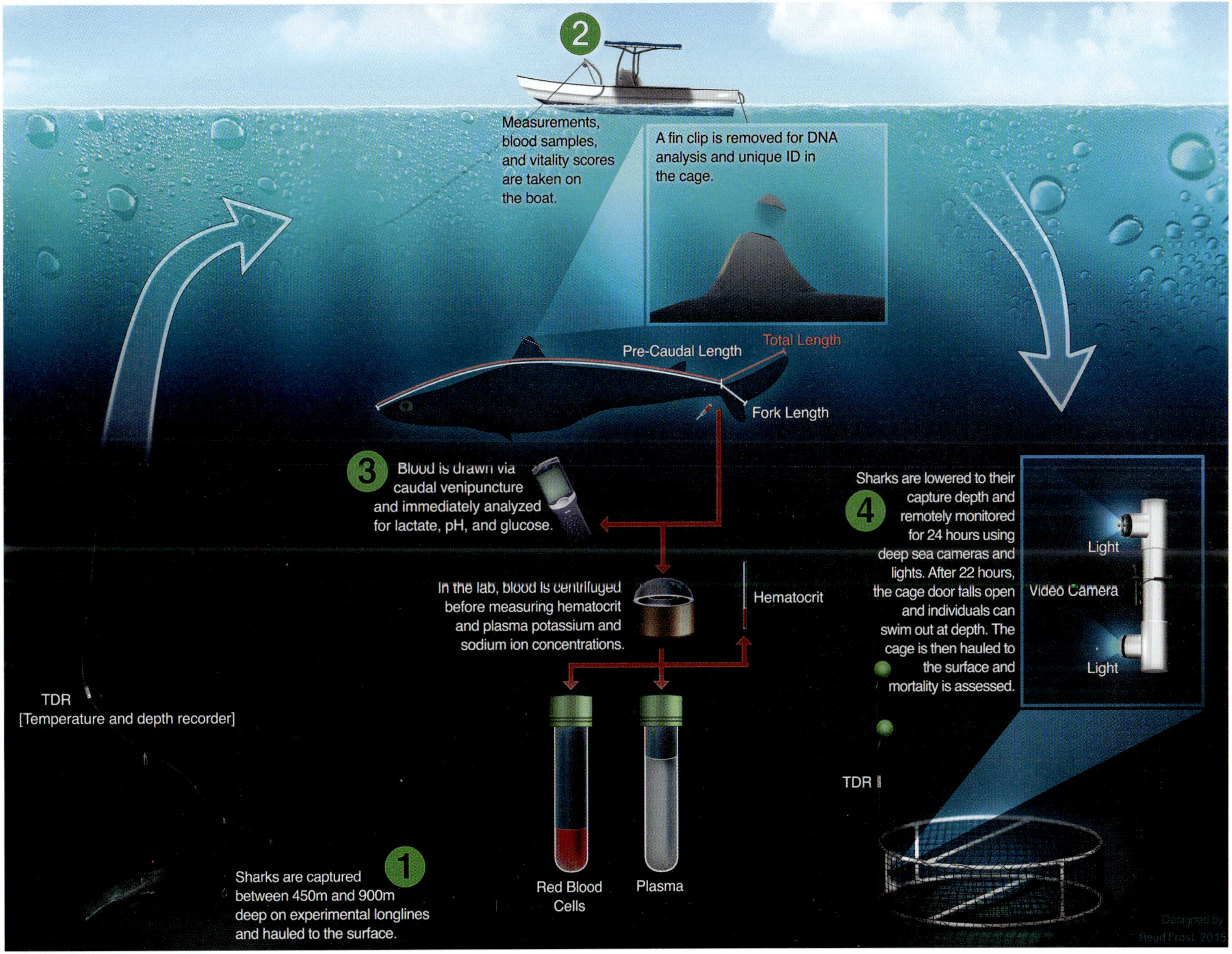

2
Measurements, blood samples, and vitality scores are taken on the boat.
A fin clip is removed for DNA analysis and unique ID in the cage.
Pre-Caudal Length
Total Length
Fork Length
3
Blood is drawn via caudal venipuncture and immediately analyzed for lactate, pH, and glucose.
In the lab, blood is centrifuged before measuring hematocrit and plasma potassium and sodium ion concentrations.
Hematocrit
Red Blood Cells
Plasma
TDR
[Temperature and depth recorder]
1
Sharks are captured between 450m and 900m deep on experimental longlines and hauled to the surface.
4
Sharks are lowered to their capture depth and remotely monitored for 24 hours using deep sea cameras and lights. After 22 hours, the cage door falls open and individuals can swim out at depth. The cage is then hauled to the surface and mortality is assessed.
Light
Video Camera
Light
TDR
Designed by
Read Frost, 2015

Figure 9.7. Author Grubbs releasing a Great Hammerhead from an experimental longline as part of a study of post-release mortality (Courtesy of Michael Scholl).

colleagues used fishery-independent surveys incorporating hook timers (magnetic devices that begin a clock when a shark bites the hook) to estimate the effect of varying hook soak time on post-release mortality in Great and Scalloped (*Sphyrna lewini*) Hammerheads (fig. 9.7). Shark surveys take on many designs depending on purpose, funding, and infrastructure (e.g., vessels, personnel). Given the conservative life histories of sharks, long-term fishery-independent surveys are critical for assessing population trends. Consistent, long-term funding to support these surveys is often the biggest challenge to their continuation. Here, we briefly describe six surveys we have been involved in or developed that illustrate this variety and the challenges.

CCU Shark Project (fig. 9.8)—Begun in 2002 at Coastal Carolina University by co-author Abel, the CCU Shark Project is a small-scale, episodic, estuarine survey started with the dual purpose of teaching students about shark research and to understand ecology and physiology of the sharks found in three NE South Carolina estuaries: Winyah Bay, Murrell's Inlet, and North Inlet. The initial research goals of the project were to (1) identify sharks inhabiting these systems; (2) describe shark population structure, distribution, and migrations and their environmental influences; (3) determine whether these systems serve as nurseries; and (4) identify human impacts. The project ended in 2025 upon Abel's retirement.

VIMS Shark Survey—The Virginia Institute of Marine Science Shark Survey is a small scale, but long-term survey. Eight fixed stations ranging from inside Chesapeake Bay to offshore Atlantic waters 30 m deep are sampled monthly with bottom longlines from June through September. Started in 1973 to monitor relative abundance of large coastal sharks, the survey's strength lies in its longevity.

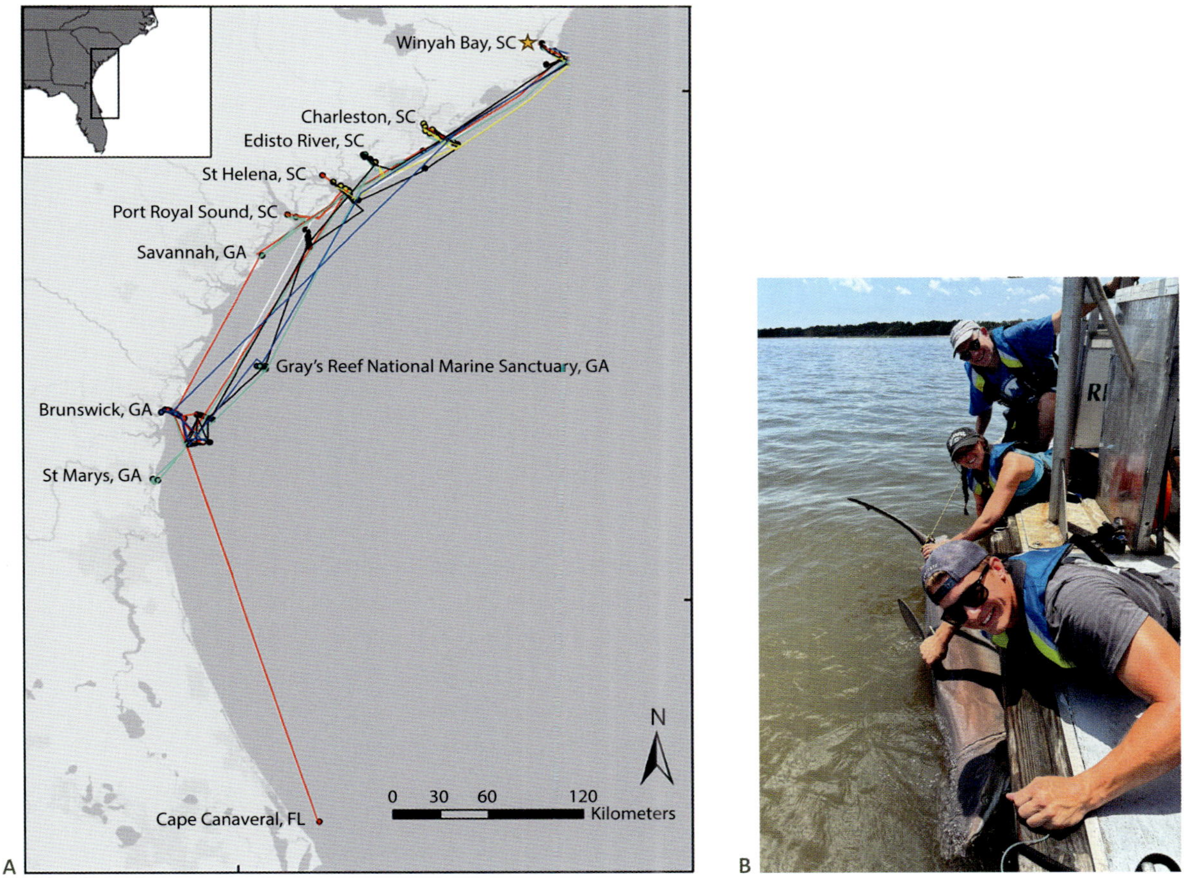

Figure 9.8. (A) Late fall to spring migration tracks of six acoustically-tagged juvenile Sandbar Sharks that exhibited heretofore unknown southern migrations from Winyah Bay, from a study conducted by the Coastal Carolina University Shark Project (Courtesy of Caroline Collatos). Esri, HERE, DeLorme, MapmyIndia, © OpenStreetMap contributors, and the GIS user community. (B) Photo from the Coastal Carolina University Shark Project.

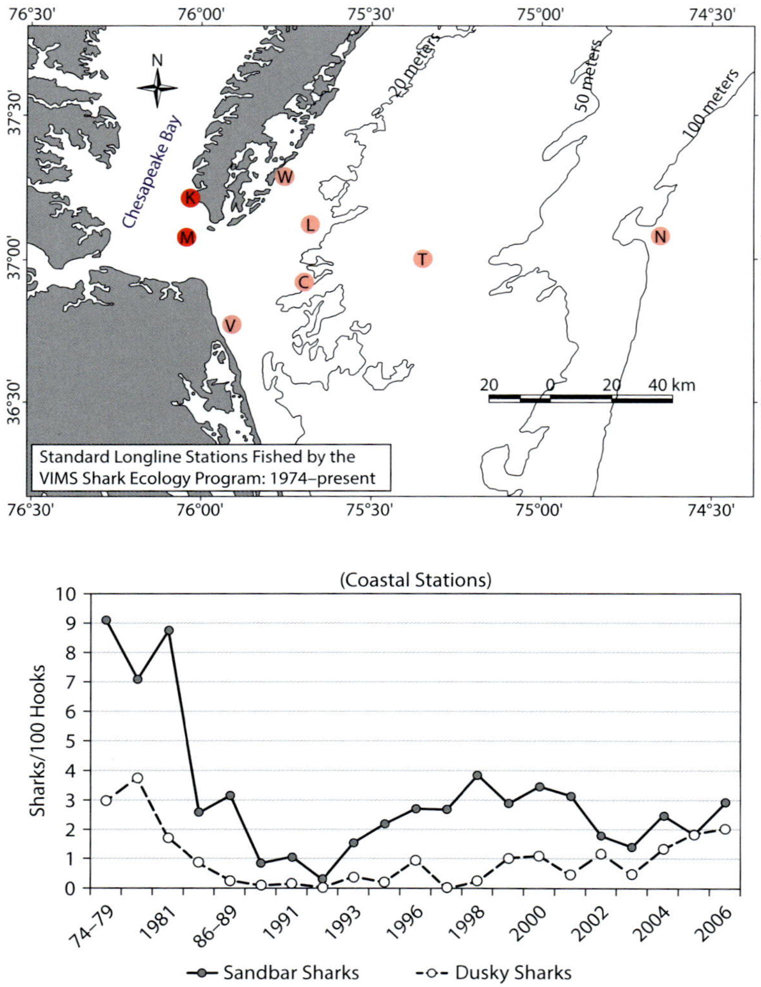

Figure 9.9. Top: Standard stations sampled in the VIMS Shark Survey. Bottom: Trends in relative abundance of the two most common large coastal sharks, Dusky and Sandbar Sharks, from the project from 1974–2006. Abundance trends reached a minimum in the early 1990s prior to the Federal Management Plan that went into effect in 1993.

The VIMS longline survey was among the primary fishery-independent data source used in the initial US Federal Fishery Management Plan for Sharks of the Atlantic Ocean implemented in 1993 and remains a primary index used in stock assessments for numerous large coastal and small coastal sharks (fig. 9.9). A second fixed station survey along with a randomized survey using smaller hooks was started at VIMS by co-author Grubbs in 1996 to examine seasonal and spatial use of Chesapeake Bay by juvenile Sandbar Sharks (see fig. 9.4).

FSUCML Big Bend Surveys—Started by co-author Grubbs in 2009 based out of the Florida State University Coastal and Marine Lab, this survey is unique in that it employs multiple gear types. The survey operates in the eastern Gulf of Mexico. Experimental longlines employing four hook sizes and gill nets employing six mesh sizes are used concurrently. Using the dual gears allows the researchers to sample all sharks from the smallest newborn small coastal species to adult large coastal sharks. In addition, the gillnet survey provides an index of prey abundance for large sharks and samples for food web analyses. Fixed stations are sampled one day per month, 12 months per year, providing data on seasonal migration and residency. A larger scale survey in the summer provides an annual index of relative abundance for numerous species and data used for large scale food web and community structure analyses. Both surveys and gear types are now used in stock assessments for coastal sharks in the Gulf of Mexico. Update: Funding was discontinued by the "Department of Government Efficiency" (DOGE) in 2025.

FSUCML Deepwater Longline Survey—Very few standardized deep-sea shark surveys exist. Started in 2011, this survey was designed to study the deep demersal (bottom-associated) sharks, bony fish, and invertebrate scavenging community in NE Gulf of Mexico in the wake of the Deepwater Horizon oil spill. The gear employs baited bottom longlines with five hook sizes and two trap types soaked for three hours. The purpose of the survey is to document changes in relative abundance, species assemblages, food web structure, and toxicological responses for sharks and other species as a function of time and space since the oil spill. It is the only survey of its kind and has become one of the largest surveys of deep-sea sharks ever conducted. Update: Funding was discontinued by DOGE in 2025.

BBFS PIT Study—The Bimini Biological Field Station in The Bahamas started a PIT tag (Passive Integrated Transponder, like the rice-grain-sized microchips implanted in dogs and cats) study of Lemon Sharks (*Negaprion brevirostris*) in 1995 (see fig. 12.6), but terminated in 2020. This was a long-term, fixed station, gillnet survey conducted at night in a semi-enclosed lagoon. It was unusual in being one of few census surveys, where the entire population of small juvenile Lemon Sharks was sampled, counted, and tagged each year. Census sampling is rare among shark surveys due to logistical constraints on capturing all individuals. Over the course of two weeks, sharks were captured, PIT tagged, and placed in holding pens. Sampling continued until no sharks were caught, and all sharks were then safely released. Because all sharks in the population less than about 3-years-old were caught each year, this survey provided one of the only empirical estimates of natural mortality in a shark population.

National Marine Fisheries Service (NMFS) Pascagoula Longline, GulfSPAN, and CoastSPAN Surveys—These are the largest shark surveys. Information can be found on the NMFS website. Update: Funding for GulfSPAN was discontinued by DOGE in 2025.

Other Methods of "Catching" Sharks

While surveys that actually catch sharks are critical for collection of many of the samples and data needed to understand the ecology of sharks and to manage their populations, there exist numerous non-capture methods of collecting data on shark distributions and habitat associations.

Submersibles

Manned and unmanned submersibles have provided valuable data on the distributions of deep-sea sharks and often provide our only *in situ* observations of these taxa. For example, most of the records for the Caribbean Roughshark (*Oxynotus caribbaeus*) have come from submersibles[5] (see fig. 3.14). Submersibles may also provide viable methods to tag deep-sea sharks at depths without exposing them to the trauma of bringing them to the surface. It is even possible to outfit unmanned submersibles with hydrophones and receivers that allow them to track tagged sharks at depths.

Camera Traps

Camera traps have long been used to document the presence of elusive wildlife in terrestrial and underwater environs. Advancements in video camera technology and economical underwater housings have made employing large-scale camera trap surveys possible even in the deep sea. These methods are obviously limited to waters that are relatively clear, but studies using baited remote underwater video (BRUV) surveys specifically to study sharks are now taking place all over the world in coral reef, pelagic, and deep-sea environments.

Drone Surveys

Similar to camera trap surveys, aerial (fig. 9.10) and underwater drones are increasingly being used to assess shark occurrence, habitat use, and relative abundance. Drones have even been used to study predator prey interactions (e.g., Great

Figure 9.10. Drone photograph of White Shark (Carlos Gauna/@TheMalibuArtist).

Hammerheads hunting Blacktip Sharks) and hunting behaviors (e.g., species-specific differences in stingrays excavating feeding pits). As drone technology expands and costs decrease, drone-based research will undoubtedly become common.

Environmental DNA

One final method to indirectly investigate the presence of sharks is the use of environmental DNA (eDNA). In concept, every living creature leaves fragments of their DNA in the environment after they are gone (e.g., skin cells, fecal matter, mucous, blood). By analyzing water samples for the presence of these fragments of genetic material, scientists can potentially assess if a species of shark was present in the area. It is important to realize that there are many caveats with these methods. Failure to detect a species does not mean it is absent. DNA degradation rates are usually unknown so if a species is detected in a water sample, there is no way to know if it was there recently or years ago. Also, since the genetic signature is in the water, currents may have carried it long distances from where the animal lived. This technology can prove valuable in detecting species presence but will never replace the need for classic surveys involving fishing since eDNA provides no indication of sex, size, reproductive status, trophic status, or any of the other parameters that require capture. However, combining increasingly sensitive genetics techniques such as Droplet Digital PCR (ddPCR) with eDNA sampling provides a powerful tool for detecting elusive or rare species and can be used to direct targeted sampling using capture surveys.

Where Do Sharks Move and Why?

Before we delve into shark movements and how researchers study them, let us address why they move at all. For that matter, why does any animal move? The answers to this question are fundamentally the same across all species and are related to the reasons fish live where they do. The objectives of most movements are related to gaining access to prey or mates while remaining physiologically comfortable and avoiding potential predators. Patterns of shark movements take place on many temporal scales in reaction to many stimuli ranging from the immediate and patterned reactions to predators, prey, or social cues to regimented seasonal migrations. Within a given day, shark activity patterns are often linked to predictable abiotic patterns, such as tidal cycles (ebb and flood) and time of day (day versus night).

As biologists, we may be interested in what drives animal movements on a daily scale and a lot of work in this arena has been geared toward delineating important nursery grounds. Not all sharks use defined nurseries, but many certainly do. In fact, it has even been hypothesized that sharks in the Paleozoic Era used nurseries. Use of nurseries may be an evolutionarily stable strategy of size-based habitat selection that minimizes predation risk and limits competition between sharks of different age classes. Some sharks, like juvenile Lemon Sharks in

Bimini, The Bahamas, establish small, well-defined home range within their nurseries whereas others, such as juvenile Sandbar Sharks in Chesapeake and Winyah Bays are essentially nomadic, ranging widely each day within the estuaries. In both cases, with growth the sharks move to adjacent, more expansive habitats due to different trophic requirements and size-based reductions in mortality risk.

Ontogenetic (i.e., over the course of their growth) expansion in spatial habitat use occurs nearly universally among animals. Activity space increases with body size in most animals, particularly carnivores like sharks. Habitat selection for very young sharks is most often a balance between maximizing accessibility to available resources (e.g., minimizing competition) and minimizing predation risk, which is often lower in shallower habitats typically used as nurseries. Shallower waters are often warmer and more productive than adjacent deeper waters, which facilitates faster growth. Predictably, species that have large birth sizes (e.g., Dusky Sharks, *Carcharhinus obscurus*) and/or fast juvenile growth rates (e.g., Atlantic Sharpnose Sharks, *Rhizoprionodon terraenovae*) often inhabit unprotected coastal areas as juveniles while those that have small birth sizes and slow growth rates typically use more protected areas such as estuaries as nurseries. Predation risk declines with growth while movement rates and foraging abilities increase. Combined with higher energetic demands and intraspecific competition, older individuals are driven to exploit additional habitats, resulting in increasing the breadth of their diet as well as their activity space.

Sharks move on much larger temporal and spatial scales than their daily activity spaces. You often hear that sharks are *highly migratory*, but what does this mean, and is it true? The term *migration* refers to predictable movements between areas, generally related to resource availability, in which the migrants are compelled to return to their place of origin. Migrations occur on tidal, diel (day-night), lunar, and seasonal cycles and over a wide range of horizontal and vertical spatial scales, but it is the larger scale seasonal migrations that draw most attention. Whereas most coastal sharks exhibit some form of seasonal migratory behavior, many deep-sea sharks examined to date do not. Even within the coastal species that do typically migrate, there can be migratory and non-migratory populations and even non-migratory contingents within a migratory population. We classify migrations by the primary driver into four main categories: *climatic*, *alimentary*, *gametic*, and *refuge* migrations.

Climatic migrations are driven by physiological tolerances of individuals to environmental factors such as temperature or salinity. Inshore juvenile habitats in tropical and subtropical regions, such as Lemon Sharks in Bimini, The Bahamas, may be occupied year-round, but in temperate climates such as Chesapeake Bay, this is not an option. Juvenile Sandbar Sharks can survive temperatures only down to about 15°C (59°F), but Chesapeake Bay drops to 5°C (41°F) in the winter. Thus, these juvenile sharks must migrate or die.

Alimentary (trophic) migrations are driven by prey availability, allowing sharks access to prey resources that are only present at certain times or only accessible due to abiotic constraints. Juveniles and adults may undergo alimentary migrations.

Gametic migrations are related to reproduction. In sharks, these may be directed toward increasing the survival of offspring by depositing eggs or newborns in habitats where they will be protected from predation and where prey densities are high, or these may be related to locating mates. Migration can facilitate aggregation of reproductively active sharks, thereby increasing the encounter rate with potential mates. It has been shown that the migratory population of Sandbar Sharks off the east coast of the United States had much higher rates of genetic polyandry (i.e., multiple fathers for a single litter) than the nonmigratory population in Hawaii. One theory to explain this posits that gametic migration in the Atlantic leads to a shorter, geographically constrained mating period, with higher encounter rates between mates and thus more multiple matings, than in Hawaii where mating takes place over broad areas for several months.

Refuge migration functions to decrease predation risk. Often, refuge habitats are unreachable by predators due to physical (e.g., water depth) or physiological (e.g., salinity) constraints. Complex habitats may also serve as refuges even in the presence of predators by allowing concealment. For example, the prop roots of red mangroves provide refuge habitat for newborn Lemon Sharks and Smalltooth Sawfish (*Pristis pectinata*) that is inaccessible to large sharks. Juveniles of many species migrate to distinct nursery habitats that provide refuge but are also highly productive forage areas; therefore, the migration has both refuge and alimentary purposes. Climatic forcings (factors) like temperature are often the proximate causes for migrations that ultimately serve to increase foraging and reproductive success or increase protection. So, migration behaviors often fall into multiple categories.

Studying Shark Movements

Determining where sharks move is central to studying their ecology in myriad ways. Movement studies can provide insight into changes in habitat use over scales from days to seasons to lifetimes. They can be used to determine important areas for feeding, mating, and giving birth as well as the habitats most susceptible to human degradation or where the sharks are exposed to fisheries capture. Together, these data are often used to designate Essential Fish Habitat in management plans or Critical Habitats for endangered species such as Smalltooth Sawfish. Shark researchers today have an array of tagging and telemetry tools at their disposal to study the movements and habitat-use patterns of their subjects over a wide range of spatial and temporal scales. These methods range from the very simple to complex.

Conventional Mark and Recapture

Conventional tagging (mark and recapture) methods are the simplest and have been used for nearly a century to study shark and ray movements. The tags employed include discs attached to each side of the dorsal fin using a stainless-steel needle, roto-tags attached to the dorsal fin (modeled after livestock ear tags), nylon dart tags anchored through the basal cartilages supporting the dorsal fin (fig. 9.11), stainless steel dart tags anchored in the muscle, as well as PIT tags discussed earlier. Except for PIT tags, which require a reader to detect their presence and identify the tagged shark, all other tags can generally be read by anyone that recaptures the shark. The tag number and the researcher's contact information are printed on the tag. Modern tagging programs often include a website and email address to report recaptures. The advantages of conventional tagging are that the tags are inexpensive, thousands of sharks may be tagged, and the recapture can be reported by anyone. If the appropriate tag is selected and applied correctly, conventional tags can be retained for the life of the shark. While numerous studies have reported recaptures more than 20 years after tagging, the record is for a School Shark (*Galeorhinus galeus*) recaptured in southern Australia after nearly 42 years.[6] The disadvantage of conventional tagging is that obtaining data on a tagged shark requires that it is recaptured and that the person recapturing the shark contacts the scientist. Though recapture rates are low (in the single digits in most cases), conventional tagging has provided a wealth of information on migration patterns, site fidelity, and even estimates of natural mortality rates. For example, tag recaptures from juvenile Sandbar Sharks tagged in Chesapeake Bay and at liberty up to 14 years were used to discern ontogenetic and sex-mediated changes in migratory patterns in this economically important species (fig. 9.12).[7] In addition, since the sizes of the tagged sharks are typically measured when caught and when recaptured, information on growth rates is obtained. Combin-

Figure 9.11. Cuban Dogfish in Eleuthera, Bahamas, with nylon dart tag (Courtesy of Lance Jordan).

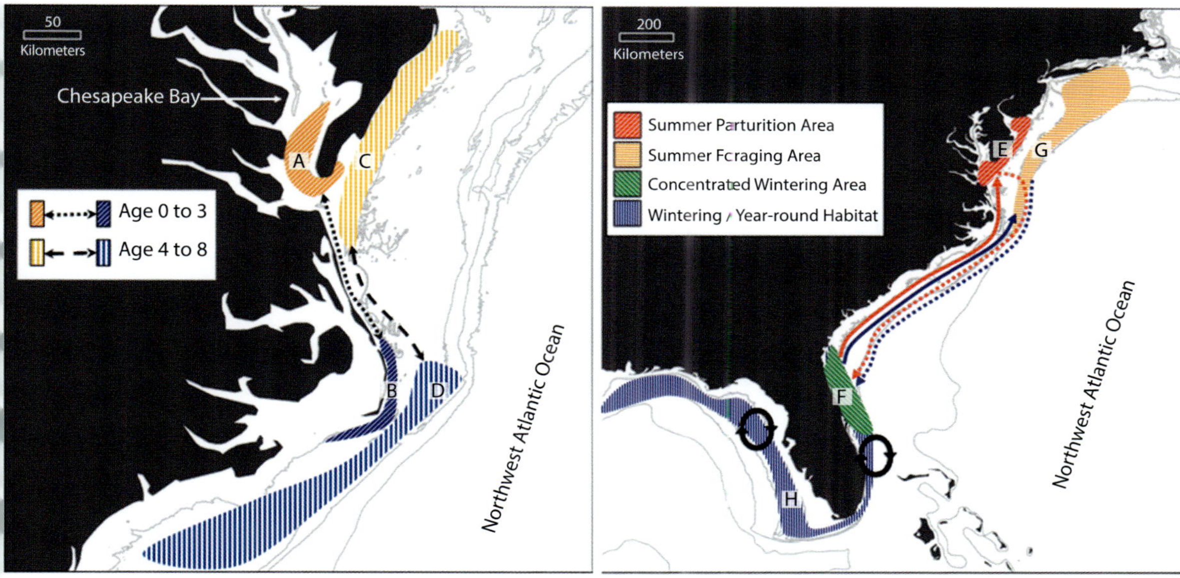

Figure 9.12. Migratory expansions in Sandbar Sharks from Chesapeake Bay based on tag recaptures. Left: Younger sharks (age 0 to 3) remain mostly within the Bay, whereas older juveniles (age 4 to 8) expand their range outside of the Bay. Right: Seasonal migrations of Sandbar Sharks from their birth and summer foraging areas (top) to the overwintering and, in some cases, year-round habitat (bottom) (Figure modified from Grubbs and Kraus 2019).

ing growth rates obtained from conventional tagging with chemical staining of hard parts, such as the vertebrae or fin spines (see pp. 353–356), is an important method in validating age and growth models.

Acoustic Telemetry

Though conventional tagging remains an important tool for studying shark movements, it provides no insight into what the shark is doing and where it is located between the initial tagging and subsequent recapture. The advent of acoustic (sound waves) telemetry greatly improved our capabilities for understanding fish movement and habitat-use patterns. The first use of underwater biotelemetry techniques on sharks dates back to 1963, when researchers tested the use of an ultrasonic (wavelengths higher than human hearing) transmitter for tracking a Scalloped Hammerhead and a Sandbar Shark off the east coast of Florida. Since then, scores of species and sharks and rays have been tracked using acoustic telemetry. The basic approach is to "tag" the shark with a transmitter that frequently emits an aural ping that can be detected using a directional hydrophone attached to a receiver. The frequency of the ping is usually 35–80 kHz, with 69 kHz being the most common frequency in use today. Humans can typically hear no higher than 10 kHz whereas sharks can only hear to around 1 kHz, so these pings are inaudible to us and to sharks. The high frequencies are inaudible to most marine animals, except for some marine mammals that produce high frequency sounds, and some clupeiform fishes (herrings and relatives) that are prey for marine mammals. However, the sound intensity of the transmitters is far too low to be audible even for those species.

Acoustic telemetry is conducted using two major approaches—active and passive tracking. In active tracking, researchers use a directional hydrophone and receiver to detect the shark's position, and they physically follow the sharks, recording locations and habitat characteristics continuously. This allows researchers to study the shark's activity space and use of specific habitats and depths at a relatively fine scale and to determine how characteristics such as temperature, salinity, depth, and habitat type influence these movements. Active tracking requires transmitters that emit aural pings between short intervals (~1 s) in order to maintain contact, which also limits their battery lives (usually weeks to a year). Tracking is usually conducted from small boats and the duration of each track is limited by the endurance of the researchers (usually no more than a few days), except in settings where multiple teams are available to rotate on and off of the track.

In passive tracking, individual animals are not followed directly but are instead detected by receivers placed semi-permanently in an array. Each time a shark with an acoustic transmitter passes within the detection range (usually 200–500 m, or 650–1640 ft), the tag number, date, and time are recorded and stored on the receiver. Researchers can then download the data from the receivers periodically. Transmitters used for passive telemetry transmit much less frequently than those used in active tracking (60–180 s intervals instead of 1 s), which permits battery lives of up to 10 years. The design of a receiver array depends on the scientific questions being asked. If interest lies in determining seasonal residency of juvenile sharks in an estuary that serves as a nursery, then a gate of receivers across the entrances to the estuary may suffice for determining the timing of the sharks entering and leaving. However, if interest lies in estimating the amount of space used within that same estuary, then an array that covers the whole estuary is needed and the receivers should be spaced such that all space is within the detection range of a receiver (i.e., ideally there is no place the shark can swim where it would not be detected). Until recently, the scale of receiver arrays was incompatible with the scale of movements of many sharks, particularly adults. In other words, it was logistically impossible and too costly for a researcher to deploy and maintain an array of receivers large enough to track shark migrations or even their seasonal use of large areas. For this reason, most passive telemetry studies until recently were conducted on juveniles or relatively sedentary species. However, over the last decade, teams of researchers have contributed to developing arrays of receivers that cover entire coastlines with hundreds of receivers. A database is maintained of all of the researchers that are tagging animals and their associated tag numbers. When a researcher downloads their receivers and detects tags that are not theirs, they send that detection data to the owner of the tag. Through these large cooperative arrays and long tag battery lives, passive telemetry has become

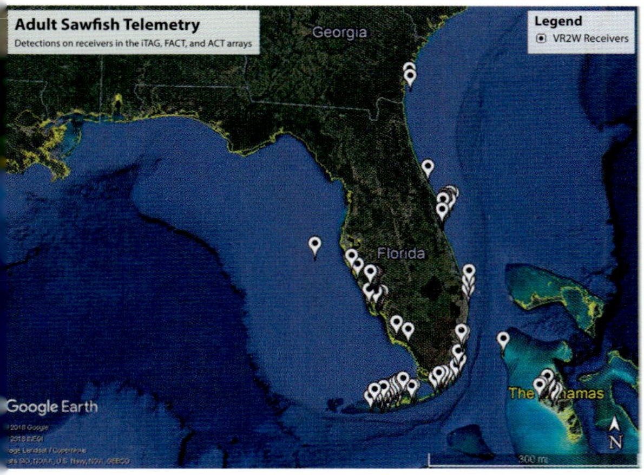

B

Figure 9.13. (A) Locations of acoustic receivers in Florida and the Bahamas that have detected adult sawfish tagged by author Grubbs and colleagues. (B) Author Grubbs and graduate student Bianca Prohaska implanting an acoustic tag through a small surgical opening in a Smalltooth (but large-bodied!) Sawfish.

a viable tool for studying the movement and migration patterns of even the largest of sharks and rays over multiple years and life stages (fig. 9.13). However, most acoustic receiver arrays are deployed in coastal zones in relatively shallow depths so they can easily be maintained from the vessel or using SCUBA, thus limiting application of passive acoustic telemetry for pelagic or deep-sea sharks.

Satellite Telemetry

The primary limitation of acoustic telemetry is that it requires repeated contact between the shark and the receivers. Whenever the shark moves out of the range of the receivers, it can no longer be tracked. The advent of satellite telemetry changed this by allowing animals to be tracked by orbiting satellites independent of the researchers. Essentially, radio transmissions are sent from the tag to satellites, as long as the tag's antenna is above the water surface. The first fish ever tracked by satellite was a Basking Shark *(Cetorhinus maximus)* tracked for 17 days off the coast of Scotland in 1984.[8] Studies of shark movements using satellite telemetry increased dramatically once the tags became commercially available. Now multiple companies specialize in satellite tag manufacturing. The size of the tags initially limited their use primarily to large species such as White Sharks, Whale Sharks (*Rhincodon typus*), and Basking Sharks and the results gave us glimpses into shark movements we never thought possible, including tracks of a White Shark moving from South Africa to Australia and a Basking Shark migrating from off Massachusetts to Brazil.[9, 10] Recently, improved design and tag miniaturization have made satellite telemetry amenable to many species, including small sharks like Bonnetheads (*Sphyrna tiburo*) and Spiny Dogfish (*Squalus acanthias*), and even skates and rays.

The advantage of satellite telemetry lies primarily in the fact that the researcher needs only to come into contact with the shark one time, to apply the tag. After that, all tracking is done through communications between the tag and satellites. Cost is the major disadvantage of satellite tags, as individual tags cost $1000–$5000 and the satellite service may cost an additional $500 per tag. Though satellite tag technology is continually advancing (e.g., more complex programming, added sensors, improved geolocation capabilities) two basic configurations exist, commonly referred to as SAT (satellite-linked transmitter) and PAT (pop-up archiving transmitter) tags.

SAT tags transmit radio signals that are picked up by orbiting satellites using calculations based on the Doppler Effect to estimate the distance to the tags. The Doppler Effect is the change in the frequency and wavelength of a wave (e.g., sound, radio, light) as the distance between a transmitter object and receiver changes. A simple example of the Doppler Effect is the apparent change in frequency you hear as a motorcycle passes by. The tires and engine of the motorcycle are transmitting sounds that are constant to the driver. But as a bystander, the sound of the motorcycle gets higher pitched as it gets closer to you and lower pitched after it passes. Knowing the radio frequency transmitted from the tag, the distance to a tagged shark can be calculated based on changes in that frequency as the satellite moves closer to or farther from the tag. Each time a signal is detected by the satellite, the change in frequency and time-lag are used to estimate the tag's location. Algorithms that use consecutive location estimates to predict future locations are used to refine the track of the tagged animal.

SPOT (Smart Position and Temperature) tags are the most commonly employed SAT tags for sharks. These have a wet/dry sensor that detects when the shark reaches the water's surface and send up signals, thus avoiding using power when the tag is submerged.

The advantages of SAT tags are the long battery life (one year or more), powerful transmitters that reach satellites 1000 km above the earth, and accurate, near real-time location estimates. This is the type of tag used to track the White Shark

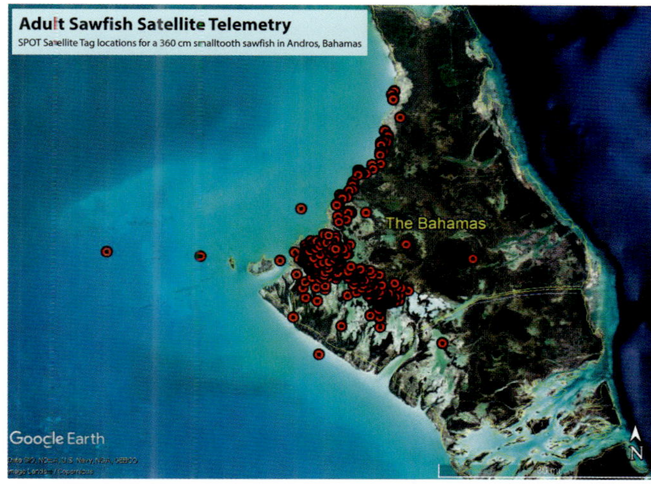

Figure 9.14. Example of the track of a Smalltooth Sawfish using a SPOT tag in the shallow mangroves of Andros in the Bahamas (Andrea Kroetz and John Carlson, unpublished data). Each red dot represents a fix uploaded to a satellite. Where did this sawfish spend most of its time?

mentioned earlier. You may be able to predict the major disadvantage of SAT tags. Recall that sharks live underwater. SAT tags are only useful for species that come to the surface frequently or live in shallow water. For this reason, these tags are usually attached to the dorsal fin or towed behind the shark with a tether. Figure 9.14 shows an example of a Smalltooth Sawfish tracked using a SPOT tag in the shallow mangrove backcountry of The Bahamas.

The accuracy of the locations produced from SAT tags using the Argos satellites varies considerably and for animals that surface infrequently, few locations may be received. However, making use of GPS satellites, which are far more numerous, can greatly reduce the error. SAT tags now have onboard GPS receivers that can provide location accuracies of 50 m or less when at the surface for about 20 s.

PAT tags may be used with animals that never come near the surface and are therefore used with a wider variety of shark species than SAT tags (fig. 9.15). PAT tags are programmed to remain on a shark for a specific time, which is usually less than a year. While attached to the shark, the tag measures and records (archives) time, temperature, depth, and light data. Once the tag release date is reached, a mechanism such as a burn wire, which predictably corrodes, detaches the tag from the attachment. The tags are positively buoyant on one end and float to the surface such that the antenna breaks the surface vertically. The tag's current (pop-up) location is transmitted to the satellites along with all the stored data. The light data recorded during the track are used to estimate the animal's location using light-based geolocation. How does this work? Day length varies depending on latitude and time of year. At the equator, there is 12 hours of light and 12 hours of dark all year long. But as you move north or south (increasing latitude), day lengths decrease in the winter and increase in the summer. This is why in Anchorage, Alaska (60 degrees North latitude) there is less than six hours of daylight in the winter and more than 18 hours of daylight in the summer. The time of solar

Figure 9.15. PAT (or PSAT) tag on a Sandbar Shark in Hawaii.

noon varies predictably by longitude. Thus, from the light data stored on the tag, we can determine the time of sunrise and sunset which gives us day length and peak daily light levels provide the time of solar noon. These two pieces of information allow us to estimate the position of the tag over the course of the track. These estimates are far less accurate than those obtained from the surface by the SAT tags. Small errors in day length estimates can cause large errors in latitude estimates, and do not forget about the pesky equinoxes when the sun is directly over the equator in the spring and the fall. During this period, day length is roughly the same all over the globe, causing huge errors in light-based geolocation estimates. However, we do have tricks to reduce these errors. We know exactly when and where the shark was tagged and where the tag popped up. We also know something about the swimming capabilities of our sharks: no shark can swim from Florida to Antarctica in one day (we had PAT locations that suggested this). We also have the temperature data recorded by the tag. By using the tagging and

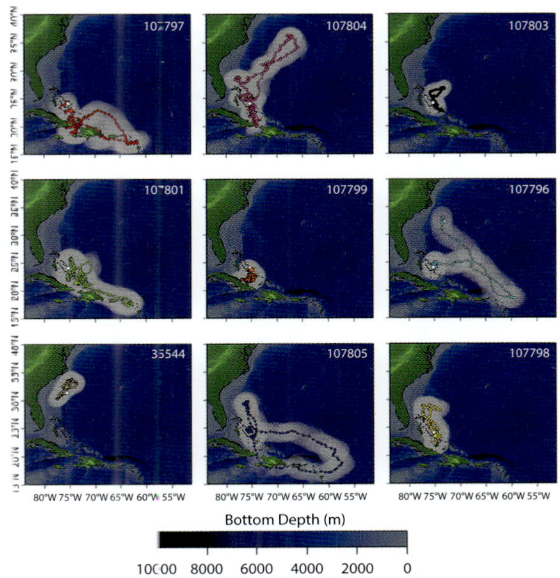

Figure 9.16. Likely tracks for nine Oceanic Whitetip Sharks tagged with PAT tags (From Howey-Jordan, L.A. *et al.* 2013. PloS one, 8(2), p.e56588).

pop-up locations, filtering out erroneous location estimates, and comparing the temperature when the shark is near the surface to sea surface temperature from satellites, we can produce a "most probable track" the shark took. However, due to the geolocation limitations, PATs are better suited for species that travel large distances than those with small home ranges. For example, figure 9.16 shows the most probable tracks for numerous PAT-tagged Oceanic Whitetip Sharks.[11]

The other great advantage to PAT tags is the archived depth data. The ability to record depth and temperature make PAT tags very useful in studying the depth-use patterns, even in deep-sea sharks where no light data are recorded to allow light-based geolocation. Figure 9.17 is an example of depth data obtained from a PAT tag on a Bluntnose Sixgill Shark. Since the tags are recording temperature as well as depth, the tagged animals can serve as incredibly detailed water samplers. These methods are also very useful in studies of post-release mortality in fisheries. If a released shark goes to the bottom after release and does not move, it likely died. In fact, several manufacturers are now making tags specifically designed to assess mortality. Since most sharks captured in fisheries are bycatch that are discarded, estimating post-release mortality is critical in determining the effects of fishery capture on populations.

Emerging Telemetry Technologies

We will conclude our discussion of methods to study shark movements by briefly mentioning emerging telemetry technologies that are allowing researchers to study shark behavior like never before. Researchers have long wished to gain more insight into what sharks are actually doing while they are being tracked. When are they chasing prey? Avoiding predators? Mating? Feeding? Resting? Some early

A

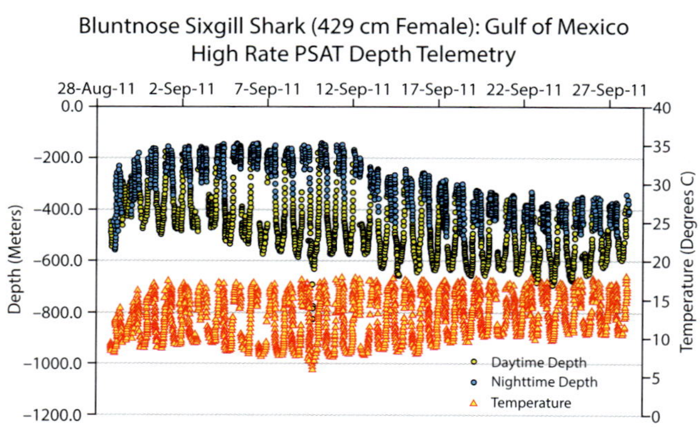

B

Figure 9.17. (A) Author Grubbs attaching a PAT tag on a Bluntnose Sixgill Shark in Eleuthera, Bahamas (Courtesy of Lance Johnson). (B) Depth and water temperature data from a PAT-tagged Bluntnose Sixgill Shark in the Gulf of Mexico, showing nocturnal migrations from colder depths to warmer surface waters (Grubbs unpublished data).

studies employed tail beat monitors to estimate when sharks were burst swimming versus sustained swimming, but these were relatively cumbersome. Tri-axial accelerometers are taking the technology leaps forward. When incorporated into archival or acoustic transmitters, accelerometers record acceleration in three dimensions (forward-back, side-to-side, up-down) in great detail. Multi-sensor transmitter tracking packages with accelerometers (e.g., daily diary[12]) can provide information on movements in terms of location and animal behavior (e.g., resting, chasing, even mating in Nurse Sharks[13]) and even energetics (see Chapter 8).

The accelerometer data can be used to determine overall dynamic body acceleration (ODBA), which has been shown to be correlated with oxygen consumption and thus energy expenditures in a wide range of animals, including sharks.[14] These archival transmitter packages can also include sensors to measure environmental parameters externally (e.g., temperature, depth) and even internal to the shark (e.g., pH). Another emerging frontier in shark movement studies is inter-animal telemetry, using what has been referred to as "business card" tags that are not only transmitters but also include receivers that can detect and archive detections from transmitters on other sharks.[15] These methods hold tremendous promise in studying schooling, aggregation, and social behaviors in sharks.

Using Genetics to Study Shark Movements and Population Structure

Tagging and telemetry provide a wealth of information regarding the movement and habitat-use patterns of sharks. From the standpoint of fisheries management, this is critical because it provides managers with data concerning the likelihood that sharks from a given population will be caught by specific fisheries. However, population genetics offers powerful tools to discriminate whether movements of sharks between two regions are interbreeding or only mixing ephemerally. The techniques employed in population genetics are beyond the scope of this book, but in short involve looking for differences in frequencies of genes and alleles (alternative forms of a gene) between groups of sharks that may reflect isolation and adaptation (local evolution) and indicate whether the groups are members of the same or different populations. Defining the boundaries of a population (or stock) is the critical first step in fisheries management.

Tagging studies may indicate that sharks in two regions rarely if ever mix, but molecular data indicate they are part of the same population. This suggests that sharks crossing from one area to the other happens too infrequently to pick up in the tagging data but frequent enough to genetically mix the populations. For fisheries management, these two groups of sharks may be managed as a single stock. This makes management difficult if a single stock crossed international boundaries, which is common for pelagic sharks. In fact, Blue Sharks have been shown to be part of one single worldwide genetic stock, hence the extreme difficulty in managing them. Conversely, tagging data may suggest two groups of sharks mix during part of their migration, but genetics data may suggest there are distinct populations that are reproductively isolated. The genetics data confirm that sharks in the two regions are distinct, they would appropriately be managed as two separate stocks. The peninsula of Florida has been shown to be a barrier to movements and gene flow to many coastal shark species (e.g., Blacktip [*Carcharhinus limbatus*], Blacknose [*C. acronotus*], and Bonnethead Sharks [*Sphyrna tiburo*]) and thus in US Atlantic and Gulf of Mexico waters these are managed as separate stocks.

Genetics can also be used as a form of tagging in numerous ways. Recaptures of animals can be identified from their genes by comparing microsatellites - short, repeated DNA sequences that are highly variable among individuals. Co-author

Grubbs sampled an adult Smalltooth Sawfish in Andros, The Bahamas that had been sampled 15 years earlier in Bimini, The Bahamas. This recapture was only determined through genetic analyses.

Through genetic analyses siblings, half-siblings, and mother-offspring can be identified and potentially tracked. In this sense, a sample from a parent can be considered the tag and a sample from the offspring represents a recapture of that parent. In Bimini, The Bahamas, female Lemon Sharks were found to return to give birth in the same area every two years for 20 years of more based only on their genetic identification from their offspring caught in the PIT study discussed earlier[16]. Taken further, using a technique called Close-Kin Mark-Recapture (CKMR) these parent-offspring and other close-kin relationships in a sample can be used to estimate population size and compared with more traditional methods.

A Whale Shark was the first shark to have its whole genome (complete DNA of an individual) sequenced, followed, unsurprisingly, by a White Shark. High quality reference genomes that represent the genetic blueprint for a species are now available for numerous species of sharks and rays. Sequencing whole genomes and assembly of reference genomes for a species are becoming less expensive and time consuming. Advanced genetic techniques using these genomes may allow estimation of changes in population size and structure over time scales of many generations to tens of thousands of years.

Life Histories of Sharks: Age and Growth

Sharks are typically slow-growing and long-lived. How slow-growing and how long-lived? Knowing this is central to understanding the ecology of shark species as well as managing them, which differs from managing bony fishes.

Reproduction of sharks as a group is more limited than that of bony fishes. Reproductive limitations are characteristic of animals classified as K-selected (as opposed to r-selected; see Chapter 1). The concept of K- and r-selection is an ecological idea based on diametrically opposed life-history, specifically reproductive, strategies. Although the idea is somewhat outdated, in part because life-history characteristics operate on a continuum and are rarely the extremes depicted in r- and K-selection, it is worth quickly reviewing these as they apply to sharks (fig. 9.18). We will revisit and apply these concepts in the following discussion.

Before fishery biologists can implement management plans for any species, they must consider the vulnerability of a species or stock (population) to overfishing. To manage a fishery (i.e., to assess this vulnerability), it is crucial to know information about the animal's life history. It is also critical in fisheries with multispecies catches to know life-history characteristics of incidental catch such that the fishery does not deplete the stocks of bycatch. These characteristics are also important in other areas (e.g., predator–prey studies). Critical life-history characteristics include:

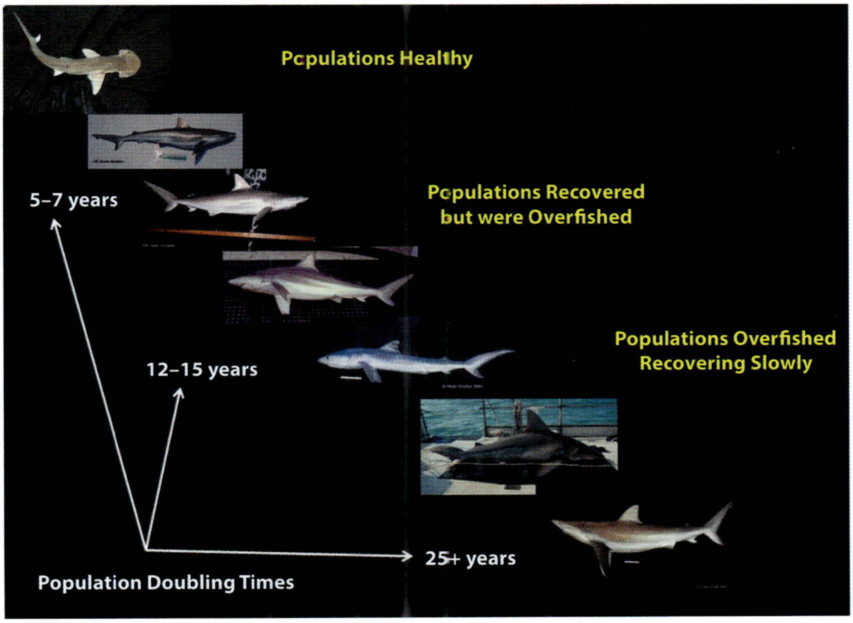

Figure 9.18. Spectrum of population doubling times for sharks along the US East Coast and a generalized view of how they have responded to fisheries and management.

- Size and age at maturity
- Size and age at first reproduction
- Fecundity, both per pregnancy and lifetime
- Frequency of reproduction
- Survival of offspring (the most difficult to accurately determine)
- Maximum size
- Maximum age

How are these characteristics determined? Let us start with age and growth. Access the web, go to Google Scholar, and search for *Age and Growth Sharks*. Surprised at the number of hits? Although not the most exciting research, obtaining these life-history characteristics is the *sine qua non*[17] of fisheries management, which is reflected in the number of studies. Further note that many of these studies are for the same species, but in different locations. We will tell you why.

How are age and growth studies done? In principle, it is fairly easy and relies on the same idea of trees depositing hard tissue seasonally, annually, or during some environmental events over the course of their entire life. In bony fishes, age is recorded primarily in calcified layers of the *otoliths*, or ear bones. Sharks lack otoliths and instead have *stataconia*, which are like squishy sand grains and which do not lay down concentric rings.

Some bony fishes (e.g., Striped Bass [*Morone saxatilis*]) can be aged by removing a few scales, which record growth. Although sharks also have scales, recall that they are morphologically different from those of bony fishes. Moreover, the

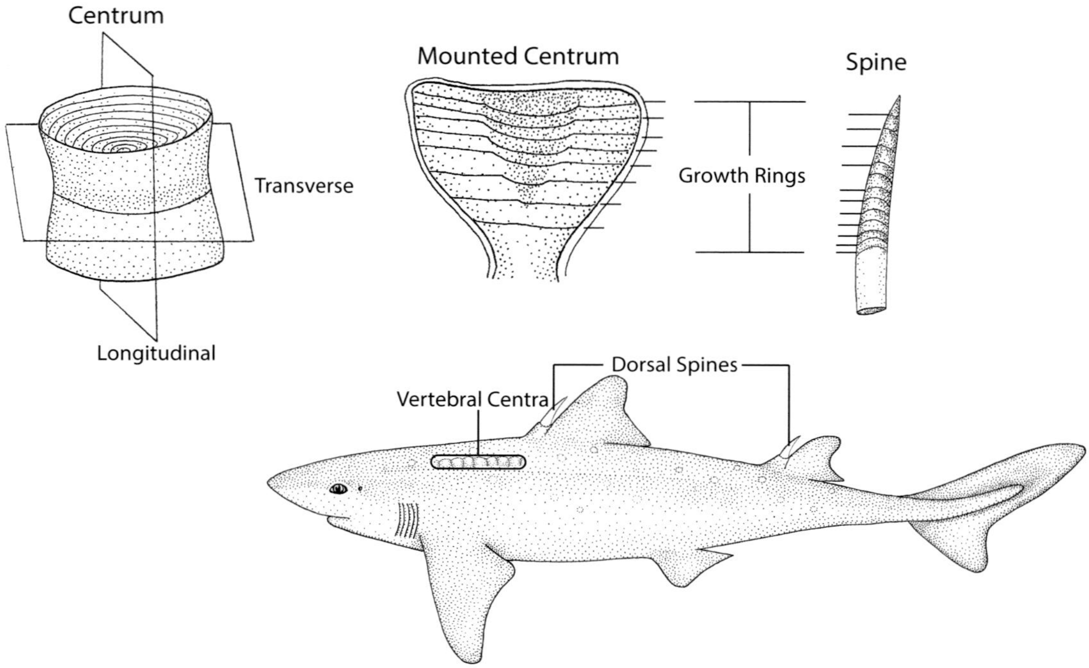

Figure 9.19. Ways to determine the age of a shark from vertebral centra and dorsal spines. Individual centrum (top) is pictured rotated 90°. See text for explanation (Redrawn from fig. 6.1 and 6.4, Goldman, K.J., 2005. FAO Fish. Tech. Pao. 474: p.76).

dermal denticles of sharks are replaced, perhaps several times over the life of the shark. Thus, the dermal denticles of sharks cannot be used for aging.

The only hard parts that can be used to age chondrichthyan fish with reliability are parts of the vertebral centra (fig. 9.18), the main structural elements of the backbone or, on some species (e.g., some Squaliformes [dogfishes], Heterodontiformes [horn sharks], and Chimaeriformes [chimaeras/ghost sharks]) the spines on one or both dorsal fins (fig. 9.19). For vertebral centra and dorsal spines, until a noninvasive reliable imaging technique is invented, the animal must be killed.

The use of spines to age sharks goes back to the 1930s, for the Spiny Dogfish (*Squalus acanthias*). In some cases (e.g., in deepwater squalids whose vertebral centra are not calcified sufficiently to yield concentric rings), it is the only method available. Using dorsal spines for aging can be problematic in older individuals of some species. First, the bands may become too tightly spaced to accurately infer age. Second, the spines of the first dorsal fin may become too worn down to be of use. Spines are removed in their entirety, including the base below the skin. Figure 9.20 shows ages of spines of a dogfish.

Vertebral centra have been used to age sharks for over 90 years. After sacrificing the shark, a section of 10 or more vertebral centra are removed, typically the largest ones in the thoracic (or trunk) region beneath the first dorsal fin. Recall that each vertebral centrum is a disc that is concave on both sides with the remnant

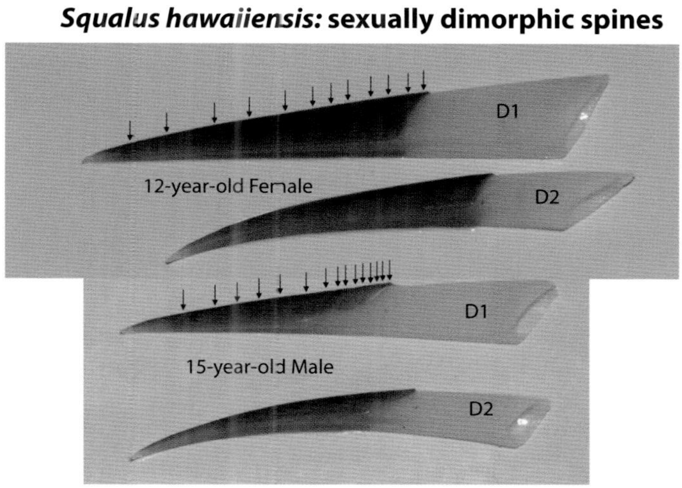

Figure 9.20. Ages from spines of male and female Hawaiian Spurdog (Courtesy of Charles Cotton).

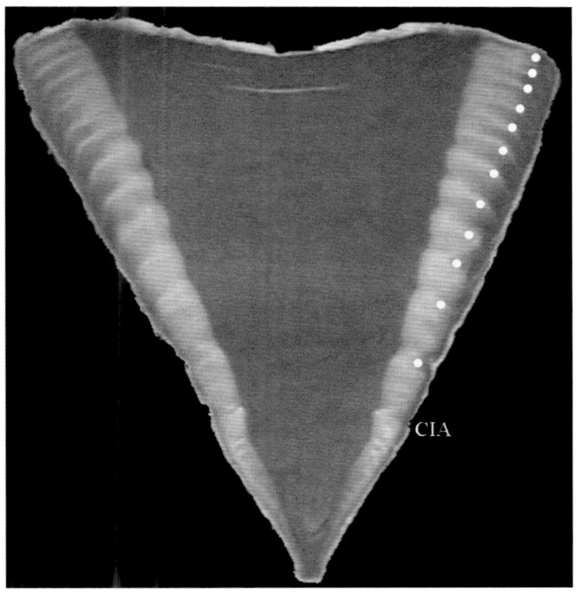

Figure 9.21. Sectioned vertebral centrum from a Sandbar Shark. CIA stands for Change in Angle and depicts the point of birth. If you estimated that the shark from which this centrum came was 12 years old, you'd be correct (Courtesy of Jason Romine).

of the notochord running through its center. The centra are then cleaned, cut with a diamond blade, and mounted. Figure 9.21 shows the bow tie, or hourglass, shape of a section through the centrum. After sanding the section down until it is sufficiently thin that light passes through it, you can examine rings; for example, the wide, light-colored rings laid down in the summer and the narrower, denser, darker rings deposited during the slower growth winter period.

The aging protocol requires considerable practice and uses multiple readers (i.e., scientists to interpret the number of rings). Additionally, it is crucial to know that the rings that are counted correspond to years and not seasons. There are two main ways to validate the time period of rings:

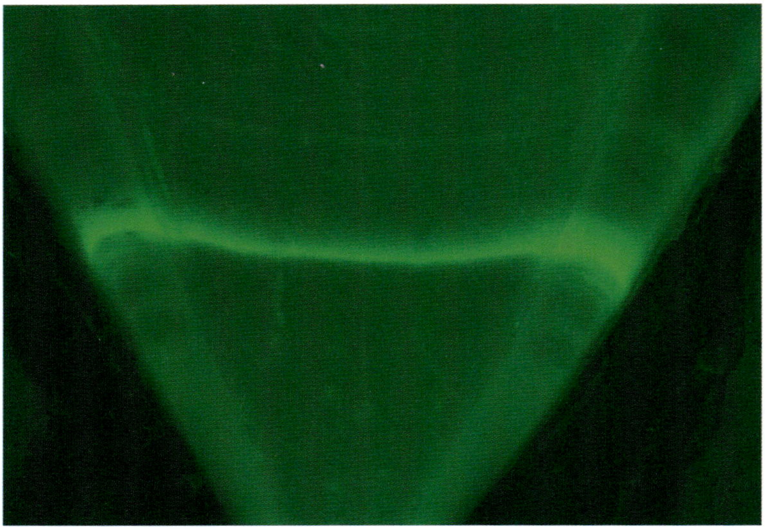

Figure 9.22. Oxytetracycline (OTC)-stained sectioned vertebral centrum from a Sandbar Shark. Green line denotes the time of the OTC injection (Courtesy of Jason Romine).

- Catch young-of-year (YOY) or neonates (newborns), measure, and tag large numbers of them, then hope that some will be recaptured three or more years later. Then remeasure and sacrifice recaptured sharks and remove centra and count rings. If the bands in the centra correspond to years at liberty, then you can safely conclude that they lay down bands annually, as well as how much they have grown.
- Beginning in the 1970s and 1980s, sharks were injected in their muscles with a chemical—oxytetracycline (OTC), calcein (a fluorescent dye), crystal violet, or silver nitrate—which stained the hard parts. Then the sharks were measured, tagged, and released. Recaptured individuals were measured, sacrificed, and then rings were counted outward from the point of stain and age and growth could then be determined (fig. 9.22).

Age and Growth: A Case Study (Sandbar Shark)

Sandbar Sharks are circumglobal in distribution. Robust age and growth models from the Virginia Institute of Marine (VIMS)[18] showed that it takes about 15 years to reach sexual maturity, and they produce about eight pups every two years. These results further showed that Sandbar Sharks in and near Chesapeake Bay are born at a precaudal length of about 45–50 cm (16–20 in) and females reach sexual maturity at about 136 cm (53.5 in). The species has a life span of about 30–35 years.

The VIMS data on Sandbar Sharks was critical to the National Marine Fishery Service's fishery management plan (FMP; see Chapter 11) for this and other large coastal shark species, a group that was overfished along the US Atlantic Coast. Among these species, Sandbar Sharks were among the most overfished from the 1980s and early 1990s, primarily for their large fins, which were popular in the shark fin soup industry, as well as for meat. Because of their low reproductive rates, population growth could not exceed 4% per year, and thus the potential for quickly rebounding from overfishing was very low.

In 1993, the FMP that was implemented put heavy quotas on landings of large coastal sharks, including Sandbar Sharks along the US East Coast, based on the life-history characteristics reported by VIMS researchers.

Representatives of the commercial shark fishing industry challenged the validity of the VIMS data, instead suggesting that reproductive rates were in fact much higher. They argued that another scientific study, one on Sandbar Sharks in captivity in Hawaii published in 1973,[19] suggested that reproduction occurred twice annually, and that the species reached sexual maturity at three years. Thus, the industry concluded that what they interpreted as draconian management policies, were in fact based on bad information. Who was right, the commercial shark fishers or the VIMS scientists?

To answer that question, another age and growth study on Sandbar Sharks was conducted in Hawaii.[20] The reason for repeating the study in Hawaii was twofold. First, it is not unusual for aquarium animals (and those in zoos), to grow faster, sometimes *much* faster, than wild counterparts. Second, the ecology of Sandbar Sharks in Hawaii differs significantly from that of those on the US Atlantic Coast. Hawaiian Sandbar Sharks live in deeper water than Atlantic Sandbar Sharks. Moreover, in Hawaii, Sandbar Sharks are year-round, nonmigratory inhabitants, whereas those on the US East Coast are migratory. Additionally, water temperature is high year-round in tropical Hawaii, and the mating season is thus protracted. Finally, Sandbar Sharks are more sparsely distributed in Hawaii than along the US East Coast.

To conduct the study, researchers (including co-author Grubbs) set demersal longlines outside Kaneohe Bay, Hawaii, and caught and sacrificed up to five specimens of each sex and 5 cm size class (45–150 cm or 18–59 in) for a total of 194 sharks. This may seem like a lot of sharks, but it is the minimum required to get robust, accurate estimates of these extremely important life-history parameters. If in fact the estimates of the industry were incorrect and were implemented anyway, which is a real possibility since the fishery management decisions included political and economic dimensions, tens of thousands more sharks would have been killed than the 194 for this study.

The results validated both of the scientific studies. Sharks do in fact mature earlier in Hawaii than in Chesapeake Bay, at 10 years for females and 8 for males (compared to 15 combined for both sexes in Chesapeake Bay), though not at 3 years as the 1973 paper asserted. Moreover, the representatives of the commercial shark fishery ignored another hypothesis from that paper—that an analysis of tooth replacement suggested the age of maturity as 10.2 and 13.1 years for males and females, respectively. Science indeed works, if used as directed.

So, even though Hawaii has no shark fishery and longlining within 80 km (50 mi) of the coast is prohibited, this study was important for Sandbar Sharks about 7900 km (4900 mi) apart. Data from a long-term tagging study of neonate Sandbar Sharks tagged in Chesapeake Bay and recaptured up to 14 years later further confirmed the initial growth models proposed by the VIMS researchers.

Life-history Theory

Life-history theory refers to understanding growth, survival, and the reproduction of species. Understanding the life history of any organism is critical for its management and conservation. The bulk of life-history research on sharks is directed toward estimating five parameters:

- Size or age at first reproduction (i.e., maturity)
- Reproductive output (e.g., fecundity, periodicity)
- Survival of offspring
- Maximum size
- Longevity (maximum age)

All of these parameters are related to *fitness*. By fitness we do not mean the ability of sharks to swim a marathon, but rather their biological or Darwinian fitness, which is defined as success in producing viable offspring that themselves survive to reproduce. The goal of any organism in terms of reproduction and life history is to maximize its fitness. One of the metrics used to estimate how resilient a population may be to fishing is the *intrinsic rate of natural increase* (lowercase *r*), a theoretical maximum value which is a function of two important aspects of its life history—its reproduction rate (*R*) and the generation time (*G*). Another important estimator of population resiliency is the *doubling time*—the time it takes a population to double in size given a stable growth rate. Doubling time is estimated by 70/% growth rate. For example, a population growing at a rate of 5% per year theoretically doubles in about 70/5 = 14 years.

The *intrinsic rate of natural increase* (or *population growth*), (*r*) is calculated as follows:

$$r = \log_e (R_0)/G$$

where,

R_0 = the average number of offspring that a female produces during her lifetime, also called the *net reproductive rate*
G = generation time (i.e., the mean period between when the parents and all their offspring are born)

This intrinsic rate of natural increase *r* in the above equation, is the "r" in r-selection. r-selected species maximize their rate of population increase. The "K" in K-selected species stands for *carrying capacity* (*K* in German).

There are three general ways to maximize the intrinsic rate of natural increase *r*:

- Minimizing generation time (e.g., mature faster)
- Maximizing fecundity (e.g., have more offspring)
- Maximizing survival (e.g., grow faster, have larger offspring)

There are limiting tradeoffs. Maturing fast generally results in small body size. The smaller the young, the higher the mortality. Therefore, mortality is minimized by having larger young. But larger young require higher maternal investments and results in lower fecundity. Small body size also results in fewer offspring unless those offspring develop external to the mother.

Most organisms lie on the continuum of r- and K-selection and not on the extremes. r-selected species are those that have evolved to maximize their rate of population increase versus K-selected species, which have evolved to live near their carrying capacity.

Sharks and their close relatives are on the K-selected side of the spectrum. r-selected species tend to be small, mature quickly, and have very high fecundities (as many as millions of eggs) with very little investment of maternal resources in each egg. An example among fish would be the anchovy. The largest fishery in the world, Peruvian Anchoveta, used for fishmeal and oil, persists because these tiny bony fishes mature in about a year and produce tens of thousands of offspring. Similarly, egg-laying in sharks and rays is likely an adaptation for increasing fecundity in small species. Skates and catsharks can produce dozens of offspring through egg laying, whereas they could produce only a few similarly sized offspring through live birth.

At the other end of the spectrum are K-selected species, which have large body size, take years to decades to reach sexual maturity, and have very low fecundity (just a few offspring at every reproductive event, which may occur every other year, or even every three years) with a large investment of maternal resources in every pup. All large sharks give live birth and again there is a tradeoff. Sand Tigers (*Carcharias taurus*) produce only two pups, but they are very large and have a high likelihood of survival. Tiger Sharks may produce 60 or more pups that are very small and are at high risk of being eaten but grow very quickly. Both strategies result in at least enough offspring surviving to replace the mother during each cycle.

Thinking of life histories in terms of fisheries, if a fishery targets tunas, like the Yellowfin Tuna, which matures at age three and produces millions of eggs annually, it takes only two years for the population of this r-selected species to double. So even if the stock is overfished, if management measures are put in place, recovery may be achieved in a few years. Conversely, a Sandbar Shark takes 15 years to reach maturity and produces eight pups every two years (i.e., on average two female pups per year). Their population doubling time is about 25 years. So, an overfished stock of Sandbar Sharks takes many decades to recover, if allowed to.

If you look at the intrinsic rate of population increase in all taxa, it is generally inversely proportional to the body weight. In other words, the smaller you are, the faster your population grows or, more appropriately for sharks, the bigger you are, the more slowly your population is going to increase.

This spectrum occurs not only between sharks and other species, but also among different sharks. For example, the Bonnethead takes three years to mature, reproduces every year, has relatively high fecundity, and has a population doubling time of six years. Thus, it is important to realize that *within* the sharks and rays, some species are more vulnerable to overexploitation than others. But it is also critical to realize that there are notable exceptions to this rule among the sharks. Many of the deep-sea sharks, like the dogfishes and gulper sharks, are around the same size as the small coastal species that grow quickly, but they have among the lowest rebound potentials due to very slow growth, long gestations, and few offspring.

Next, we will discuss the management of sharks along the US East Coast. Now, many shark species in this region are either prohibited from harvest (19 species) or managed with species-specific quotas. However, for many years, sharks were managed as three major groups—Large Coastal Sharks, Small Coastal Sharks, and Pelagics. The reason for these divisions, in part, is that members of each group tend to have different life histories. For example, in the Small Coastal Shark category, Bonnethead, Blacknose Shark (*Carcharhinus acronotus*) and Atlantic Sharpnose Shark (*Rhizoprionodon terraenovae*) all have population doubling times in the 5- to 10-year range. Members of the Large Coastal Shark group, on the other hand, like the Lemon, Sandbar, and Dusky Sharks, have population doubling times in the 20- to 30-year range. Pelagic sharks, including the Blue and Oceanic Whitetip Sharks, are somewhere in the middle.

Large Coastal Sharks were further divided into *ridgeback* sharks and *non-ridgeback* sharks (see Chapter 3) in part because of life-history differences that necessitate that they be managed separately. Historically, the two most-frequently caught Large Coastal Sharks are the Sandbar and Blacktip. The Sandbar, which has a long population doubling time, is a ridgeback. The Blacktip Shark is a non-ridgeback and it matures early and thus has a population doubling time that is half a Sandbar's. Managing them both as Large Coastal Sharks without regard for these life-history differences would most likely be problematic for recovery of Sandbar Sharks.

Complicating management even further, within a species there can be variability. Consider that the age of maturity of Sandbar Sharks for populations in the Indian Ocean, NW Atlantic Ocean, and the Pacific Ocean off Hawaii is about 16, 15, and 10 years, respectively.

One of the major impediments to manage sharks in the 1980s was trying to do so like bony fishes were managed. In reality, sharks should have been managed more like whales and other cetaceans, or even elephants, because the intrinsic rates of population growth of sharks, like Sandbar Sharks, are more similar to these mammals, or even sea turtles, than they are to bony fishes. The annual rates of population growth for the African elephant, loggerhead sea turtle, and most larger species of sharks are in the ranges of 4.0–7.0%, 2.0–6.0%, and 1.7–6.9%, respectively. Managing sharks as bony fishes, whose annual rates of population growth are much higher, would lead to overfishing in most cases.

Trophic Ecology

Sharks are known as opportunistic predators, and prey selection often varies on multiple time and space scales. The diets of most species are diverse, and they can switch prey types in response to changes in prey abundance or species composition. The hypothesis that sharks consume the most abundant prey available has been supported by numerous studies, however, there is also plenty of evidence that some shark species are specialists on certain prey types. For example, Horn Sharks (*Heterodontus francisci*) feed primarily on hard-shelled molluscs and crustaceans, Dusky Smoothhounds feed mostly on a few species of crabs (primarily recently molted crabs), and Frilled Sharks (*Chlamydoselachus anguineus*) eat mostly squids.

How Do We Know What Sharks Eat?

There are multiple approaches to studying trophic ecology, but two dominate the literature: gut content analyses and analysis of stable isotope ratios. Firstly, a lot of what we know about the trophic ecology of sharks comes from actually looking at stomach contents and identifying the prey items (fig. 9.23). Typically, the stomach contents are removed in the field, preserved or frozen, and then identified in the lab to the lowest possible taxon. This is a meticulous process that requires a deep knowledge of taxonomy and identification of the species that could be in the potential prey base and, to be done properly, it requires access to scores of stomachs from both sexes across all shark life stages. The advantage is that prey can

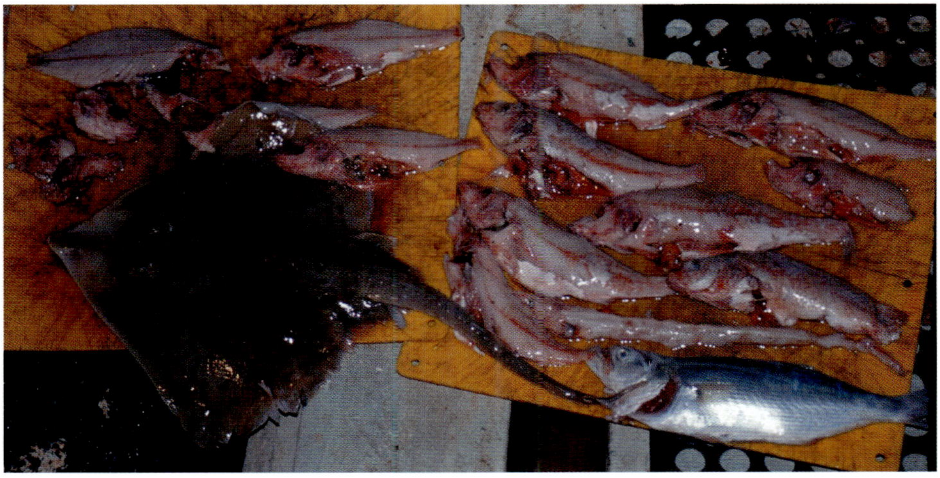

gure 9.23. Gut ntents of Sand ger, whose diet ludes fishes that y be swallowed ole, like Atlantic akers, a efish, and a arnose Skate.

potentially be identified to species and the relative contribution of each prey species can be quantified. Stomach content analysis remains our most robust method for studying the actual diet of sharks, but it has disadvantages beyond the time, skill, and labor required. Though some studies use stomach lavage with water to flush contents from live sharks that can then be released, most samples are obtained from dead sharks, typically obtained from fisheries or from sharks that have been used in age and growth studies. Many studies are hampered by the prevalence of empty stomachs, since the hungriest sharks take baited hooks. Stomach contents also only tell us what the shark was eating in the hours to days before being caught, so samples are needed over longtime periods to robustly quantify the diet.

Secondly, analysis of stable isotope ratios is based on the adage "you are what you eat." Recall that *isotopes* are multiple forms of a chemical element that vary by the number of neutrons in the nucleus. The more neutrons, the heavier the isotope. These isotopes can be generally considered either *radioactive* (which spontaneously decay over time; that is, they are *unstable*) or *stable*. The element Carbon, for example, has three naturally occurring isotopes, of which two are stable: Carbon-12 (^{12}C or C-12), with six protons and six neutrons, and ^{13}C, with six protons and seven neutrons. The $^{13}C/^{12}C$ ratio[21] (conventionally reported as heavy isotope/light isotope) is about 1/93.

For analysis of stable isotope ratios, the number in which we are interested would be calculated differently and reported as $\delta^{13}C$ (pronounced *delta-13-C*), whose units are parts per thousand (‰, ppt, or per mil). $\delta^{13}C$ is calculated by taking the above ratio, $^{13}C/^{12}C$, dividing it by a standard, then subtracting 1, and then multiplying by 1000.

Carbon, Nitrogen, and Sulfur isotopes are most commonly used for aquatic diet studies because they are *tracers* of different aspects of trophic ecology. Carbon stable isotopic ratios provide a good idea of which primary producers are at the base of the food chain. Sulfur ratios give insight into whether the animal is feeding in a benthic or pelagic food web. Nitrogen isotopes are most useful for determining a shark's trophic level (how high in the food chain it feeds).

For marine plants at the base of the food web, $\delta^{13}C$ values for seagrass, macroalgae, and phytoplankton are, respectively: -11‰, ≅-17‰, and ≅-22‰. These values change, or *fractionate*, at the low rate of only ~1‰ per trophic level. The minus sign indicates that the heavier isotope is *depleted* as it rises through trophic levels.

Nitrogen isotopes fractionate at a rate of 3.0–3.4‰ per trophic transfer, though there is evidence that this fractionation shrinks as you move to higher trophic levels. Increase in the fractionization ratios is known as *enrichment*. One complication is that Nitrogen isotopic ratios also vary by Carbon source, so Carbon isotopes are also needed to interpret the Nitrogen isotopes.

From our own work, a Cownose Ray (*Rhinoptera bonasus*) feeding on bivalves, which are primary consumers of phytoplankton, has a signal of $\delta^{13}C = -20$‰ and $\delta^{15}N = 10$‰. In contrast, an adult Blacktip Shark feeding on fishes and small sharks in a seagrass-dominated system has a signal of $\delta^{13}C = -13$‰ and $\delta^{15}N = 16$‰.

Confusing as it may seem, the advantages of stable isotopes are that the tissues (e.g., muscle, blood, skin) can be collected in relatively safe and noninvasive ways, and the shark can then be released. The results provide a signal of diet that is constrained only by the turnover time of the particular tissue (e.g., days for blood, weeks to months for muscle). The primary limitation of this method is its lack of specificity; that is, the results reflect the signal from all of the prey items eaten and do not provide insight into the actual prey species.

Additional methods of assessing shark diets that are gaining traction is the use of DNA barcoding to identify the prey. This can involve genetically identifying the species in the stomach contents or swabbing the cloaca to obtain fecal tissue that contains the DNA of prey that have been eaten and digested. Think of this as similar to the eDNA techniques discussed earlier for detecting the DNA of species present in a water sample. The only difference is the use of fecal DNA to elucidate diet. The advantages of these techniques are that samples are collected non-lethally and prey in the stomach or fecal matter can potentially be identified to species. The disadvantages are that only prey species that have reference DNA barcodes available can be detected and the relative contributions of the species to the diet (i.e., number and biomass) can't be determined. In addition, there detected species in the sample may not only include the prey of the shark, but also the species those prey species were feeding on (i.e., prey of the shark's prey).

We can also study shark trophic ecology through a functional morphology lens. Variation in jaw and tooth morphology (shape) reflects the varied trophic ecology of sharks (see fig. 4.19). Many familiar sharks like Blacktips, Sandbars, and Lemons have narrow-cusped lower teeth for grasping prey, whereas the upper teeth are slightly wider with lateral edges that allow them to slice prey into pieces or grasp and swallow whole prey. Co-author Grubbs saw this when he had his thumb nearly completely removed by a juvenile Sandbar Shark he was tagging. You might say he observed this tooth morphology in action first*hand*. White and Bull Sharks have triangular upper teeth for cutting large chunks from larger prey. Of course, the dalatiid sharks (e.g., Cookiecutter, *Isistius brasiliensis*) take the opposite approach and have small grasping teeth in the upper jaw and triangular cutting teeth on the lower jaw for carving melon-ball chunks out of unwitting prey. Tiger Sharks and Sixgills take flesh removal to extremes, having wide multicusped teeth in both jaws and relative weak jaws that are adapted to bending across the body of large prey (e.g., sea turtles, dead whales). They then spin their bodies or use strong jaws to carve out huge chunks of flesh. Dogfish sharks actually feed similarly and have oblique bladed teeth in both jaws that lie close together forming a single knifelike blade in both jaws, allowing them to cut small prey in half or cut small pieces from large animals while scavenging.

At the other end, Shortfin Mako and Sand Tigers concentrate on prey they can swallow whole and thus have very narrow cusps on the teeth of both jaws. The Frilled Shark has tri-cusped teeth angled backward to grasp and swallow soft-bodied prey, and like the jaws of a python, once the prey enters the mouth, there is no backing out. Co-author Grubbs also witnessed this when he had to call for assistance when his thumb became stuck in the mouth of a dead Frilled Shark that had been preserved in a museum for decades. (At this point, you would be justified by questioning why co-author Grubbs seems to always be placing his thumb in proximity to the mouths of sharks.)

There are many shark species with specialized dentition adapted to some level of durophagy (eating hard prey). Many smoothhounds have pebble-like teeth similar to many stingrays for crushing relatively weak shelled prey like shrimp, crabs, and small clams. In the extreme, the Horn Sharks have cusped teeth up front, molariform teeth in the back, and hypertrophied jaw muscles for crushing snails, urchins, and crabs. Nurse Sharks have pavement-like teeth with pointed cusps that are used to crush and tenderize large prey like lobsters until they can be broken into pieces small enough to swallow.

In addition to selecting specific prey taxa, some sharks may also select large or energy-rich prey. For example, one study showed Basking Sharks selected areas to filter feed where large copepod species were concentrated and smaller, less energy-rich species were less dense. Another study showed that when offshore, Shortfin Makos fed primarily on squid but when they moved inshore, they fed heavily on oily energy-rich bluefish (*Pomatomus saltatrix*) even though there were high abundances of squid prey inshore.

High diversity of sharks in a given region may by supported by opportunistic versus specialized diets. This constitutes a kind of resource partitioning between species (also known as *interspecific specialization*), which refers to dividing limited resources in ways that minimize competition for them.

High densities of sharks can also be supported by ontogenetic niche shifts of the same species (*intraspecific specialization*). Just as habitat use changes with growth and age, sharks undergo changes in diet as they grow. Since mouth gape increases with growth, which allows consumption of larger prey, the trophic position of predators, including sharks, usually increases with age. Larger sharks also eat larger and faster prey than juvenile sharks, but still consume many of the same prey taxa that were consumed as juveniles.

It is often estimated that sharks consume 2–3% of their body weight per day, but this also varies depending on the energy demands of a specific shark species or life stage and the energy content of the prey. The daily ration of Sandbar Sharks was estimated to decrease over the first few years of life, but the energy demand increased, which required consumption of prey with higher caloric content. So as juvenile Sandbar Sharks grew, they consumed larger and more energetic prey.

Both prey density and accessibility are likely important to habitat selection by sharks. Recall the ontogenetic expansion of habitat use by Sandbar Sharks discussed earlier. Even over the first several years of life, the shift by juvenile Sandbar Sharks to deeper habitats is accompanied by a shift from consuming crustaceans like swimming crabs and mantis shrimp, to bony fishes and skates. The shifts in diet that occur with growth in Tiger and White Sharks are very dramatic. Both species feed mostly on teleost fishes as juveniles. But as adults, Tiger Sharks commonly eat elasmobranchs, sea turtles, birds, and mammals. For adult White Sharks, marine mammals become a major component of their diet. The shape of the teeth of White Sharks even undergoes an ontogenetic change that facilitate the dietary shift.

Roles of Sharks in Ecosystems

We end this chapter by addressing the ecological roles of sharks. The first studies to assign trophic levels to sharks (149 species in eight orders and 23 families), skates (60 species), and stingrays (67 species) as groups were published only in 1999,[22] 2007,[23] and 2013,[24] respectively.

These studies validated the perception that most sharks are upper-level predators and their food webs are composed of at least four trophic levels. Overall, sharks were considered *tertiary consumers* with a mean trophic level > 4 (see fig. 9.3). Trophic level generally correlated positively with body size, although large planktivorous sharks had low trophic levels (3.2 and 3.4 for Basking and Megamouth Sharks, respectively). Carcharhiniform, hexanchiform, and lamniform sharks exhibited the highest mean trophic levels, as high as 4.7 for the Broadnose Sevengill Shark (*Notorynchus cepedianus*), 4.5 for the White Shark, and 4.4 for the Sand Tiger and Bramble Sharks (*Echinorhinus brucus*). In addition to the planktivores, on the low end of the trophic level scale were the Zebra Shark (*Stegostoma fasciatum*) at 3.1 and the Horn Shark at 3.2.

Thus, large sharks that feed on fish, squid, and mammals are tertiary consumers (trophic level 4) and are at the top of the food chain and in some cases serve as apex predators. Recall, however, that most sharks are not large, but are less than 100 cm (3.3 ft) long as adults. Many of these species as well as large species that feed on invertebrates are secondary consumers (i.e., *mesopredators*, trophic level 3).

It is also critical to understand that large sharks were not always large. Juveniles of sharks that are apex predators as adults are mesopredators as juveniles. Many coastal and deep-sea sharks undergo an ontogenetic shift of more than one trophic level. Since the majority of shark species are small, and all sharks start life small, then by sheer numbers, the vast majority of sharks swimming the world's oceans are not apex predators, or even consumers at trophic level 4. Most sharks are mesopredators.

This means the ecological roles of sharks in marine ecosystems are far more complex than the simplistic notion of them as apex predators influencing marine food webs through top-down forces. They also play critical roles in the middle of the food webs. Even mesopredators can have structuring roles on invertebrate populations and communities. In many cases, sharks that are at trophic level 4 (e.g., reef sharks) function as critical members of an assemblage of shark and teleost species that together occupy the highest trophic level in a system. A recent study[25] found that Bull Sharks and Smalltooth Sawfish share the top predator space in mangrove estuaries. In some specific cases with simple food webs, adults of a single species of shark may be the apex predator. We are tuned to assume that marine food webs should exist in simple linear ways with a single apex predator, as has been depicted in many terrestrial food webs. But marine food webs are often so complex that there is functional redundancy at all trophic levels, including the highest-level predators. The idea that sharks are apex predators is akin to saying all snakes are apex predators because anacondas exist. Sharks are highly connected predators that play complex roles in marine, and some freshwater, food webs.

Concluding Comments

We began the chapter with a list of the subdivisions of ecology. The overarching goal of ecology is to assemble these reductive views into the big picture of the ecological roles of sharks as some of the questions we raised earlier are answered. Although the gains in our understanding of sharks in their ecosystems are substantial, there remains much to be learned, e.g., the ecology of deep-sea sharks, or the impacts of anthropogenic impacts like climate change and overfishing on shark ecology. We discuss the latter in chapters 11 and 12.

NOTES

1. Your first two co-authors both hail from the Southern US, where the indigenous grocery store used to be the neighborhood *Piggly Wiggly*.
2. Grubbs, R.D. and Musick, J.A. 2007. In: McCandless, C.T. et al. (eds.). Shark nursery grounds of the Gulf of Mexico and the East Coast waters of the United States. Am. Fish. Soc. Symp. 50: 63–86.
3. NMFS. 1999. *Final fishery management plan for Atlantic tuna, swordfish, and sharks.* Silver Spring.
4. Talwar, B. et al. 2017. Mar. Ecol. Prog. Ser. 582: 147–161.
5. Messing, C.G. et al. 2013. Proc. Biol. Soc. Wash. 126: 234–239.
6. Coutin, P. 1992. Austr. Fish. 51: 41–42.
7. Grubbs, R.D. and Kraus, R.T. 2019. In: Choe, J.C. (ed.). Encyclopedia of animal behavior 3: 553–563. Elsevier, Academic Press.
8. Priede, I.G. 1984. Fish. Res. 2: 201–216.
9. Bonfil, R. et al. 2005. Science 310: 100–103.
10. Skomal, G.B. et al. 2009. Cur. Biol. 19: 1019–1022.
11. Howey-Jordan, L.A. et al. 2013. PloS One 8(2): e56588.
12. Wilson, R.P. et al. 2008. Endanger. Species Res. 4: 123–137.
13. Whitney, N.M. et al. 2010. Endanger. Species Res. 10: 71–82.
14. Lear, K.O. et al. 2017. J. Exp. Biol. 220: 397–407.
15. Holland, K.N. et al. 2009. Endanger. Species Res. 10: 287–293.
16. Brooks, J.L. et al. 2016. J. Fish Biol. 88: 2067-2074.
17. *Sine qua non* is Latin for *that without which there is nothing* (i.e., essential element).
18. Sminkey, T.R. and Musick, J.A. 1995. Copeia 4: 871–883.
19. Wass, R.C. 1973. Pac. Sci. 27: 305–318.

20. Romine, J.G. et al. 2006. Env. Biol. Fishes 77 229–239.

21. Note that "$^{13}C/^{12}C$" *is* a ratio and we are being redundant by writing "$^{13}C/^{12}C$ ratio."

22. Cortés, E. 1999. ICES J. Mar. Sci. 56: 707–717.

23. Ebert, D.A. and Bizzarro, J.J 2007. In: Ebert, D.A. and Sulikowski, J. (eds.). Biology of skates. Springer Science. 115–131 pp.

24. Jacobsen, I.P. and Bennett, M.B. 2013. PLoS One 8(8): p.e71348.

25. Poulakis, G.R. et al. 2017. Endanger. Species Res. 32: 491–506.

10 / Behavior and Cognition

CHAPTER OUTLINE

Introduction

Let us be honest—sharks have never had the greatest reputation. Many people today still consider them solitary, mindless, ocean wanderers who exhibit limited behavioral repertoires and cognitive abilities. However, this view could not be further from the truth. Sharks have evolved extraordinary behaviors that set them apart as arguably one of the most successful predators to inhabit the planet. Whether finding food, hiding from predators, or searching for a mate, sharks are always behaving!

In this chapter, we walk you through the behavioral diversity of sharks, revealing insights from their:

- Social lives—Do Scalloped Hammerheads (*Sphyrna lewini*) displace each other in a school by shaking their heads?
- Cognitive capacities—Can reef-dwelling Bamboo Sharks learn how to navigate a maze?
- Navigation—Can Lemon Sharks (*Negaprion brevirostris*) return to where they were born to give birth themselves?
- Predatory behavior—Do White Sharks (*Carcharodon carcharias*) use the sun to launch cryptic attacks on unsuspecting seals?
- Predator avoidance—Is there a shark that swells up like a balloon to avoid being eaten?

Group Living

There is great diversity in how, when, and why sharks form groups: think resting catsharks in labyrinth-like caves, daytime schooling Scalloped Hammerheads over seamounts (fig. 10.1), or filter-feeding giant Basking Sharks (*Cetorhinus maximus*) congregating in the open ocean.

Like other animals, sharks come together frequently, sometimes in large numbers, but these groups vary in their organization, behavior, and the environmental or ecological conditions under which they form. However, before we begin exploring this phenomenon, it is helpful to distinguish between *aggregations* formed when sharks are drawn together because of a common attractive resource or constraint (e.g., food or shelter) and *social groups* that result from individuals being drawn to one another (e.g., influenced by social interaction).[1]

Aggregation

Large numbers of sharks always draw great interest. Show people a photo of a Bull Shark (*Carcharhinus leucas*) and their response is "that's neat," but show them

Figure 10.1. Scalloped Hammerheads schooling at Cocos Island, Costa Rica (Courtesy of Shmulik Blum).

40 Bull Sharks and their response is more like "Wow! That's incredible!" and they sit up and pay much more attention. The spatial distribution of sharks has always fascinated researchers too, and there is widespread evidence of single and multi-species aggregations across the 550+ species that roam our seas. Food is a primary driver of aggregations, especially when it is clumped (as opposed to dispersed) in distribution, as *everyone* comes to exploit the plentiful resource.

One of the most extraordinary examples of sharks aggregating for food is the massive numbers of Whale Sharks (*Rhincodon typus*) that come together during the summer months off Mexico's Yucatan Peninsula (see fig. 3.23). Up to 420 Whale Sharks have been observed in a single aerial survey, all gathered in an approximately 18 km^2 (7 mi^2) elliptical patch of ocean, feeding on dense patches of pelagic fish eggs.[2] This species is regularly seen feeding together throughout the world, and most research focuses on learning about the spatio-temporal dynamics of aggregation sites with a view to conserving them. For Whale Sharks, it is unclear whether such gatherings also facilitate exchange of social information. In Basking Sharks (*Cetorhinus maximus*), for example, a similar planktivorous ocean wanderer, torusing behavior (i.e., circling in a donut-shape) recorded post-feeding, was shown to serve a secondary reproductive function.

Another incredible sight to behold is the seasonal mass aggregations of migrating Blacktip Sharks (*Carcharhinus limbatus*) off West Palm Beach, Florida (see fig. 1.19). Aerial surveys conducted throughout the year saw huge spikes in the number of Blacktips during the winter months (temperatures below 25°C or 77°F), with up to 800 sharks per km^2 (2072 per mi^2).[3] It was hypothesized that the sharks gathered to feed on spawning baitfish; however, these Blacktip Sharks also performed coordinated, school-like, evasive maneuvers when attacked by large Great Hammerheads (*Sphyrna mokarran*), suggesting an antipredatory function to the aggregations as well.

The tendency to group is important throughout life stages. Juvenile sharks of many species, particularly in the early stages of their lives, aggregate in nursery areas (often at the same time), presumably to avoid predators and search for food. Newborn Blacktip Sharks, for example, frequently aggregated in specific areas during daylight hours before dispersing at night in a small bay near Tampa, Florida.[4]

Within these nursery areas, more localized responses to environmental factors lead to the formation of aggregations. For example, juvenile Lemon Shark movements are tidally linked in the Bahamas and Brazil, with sharks aggregating at high tide in the back of mangroves or pools, presumably to avoid larger patrolling predators that have gained access to habitats closer to shore as the tide rises. Lemon Sharks also aggregate in mass off the beaches of Cape Canaveral, Florida, despite this constituting a major shift in habitat from mangroves to the high-energy surf zone, providing further evidence that this behavior is adaptive across habitats and regions.[5]

Shifting to the pelagic environment, Blue Sharks (*Prionace glauca*) provide an interesting example of aggregations, with juveniles segregating. Dense aggregations of young males (~70 cm or 28 in, in length) form at oceanic seamounts, while females remain in coastal habitats as they mature.[6] Interestingly, at the same life stage, another pelagic shark species, the Silky Shark (*Carcharhinus falciformis*), is regularly seen in mixed size and sex groups aggregating underneath floating aggregatory devices (FADs), buoys, or natural logs in the open ocean (fig. 10.2). Unfortunately, this species' propensity to gather under FADs sees them regularly caught as bycatch in purse seine net fisheries that target tuna (see chapter 12). Recent statistical modeling suggests an important social component to these aggregations. Thus, studies exploring the mechanisms underpinning this species' aggregatory behavior could prove particularly important for future conservation efforts.

The deep-sea is a challenging habitat to study, for obvious reasons. However, imagine attempting to conduct behavioral trials investigating grouping behavior 500 m (1640 ft) or more below the surface? It probably comes as no surprise to

Figure 10.2. Silky sharks aggregating under ocean buoy, Andros Island, Bahamas (Courtesy of Annie Guttridge).

learn that much of our understanding of the associations and aggregations of deepwater sharks is inferred from catch composition in fisheries. A study[7] in New Zealand explored these patterns among 14 chondrichthyan species (10 sharks and 4 chimaeras) captured across a 27-year period in 5165 bottom trawls. Results indicated that not all species engaged in aggregative behavior, but those that did showed "patterns of sex- and size-specific associations, which varied with catch density." Interestingly, like other sharks that show sex segregation at maturity, "adult females were highly associated with other adult females", and "adult males were consistently associated with each other."

Some species of sharks have been shown to segregate into small groups or packs in proximity to females. In Whitetip Reef (*Triaenodon obesus*) and Nurse Sharks (*Ginglymostoma cirratum*), multiple males have been observed to pursue and attempt to mate with females. Female sharks are often bigger-bodied than males (i.e., they are sexually dimorphic), thus this gang method might increase their chances of successful mating, particularly as larger females can probably outmuscle and deter a single male. Whether this pack-type tactic is common among sharks in general is unknown.

This idea of female sharks aggregating (i.e., sex segregation) to avoid male harassment is not uncommon. Mature or near-mature female Leopard Sharks (*Triakis semifasciata*) aggregate in shallow, warm waters during the summer months off the coast of California, US. More than hundred female Port Jackson Sharks (*Heterodontus portusjacksoni*) were observed in a mass aggregation in waters near Melbourne, Australia. Female Silky Sharks that exhibited fresh mating wounds (30% of 210 individuals photographed) were encountered in large groups (of at least 50 individuals) off the southern tip of the Baja California Peninsula, Mexico. Although it seems that the males are to blame for these large, female only aggregations, some researchers have speculated that the warm water may also influence these aggregations, possibly enhancing gestation and embryonic development.[8]

Social Groups

Sharks are social, but how do we know when sharks are socializing? Is it proximity to another shark or physical contact? Or some form of coordinated behavior? Or all the above? Just over 50 years ago, Stewart Springer identified that "some shark populations exhibit complex behavior that constitutes part of their social organization" and that based on anecdotal evidence from aerial surveys, "large sharks (and rays) are often in groups and not randomly distributed."[9] A few years later, Samuel Gruber and Arthur Myrberg developed an ethogram (i.e., a catalog of behaviors) of social interactions for captive Bonnetheads (*Sphyrna tiburo*), including definitions of *following*, *give-way*, and *circling*[10] (fig. 10.3). Using these behaviors, they showed a clear dominance hierarchy existed among 15 captive sharks, with the largest females leading the most groups and deferring the least in head-to-head (also called *give-way*) encounters.

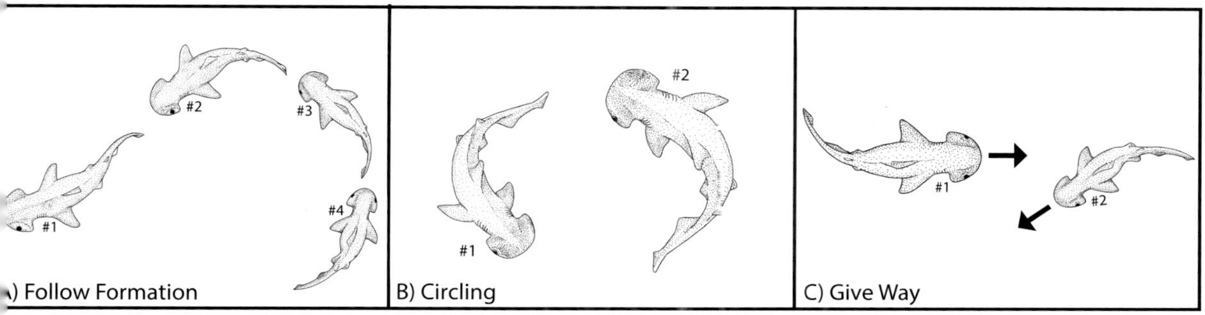

Figure 10.3. Social interactions of Bonnetheads (Adapted from Myrberg & Gruber, 1974).

This ethogram of behaviors and results was a tantalizing insight into the social lives of sharks and has since formed the basis of research examining the functions and mechanisms of shark sociality. One study[11] conducted in Bimini, Bahamas over two years documented the grouping behavior of 38 juvenile Lemon Sharks in a shallow-water mangrove inlet. Daily observations of following and circling behavior revealed repeatable social interactions with size-matched and often related individuals. Groups were typically observed during high tide when larger predators could access the habitats closer to shore, suggesting an antipredatory function. This makes perfect sense, since more sharks equals more senses to detect potential predators, and being part of a group means a lower probability of being the one eaten. Through a series of creative semi-captive experiments, the research team went on to explore the mechanisms underpinning group formation, finding that juvenile Lemon Sharks were attracted to other Lemon Sharks (presumably to form groups), preferred to socialize with sharks of similar sizes, and even had preferences for familiar individuals, as well as social personalities.[12]

Concurrent to these discoveries, a research group explored sociality and refuge behavior in the Small-spotted Catshark (*Scyliorhinus canicula*) in the United Kingdom. This benthic species exhibits a very different type of social behavior that involves close or tactile resting in caves or under ledges, almost like roosting insects. Yet despite this obvious difference in sociality between Small-spotted Catsharks and Lemon Sharks, striking similarities existed in the results. For example, like young Lemon Sharks in the Bahamas, juvenile Catsharks showed repeatable associations (e.g., resting with the same individuals), were consistent in their position within the social network (i.e., had a social personality), and familiar sharks (e.g., those kept in the same tank together) formed more groups of greater size than non-familiar groups.[13] Adult Catsharks formed single sex groups, with females resting in labyrinth-like cave systems and, interestingly, males regularly visiting outside, presumably waiting for an opportunity to court a female. Follow-up captive experiments found that the introduction of a male to a tank of females caused disruption to the social network, confirming that sexual harassment can have implications on group behavior in sharks (fig. 10.4). Out on the coral reef,

Figure 10.4. Small-spotted Catsharks resting together during lab experiments (Courtesy of Paul Naylor).

Blacktip (*Carcharhinus melanopterus*) and Grey Reef Shark (*Carcharhinus amblyrhynchos*) social groups were also characterized by non-random associations with both species showing evidence of community structure—a first for sharks! For the Blacktips, sex and size differences seemed to be driving these preferences with kinship (i.e., relatedness) not being a predictor of affiliations among sharks. Amazingly, through the use of acoustic tracking methods, some of the Grey Reef Shark associations were shown to occur over 4-year periods, within fixed areas of reef habitat, which may be an opportunity for sharks to share information. Indeed, the species are recognized for their use of body postures (e.g., arched back, lowered pectoral fins, raised snout and exaggerated swimming) to communicate with each other.[14]

Arguably the most well-known example of shark grouping behavior is the polarized schools of Scalloped Hammerheads. It is an astonishing sight to witness hundreds of large sharks (approximately 2 to 3 m or 6.6 to 9.8 ft in length) showing highly coordinated movements, all within a body length of each other. In the 1970s and early 1980s, shark ethologists Peter Klimley and Don Nelson spent hours free diving among Scalloped Hammerhead schools, recording their behavior in great detail.[15] The results were fascinating! Schools that formed during daylight hours were predominantly female in composition, with larger individuals typically remaining at the top, in more central positions, asserting dominance over smaller sharks through posturing (e.g., body tilting, head shaking, and accelerating) or direct contact (e.g., bumps, hits). They also witnessed a behavior they termed *corkscrewing*, defined as a 360° rotation around the longitudinal axis, used to displace another shark. The function of the schools was difficult to interpret because there are numerous factors that influence school formation. Here are some possible explanations: *Anti-predatory?*—few predators would take on Scalloped Hammerheads; *Foraging?*—no observations of feeding behavior; *Courtship?*—few males present and no mating scars, or attempts observed; *Avoidance/refuging?*—females schooling to avoid costly mating attempts by males; *Social?*—like other animals that are attracted to each other, maybe Scalloped Hammerheads are too.

More recently, Basking Shark social groups off the eastern North Atlantic Ocean were observed forming mesmerizing "circles in the sea." Underwater and aerial recordings of 6–23 likely mature non-feeding groups were analyzed, revealing that individuals were layered vertically from the surface to ~16 m (50 ft) depth, engaged in close-following, echelon and parallel-swimming behaviors.[16] Further, some individuals observed to breach, interacted through fin-to-fin or fin-to-body contact, and rolled to expose their undersides to following sharks. Taken together, these results strongly suggest a courtship function to torusing and highlight the importance of productive thermal front habitats as a key driver of feeding aggregations and courtship reproductive behavior for the Endangered Basking Shark.

Evidence of intraspecific communication among sharks (like the Scalloped Hammerheads discussed earlier) suggests more complex social lives and indicates that information transfer could be an important function of their social behavior. Indeed, White Sharks (*Carcharodon carcharias*), that are typically known for solitary, more nomadic lifestyle, inadvertently share information on the location or remains of large prey by associating with other individuals, eavesdropping on each other's predations.

Evidence of more coordinated predatory behavior comes from one of the most archaic species, the Broadnose Sevengill Shark (*Notorynchus cepedianus*; fig. 10.5). These sharks have been observed to feed together, where a group slowly encircles a larger prey item (e.g., a seal), finally converging in on it for the kill.[17] Clearly this pack-type behavior provides the opportunity to access prey items that would not be possible for an individual. Further examples of group attacks have been observed in reef-dwelling Whitetip Reef (*Triaenodon obesus*) and Grey Reef Sharks trapping and chasing fish among coral heads, Bronze Whalers (*Carcharhinus brachyurus*) and Blacktips blitz-attacking mass sardine schools, and juvenile Lemon and Blacktip Reef Sharks herding schools of baitfish toward the shore and then feeding on them.[18] Interspecific foraging has also been proposed for why Oceanic Whitetip Sharks are regularly observed to follow Pilot Whales (*G. macrorhynchus*). A more surprising association was recently documented from a camera tag deployed on an adult Nurse Shark off the Dry Tortuga's, Florida.[19] The shark appeared to directly follow Bottlenose Dolphins (*Tursiops truncatus*) for ~ 15 mins, performing gliding descents behind the dolphins back to the substrate. The shark repeatedly swam through clouds of sand that were produced from dolphins that were exhibiting crater feeding behavior. Anecdotally, similar behaviors have been observed in the Bimini Islands, The Bahamas, with Nurse Sharks observed trailing Dolphin pods engaging in feeding behaviors. So, in this instance perhaps the roles are reversed with the smart dolphins being exploited by the slow, sluggish nurse shark!

These wild observations provide compelling evidence that sharks are capable of social learning (learning from other individuals); however, only recently has this been demonstrated experimentally using novel food tasks. Juvenile Lemon

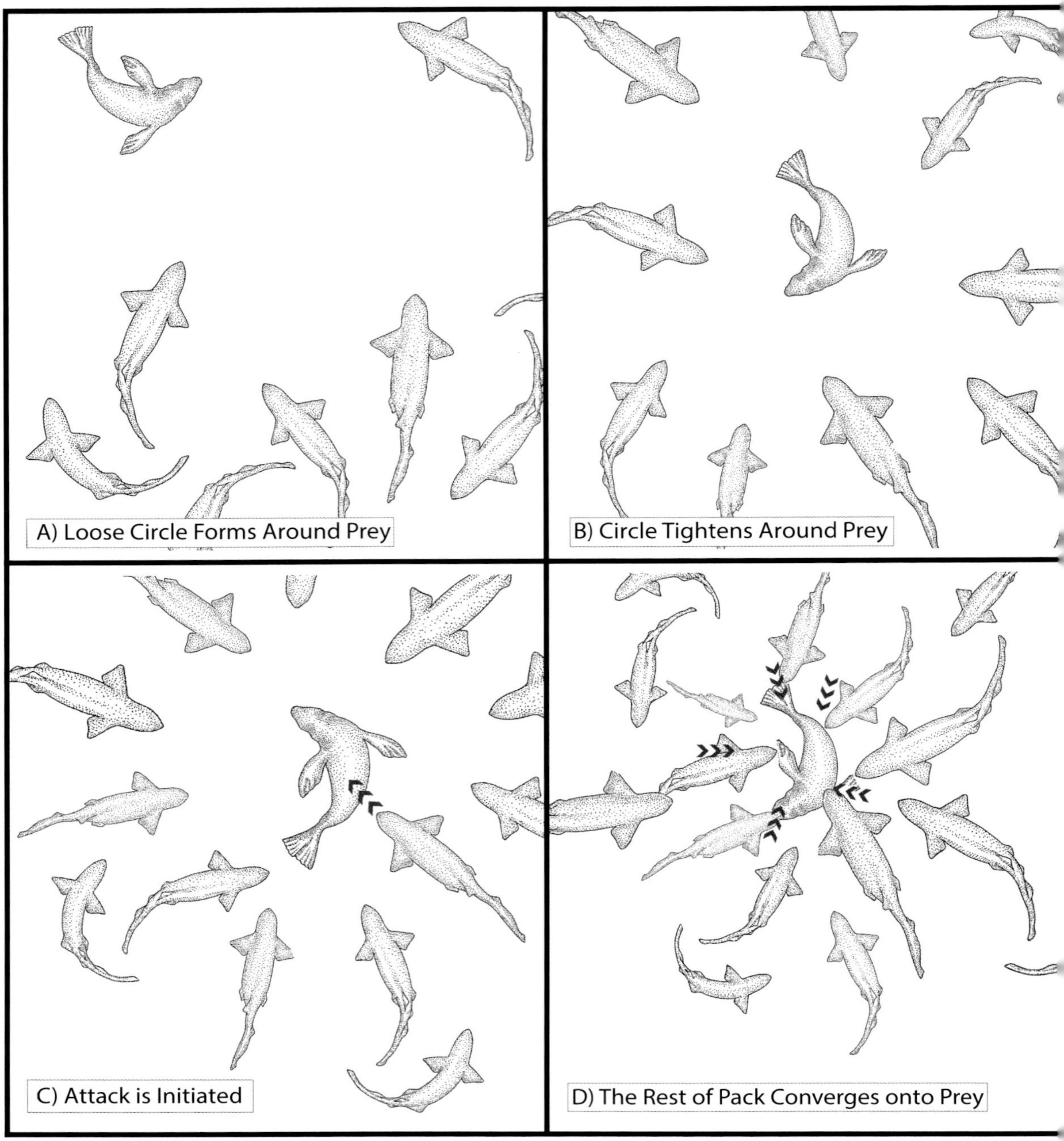

Figure 10.5. Sevengill Sharks hunting as a pack to feed on a Cape Fur Seal in St Helena Bay, South Africa (Adapted from Ebert, 1991).

Sharks were required to enter a start zone and make physical contact with a target to receive a food reward. Juvenile Port Jackson Sharks were trained to solve a simple choice test and swim through the correct door to access food. For both species, naïve (inexperienced) sharks were then paired with either trained or untrained partners and their performance in the task was compared. Port Jackson Sharks that were trained with demonstrators were far more likely to learn the task and Lemon Sharks working with trained partners completed the trials quicker and used more task-related behaviors (e.g., bumping a target).

Understanding the patterns, mechanisms, and functions of shark grouping behavior is particularly important given the conservation status of many elasmobranch fishes, especially as concentrations of sharks are more susceptible to exploitation through spatially focused fisheries.

Learning and Memory

Can you train a shark? Sharks live in a watery world where the habitats and occupants are subject to year-by-year, seasonal, day-by-day, or even moment-to-moment variation. Having the ability to retain information about where to find food or learn an association with a stimulus, such as a visual feature of an approaching predator, is adaptive and can ultimately lead to enhanced survival. The general principles of learning can be described in two forms: *associative learning*, which occurs when an association or relationship between two events is established (e.g., operant, classical conditioning, and observational learning) and *non-associative learning*, which occurs as a result of the presentation of a single stimulus (e.g., habituation and sensitization).[20] These terms, developed by comparative psychologists, have provided a useful conceptual framework for researchers working on learning in a variety of animals and are used here to map out our knowledge of shark learning abilities.

Associative Learning

The capacity of sharks to learn was first demonstrated experimentally by Eugenie Clark and colleagues in the late 1950s, using an operant conditioning regime.[21] Two adult Lemon Sharks were trained to bump an underwater target, on hearing a submerged bell ring, to receive a food reward. Training was in accordance with positive reinforcement, where the consequence of bumping a target (correct voluntary behavior) produced a food reward. Sharks were trained in a 6-week period and retained a strong response even after a 10-week absence from exposure to the stimuli. Soon after this study, Samuel Gruber and Neil Schneiderman reported the first authoritative account of classical conditioning, again using the Lemon Shark as their study species. Of course, this learning process is most well-known from Ivan Pavlov's work with dogs; however, instead of using food (a positive/reward stimulus), sharks were restrained while exposing them to a light flash that was paired with an electric shock (negative/aversive stimulus), producing an

Figure 10.6. Lemon Shark secured during aversive classical conditioning experiments. See text for explanation.

eye-blink conditioned response (not salivation, like in Pavlov's dog experiment; fig. 10.6). After about 60 trials, sharks exhibited a reliable conditioned response (e.g., eye blink on light presentation), indicating they had learned the association. This finding was comparable to other vertebrates tested in a similar paradigm, suggesting sharks were able to learn discriminative tasks as rapidly. So, sharks are not just swimming noses after all!

Since these early landmark demonstrations, classical and operant conditioning paradigms have been used effectively to explore a variety of shark cognitive abilities. A team led by Vera Schluessel in Germany, working with captive bred Grey Bamboo Sharks (*Chiloscyllium griseum*), has pioneered research examining recognition and discrimination abilities.[22] Through a series of elegant experiments, her team showed that sharks could discriminate between symmetrical and asymmetrical shapes, contrasts but not colors, stationary two-dimensional objects, and between moving objects ranging from moving circles to differently moving organisms. She even found that these sharks could perceive optical illusions, count (i.e., discriminate quantities) and most recently they outperformed cichlids (a species of teleost fishes) in a serial reversal learning task using two-dimensional objects varying in color and form. She describes some of her groundbreaking work in *Research, Conservation, and Education Focus* Box 10.1. For a shark that lives in complex, visually well-structured environments, such as coral reefs, these abilities are probably extremely advantageous. It will be revealing to test other species that occupy different habitats to see how capabilities vary between sharks.

Concurrent to these studies, Culum Brown, an expert in fish cognition, has reported some surprising findings on the Port Jackson Shark, found off the coast of Australia. If you have ever visited an aquarium and seen this species, you would likely not have paid them much attention, as they barely move and show little to no interest in anything. But underneath this rather uninteresting demeanor, there are hidden smarts. Culum's team taught juvenile Port Jackson Sharks a classical

BOX 10.1

Research, Conservation, and Education Focus

Nothing Fishy Going On: Sharks Can Put Two and Two Together

Vera Schluessel

Bonn Institute for Organismic Biology
University of Bonn
Bonn, Germany

In my lab, we have been researching cognitive abilities of Grey Bamboo Sharks (*Chiloscyllium griseum*) and Ocellate River Stingrays (*Potamotrygon motoro*) for many years, comparing them to each other as well as teleosts. Topics are diverse, including visual discrimination and categorization of two-dimensional and three-dimensional objects, same-different and matching-to-sample tasks, serial reversal learning, numerical abilities, perception of illusory contours, symmetry, color and movement, optical illusions, memory, and spatial orientation. There are large inter- and intraspecific differences in learning performance of both sharks and stingrays, with some recurring trends likely related to different lifestyles and requirements.

Occasionally, there is a study that you look back on and wonder how you found the courage to do it, how you got the animals to "get it" and how in the world they performed as wonderfully as they did. Funnily enough, when you start, you don't think about any of that at all. I have many studies, but two continue to amaze me.

A few years ago, I read a paper on honeybees that could do math,[1] i.e., successfully add and subtract "one" in the number space from one to five, and wondered, if the sharks and stingrays at our Institute could do that as well. As with so many things, no one had previously bothered to look at "mathematical skills" in sharks before. To begin, we tested if quantities of objects presented either on a screen or laminated cards, could be discriminated solely on the basis of numerical information, with non-numerical factors, such as size or area being controlled for.[2] In a procedure known as operant conditioning, individuals were repeatedly presented with two stimuli in a pseudo-random order on either side of a divider, featuring a number of two-dimensional geometric symbols. Choosing the side with the stimulus displaying the higher number of symbols led to a food reward while choosing the alternative stimulus with the lower number of symbols did not. In daily sessions conducted over several months of training, sharks and rays learned to differentiate between various ratios such as 5:2, 7:5, or 4:1 and successfully extrapolated learned knowledge onto new ratios. Presenting the animals with many number combinations showed that both species preferred using a relative strategy (always choosing the larger quantity) over an absolute strategy (e.g., always choosing the trained number, such as "5"). Ratios that differed by only one item, were only successfully discriminated up to 4 (such as 3:2 or 4:3).

As there was great intraspecific variation, we decided to go one step further and train stingrays—similarly to the honeybees—in a basic arithmetic task involving numbers up to five.[3] On each trial, the stingrays had to perform either an addition or a subtraction, based on presenting them with two-dimensional objects of two distinct colors. The color signaled a particular arithmetic process (i.e., blue-addition, yellow-subtraction). Stingrays excelled at this task; they actually counted and not just memorized the symbols, and also learned to add and subtract exactly "one" instead of just choosing the highest or lowest number presented. I think looking back at the last 15 years of my studying fish cognition, that was my most surprising finding. While going through months and months of training, always being able to solve any of the tasks presented correctly by picking the highest or the lowest number, the stingrays here went out of their way and "added or subtracted exactly one" when they would have been just fine by sticking to that previously learned rule. But they recognized during training that "more" or "less" was always "one more" or "one less" and when given the option of picking "one more/less" vs. "two more/less," they chose the former. They paid attention to the (unnecessary but consistent) detail; they learned more than they would have needed to. I was thrilled; in fact, I still am.

The second study looked at categorization of two-dimensional objects. We had just shown that sharks could categorize "fish" and "snail," choosing the former over the latter irrespective of many structural, color, or design differences.[4] That was pretty cool, but we decided to go one step further and check if sharks could also categorize symmetrical and asymmetrical symbols, which were identical in all characteristics (volume, size, color, etc.) but symmetry.[5] To cut a long story short, the sharks did fabulous. After learning to pick a symmetrical over an asymmetrical symbol in three single pair-experiments (in another operant conditioning task, identical to the one described before), the sharks succeeded in choosing any symmetrical over any asymmetrical form, even if new symbols were presented on each and every trial. They somehow understood what we wanted them to do, learned the rule, paid attention to very small differences and details and did not shy away from the ridiculously abstract images we presented them with.

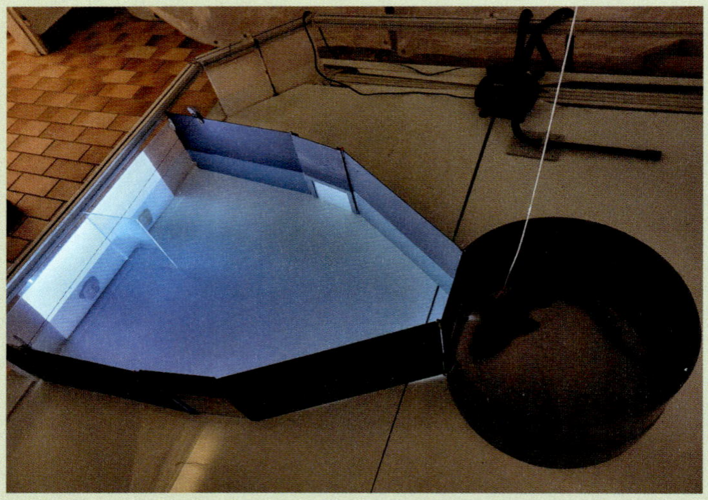

Overview of the experimental tank. There is a starting compartment holding the animal before each trial, from which it enters into the experimental arena. The decision area is on the far side of the starting compartment, featuring the two stimuli (here faces) to be discriminated. These are shown 10 times per session (one session/day) in a pseudorandom order to prevent sharks associating a stimulus with a particular side of the tank. Choosing the correct stimulus by pressing the snout against it elicits a food reward.

Finally, a more recent task that sharks were really good at: they outdid both stingrays[6] and Malawi cichlids in serial reversal learning.[7] Again, sharks learned to pick a star over a circle in operant conditioning trials. Once the task was learned, the food association was switched to the alternative stimulus (circle) and then repeatedly switched again, to observe if individuals would decrease the time needed, i.e., reduce the number of errors over consecutive reversal phases to consistently pick the correct stimulus. This is exactly what happened—the sharks took about the same time to learn the task as stingrays and a bit longer than cichlids, but they reversed much quicker than the other two and were the only species to demonstrate progressive improvement over many reversal phases—again indicating rule learning.

1. Howard, S.R. *et al.* 2019. Sci. Adv. 5: eaav0961.

2. Kreuter, N. *et al.* 2021. Animals 1(9): 2634.

3. Schluessel, V. *et al.* 2022. Sci. Rep. 12: 3894.

4. Schluessel, V. and Düngen D. 2015. Anim. Cogn. 18: 497–507.

5. Schluessel, V. *et al.* 2014. Anim. Cogn. 17: 1187–1205.

6. Daniel, M. and Schluessel, V. 2020. Anim. Cogn. 23: 109–119.

7. Schluessel, V. *et al.* 2023. Behav. 160 (11–14): 1259–1281.

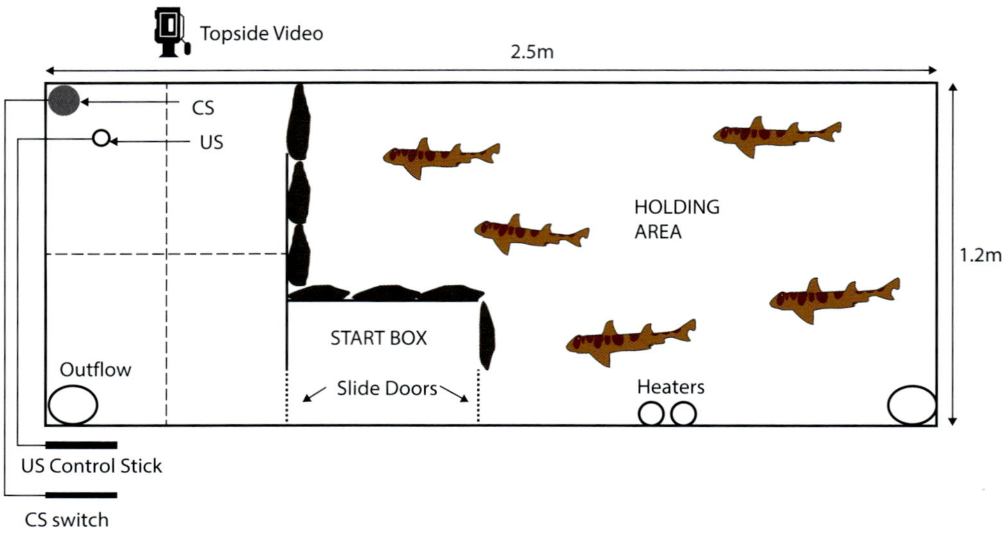

Figure 10.7. Port Jackson Shark classical conditioning setup. See text for explanation. CS = Conditioned Stimulus; US = Unconditioned Stimulus.

conditioning paradigm using an underwater bubbler paired with a food reward. Most sharks reached the learning criterion, biting at the bubbles, within 30 trials, with some remembering the association for 40 days. The researchers also increased the difficulty of the task by adding a 10-second time interval between the bubbles and food reward, and despite this more cognitively demanding task, some of the sharks were still able to learn the association (fig.10.7). A different study found that sharks that were trained at a higher frequency learned the task faster, with little influence by the size of the reward. Putting these results into the context of shark diving and feeding, the team concluded that reducing the frequency of reinforcement might reduce the likelihood of sharks making associations with food. Another study[23] investigated whether Port Jackson Sharks could learn about sounds. Low frequency sounds are known to be attractive to sharks (see chapter 5), and they paired a musical stimulus (jazz) with a food reward. Five of eight sharks learned the association, swimming to the correct corner when presented with the music. Taken together, these findings suggest that benthic shark species, although regularly perceived as lazy and slow, are capable of learning just as effectively as their more active larger-brained relatives.

Non-associative Learning

In the late 1960s and early 1970s, field studies investigating shark acoustic detection abilities were conducted using artificial low-frequency sounds (see chapter 5). The behavioral responses of sharks to these sounds were recorded and a simple form of learning known as *habituation*, a decrease in response to stimulus after repeated presentations, became apparent. Fewer sharks were observed to respond to the sound stimulus and their response intensity was seen to decrease within minutes or even seconds after their initial approach to the speaker. This simple form of learning acts as a central filter mechanism, which means sharks know to ignore recurring stimuli that are irrelevant, and which might waste important time and energy.

Another context where habituation has been discussed regularly is when people SCUBA or free dive with sharks—this idea was introduced by Don Nelson in the 1970s during his pioneering research on shark behavior.[24] Of course, these days, in addition to repeated exposure to divers, many operators also provide frequent, predictable (e.g., in time and location) food incentives (i.e., inadvertently conditioning sharks; as discussed earlier) to sharks to enhance the visual spectacle for their guests. As a result, changes in behavior have been documented for various species, from increased shark numbers to greater residency time and speed of arrival to the dive boat. Indeed, some Great Hammerheads (a huge female called 'Gaia' and a male called 'Atlas') in Bimini, The Bahamas and Bull Sharks in Fiji have returned annually to their respective dive sites for up to a decade or more, and despite absences of up to 18-months during COVID-19 anthropause, sharks returned to feeding behavior rapidly, suggestive of long-lasting cognitive capacities for both species.

To explore habituation in a controlled setting, one study exposed 11 captive juvenile Port Jackson sharks individually to the smell of squid for 4 min. The researchers repeated this three times a day for 21 days without a food reward. It didn't take the sharks long to figure out that despite the squid smelling great, there was no end result; in fact the sharks' responses declined within and across both daily exposures, suggesting habituation on both short and long-term scales. Anecdotally, a similar response has been noted for White Sharks in the Neptune Islands, Australia, where sharks that did not receive food rewards (i.e., tuna heads on lines) began to ignore the dive operators' chum slick. More recently, evidence of habituation has been recorded toward electric stimuli. Unrewarded Small-spotted Catsharks exposed to artificial, prey-type electric fields showed reduced responses with eventual cessation, and White Sharks presented with low voltage electric fields (during deterrent studies) showed habituation over short time scales. Amazingly, Clearnose Skates (*Rostroraja eglanteria*), Small-spotted Catsharks, and Bamboo Shark embryos also showed reduced responses to electric stimuli that simulated a passing predator, with Bamboo Sharks taking 30 to 40 min to recognize previously presented stimuli versus 10 to 15 min for Skates and Catsharks (See chapter 5 for electroreception discussion). One study of personality traits in juvenile Lemon Sharks even found evidence for individual differences in the rate of habituation, and a relationship between this and personality, with faster explorers of a novel open-field-test showing more rapid habituation.[25]

Given these types of studies and the growing literature that showcases the diverse cognitive functions that sharks are capable of, we believe the days of people considering sharks as dumb and mindless are in the past.

Orientation and Navigation

Figuring out where to go, or what route to take, is hard enough with a smart phone and an app calling out directions. Now imagine navigating coastal habitats with intricate mangrove systems, homogenous sand flats, tidal ranges, or traveling across endless blue water searching for a profitable restaurant, um, we mean food patch. There are changes in temperature to consider, current, light intensity, and sounds coming from all directions. Then, once you have finally figured out how to get from A to D, via C and B, a predator comes out of nowhere and makes a meal of you. The challenges are daunting, and sharks, like many other ocean inhabitants, have evolved a suite of incredible senses to help them track both abiotic and biotic changes with some considerable accuracy (see chapter 5).

There are many impressive feats of navigation by sharks. For example, Scalloped Hammerheads disperse from islands and seamounts at night to the surrounding pelagic environment, where they will forage. Salmon Sharks (*Lamna ditropis*) make rapid directed migrations between the two most productive ecoregions of the eastern North Pacific; Spiny Dogfish (*Squalus suckleyi*) migrations of up to 7000 km (4400 mi) provide evidence for the trans-Pacific connection of stocks; even Greenland Sharks (*Somniosus microcephalus*), a species known for its sluggish, slow moving lifestyle, migrate from Greenland 1000+ km (600+ mi) off the continental shelf in abyssal waters presumably to mate or pup; and Basking Sharks track the daily change in vertical abundance of zooplankton, adjusting their vertical migration depending on prey behavior and ocean habitats.

Some of the early investigations of shark navigation suggested that they used geomagnetic orientation, particularly during directed pelagic movements. Peter Klimley, in his seminal work on Scalloped Hammerheads, mapped the associated magnetic gradients within the region of his study site in Espirito Santo, Baja, Mexico, proposing that Hammerheads were able to orient by comparing the intensity of magnetic stimuli (i.e., *tropotaxis*). Support for this hypothesis has emerged from lab-based experiments in Hawaii with Sandbar Sharks (*Carcharhinus plumbeus*), where they were successfully conditioned to magnetic stimuli within the range of values that Klimley previously hypothesized[26] and more recently where researchers exposed 20 juvenile, wild-caught Bonnethead Sharks to magnetic conditions representing locations hundreds of miles from where the sharks were captured. The team showed for the first time that sharks use the Earth's magnetic field to orient toward home—meaning that sharks can use a magnetic map to derive positional information! (See chapter 5 for detailed discussion of magnetoreception).

Other researchers have explored the contribution of olfactory stimuli (i.e., chemical cues) to navigation by displacing sharks offshore from their typical home ranges. Leopard Sharks rendered anosmic (nares temporarily blocked such that the sense of smell was masked) were displaced 9 km (5.6 mi) from their nearshore habitat and tracked on release. Individuals that could not smell made circular movements on return and ended further from shore than control sharks that exhibited more directed movements.[27]

This idea of being able to return *home* (i.e., homing) was tested with juvenile Lemon Sharks in Bimini, Bahamas, where 31 of 32 individuals that were displaced up to 16 km (9.9 mi) from their capture sites returned to their specific, spatially limited home ranges.[28] A shark displaced in the Gulf Stream, a powerful, warm western-boundary current, with 1000+ m (3300 ft) depth, miraculously managed to make it back home. Whether these young sharks imprint (like salmon or turtles) on chemical clues or the geomagnetic signature of their home habitat is unknown, but recent evidence of adult Lemon Sharks returning to their own site of birth to pup themselves, some 14 years later, clearly demonstrates that these homing abilities are retained into adulthood and play an important role in their life history. Indeed, both adult Lemon and Nurse sharks monitored in Bimini, The Bahamas and Dry Tortugas, Florida showed long-term site fidelity to pupping and breeding grounds, respectively. How many of the 118 Nurse Sharks tagged from 1993 to 2014 returned in subsequent years? At least 80 (68%) came back during the June–July mating season. Some individuals returned for periods of up to 28 years! For female Lemon Sharks, some were shown to pup (i.e., through genetic analysis, linking baby sharks to mothers) in the system across 10 different mating cycles, a 20-year time period.

Navigation and orientation are both dependent on the formation, storage, and retrieval of spatial memories, which permit repeated visits to fixed points in the environment. In the early 1990s, researchers working with Port Jackson Sharks in Australia speculated that this species was able to remember features of the reef, potentially having some form of mental map. After being displaced 3 km (1.9 mi) away, within 48 hours the sharks were observed to rest in the same reef sites.[29] More recently, scientists analyzing movement tracks of Tiger (*Galeocerdo cuvier*) and Galapagos Sharks (*Carcharhinus galapagensis*) in Hawaii concluded that the former may use cognitive spatial maps for orientation while moving between different foraging areas.[30]

Using an experimental platform designed for research on cognition, Vera Schluessel and colleagues recently explored the spatial learning and memory capabilities of Bamboo Sharks. Experiments used a two-choice T-maze setup, where sharks were trained to retrieve food from a specialized feeding apparatus and solve spatial tasks. After many weeks of training, all eight sharks were able to remember a feeding location(s), either by memorizing a particular turn response and/or with the help of external visual landmarks (place learning) and were still able to successfully perform the task after various experimental breaks (up to six weeks).[31] One recent experiment investigated the ability of juvenile Lemon Sharks in a semi-controlled environment to learn time–place associations. It may be an advantage to learn about consistent spatiotemporal patterns in food availability, which could increase foraging efficiency. To the team's surprise, despite feeding sharks in two distinct locations for 41 consecutive days, monitoring their progress along with untrained, control individuals, they were unable to learn the association. It is hard to know why the sharks did not learn this seemingly simple association. Perhaps the sharks picked up the association between the arrival of the feeders instead of the time of day or other confounding factors, such as the size of the pen or being exposed to outdoor conditions meant they were unable to learn.

Nevertheless, these experiments provide an exciting first indication that sharks can form cognitive spatial maps; however, more elaborate experiments should be conducted in the future to explore the phenomenon further.

Predator Behavior

To us, sharks live in a visually concealing world, which makes it extremely challenging to directly observe them. Thus, in comparison to terrestrial predators or smaller-bodied fish, our understanding of their predatory behavior is generally poor. Of course, when most people think about shark predation, they immediately visualize a White Shark at full tilt, launching a seal into the air. And yes, this is a brutal, iconic event to watch, but what makes it even more extraordinary is when you consider the details; that is, how the White Shark got to that moment to launch an attack, resulting in a successful, profitable meal. Beyond the exquisite senses and physical attributes that sharks have evolved, there are tactics and strategies to consider, and split-second decisions that can lead to success or failure. For example, from where should an attack be launched? What depth, habitat, or angle? What time of day or conditions are most effective? Which hunting mode or movement pattern should be used for which prey species or habitat characteristics?

For many sharks, these questions are almost impossible to answer, but the visual nature of White Shark predation events, coupled with their proximity to shore (due to their preferred prey, pinnipeds [seals and sea lions], aggregating on rocky islands) has provided a unique opportunity to study this predator–prey relationship in detail. Early examinations of White Shark predatory behavior were undertaken by Tim Tricas and colleagues in Australia, on the aptly named *Dangerous Reef*.[32] They described behavioral responses to surface and mid-water baits, which included terms such as *surface charge* or *side roll* and, based on these observations and examinations of wounded pinnipeds, suggested a typical White Shark attack scenario. This included an ambush, stealth-type strategy, searching for a silhouette with an attack made from beneath (from depth) or from behind the unsuspecting prey. For large prey items, such as Elephant Seals, they proposed that a *bite-and-spit* strategy might be used to avoid injury, leaving the seal to bleed to death or go into shock after an attack. Interestingly, in other locations, White Sharks were not observed to use this tactic for the less dangerous, juvenile Sea Lions, indicating that they altered their predatory tactics depending on the prey species they were targeting.

Since this early investigation, behavioral ethograms of White Shark natural predation events have been developed describing interactions with pinnipeds at

Seal Island, South Africa and the Farallon Islands, California.[33] These descriptions generated 20 and 24 distinct behavioral units respectively, and included behaviors such as *polaris attack, surface lunge, lateral head shake,* and *subsurface carry.*[34] At both study sites, the highest rates of foraging success by White Sharks occurred in the hours following dawn. This peak was likely due to a combination of the presence of pinnipeds (the primary prey item) and optimal ambient light conditions for attacking at the surface. Attacks were launched in specific locations (presumably optimal), with larger sharks showing greater success rates and more restricted search areas, indicative of learning. Changing abiotic (i.e., sunlight, wind, currents) and biotic (i.e., presence/absence of competitors, anti-predatory tactics of prey) factors clearly creates a dynamic, complex predator–prey relationship, with White Sharks needing to adapt and learn features of their environment in order to be successful.

Clearly, much of our knowledge of White Sharks' predatory behavior comes from the last few moments before a seal becomes a meal with little description of the underwater component. However, the latest studies use miniaturized camera tag technology with sensors that can measure a suite of parameters (e.g., 3D acceleration, pressure, temperature). Tag packages are attached to the shark's dorsal fin providing a unique shark-eye view, as well as high resolution data for analysis (fig. 10.8). In South Africa, breaching behavior was recorded to begin at depths up to 66 ft (20 m), with a short, 7 to 16 s ascent to the surface. During this ascent, the pitch angle increased by ~ 30° with tail-beat frequency and swim speed increasing as much as 6.5-fold.[35] One shark was even recorded rapidly adjusting its swim speed and angle during a sharp turn at the breach's start. Another study[36] in the region observed White Sharks to move throughout densely kelp-covered areas (4 hrs of 28 hrs of footage), navigating channels and pushing directly through fronds. Increased turning and activity rates were recorded, and seals displayed evasive behaviors (e.g., blowing bubbles, hunkering down), providing the first evidence of predator–prey interactions and repeated use of this habitat for the species. This sub-surface predatory behavior (confirmed from video and acceleration profiles) was also highlighted at the Neptune Islands, Australia occurring at various depths (0–53 m). Long-nosed Fur Seals (*Arctocephalus forsteri*) in this region are known as benthic divers, foraging along the bottom so it seems reasonable that the White Sharks have adapted their hunting strategies to optimize capture success. Although no actual predation attempts were recorded in this region, White Sharks in Guadalupe Island off the coast of Mexico were recorded to bump and attack an autonomous vehicle at depths of up to 90 m. One attack revealed a rapid vertical approach from depths well below the autonomous underwater vehicle (AUV) toward the back-lit silhouette, a strategy typically used on surface moving seals. The idea that White Sharks attack in low light levels or show behavioral flexibility in their exploitation of environmental features is fascinating and led White Shark researcher Charlie Huveneers to experiment with baited lines at a known aggregation site in Australia. He found that through

Figure 10.8. Great Hammerhead quipped with multi-sensor tag package in Bimini, Bahamas (Courtesy of Eugene Kitsios).

the course of a sunny day, White Sharks moved their attacks so that the sun was behind them and from morning to afternoon, they apparently reversed their direction of approach along an east-west axis. This could be considered a behavioral modification to improve crypsis or a better method for the sharks to track their prey visually.[37] Either way, it is impressive!

More recently, a study linked White Shark hunting modes to predation attempts/successes on seals at the surface. Hunting modes were defined as: 1) active search/patrolling when a shark is moving through its environment searching (either directed or random) for prey; 2) sit-and-wait, or area-restricted-search (ARS), which sees the shark waiting (albeit still very slowly moving) for prey to cross the boundary of its strike space.[38] To undertake such a task, the team actively tracked White Sharks with electronic tags, following them on multiple days throughout a month. Results revealed that White Sharks ambushed seals successfully at the surface using both modes, but preferred ARS in the mornings with clear sex and individual differences detected. Interestingly, Charlie's team in Australia also suggested White Sharks in Australia were using this "sit and wait" strategy. They documented swim speeds slower than the speed that minimizes the cost of transport, which was hypothesized to optimize encounters with their prey. Another large predatory shark, the Tiger Shark (*Galeocerdo cuvier*) showed a similar slow-moving strategy when stalking turtles on the reef. Footage from Tiger sharks equipped with camera-tag packages off Ningaloo Reef, Western Australia were analyzed using a motion detector based on the sharks' visual capabilities to estimate the strength of visual motion signals generated by the turtles. The team found that the Tiger Sharks' energy expenditure decreased after interactions with more visible turtles; however, their movements were more tortuous (twists and turns). They concluded that the Tiger Sharks may potentially stalk turtles that produce greater motion cues providing further insights into how these incredible predators hunt.

Let us move to couch-potato sharks, those that spend most of their day on the seabed. Pacific Angel Sharks (*Squatina californica*) also used ambush tactics by rapidly lunging from the seafloor on oblivious demersal fishes. Resting Angel Sharks were usually located next to reef patches or rock-sand interfaces. These sites serve as refugia for a variety of fishes and no doubt improved encounter rates with potential prey. Sharks showed reuse of ambush sites across the study which was discussed in the context of an ARS hunting mode and models of search behavior.[39]

Traditionally, using sharks as models of behavior to test theoretical predictions (e.g., optimal foraging) has proved challenging. However, with the huge improvement in tracking, remote sensing, and data logging technologies, alongside data processing and statistical modelling, it is becoming increasingly possible to conduct hypothesis-driven behavioral studies on elasmobranchs.

Behavioral ecologist David Sims has pioneered this type of approach, with sharks as models of foraging behavior. His team's work over the past two decades has examined pelagic shark movements (e.g., Basking and Blue Sharks) via satellite tracking to examine how their movement patterns conform to optimal search theory (e.g., *Lévy-flight foraging hypothesis*, a type of random movement with many short moves and fewer long-distant displacements; or *Brownian motion*, which are random movements).[40] Results revealed that Lévy (47%) and Brownian (21%) movement patterns were both important strategies for open ocean sharks, and that environmental context impacted their use. In one study,[41] an NE Atlantic Blue Shark which was followed across both productive shelf waters and less productive deep waters, displayed dive patterns consistent with Brownian movements when in the former and Lévy flights in open ocean habitats.

Since this pioneering research, others have explored the use of *directed walks* (where the shark moves to a known goal) to examine movement patterns of sharks at different spatial scales, and to determine if they were using food patches. Active tracking data for Tiger, Thresher (*Alopias vulpinus*), and Blacktip Reef Sharks, of varying life stages, revealed a variety of movement strategies both between and within species.[42] Tiger Sharks performed directed movements to locations 6 to

8 km (3.7 to 5 mi) away, whereas Blacktips did not show oriented movements. Another species showing directed movement was Thresher Sharks, at scales of 0.4 to 1.9 km (0.2 to 1.2 mi). Adult threshers oriented at greater scales than juveniles. These results suggest that learning improves their ability to perform directed walks.

Some sharks have impressive weapons at their disposal that aid them in prey capture. Take the Pelagic Thresher (*Alopias pelagicus*) for example, with its huge scythe-shaped tail that is half the length of its body. Recent video recordings taken in Cebu, Philippines showed that Thresher Sharks had remarkable control of the upper caudal lobe, using it to target bait fish in dense schools by slapping and stunning the fish. Amazingly, tail-slaps occurred with such speed (fastest recorded = 21.8 m per s, almost 50 mph) and force that they may have caused dissolved gas to come out of solution-forming bubbles.[43] In addition, there are various other reports suggesting Thresher Sharks hunt in groups, corralling bait fish ready to tail-whip.

Let us switch ends of the body to the head. Probably the most obvious example of sharks with head weaponry are the Sawsharks (Order Pristiophoriformes), with their elongated rostrum shaped like an isosceles triangle and teeth lining the edges (fig. 10.9). Although there are no direct observations of Sawsharks using their rostrum in the wild, studies of the microwear surface of the rostral teeth of the Common Sawshark (*Pristiophorus cirratus*) suggests that it uses it to capture prey (though not necessarily to impale them) as well as in defense.[44] A captive study on Freshwater Sawfish (*Pristis microdon*) described 17 different behaviors in which Sawfish used the saw, first to sense prey-simulating electric fields, and second to capture prey. When prey was encountered in the water column, the Sawfish used lateral swipes of the saw to attack. This resulted in splitting a fish in half, impaling it on the rostral teeth, stunning it, or sweeping it onto the substrate.[45] Another shark with a weapon-shaped head is the Hammerhead, where the head was used to hit a fleeing stingray and pin it to the bottom, ready for consumption (see chapter 4).

Other sharks may lure prey to themselves. The bearded appearance (branching, freaky-looking dermal lobes around the head) of the Tasseled Wobbegong Shark (*Eucrossorhinus dasypogon*; fig. 10.10), in combination with cryptic coloration, attracts small unsuspecting fish to refuge. Observations in captivity have also revealed that when it detects a potential prey nearby, this shark slowly waves its tail back and forth, which resembles a small fish, complete with a dark eyespot at the base. Bigger fish come in to investigate and "snap . . . dinner time!" Although not active luring, like the Wobbegong earlier, the upper jaw of the Megamouth Shark (*Megachasma pelagios*) has luminescent tissue which might attract shrimp and other prey, providing an easy meal for this species.[46]

Ethologist Arthur Myrberg in the early 1990s also suggested that the spot patterns on the fins of the Oceanic Whitetip Sharks (*Carcharhinus longimanus*) were

Figure 10.9. Shortnose Sawshark off Merimbula, NSW, Australia (Courtesy of Tristan Guttridge).

a form of *aggressive mimicry*. From a distance, the white blotches resemble baitfish, which act as a visual lure for fast-moving prey that the shark can more easily capture. Oceanic Whitetips are regularly seen in pairs or small groups, which would likely enhance this visual aid.[47] More recently, observations were made of associative behavior between Sand Tigers (*Carcharias taurus*) and Round Scad (*Decapterus punctatus*) at Frying Pan Shoals, off the North Carolina coast, in which the shoal of scad surrounds the shark (fig. 10.11). This shoal provides camouflage for the shark, and offers up new opportunities for feeding, as the predators of the scad are the prey of the shark. Whether the sharks are actively seeking the scad schools for this predatory camouflage is unknown, but this type of mutualistic behavior (the scad apparently benefit as well as they can use the shark as a refuge) provides yet another interesting predatory tactic.

Figure 10.10. The bearded appearance of the Tasseled Wobbegong Shark (Courtesy of Annie Guttridge).

Predator Avoidance

It may come as a surprise to learn that sharks are regularly prey themselves; in fact, most sharks live in constant fear of being eaten by another shark, and even some of the *baddest* sharks on the planet employ antipredatory behaviors to ensure they do not end up as someone's lunch (e.g., Orcas [*Orcinus orca*] are often in the news now hunting and killing White Sharks off the coast of South Africa). Of course, having a large body size restricts the list of predators to worry about, but this does not limit the range of behaviors that sharks use—from hiding under crevices, puffing up like a balloon, to counterillumination and even feigning death.

Figure 10.11. Video frame grab of a Sand Tiger shark surrounded by scad school (Courtesy of Erin Burge).

Probably one of the most common (and obvious) strategies to avoid being eaten is to hide or take refuge in a habitat that conceals you or is inaccessible to larger predators. Knowing which habitats are safe and when to use them likely comes from experience, and no doubt many young sharks' lives are a series of near misses. Juvenile sharks of many species are known to use shallow-water nursery habitats; for example, the use of coastal bays with low turbidity by juvenile Scalloped Hammerheads (despite low resources), Blacktip Sharks showing strong fidelity to a small bay fringed by mangroves, and Bull Sharks' use of some low salinity estuaries. At a finer scale, even within these habitats, sharks use their features to aid in predator avoidance; for example, Nurse Sharks wedge themselves under rock crevices and juvenile Lemon Sharks use the complex mangrove root structure more when in the presence of a predator (fig. 10.12).[48] As discussed earlier in the chapter, often many young sharks will form groups, which likely sees them benefiting from greater vigilance, encounter dilution (lower predation risk), and predator confusion.

Reducing activity levels (i.e., movement speed or direction) is another strategy that can help avoid being detected by a predator. Although it is difficult to disentangle physiology (i.e., bioenergetics) from behavior, it is likely that resting or slow swimming while refuging is an effective antipredatory tactic employed by many sharks. Indeed, Whitetip Reef Sharks spend much of their days huddled in reef overhangs; juvenile Port Jackson Sharks rest in seagrass beds for long periods of

Figure 10.12. A juvenile Lemon Shark patrols the mangroves in Bimini, Bahamas (Courtesy of Annie Guttridge).

the day (fig. 10.13); whilst adult Scalloped Hammerheads form huge, polarized schools at seamounts, slowly swimming together. Does this inactivity mean sharks sleep? Well, a team from Australia demonstrated for the first time that resting Port Jackson and Draughtsboard Sharks (*Cephaloscyllium isabellum*) were technically "asleep." Inactive sharks of both species required significantly higher levels of electric stimulation before they showed a visible response, which fulfills a key criterion for separating sleep from other quiescent states. Whether other species that are required to swim (i.e., obligate ram ventilators) to breathe go into a similar sleep-like state is questionable; however, footage of a White Shark at depth showed on Discovery Channel seemingly in a slow-moving trance provides some interesting evidence to consider.

Some sharks, even as adults, will bury themselves in the substrate to remain undetected (e.g., Angel Sharks, *Squatina squatina*) and many have evolved cryptic coloration to blend into the habitats in which they reside. The Ornate Wobbegong Shark (*Orectolobus ornatus*), or *shaggy beard* in Australian Aboriginal language, uses its incredible camouflage (e.g., mottled appearance and small

Figure 10.13. Whitetip Reef Sharks resting during the day under a reef ledge at Cocos Island, Costa Rica (Courtesy of Matt Potenski).

weed-like whisker lobes) to remain invisible to predators and it also selects day-time resting positions with high complexity and crevice volume. These locations are not good areas for prey availability, and thus were likely chosen to avoid predators.

Amazingly, even as embryos, some sharks (e.g., Grey Bamboo and Small-spotted Catsharks) show evidence of antipredatory behavior—they can detect electric fields that mimic those of predators and respond by stopping their respiratory gill movements. Despite being restricted to such a small space and thus being vulnerable to predators, these embryonic sharks can identify potentially dangerous stimuli and react with an avoidance response that is innate (see chapter 5). In addition, Bamboo Sharks, only 10 to 12 cm (4 to 4.7 in) long at birth, are extremely vulnerable to predation and have a distinctive pattern of high contrast banding that mimics the coloration of unpalatable or venomous prey such as sea snakes, in the process avoiding predation (this phenomenon is known as *Batesian mimicry*).[49]

Some sharks are actually able to change color to match their background. Experiments conducted on Small-spotted Catsharks and three species of skates demonstrated a darkening or lightening of skin color when placed in either black or all white tanks, thus effectively matching to the background of their environments. Such an ability would likely help them avoid detection by predators.

Another form of predator avoidance comes from the deep-sea, where camouflage by counterillumination is used by species within the kitefin and lantern sharks. These sharks can switch on their luminescence in response to overhead illumination, which makes them invisible to predators. This ability likely helps with stalking prey too!

Another fascinating capability that has recently surfaced is that some sharks can glow in the dark (i.e., *biofluoresce*). The Swell Shark (*Cephaloscyllium ventriosum*) and Chain Catshark (*Scyliorhinus retifer*) were found to exhibit bright green fluorescence, and incredibly, could visualize their own fluorescent patterns, which could be useful for conspecific communication and possibly avoiding predators that lacked similar visual capabilities.[50]

When a predator is about to strike its prey, the options for escape are limited to either *flight* or *fight*. Some species have evolved morphological characteristics to aid in defense; for example, horn sharks have spines anterior of their dorsal fins that may prevent a predator swallowing them, resulting in ejection and ultimately survival. The eponymous Swell Shark can inflate to almost double its size through taking air into its cardiac stomach. This increase in size likely intimidates a predator or tricks them into thinking the prey is too large for consumption. Of course, stingrays are equipped with a venomous barb that can be used to defend against predators as well as harassment from mates, although a recent study found that the venom from the Atlantic Stingray (*Hypanus sabina*) was low in toxicity and metabolic cost compared to other venomous species, suggesting that defense is not the primary purpose of the spine.[51] One Great Hammerhead was found with 96 barbs in its mouth, so how effective they are for defense seems questionable? Other rays may use their electric organ discharges (EODs) in predator defense. Interestingly, the Lesser Electric Ray (*Narcine brasiliensis*) did not use it for foraging, indicating that it was primarily for predator defense or conspecific communication.[52]

Threat displays can also be an effective means of warning off predators. Many sharks have been observed to use agonistic behaviors, such as depressed pectoral fins, arched back, stiff or jerky movements, and jaw gaping. In the 1980s, ethologist Don Nelson examined the intensity and frequency of these behaviors by approaching Grey Reef Sharks in a submersible. He observed that displays were common, often escalating to full-on attacks when sharks were cornered. Bull Sharks, when followed in the shallows, will also exhibit similar behaviors, beginning with flight, then exhibiting the stereotypical pectoral fin depressing, S-shaped swimming followed by the final, last-resort defensive yet powerful attack. Most predators prefer the element of surprise during attack and want to avoid injury at all costs, thus the behaviors described above are likely used to signal a predator—"you've been spotted, and if needed I'm ready to attack to defend myself."

Lastly, probably one of the most underreported but fascinating antipredator strategies is death feigning. This bizarre behavior is referred to as tonic immobility (TI) in the shark literature (fig. 10.14), and in other animals is known to be adopted late during a predation event, and frequently following a predator's making

Figure 10.14. Dr. Samuel "Sonny" Gruber placing a juvenile Tiger Shark in tonic immobility (Courtesy of Bob Crimian).

physical contact. A handful of studies have explored the onset and timing of TI in juvenile Lemon Sharks, with one examining stress via blood chemistry parameters.[53] Whether this strategy is used by sharks as a last-ditch attempt to confuse a predator is unknown. Most likely, it serves multiple functions including being a component of courtship and copulation (see chapter 6).

Concluding Comments

It is an incredibly exciting time to be a scientist exploring shark behavior!

There are still so many behaviors that we have only just touched the surface of. Continuing advancement of biologging and camera tools, alongside enhanced remote sensing and data processing, will pave the way for a better understanding of shark behavior, which will ultimately improve their conservation and survival.

NOTES

1. For further discussion, see McInturf, A. G. et al. 2023. ICES J. Mar. Sci. 80(6), 1551–1566.
2. de la Parra Venegas, R. et al. 2011. PLoS One 6(4): p.e18994.
3. Kajiura, S.M. and Tellman, S.L. 2016. PloS One 11(3): p.e0150911.
4. Heupel, M.R. and Simpfendorfer, C.A. 2005. Mar. Biol. 147: 1239–1249.
5. Reyier, E.A. et al. 2008. Fl. Sci. 2008: 134–148.
6. Litvinov, F.F. 2006. J. Ichth. 46: 613–624.

7. Finucci, B. et al. 2019. ICES J. Mar. Sci. 75: 1613–1626.
8. Hight, B.V. and Lowe, C.G. 2007. J. Exp. Mar. Biol. Ecol. 352: 114–128.
9. Springer, S. 1967. In: Gilbert, P.W. et al. (eds.). Sharks, skates and rays. Johns Hopkins U. Press. 149–174 pp.
10. Myrberg Jr., A.A. and Gruber, S.H. 1974. Copeia 1974: 358–374.
11. Guttridge, T.L. et al. 2011. Mar. Ecol. Prog. Ser. 423: 235–245.
12. Finger, J.S. et al. 2018. Behav. Ecol. Sociobiol. 72: 17 and Keller, B.A. et al. 2017. J. Exp. Mar. Biol. Ecol. 489: 4–31.
13. Jacoby, D.M.P. et al. 2012. J. Fish Biol. 81: 1596–1610.
14. Jacoby, D. M. et al. 2016. J. Royal Soc. Interf. 13(124): 20160676.
15. Klimley, A.P. and Nelson, D.R. 1981. Fish. Bull. 79: 356–360.
16. Sims, D. W. et al. 2022. J. Fish Biol. 101(5): 1160–1181.
17. Ebert, D.A. 1991. S. Afr. J. Mar. Sci. 11: 455–465.
18. Jacoby, D.M. et al. 2012. Fish Fish. 13: 399–417.
19. White, C.F. et al. 2022. Anim. Biotelem. 10(1): 35.
20. Lieberman, D.A. 1993. *Learning: behavior and cognition*. Thomson Brooks / Cole Publishing Co.
21. Clark, E. 1959. Science 130: 217–218.
22. Schluessel, V. 2015. Anim. Cogn. 18: 19–37.
23. Vila Pouca, C. and Brown, C. 2018. Anim. Cogn. 21: 481–492.
24. Nelson, D.R. 1977. Am. Zool. 17(2): 501–507.
25. Finger, J.S. et al. 2016. Anim. Behav. 116: 75–82.
26. Anderson, J.M. et al. 2017. Sci. Rep. 7: 11042.
27. Nosal, A.P. et al. 2016. PloS One 11(1): p.e0143758.
28. Edrén, S.M.C. and Gruber, S.H. 2005. Env. Biol. Fish. 72: 267–281.
29. O'Gower, A.K. 1995. Mar. Freshwater Res. 46: 861–871.
30. Meyer, C.G. et al. 2010. Mar. Biol. 157: 1857–1868.
31. Schluessel, V. and Bleckmann, H. 2012. Zool. 115: 346–353.
32. Tricas, T.C. 1985. Mem. S. Calif. Acad. Sci. 9: 81–91.
33. Klimley, A.P. 1994. Am. Sci. 82(2): 122–133.
34. Martin, R.A. et al. 2005. J. Mar. Biol. Assoc. U.K. 85: 1121–1136.
35. Semmens, J. M. et al. 2019. Mar. Biol. 166: 1–9.
36. Ryan, L. A. et al. 2022. J. Exper. Mar. Biol. Ecol. 553: 151764.
37. Huveneers, C. et al. 2015. Am. Nat. 185: 562–570.
38. Towner, A.V. et al. 2016. Func. Ecol. 30: 1397–1407.
39. Fouts, W.R. and Nelson, D.R. 1999. Copeia 1999: 304–312.
40. Sims, D.W. et al. 2008. Nature 451: 1098.
41. Humphries, N.E. et al. 2010. Nature 465: 1066.
42. Papastamatiou, Y.P. et al. 2011. J. Anim. Ecol. 80: 864–874.
43. Oliver, S.P. et al. 2013. PLoS One 8(7): p.e67380.
44. Nevatte, R.J. et al. 2017. J. Fish Biol. 91: 1582–1602.
45. Wueringer, B.E. et al. 2012. Current Biology 22: R150-R151.
46. Nakaya, K. 2001. Bull. Fish. Sci. Hokkaido Univ. (Japan) 52: 125–129.
47. Myrberg Jr., A.A. 1990. J. Ex. Zool. 256: 156–166.
48. Heupel, M.R. et al. Mar. Freshwater Res. 70: 897–907.
49. Kempster, R.M. et al. 2013. Plos One 8(1): p.e52551.
50. Gruber, D.F. et al. 2016. Sci. Rep. 6: 24751.
51. Enzor, L.A. et al. 2011. J. Exp. Mar. Biol. Ecol. 409: 235–239.
52. Johnson, R.H. and Nelson, D.R. 1973. Copeia 1973: 76–84.
53. Brooks, E.J. et al. 2011. J. Exp. Mar. Biol. Ecol. 409: 351–360.

PART IV

Fisheries, Climate Change, and Other Human Impacts

11 / Fisheries

CHAPTER OUTLINE

Introduction

Chances are you are far removed from shark fisheries and the products of these fisheries. You may never have caught a shark or eaten one. You likely have not consumed shark fin soup or worn shark-leather products. Most readers will never have seen a commercial shark fishing operation, especially at sea catching sharks. On the other hand, you are very likely aware that fisheries for sharks exist. But for which sharks? And where? And by what methods? And are all shark fisheries unsustainable? Is the shark fin industry responsible for killing 100 million or more sharks annually, as has been widely reported in the media? In what ways are shark fisheries managed differently than fisheries for bony fishes?

Uses for Sharks and Batoids

The use of sharks and batoids by humans is recorded from over 5000 years ago. Data on the annual catches of chondrichthyans (sharks, batoids, and chimaeras) are collated by the Food and Agriculture Organization of the United Nations (FAO), based primarily on reports by regional organizations and member nations. Globally, harvest of sharks peaked around the year 2000, though exploitation of shark stocks for fins, meat, and other shark products, including their liver, cartilage, skin, teeth, jaws, and other body parts, continues (fig. 11.1).[1] Live sharks are used in aquaria and as ecotourism draws.

Fins

Since the 1980s, shark fins (figs. 11.1 and 11.2) have been the most economically valuable part of the shark and are still one of the most profitable. They are used mostly for shark fin soup, which is considered a delicacy and status symbol in East and Southeast Asian countries that include China, Hong Kong, Taiwan, Singapore, Malaysia, Vietnam, and Thailand.

Figure 11.1. Shark Products: shark fin soup, dried shark fin, and shark oil or squalene capsules (marinethemes.com/Kelvin Aitken).

A

B

C

Figure 11.2. (A) Shark fins (top row) in a dried seafood shop in Hong Kong. The price per kg for the most expensive fins in the photo, 3480 Hong Kong Dollars per kg, is about $200 per pound. (B) Finned carcass of Silvertip Shark (marinethemes.com/ Kelvin Aitkin). (C) Blue Sharks with fins removed in a market in Cadiz, Spain.

It is important that a distinction be made between illegal finning and the legal shark fin trade. The former involves removal of the fins from the carcass, most often immediately upon capture when the animal is still alive, and disposal at sea of the less valuable finless carcass. The legal shark fin trade involves fins from legally caught sharks which typically are brought back to port whole and are used for meat and other products. Numerous countries[2] have various kinds of restrictions or bans on finning or on shark fin soup.

All fins on a shark can be used except the upper lobe of the caudal (recall that the vertebrae extend all the way to the tip of the tail). Following a week or more of processing, what is left of the fins is the key ingredient, the skeletal elements called *ceratotrichia*. Ceratotrichia, which are composed primarily of an elastic protein similar to the fibrous protein keratin of human hair, taper from a round base to finer strands. In the industry, ceratotrichia are called *fin needles*. The ceratotrichia are boiled and take on the appearance of colorless cellophane noodles, which are then ready to be made into soup (fig. 11.1).

Species of choice in the shark fin industry include carcharhinids, lamnids, sphyrnids, and even pristids. The costs per pound for wet and dried fins varies among species and are also based on how skillfully the fins have been removed and handled. Prices can rise as high as $495/kg ($225/lb) retail, but fishers typically receive only around $16/kg ($7.50/lb) off the boat.

The global trade for shark and ray fins peaked at just over of 20,000 tonnes (44 million lbs) in 2003, around the same time landings peaked for sharks and rays. By 2012, fin exports had dropped by about 50%, to less than 10,000 tonnes (22 million lb).

Peak global value for the shark fin trade was about $350,000,000. Over the period from 1993–2002, Hong Kong was the world's biggest trader of shark fins, with about 40% of the global total, followed by China, Indonesia, Singapore, and Taiwan. More recently (2013–2022), Singapore has become the largest trader of shark fins, followed by Hong Kong, Spain, Malaysia, and Indonesia. Commercial fishers from Spain and Indonesia are responsible for the largest shark catches for the shark fin industry.

In the United States, it is now illegal to sell or trade shark fins, even in fisheries that are operating sustainably. Instead of full use of the carcass, the fins must be discarded. How do you feel about this policy?

Meat

While shark meat has always been eaten in developing coastal and island countries, it was not until capture fisheries of many bony fishes began to decline in the 1980s that fresh shark meat became more acceptable as an alternative to swordfish, tuna, etc. in wealthier countries.

Shark steaks can be found in seafood markets as both fresh and frozen products. Among the biggest single uses is as a substitute for demersal (bottom) fish cod in the European and British fish and chips industry (fig. 11.3).

B

Figure 11.3. Spiny Dogfish in (A) net and (B) on culling table. These sharks most likely are destined for the European and British fish and chips market (Courtesy of NOAA).

Like fins, the trade for shark and ray meat increased in the early 2000s but has decreased in recent years. World exports of shark meat, a category that includes rays, increased from about 70,000 tonnes (154 million lb) to over 124,000 tonnes (273 million lb) in the decade between 2003 and 2012, but decreased to about 80,000 tonnes (176 million lb) by 2022. The largest importers of shark meat in 2022 were Brazil, Spain, Portugal, Republic of Korea, and Italy. Brazil's imports comprise coastal and pelagic sharks including blue sharks, and a mix of blue sharks and deepwater squaliform sharks dominate European imports. Korea's imports are predominantly rays, which are used as substitutes for scallops.

The value of shark and ray meat increased rapidly beginning in 2000, driven in part by the heavy use of shark meat in fish and chips in Europe. The value of shark and ray meat now equals or exceeds the value of shark fins globally, peaking at nearly $450,000,000 in 2011.

Liver

The largest organ in sharks is the oil-rich liver (see fig. 7.5). It has been found to constitute as high as 19.2% of body weight in the Kitefin (*Dalatias licha*), to as

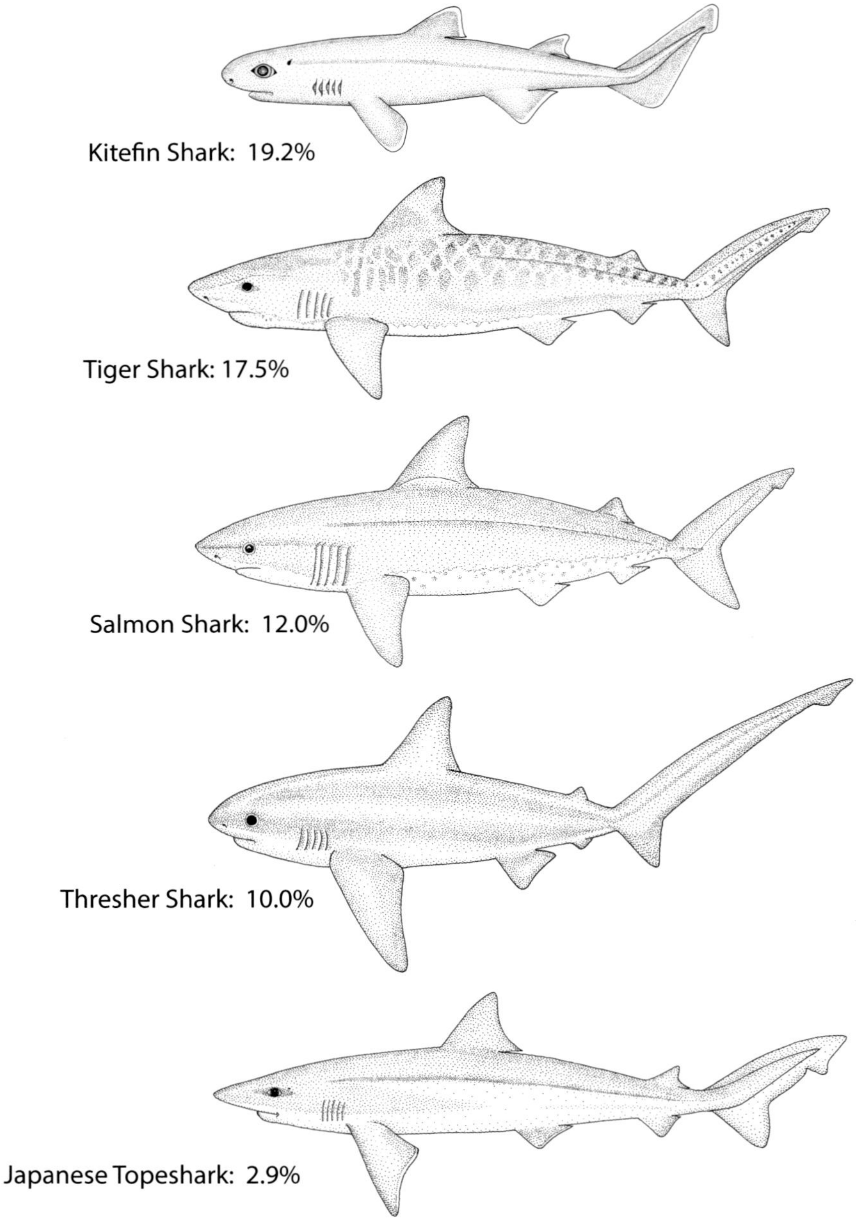

Figure 11.4. Percentage of body weight of the livers of selected sharks.

low as 2.9% in the Japanese Topeshark (*Hemitriakis japanica*; fig. 11.4). The weight of the liver of an adult human is about 2–4% of body weight.

In the late 1930s, vitamin A, a fat-soluble compound important to vision, the immune response, reproduction, and cellular communication, was found in high concentrations in sharks. When supplies of cod liver oil, then the major vitamin A source, were cut off during World War II, fisheries for sharks exploded in North America and Australia, particularly for the School (or Tope) Shark (*Galeorhinus galeus*) off California, and the Spiny Dogfish (*Squalus acanthias*) off British Columbia. Both of these species were relatively abundant and had high vitamin A levels. To make use of the sharks and rays captured for the livers, the US government instituted a campaign to encourage Americans to eat sharks and rays (fig. 11.5). Synthetic vitamin A was manufactured in the early 1950s, after which the shark-derived form was no longer needed.

Shark liver oil (fig. 11.1) is used in a variety of products, including as a lubricant in the textile industry, cosmetics, health products, fuel oil for lamps, and to protect boat hulls. World trade in shark liver oil is small. The highest recorded production was the export of nearly 1,000 tonnes (2.2 million lb) in 1985, mostly from Portugal. Recent exports of shark liver oil have been less than 30 tonnes (66,000 lb) per year.

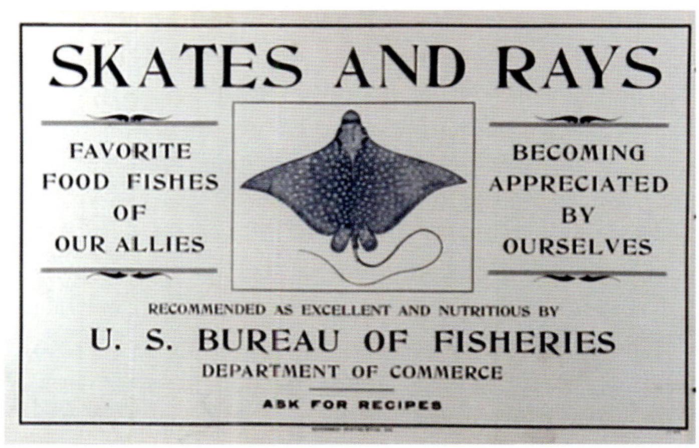

Figure 11.5. US Department of Commerce posters encouraging fishing for sharks and rays.

Cartilage

Shark cartilage is widely used as a health supplement, although the health benefits are not always supported by sound, science-based evidence. These so-called benefits include protecting against or curing eczema, ulcers, hemorrhoids, arthritis, and other diseases and disorders, most notably cancer (see chapter 1).

Shark cartilage is sold as powder, capsules, creams, and tablets, often labeled *shark cartilage* or *chondroitin*. The latter is used as a dietary supplement for treating osteoarthritis in humans, dogs and cats, although there is conflicting evidence of its efficacy. Chondroitin is more often sourced from cows (bovine) or pigs (porcine) than from sharks. Accurate records of the shark cartilage trade are lacking.

Other Products

Other products from sharks and rays include skin, gill rakers from Mantas, jaws and teeth, rostra from Sawfish, and intact, preserved neonate sharks or embryos.

Shark skin is used mostly to make shark leather. The skin can be used for most consumer products where cow leather is used (e.g., furniture, wallets, purses, shoes, book bindings, belts, etc.). Shark skin has also been used as sandpaper and for polishing wood.

Gill rakers from rays of the family *Mobulidae* (mantas) are used in traditional Chinese medicine, purportedly to aid in the immune response as well as to cure or prevent other diseases or disorders.

Shark jaws and teeth, rostra from sawfish, and preserved shark embryos and juveniles are bought and sold at trade shows, markets, and online. All are legal to purchase, except for species like sawfishes and others that are on the US Endangered Species list or are present in Appendix I of CITES (the Convention on International Trade in Endangered Species of Wild Fauna and Flora), which prohibits their trade.

A Primer on Fisheries

The purpose of any capture fishery (as opposed to aquaculture) is to catch targeted species. For any species in a fishery, there is a *maximum sustainable yield* (MSY), which is defined as the highest catch (typically expressed as weight more than numbers) of that species that can be removed annually without irreversibly damaging the population. Note that no attention to impacts on nontarget species or the ecosystem is included in this concept.

Maximum sustainable yield occurs at the point on the growth curve where population growth rate is highest (fig. 11.6). Mathematically, the MSY should be half of the *carrying capacity*, the maximum number of individuals of a given species that an ecosystem's resources can support in the long term without significantly depleting or degrading those resources. In reality, it is not that simple. For fishes like tunas with very fast-growing populations, MSY may be at half or even

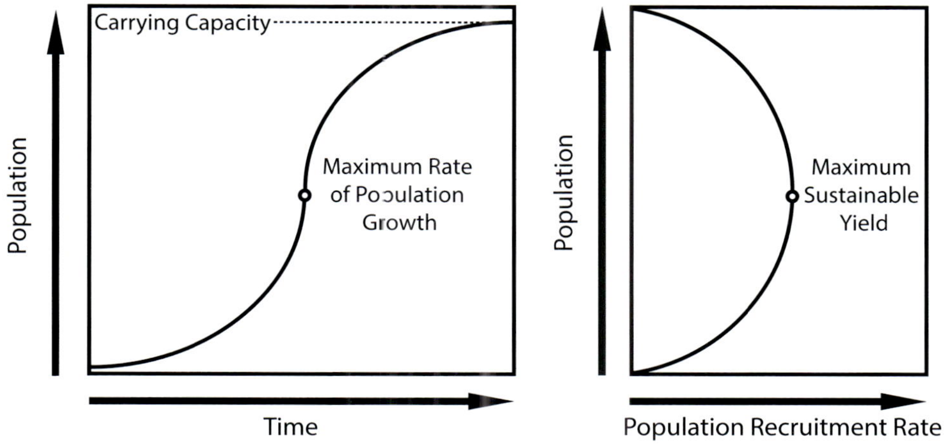

Figure 11.6. Graph of Maximum Sustainable Yield.

more of the carrying capacity. However, for species, like sharks, that take many years to mature, MSY is often far below half the carrying capacity. If the catch of any species is too high (i.e., above the MSY), then resulting population decreases reduce the future yields, or the individuals caught become too small.

Overfished versus *overfishing*—these two terms may seem synonymous but mean very different things when assessing the sustainability of a fishery. *Overfished* refers to the numbers or biomass of fish in the stock. If more sharks are harvested than the MSY, the number of females in the population will be too low to produce the offspring needed to support that MSY, and hence the population is overfished. *Overfishing* refers to the rate of fishing mortality. If this mortality is too high, the population will decline and it will eventually become overfished. When we assess the status of a shark stock, we must consider both concepts. The goal is a stock that is not overfished (the population is large enough to support MSY) and no overfishing is occurring (the population is not declining). This is the case in the most recent stock assessments for US Atlantic Blacktip Shark and Scalloped Hammerhead stocks. In a stock that is not overfished, but overfishing is occurring, the fishing mortality must be decreased, or the stock will become overfished. A stock that is overfished, but no overfishing is occurring, is essentially recovering and on the trajectory to no longer be overfished. This is the case for the most recent assessments for Great Hammerhead and Sandbar Shark stocks in US Atlantic waters, though Sandbar Sharks are due for reassessment and may now be recovered. The worst case is a stock that is overfished and overfishing is occurring, because this is a stock that is not recovering, which was the case for the last assessment of US Atlantic Dusky Sharks.

Fisheries can be classified by the type of catch, the type of gear, the geographical location, as well as position in the water column (surface or bottom, shallow or deep) and the scale (no pun intended) of the fishery.

Fisheries can be either *subsistence, artisanal, recreational,* or *commercial*. Subsistence fisheries are those whose catch is typically for the consumption of the fisher and his or her extended family. Subsistence fishers use traditional techniques and small vessels without advanced technology and are small-scale. Even though the catch of every subsistence fisher is small, there are enough of them globally to contribute to overfishing of some species, although their impact is not fully known since most subsistence fisheries are not monitored and their catches are thus unreported.

Artisanal fisheries are intermediate between subsistence and commercial fisheries in practically every aspect. They provide seafood to their local village or town but usually not for export, and most of their catch is used as caught and is not heavily processed. The gear of artisanal fishers is typically more modern than that of subsistence fishers, but artisanal fishers still use traditional techniques as well. The impact of artisanal fisheries is also difficult to assess but is often substantial.

Recreational fishers fish for leisure and while many eat their catch, they typically do no need to fish for their income or subsistence. Most often the equipment used includes rod and reel, spear guns, or cast nets. Recreational fisheries can contribute significantly to local economies, and for-hire charter boat industries prosper by providing recreational fishing to those without their own vessels, means or expertise. Given the large number of anglers involved in recreational fisheries, the mortality, even in fisheries where most captured fish are released, can rival commercial fishing.

Finally, the most advanced, largest-scale fishery is commercial/industrial. Vessels may be quite large, and are well-equipped, often with technologically sophisticated instruments for finding fish. The catch may be brought back to port for processing or is processed and frozen at sea. The capital investment is high, as is the impact on target and nontarget species and marine ecosystems. Commercial fisheries are the most closely monitored of the three types, but illegal and undocumented fisheries occur with alarming regularity. It has been estimated that up to 50% of shark and ray landings may go unreported. Most of the shark and ray fisheries we present in this chapter are commercial.

Ways to Catch Sharks in Fisheries

There are four major categories of gear on which sharks and batoids (and other fishes as well) are caught, either intentionally or as bycatch (untargeted or unwanted catch; fig. 11.7).

Facing page

Figure 11.7. Ways to catch sharks as well as sources of shark and ray bycatch (Marc Dando. Used with permission from Save Our Seas Foundation © 2016).

SOURCES OF SHARK AND RAY BY-CATCH

Globally, the number of sharks and other elasmobranchs captured as by-catch (species not targeted) is probably greater – perhaps by several times – than the number harvested in fisheries. The greatest biomass of sharks captured as by-catch occurs in pelagic fisheries using long-lines, purse seines and drift gill nets. By-catch in these fisheries is of great concern for the populations affected due to the magnitude of mortality. However, these fisheries affect relatively few species, with catches being dominated by a handful of pelagic sharks like blue sharks and silky sharks. In contrast, fisheries employing gill nets, bottom trawls and long-lines in the coastal zone often affect a great many species of sharks, skates and rays and the biomass of discard mortality is often grossly underestimated. The discard biomass in bottom-trawl fisheries in continental shelf waters is often many times the biomass of the targeted catch. Many deep-sea sharks have life histories that render them much more vulnerable to overfishing than their coastal counterparts, and deep-water trawl and long-line fisheries now operate to the maximum known depth at which sharks occur. Deep-sea edge habitats such as submarine canyons and sea mounts concentrate biomass and biodiversity and often have unique animal communities, including isolated shark populations that can be quickly depleted by relatively small fisheries.

As the human population has grown and coastal marine resources have become fully exploited, and in some cases overfished, fisheries have spread farther from shore and into deeper regions of the world's oceans. Sharks and rays are now subject to harvest and by-catch by fisheries operating throughout their depth range from coastal rivers and estuaries to the deep continental slope more than 3,000 metres deep. The major fisheries responsible for shark and ray by-catch are illustrated and typical species that are exposed to each fishing gear are represented below.

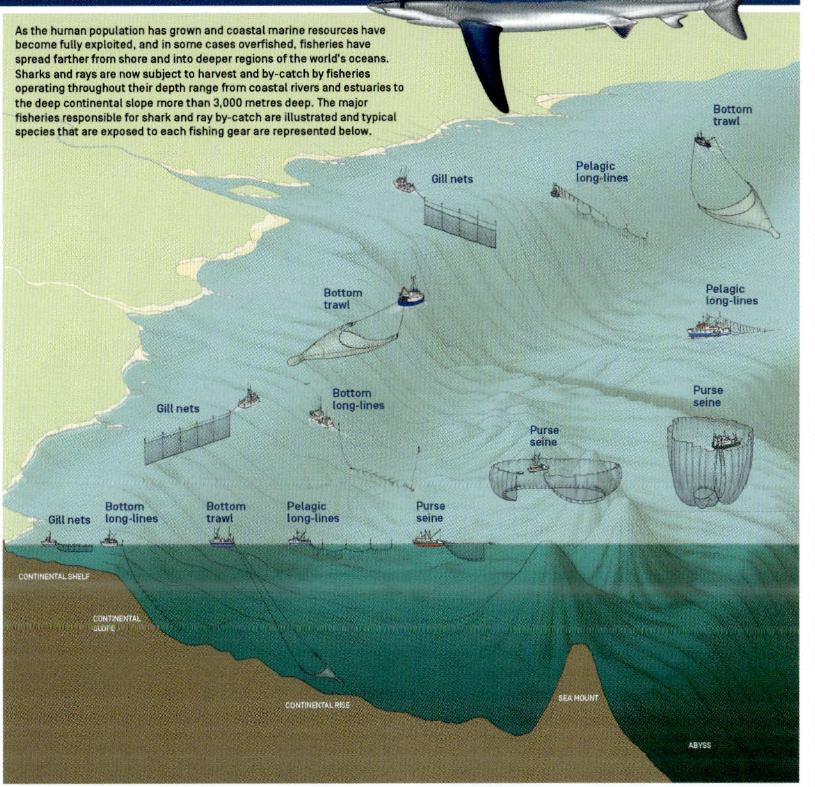

Gill nets

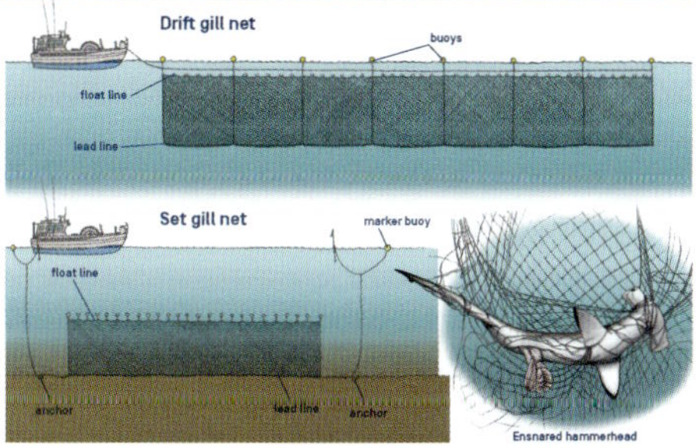

Bottom long-line

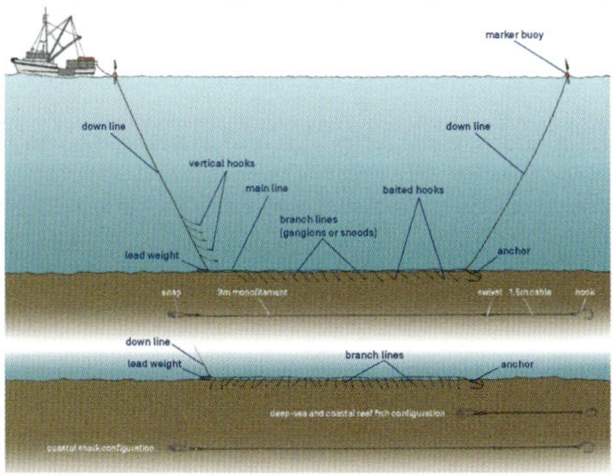

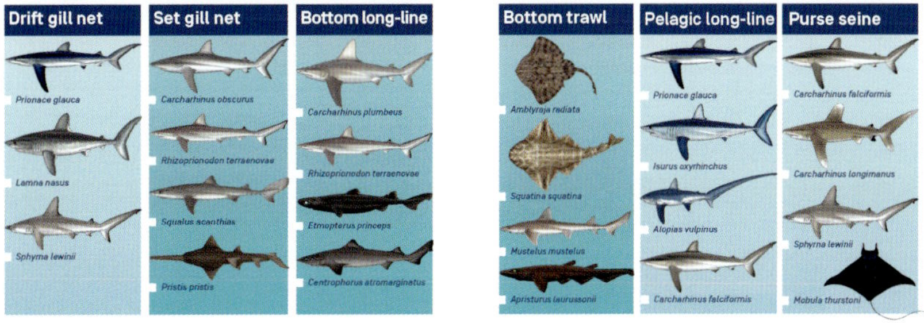

Bottom trawl

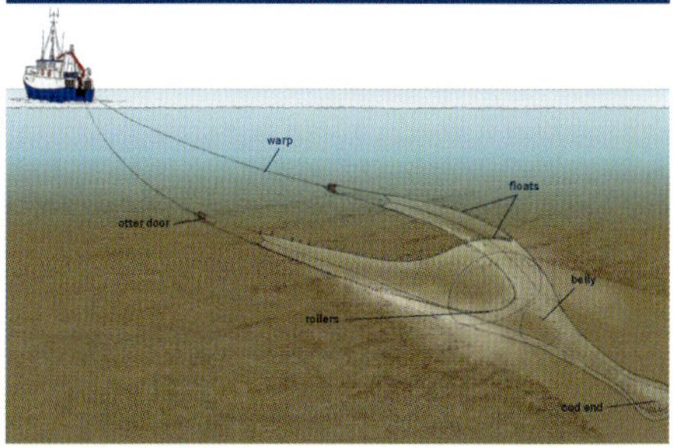

Pelagic long-lines

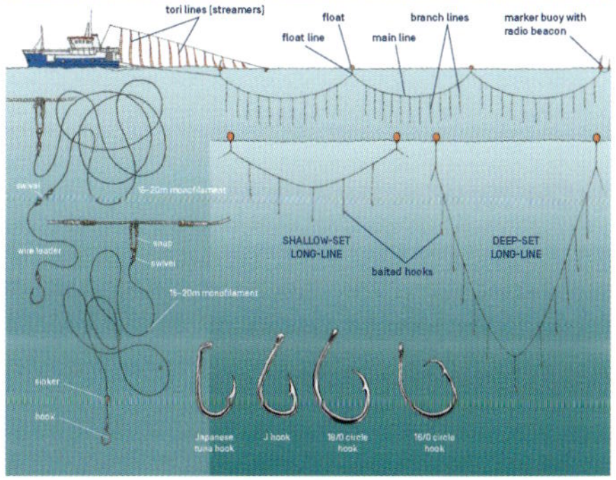

Purse seine

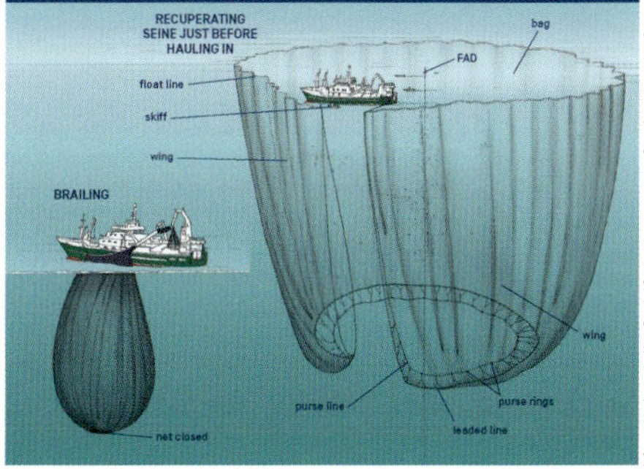

Infographic design by Marc Dando | Content interpretation by Dean Grubbs and Marc Dando

- *Active entrapment* (seines, purse-seines, trawls)—targeting fish like pollack, cod, flounder, tuna, shrimp, skates, and sharks. Trawl catches are typically multi-species except in some pelagic fisheries for small schooling fishes like sardines.
- *Hook and line* (trolling, longlines)—targeting pelagic fish like Mahi-mahi, tuna, grouper, snapper, Spiny Dogfish, and other sharks.
- *Passive entrapment* (trap nets [tonnara], pound nets)—catching tuna and sea-bass as well as estuarine species.
- *Entanglement gear* (fixed and drift gillnets, trammel nets)—targeting tuna, mackerel, estuarine fishes, and Spiny Dogfish. A variety of carcharhinid (e.g., Atlantic Sharpnose [*Rhizoprionodon terraenovae*], Blacktip [*Carcharhinus limbatus*], Blacknose [*C. acronotus*], Finetooth [*C. isodon*]) and Bonnetheads (*Sphyrna tiburo*) are targeted in the SE US gillnet fishery. Thresher sharks are caught in the Pacific Ocean using gillnets.

Overview of Shark Fisheries

Globally, 70% of annual catches of chondrichthyans are sharks and about 30% are batoids. Chimaera fisheries represent an insignificant proportion of the chondrichthyan catch, and thus we ignore them in our analyses in this chapter. Finally, the data reported here represent landings only and include only bycatch that was landed and not discarded at sea or unreported.

As a group, elasmobranch fisheries are less than 1% of the total marine capture fishery production. The catch of sharks and batoids reached a maximum of approximately 895,000 tonnes (1,970,000,000 lb) in 2003, representing a tripling of the catch rate of 1950 (fig. 11.8A). This increase was due to several factors, including increased demand, the development of global international trade, and a transition to more modern, industrial-type fishing methods.

One caveat of shark fisheries is that reporting of catches is notoriously bad, in part because identification to species level is low or inaccurate, and about one-third of the elasmobranch catch is lumped together as sharks, rays, skates, etc.

Since the 2003 peak, reported landings have declined. For 2022, the shark and batoid catch was 594,000 tonnes (1.3 billion lb), or about 0.6% of total global fish catch (excluding aquaculture) of 92,896,000 tonnes (204 billion lb).

What accounted for the decrease in landings? Globally, the principal explanation is declines of shark and batoid stocks associated with overfishing. Additionally, although to a lesser degree, landings were reduced by the implementation of effective management that included fishery management plans, harvest quotas, finning bans, establishment of marine protected areas, and so on.

This contrasts with the explanation for decreased shark and batoid landings in the United States, where catch reductions are best explained predominantly by implementation of fishery management plans (FMPs) and aggressive enforcement of quotas, although some stocks have been and still remain depleted due to overfishing.

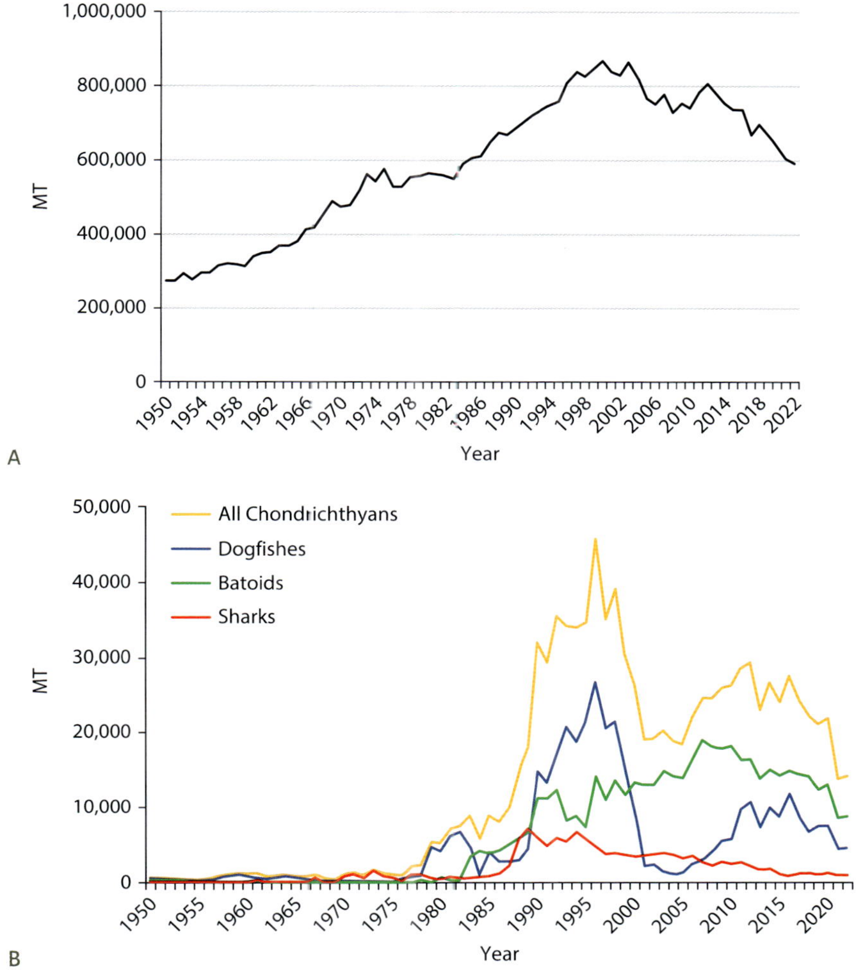

Figure 11.8. (A) Graph of global shark, batoid, and holocephalans (chimaeras) landings combined from 1950–2022. (3) US Western Atlantic landings. *All chondrichthyans* category includes non-dogfish sharks, dogfishes, batoids, and holocephalans (chimaeras). *Sharks* category excludes dogfish.

In the United States, catch of sharks and rays peaked in 1992 at 54,093 tonnes (119,255,000 lb). By 2001, the landings had declined to 21,837 tonnes (48,142,000 lb) but rose subsequently with landings of 27,969 tonnes (61,661,000 lb) in 2022. Nearly 50% of US chondrichthyan landings are from Pacific waters, but 98% are skates.

In US Atlantic waters, shark and ray landings increased rapidly in the 1980s and have fluctuated, reflecting changing fisheries and associated management (fig. 11.8B). The major groups landed have changed dramatically as well. For example, in 1995, total US Atlantic landings of chondrichthyans were 62% "dogfishes," 17% coastal and pelagic sharks, and 21% skates. By 2020, these landings shifted to only 7% "dogfishes," 17% coastal and pelagic sharks, and 76% skates and rays, but by 2022 shifted to 32% "dogfishes," 6% coastal and pelagic sharks, and 62% skates and rays. These shifts are discussed in more detail under the Contemporary Case Studies later in the chapter.

Globally, about 40–50% of elasmobranch fisheries are in the Western Central Pacific (> 130,000 tonnes or 286 million lb), and the Eastern and Western Indian Ocean (about 120,000 tonnes each or 265 million lb each). The elasmobranch fishery in the NW Atlantic is only about 40,000 tonnes (88 million lb). The catch for all of Europe is about three times that of the United States.

The top five countries with the highest landings of sharks and batoids, with a combined catch of about 245,000 tonnes (772 million lb) are Indonesia, Spain, Mexico, India, and United States. These five, plus Taiwan, Japan, Malaysia, Argentina, Portugal, New Zealand, France, and Peru make up about 60% of the shark and batoid landings worldwide. The remaining 40% are from small countries, mostly island nations or poorer countries in Africa, and mostly from artisanal fisheries which, as we already stated, are difficult to get good data from. This remains a vexing problem.

In last half of the twentieth century, another major problem has been island nations and smaller poor countries selling fishing rights in their Exclusive Economic Zone (EEZ) to big fishing nations like Korea, Japan, India, Spain, etc. The smaller countries typically receive a one-time payout for these rights. The countries who purchase them thus have no incentive to sustainably fish these stocks. In fact, the incentive is exactly the opposite—come in, exploit, overfish, and wipe out stocks, then move on. More recently, there has been a push in Indo-Pacific countries to fish their own stocks, the idea being that this would serve as an incentive to sustainably fish. To that end, in 2003, the governments of Australia and New Zealand funded a comprehensive manual on longline fishing[3] for pelagic fishes that was distributed by Indo-Pacific countries to teach coastal residents to fish for themselves.

For insight on the problems and solutions of shark fisheries conservation in countries outside of the Western world, in this case Sudan and Bangladesh, from the perspective of resident conservation scientists, see *Research, Conservation, and Education Focus* Boxes 9.1 and 11.1, respectively.

Deep-sea Fisheries

In places where deep-sea fisheries operate, sharks often become quickly depleted. Recall that deep-sea sharks are even more susceptible to overfishing than others, as they have among the most conservative life histories (e.g., slow maturation and low fecundity). A 2022 review of deep-sea shark biology and threats,[4] highlighted four regions of the world where industrial deep-sea fisheries are of most concern for deep-sea chondrichthyans—the Eastern Atlantic Ocean including the Mediterranean Sea; the SW Atlantic (primarily off Uruguay,

BOX 11.1

Research, Conservation, and Education Focus

How Fishers' Knowledge and Observations Can Transform Shark Sciences

Alifa Bintha Haque

Department of Zoology
University of Dhaka, Bangladesh
Founder and Lead, Bengal Elasmo Lab Collaboration

Sharks and rays are at risk of extinction globally; however, studies are immensely limited in the Global South. For instance, even though artisanal fisheries in the Bay of Bengal contribute to the high global fishing pressure on sharks and rays, the area is one of the most data-poor regions of the world. This region's diversity, abundance, habitats, threats, and conservation requirements still need to be adequately assessed. Traditional and scientific data collection and at-sea research can be inherently resource-intensive and sometimes guided by Western approaches, inappropriate for the context of the geography or the fishing communities of our region. The realities are the same for conservation actions. In these regions, species conservation at sea and fisheries are challenging to be a livelihood for fishers dealing with poverty, climate change, lack of social safety nets, and limited decision-making power. Moreover, for decades, the knowledge fishers gained through hundreds of years of experience in fishing and handling species of conservation interest has been neglected and only recently got traction through different social sciences approaches.

As long as sharks and rays remain "unimportant species" to a developing country, as they may not have substantial commercial value what with the area dealing with other social and economic priorities, resources to fund expensive at-sea projects will remain limited. Hence, there is a need for low-cost, low-tech, and innovative solutions to keep the research and conservation actions viable in the long run. Here, ethnoecology (how different people living in different locations understand the ecosystems around them) and citizen science can play an important part. What if we consider each fishing vessel as one ocean data point? Then, there might be no space left for us to explore. What if we have the "thinking approaches" and "implementation tools" to gather and understand these fishers' knowledge to answer critical ecological/conservation questions not only dealing with socio-ecological issues but inherently scientific issues, such as habitat mapping, ecological base-line creation, ecology, and biology of species in time and space? There will always be questions about reading "personal bias," granularity, and authenticity of data, which I believe can be overcome only by "doing" the science and enhancing our capacities to improve methods, analyze, and comprehend these better.

For instance, we examined the fishing footprint (three-dimensional mapping, magnitude, and location of the artisanal fishing activity), seasonality, and fishing efforts (annual fishing months and days) for seven different fishing gears within the Bay of Bengal, Bangladesh region utilizing > 1000 interviews with fishers and workshops facilitating participatory mapping (ethnoecological methods). Spatially explicit maps, incorporating local socio-ecological knowledge, were created to address conservation and management challenges by area-based management actions in data-limited situations. We identified potential areas of concern. The overlap of critical habitats of different species with fishing footprints varies, showing the potential for nuanced area-based management. To establish ground truth and eliminate biases, we have used low-tech mobile phones with an open-access application to track fishing trips (citizen science and GPS tracking methods) and identify fishing routes, at-sea catches, and fishing hotspots. This is the first attempt to track fishing vessels, map fishing grounds at this scale, and collect at-sea data in Bangladesh, bridging the gaps of a very data-poor fishery system. We have used ethnoecological and citizen science methods to create the most extensive datasets of shark and ray landings for over 90 species in coastal Bangladesh, a project running for eight years now with very low costs and empowering locals. We identified Critically Endangered sawfish catch hotspots and Important Sharks and Ray Area (ISRA), devised gear-specific live release protocols, and so on.

Transdisciplinary frameworks and local ecological knowledge help the science of, and can advance and foster meaningful real-world change in, marine conservation and fisheries management. This approach bridges the gap between scientific rigor and local communities' practical and lived realities, particularly in the Global South. Engaging fishers, traders, and government agencies in co-creating knowledge is robust and actionable. This blend of academic rigor and community engagement is essential in navigating the multifaceted challenges of species conservation, ecosystem preservation, and human well-being by integrating local ecological knowledge with advanced tools such as real-time data gathering, habitat mapping, and technological innovations.

From an academic standpoint, these methodological approaches challenge the limitations of traditional conservation approaches, which often fall short in Global South contexts due to their top-down structure, reliance on externally imposed regulations, and lack of attention to sociocultural dynamics, particularly the understanding of local knowledge and decision-making systems. These conventional approaches typically prioritize ecological science and regulatory mechanisms, overlooking the complex interconnections between local communities and their environments,

Participatory mapping in Dublar Char, Sundarbans 2021.

which leads to low compliance and the marginalization of fishers. In contrast, this research introduces a transdisciplinary framework that combines the strengths of Western conservation science, fisheries management, and social science with meaningful, community-driven engagement. This allows us to address critical research questions in data-deficient settings by integrating local ecological knowledge and participatory governance structures. The result is a more comprehensive and equitable model for conservation decision-making, rooted in both scientific rigor and the socio-economic realities of fishing communities, which improves conservation outcomes for threatened species and tackles the socio-political imbalances embedded in conventional methods.

Argentina, and Brazil); SE coast of Australia; and the Indian Ocean off India and the Maldives.

A study[5] in New South Wales, Australia, revealed declines of more than 90% in a large suite of deep-sea sharks between the late 1970s and the late 1990s as a result of bycatch in a deepwater trawl fishery primarily for redfish and gemfish. In the Maldives during the early 1980s, a deep-sea fishery targeting gulper sharks for liver oil led to steep declines in landings by the early 1990s. By the mid-2000s, deep-sea shark populations were overfished in the eastern Atlantic. Off the Maldives, gulper sharks declined dramatically following the development of a fishery in the early 1980s for shark liver oil.

A 2024 study[6] found that one in seven (14.1%) deep-sea shark and ray species are Threatened, driven by fisheries for shark liver oil and meat. One-third of gulper shark species, a group prized for liver oil, were assessed as Threatened. Almost 2/3 (62%) of deep-sea species were assessed as Least Concern, likely driven by three large families of small, egg-laying chondrichthyans with high fecundity (softnose skates, catsharks, and deepwater catsharks).

Life-history data are available for only a very small percentage of deep-sea sharks, rays, and chimaeras. What information there is, however, suggests that many of the live-bearing deep-sea taxa (such as gulper sharks, dogfishes, sleeper sharks, and some lantern sharks) have extremely conservative life histories. Females reach sexual maturity slowly, and while interspecific variation exists, the age at maturity can be decades long. Fecundity is often limited to fewer than 10 pups (sometimes only one), and gestation periods are among the longest of any vertebrate (e.g., perhaps 3.5 years in the Frilled Shark [*Chlamydoselachus anguineus*]). Reproduction is often continuous and asynchronous. As a result, these species have some of the longest population doubling times of any chondrichthyan, and therefore the lowest rebound potentials if overfished. Unfortunately, most of the species with such conservative life histories have not been assessed for the IUCN Red List due to a lack of information necessary for a reliable assessment.

Concerns over the extremely long recovery trajectories for deep-sea sharks have resulted in some efforts to manage their stocks and limit harvest in certain regions. In the Maldives, for example, all shark fisheries, including the deep-sea demersal fishery targeting gulper sharks, have been closed. However, the closure of targeted shark fisheries is likely to have an insignificant effect on deep-sea chondrichthyan stocks as long as there are deep-trawl fisheries that capture and discard these fish as incidental bycatch. However, the European Union (EU) implemented numerous regulations to halt overfishing of deep-sea chondrichthyans, including a 2007 prohibition on directed fishing for deep-sea sharks, a 2009 ban on gillnets set deeper than 600 m, and a 2016 ban on trawls deeper than 800 m. Such regulations by the EU and elsewhere have had positive effects on overfishing. However, we know very little about the survival rates of deep-sea sharks that are released after capture, but the evidence suggests, particularly for small species, that very few survive, even if they are released relatively unharmed.[7]

Closing areas of deep sea to trawling, such as implemented by the EU, may offer the only effective management mechanism for limiting mortality in these inherently vulnerable species.

Unsustainable Fishery Practices

Other issues impact elasmobranch populations and complicate managing them. These include bycatch, multispecies fisheries, varied life histories, and species that may be rare, poorly described, or imperiled.

Bycatch (fig. 11.7) includes juveniles of the target species that are too small to eat, as well as nonfood or nontargeted species. In some circles, bycatch is also known as *bykill* because, while some incidental catch is returned to the water and may survive, many, perhaps most, will die in nets or on longlines or are returned to the water dead or dying.

As many as 25,000 non-recreational fishing boats ply US waters and there are an estimated 4.6 million fishing vessels globally, most less than 24 m (79 ft) in length, and most in Asia (3.5 million) and Africa. Thus, bycatch is a big issue affecting many species, including sharks.

Few fisheries catch only a single species. If the goal of fisheries was overall sustainability of marine ecosystems and their inhabitants, then fisheries should be managed to ensure viability of populations of the most vulnerable taxa encountered, whether bycatch or targeted, rather than the sustainability of the most valuable stock. For example, if a fishery catches tuna, sea turtles, and thresher sharks, then it should be managed, in decreasing order of conservation importance, for thresher sharks, sea turtles, and lastly, the targeted species, tuna. However, as you might have concluded, this rarely happens, since the targeted species is the most economically valuable. Thus, bycatch is often ignored.

A 2008 study[8] estimated the source of the major threats for different at-risk marine vertebrate taxa (cetaceans, sharks, sea turtles, and seabirds). Bycatch was considered the principal current threat for all the groups, followed by directed harvest (i.e., targeted fisheries), habitat degradation and destruction, pollution, and aggregated human disturbance (persecution, noise pollution, and so on).

Most shark and other fisheries are not perfectly *clean* (i.e., the ratio of discards to landings is not zero). A fishery that is perfectly clean catches only the targeted species of marketable size. Fisheries for anchovies, herring, sardines, and menhaden are typically the cleanest, with only 1% by weight bycatch, because the targeted species are found in massive, single-species schools that are easily surrounded and caught by purse-seines. At the other end of the spectrum, the *dirtiest* fisheries are bottom trawls for shrimp and demersal (bottom-associated) fishes like cod, where the targeted catch is less than one-third of the total—i.e., most of the animals captured are discarded as bycatch. Bottom trawling has been likened to fishing with bulldozers and ocean clear-cutting and all species not capable of outswimming the trawl may be caught.

Even relatively clean fisheries, because of their immense scale, can have a large shark bycatch. The bycatch may be of concern *even if sharks are not caught,* because it may include the forage base (food source) for sharks and other species, and thus the bycatch may have indirect consequences for species not captured, or for the ecosystem.

Not all sharks and batoids are equally impacted when caught as bycatch. Those with high reproductive potential (e.g., early maturity, many offspring, short gestation) may be able to withstand higher levels of bycatch mortality, and rebound more quickly, than those with more conservative life-history characteristics (e.g., lower fecundities, longer gestation periods, later maturities), which make populations of these species more vulnerable.

Consider the sawfishes (family Pristidae), batoids which are considered to be among the most imperiled of all chondrichthyans. In the United States, the Smalltooth Sawfish (*Pristis pectinata*) was the first native marine fish to be listed as Endangered under the US Endangered Species Act.

Even though Smalltooth Sawfish are fully protected in US waters, they are still caught incidentally in shrimp trawls along Florida's coastline, and this is believed to be the main source of mortality for adults. The targeted species, shrimp, mature in less than a year and their reproductive potential is high, producing hundreds of thousands of offspring per shrimp annually, whereas the Smalltooth Sawfish takes 8–10 years to mature and produces only about 10 offspring, probably every other year. At this rate, only 10–15% of the sawfish population is replaced annually.

At the other end of the spectrum is the Atlantic Sharpnose Shark (*Rhizoprionodon terraenovae*), which experiences extremely high bycatch mortality in US trawl fisheries. However, because it matures quickly (three years) and has a high reproductive rate, its population has remained strong. Similarly, in the NW Atlantic, the Barndoor Skate (*Dipturus laevis*), whose fecundity is high (about 50 eggs per year), has recovered quickly from high levels of bycatch mortality.

On the other hand, deep-sea squaliform sharks have some of the most conservative and vulnerable life histories. Some gulper sharks (family Centrophoridae) have extremely low reproductive potentials (> 30 years to mature, 1–4 offspring, at least 2-year gestation period). It takes 30 to 60 years for their population to double.

Following is a summary of shark and batoid bycatch, based in part on an article by co-author Grubbs in *Save Our Seas* magazine in 2016 (fig. 11.7).

Trawl Fisheries

Trawls are 10–130 m (33–426 ft) long, funnel-shaped nets that are towed behind one or two fishing boats (fig. 11.9). Trawls collect virtually anything in their paths that cannot actively avoid the net or that do not slip through the mesh. As the trawl moves forward, its contents are pushed to the rear, or *cod end* of the net.

Figure 11.9. Dead Basking Sharks on the deck of a bottom trawler in the North Atlantic (Iakov Kalinin/Shutterstock).

Trawling is practiced worldwide on virtually every different bottom type[9] and in mid-water as well. According to the US National Academy of Sciences,[10] repeated trawling and dredging result in harmful changes in benthic communities, some likely permanent. Saturation trawling, in which the net is repeatedly fished in an area until virtually no fish or shrimp are left, has been compared to clear-cutting a forest. Approximately 100,000 km^2 (38,000 mi^2) of terrestrial forest are clear-cut annually, whereas an area 150 times as large is trawled.

Trawls that are fished mid-water, also known as *pelagic trawls*, catch squid, sardines, pollock, and other schooling fishes and have relatively low bycatch. Some pelagic squid fisheries, however, often catch squid predators like Crocodile Sharks (*Pseudocarcharias kamoharai*) or Sharpnose Sevengill Sharks (*Heptranchias perlo*).

Bottom-trawls are generally dirty, damaging marine habitats and taking excessive bycatch, as much as 90% of the trawl contents, which are sometimes called *trash fish* by fishers.

Mortality often approaches 100% for organisms trapped in trawls because trawls can be pulled for several hours.

Shrimp trawl fisheries in the United States catch numerous species of small coastal sharks, including Atlantic Sharpnose, Bonnethead (*Sphyrna tiburo*), and

Blacknose Sharks. Stock assessments for these species indicate shrimp trawl bycatch dwarfs all other sources of fishing mortality.

Bottom-trawl fisheries may be even more harmful to skates and demersal sharks. In the NE Atlantic and Mediterranean, declines in the populations of highly endangered Angel Sharks (family Squatinidae) and guitarfishes (Rhinobatidae) have been attributed to bottom-trawl groundfish fisheries.

Some bottom-trawl fisheries target elasmobranchs and catch other elasmobranchs as bycatch. Two species of skate (Winter Skate [*Leucoraja ocellata*] and Little Skate [*Leucoraja erinacea*]) are targeted in the NE US for their wings (pectoral fins) for human consumption and as bait for lobster traps. Five other species are caught as bycatch, one of which (Thorny Skate, *Amblyraja radiata*) has become severely depleted by this fishery.

Squaliform (dogfish) sharks and catsharks (family Scyliorhinidae) constitute significant bycatch in deep-sea bottom-trawl fisheries. The former are often utilized (for meat and livers), but their populations can be quickly depleted. Gulper Sharks, dogfishes, sawsharks, Angel Sharks, and Sevengill Sharks, all experienced population declines as a result of deep-sea trawling off the coast of New South Wales, Australia.

There has been little research into how the bycatch of sharks in trawl nets can be reduced. Although methods to reduce sea turtle bycatch have been developed successfully—and may also be moderately effective for large batoids such as stingrays—it is unlikely that they could be modified for small coastal and deep-sea sharks. In the case of the former, the effect of limiting tow time on post-release survival should be explored. However, restrictions on the type of trawling gear used and the closure of specific areas to trawling may be the only viable mechanisms for reducing elasmobranch bycatch.

Gillnet Fisheries

Gill nets rely on fishes swimming into them and getting entangled. Their rates of bycatch are very high, especially for nets with smaller mesh.

Like bottom-trawl fisheries, the bycatch in gill nets comprises a wide variety of species (fig. 11.10); some researchers estimate that globally gill nets are responsible for more bycatch mortality of marine mammals, sea turtles, and sharks than any other gear. High seas drift gillnet fisheries, such as those for flying squid and salmon in the North Pacific, which have been criticized for their extremely high rates of marine mammal and seabird bycatch, but their rates for shark bycatch are no less extreme. It has been estimated that approximately two million sharks, primarily Blue and Salmon Sharks, were caught in the squid drift-net fishery in the North Pacific in 1990 alone.

Gillnet fisheries for coastal fishes also catch large numbers of stingrays and small coastal sharks such as sharpnose and smoothhounds, as well as juveniles of large coastal species like Blacktip and Bull Sharks (*Carcharhinus leucas*). A recent analysis of bycatch in drift gill nets for Spanish Mackerel, and in sink gill

Figure 11.10. Dead shark in an abandoned, or ghost, gill net (NOAA Fisheries).

nets for drums and Spanish Mackerel along the SE US, found that more than 20 species of sharks and rays were caught, including prohibited species such as Atlantic Angel Sharks (*Squatina dumeril*), Dusky Sharks (*Carcharhinus obscurus*), Sandbar Sharks (*C. plumbeus*), and Manta Rays. Juvenile Sandbar and Dusky Sharks taken as bycatch in gillnet fisheries significantly hindered recovery of these species.

A ban on the use of gill nets, particularly in locations where there are vulnerable species or life stages, is clearly the most straightforward way to reduce bycatch. Following the ban on gillnet fisheries in Florida 30 years ago, researchers are now reporting signs that the Smalltooth Sawfish population is beginning to recover. An emerging problem is the use of gill nets by artisanal fishers in developing nations, which are difficult to regulate. These can have major conservation implications for species that are endemic or have small regional distributions, such as river sharks (*Glyphis* spp.) and freshwater rays.

Longline Fisheries

Longlines are an effective fishing gear that can be employed across various habitats from rivers to shallow coastal waters, the open ocean and the deep sea (fig. 11.7). Unlike trawls and gill nets, longlines are selective for species that can be lured to take a baited hook, which include most predatory fishes.

A longline consists of a mainline to which a series of branch lines (also called *gangions* or *snoods*) is attached. Each branch line terminates in a baited hook. The mainline generally ranges from hundreds of meters long with only 20 or so hooks in some near-shore and deep-sea fisheries to more than 100 km (60 mi) long with thousands of hooks in pelagic fisheries.

Pelagic longlines are not anchored but are set adrift and marked with highfliers (floats with a radar reflector, a strobed light, and possibly a radio transmitter) to locate the ends. They employ a combination of floats, weighted branch lines, and varied branch line lengths to reach the depths of the targeted species. Bottom-set (demersal) longlines are anchored at both ends and marked by a buoy at one or both ends. The branch lines for pelagic longlines may be as much as 10–20 m (32–64 ft) long, whereas the branch lines for demersal longlines generally range from only 20 cm (8 in) to 3 m (10 ft) long.

Pelagic longline fisheries are often seen in a negative light by environmental groups because of their bycatch, although they are relatively clean when compared to bottom-trawl and gillnet fisheries in terms of their discard-to-catch ratio. In US pelagic longline fisheries, for example, only 3–15% of the catch is discarded. However, as in trawl fisheries, the disparate life histories of target and bycatch species are a concern in longline fisheries. Targeted pelagic fishes, such as tunas and Mahi-mahi, mature early and produce many offspring, resulting in population doubling times in the order of two years, whereas the bycatch often comprises charismatic species, such as pelagic sharks and sea turtles, that have much more conservative life histories and population doubling times that may be much greater than those of the targeted species.

Sharks are often the dominant bycatch in pelagic longline fisheries. In the tuna fishery in the W. Tropical Pacific, shark bycatch has been shown to be relatively high—approximately one shark for every two tunas caught. However, a comparison of many pelagic longline fisheries has suggested that shark bycatch rates were lowest in high seas fisheries that fish deeper, targeting tunas (such as those of Japan, Fiji, and Hawaii), and were highest in fisheries that fish shallower or closer to shore (such as Chile's Mahi-mahi and Hawaii and Chile's swordfish fisheries).

Although the number of individual sharks caught may be very large, relatively few shark species are affected by pelagic longlines in comparison to bottom-trawl fisheries. Globally, the Blue Shark is the dominant species caught in pelagic longline fisheries, followed by the Silky Shark (*Carcharhinus falciformis*) and the Oceanic Whitetip (*C. longimanus*). Most of the sharks that end up in the international fin trade are bycatch in pelagic longline fisheries and there is concern that these shark species may become depleted.

In most pelagic longline fisheries, the Blue, Silky, and Oceanic Whitetip Sharks are reported being discarded alive. Marketable species such as the Shortfin Mako (*Isurus oxyrinchus*) and the three threshers are typically sold (if allowed) and, given their conservative life histories, are of great management concern. It is also important to recognize that some pelagic fisheries retain all the sharks taken as bycatch. For example, the Swordfish fishery off Uruguay, which has an extremely high rate of shark bycatch, typically markets more than 95% of the Blue Sharks caught.

Several mitigation measures to reduce shark bycatch on pelagic longlines have been assessed, though their effectiveness often depends on the species involved. The use of monofilament instead of steel leaders and of squid as bait instead of fish have been shown in some sharks to reduce the number caught, with relatively little—or even a positive—effect on the catch rate of the targeted tunas and swordfish.

The depth of the hooks is often an important consideration. The majority of pelagic sharks, such as Silky and Oceanic Whitetips, spend most of their time in the upper mixed layer shallower than 100 m (330 ft). Increasing the depth of hooks to 150 m (500 feet) has been shown to reduce bycatch rates significantly for these and most other shark species. However, Blue Sharks, Bigeye Threshers (*Alopias superciliosus*), and Shortfin Makos make daily excursions to depths of 400 m (1320 ft) and more. For them, factoring in both time of day and depth when deploying hooks may reduce bycatch. More research needs to be undertaken to determine what influences shark bycatch rates in pelagic longline fisheries so that additional mitigation measures can be identified.

At-boat and post-release mortality rates for sharks caught on pelagic longlines vary widely, depending on the species. Some sharks, such as the Smooth (*Sphyrna zygaena*), Scalloped (*S. lewini*), and Great Hammerheads (*S. mokarran*) and all the threshers, suffer at-boat mortality rates of at least 25% and sometimes more than 50%, and it is likely that most of the sharks released alive do not survive. In contrast, the at-boat mortality rate of Blue, Silky, and Oceanic Whitetip Sharks—the three species most often caught on pelagic longlines—as well as of makos, is only 5–20%, and the few data available suggest that post-release survival may be quite high.

Demersal, or bottom-set, longlines are those whose hooks lie directly on the sea floor. Many targeted shark fisheries deploy demersal longlines as their primary gear, but there are also a number of fisheries using the same gear for bony fishes, such as snappers and groupers, that take significant shark bycatch. In fact, the bycatch on demersal longlines can be 50% or more of the overall catch—a much higher overall bycatch rate than that of pelagic longlines.

Demersal longline fisheries likely take more shark species as bycatch than do all other fisheries combined. In the Gulf of Mexico, the two most common bycatch species taken by the US demersal fishery for groupers and tilefish are often Cuban Dogfish (*Squalus cubensis*) and Blacknose Sharks. It is reported that more than 95% of these sharks are released alive, but a recent study[11] has shown that about half of Cuban Dogfish would likely have soon died, even if they were released in a healthy condition. In addition, on some vessels, the hook is ripped from the shark's mouth, breaking the jaw, which probably increases the post-release mortality rate.

The deepest fisheries in the world use demersal longlines, such as those targeting Patagonian Toothfish (marketed as *Chilean Sea Bass*), down to 3000 m (9840 ft) and Greenland Halibut, grenadiers, hake, and ling (2000 m, or 6560 feet). Shark bycatch in these fisheries can be very high for species like Portuguese Dogfish (*Centroscymnus coelolepis*), lantern sharks (Etmopteridae), gulper sharks (Centrophoridae) and their relatives, as well as for numerous species of catsharks (Scyliorhinidae). Some of these species are kept and sold for their livers and meat, but most small sharks are discarded at sea. Mortality is probably 100% in these fisheries as not only are the sharks unable to survive being retrieved from great depth, but their jaws are broken on landing by the auto-line retrieval system that pulls the hooks through a set of steel rollers.

Even in targeted shark fisheries that are reasonably well-managed, bycatch is a major concern. The shark longline fishery in US Atlantic waters targets large coastal sharks and is regarded as one of the best-managed fisheries of its kind in the world. However, more than half of its catch can include small coastal sharks, such as the Atlantic Sharpnose, that may not be marketed, and most are discarded dead. Moreover, species that are prohibited because they are overfished (such as Dusky Sharks) or endangered (Smalltooth Sawfish) are also taken as bycatch. Fortunately, many of these are quite resilient and probably survive capture.

However, a challenging current problem in US Atlantic large coastal shark fishery is the capture of hammerhead species. Capture mortality is very high in these species, therefore capture prohibitions are ineffective in curbing mortality in a multi-species fishery. A recent study[12] showed that 50% of hammerheads are dead after three hours on the hook in this fishery. Co-author Grubbs studied the survival probability of hammerheads released alive and determined that after about 60 minutes on the hook, the likelihood of death increases dramatically.

Purse-seine Fisheries

The largest fisheries in the world are purse-seine fisheries that target small pelagic fishes such as anchovies, herring, and menhaden (fig. 11.7). Historically these fisheries have been relatively clean in terms of bycatch as they typically target dense, single-species schools of fish. Industrialized purse-seine fishing for larger pelagic fishes such as tunas has developed relatively recently, and purse-seine nets are now responsible for about 70% of all tuna landings worldwide.

Purse-seine nets are deployed in the open ocean over deepwater. Those used for tunas are typically 1000–2000 m (3280–6560 ft) long and about 300–650 m (985–2130 ft) in diameter, and as much as 200 m (650 ft) in depth. The seine has a float line and a lead line, and it is positioned to encircle a school of fish, with one or two skiffs pulling it from the larger harvest vessel. Once the float line circle is closed, the opening at the bottom of the net is closed, or pursed, by cinching the lead line and thus preventing fish from escaping. The net volume is reduced, and the catch is hauled aboard the harvest vessel, removed from the net, and placed in the ship's hold. Elasmobranch bycatch in purse-seine fisheries is mainly

Silky Sharks, Oceanic Whitetips, and mantas, with other species caught episodically (fig. 11.11).

Floating objects that drift around oceanic gyres, such as trees or logs that have been swept into the sea from rivers, or man-made, like lost fishing nets or pieces of wrecked vessel, tend to attract large communities of organisms, as many as 300 fish species, that seek refuge. Pelagic predators such as tunas and sharks in turn are attracted to these floating objects and the potential prey they harbor. Bycatch rates in nets set on these floating objects are much higher than on other sets. The most vulnerable component of this bycatch is pelagic sharks and mantas, as well as sea turtles.

Fishers also exploit the tendency for tunas to gather at floating objects by deploying purpose-built fish aggregating devices (FADs) to attract them (see fig. 10.2). These FADs may be floating or anchored and as simple as palm fronds tied together or as complex as large structures with radio or satellite locator beacons and integrated sonars that enable fishers to estimate the biomass of aggregated fish. Pelagic sharks, big and small pelagic fishes, and large amounts of undersize tuna discards make up most of the high bycatch of FADs.

Figure 11.11. Blue Shark caught as bycatch in a purse seine (Tara Lambourne//Shutterstock).

An interesting case study on the complexities of fisheries management, as well as the effects of well-meaning but misguided environmental activism, concerns tuna purse-seine fisheries in the Eastern Tropical Pacific (ETP). There, tunas associate with dolphins, which is thought to benefit the tunas, which can take advantage of the dolphins' superior prey-finding abilities. Once these assemblages are located, seines termed *dolphin sets* are then positioned around the combined schools.

The targeting of dolphin-associated tunas generated significant controversy in the 1960s due to the high rate of dolphin mortality in purse-seines and estimates of rapidly declining dolphin populations. However, by the late 1980s, dolphin mortality dropped substantially after the fishery developed its own technique to reduce dolphin bycatch.

Although dolphin mortality was already in a steep decline, the *dolphin-safe* tuna labelling campaign was launched, which led to laws forcing US canneries to adopt dolphin-safe policies in 1990, which required them to buy only tunas that had been caught by methods that did not involve encircling dolphins. By 1993, dolphin mortality had decreased to almost zero, thanks to the procedures developed by the fishery prior to dolphin-safe labelling, and it has remained extremely low for more than 20 years.

In an unintended consequence, however, dolphin-safe labelling led the ETP purse-seine fleets selling tuna to US markets to develop the use of FADs; by 2009, 95% of floating objects used by purse-seiners were FADs. Bycatch increased from 0.5% of the catch in dolphin sets before 2009 to nearly 10% using FADs and included nearly 3500 sharks in the former but > 35,000 sharks in the latter. Capture of one dolphin in dolphin-associated purse-seine nets was equal to the bycatch of approximately 25 sharks and more than 900 bony fishes in FAD-associated purse-seines. FAD-associated purse-seine fisheries now account for more than half the global landings of tuna.

Sharks and mantas are the most vulnerable FAD bycatch because of their life histories. In some regions, more than 20 shark species are caught in purse-seine nets. Of these sharks, Silky and Oceanic Whitetip Sharks dominate worldwide, followed at some distance by Scalloped and Smooth Hammerheads.

In most purse-seine tuna fisheries, sharks caught must be released alive. However, these sharks are mostly pelagic and, as obligate ram ventilators, must swim constantly so that sufficient water passes over their gills to oxygenate them. Thus, they tend to be quite fragile and soon die if they are not actively swimming. For this reason, pelagic sharks caught on longlines have a relatively high survival rate, whereas the rate for sharks taken by purse-seine fisheries is likely very low. Numerous recent studies[13] examined post-release mortality in silky sharks caught in tuna purse-seines set around FADs and all estimated the total mortality to be 80% or more.

The problem of bycatch in purse-seine fisheries using drifting FADs is compounded by the fact that the structure of the device itself causes additional mortality. In many drifting FADs, an old fishing net hangs down from the structure and marine life, including sea turtles and sharks, becomes entangled in it. These deaths are not included in fishery bycatch estimates, but they can be substantial. A recent study[14] conducted in the Indian Ocean estimated that the mortality of Silky Sharks entangled in FADs may be 5 to 10 times the actual bycatch in the associated purse-seine fishery.

Bycatch and how to reduce it is one of the most difficult and complex issues faced in fisheries management. Relatively few fisheries exist where bycatch is not a major concern, and it is particularly troubling when it includes elasmobranchs whose life histories are more conservative than those of the targeted catch, making them more vulnerable to overfishing. As you may now realize, the real cost of seafood often exceeds the monetary cost of the product.[15]

Historical Directed Shark Fisheries

Shark fisheries in the early to mid-20th century have provided historical context to what can be expected if the life-history characteristics of the targeted species are not considered. We give you two examples of classic historical case studies here.

Porbeagle

The most famous, perhaps infamous, historical case is the Porbeagle (*Lamna nasus*) fishery in the NE and NW Atlantic.[16] Porbeagles (family Lamnidae) are a cold-temperate shark found across the North Atlantic and in the southern hemisphere. They reach about 3.5 m (11.6 ft). Females mature at about 13 years and have about four pups every other year.

The commercial fishery for Porbeagles in the NE Atlantic began in the 1920s, with annual harvests fluctuating between zero and nearly 4,000 tonnes (8.8 million lb). Norwegian longliners initiated a fishery for the NW Atlantic Porbeagle stock in 1961, a population that was considered virgin (i.e., unexploited). Reported landings for 1961 were 1924 tonnes (4 million lb). By 1964, at its peak, the landings had reached 9360 tonnes (20.6 million lb).

The fishery collapsed in 1967, and by 1970, total reported catch in that fishery was below 1000 tonnes (2.2 million lb), and the average size of the sharks in the fishery had declined. In the 1970s and 1980s, a smaller, more sustainable fishery for the species of about 250 tonnes (550,000 lb) continued in the Faroe Islands of the North Sea.

The species began to very slowly recover until the early 1990s, at which time Canada and the Faroe Islands increased their fishing effort, although not to the 1960s levels. Still, the Porbeagle population in the North Sea could not sustain the fishery. Porbeagles were listed under CITES appendix II in 2014 which required all countries fishing the North Atlantic to release Porbeagles. As a result, landings of both the NW and NE Atlantic Porbeagle stocks have been near zero since 2015. Directly due to these measures, the latest stock assessments indicate overfishing is no longer occurring, though both the stocks remain overfished (i.e., they are recovering but have not recovered). This example demonstrates how decades are required for a species to recover from the development of unregulated fisheries. The IUCN considers the Porbeagle Critically Endangered in the NE Atlantic Ocean and Mediterranean Sea, Endangered in the NW Atlantic, and Vulnerable globally.

California and Australian Soupfin (Tope) Shark

The fishery for Soupfin Sharks (*Galeorhinus galeus*), also known as Tope, in the 1930s and 1940s in California was the biggest US west coast shark fishery. There was also a fishery for this species at about the same time along the southern coast of Australia. What drove this fishery was the need for a source of vitamin A when cod imports from Europe stopped during World War II. The livers of this species had among the highest potency vitamin A of any shark species.

Soupfin Sharks were caught using large-mesh, bottom-anchored gill nets or longlines. Prior to the increased demand for vitamin A, the fishery was a small one for fresh fillets, fins for the Asian market, or for rendering for fertilizer. The average annual catch of Soupfin Sharks between 1930 and 1936 was under 272,000 kg (600,000 lb). Data on landings of Soupfin Sharks between 1936 and 1941 have been difficult to locate. In 1939, landings of all sharks in California were about 4170 tonnes (9.2 million lb). Soupfin Sharks comprised about 52% of the catch from 1941–1944. Using that figure, landings for Soupfin Sharks at the peak were over 2040 tonnes (4.5 million lb). Though Soupfin Sharks are a medium-sized coastal species, they have a triennial reproductive pattern (reproducing every three years), which renders them particularly vulnerable to overfishing. The California fishery collapsed by 1946, partially due to declining stocks associated with overfishing as well as the development of synthetic vitamin A, which obviated the need for the fishery.

It is worth noting that since this fishery developed so fast and prices for livers was so high, virtually anyone with a boat began fishing for Soupfin Sharks. A story by the *Santa Barbara* (California) *Independent*[17] quotes locals Ralph Hazard and Red Allen: "'It took about two or two-and-a-half years, but we cleaned them all out,' said Hazard. Allen added, 'And we weren't taking only soupfins.' By 1943, sharks in the Santa Barbara Channel 'were just about all wiped out.'"

CASE STUDIES

Contemporary Case Studies

US Atlantic Shark and Ray Landings

The development and management of commercial fisheries for sharks and rays in US Atlantic waters offer examples of a suite of contemporary case studies illustrating a pattern of serially depleted stocks, subsequent management plans (often in response to campaigns from environmental advocacy groups), and stock rebuilding or recovery. Prior to the Fishery Conservation and Management Act of 1976 (currently the Magnuson-Stevens Reauthorization Act), foreign fleets frequently exploited shark stocks in US waters but US shark fisheries were very small, and the sharks harvested were primarily Spiny Dogfish caught primarily for their livers for the production of vitamin A and liver oil. In the early 1980s, in response to plummeting US stocks of groundfish-like cod off New England, fisheries managers encouraged the development of fisheries targeting Spiny Dogfish to supply *fish and chip* markets in Europe.

Also, during the 1980s, directed fisheries for skates, primarily for use as bait in lobster traps, were also developing in the NE United States. In addition, the lifting of US trade limits with China in the late 1970s opened up a huge market for US fishery products, including shark fins, leading to the development of directed fisheries for coastal and pelagic sharks. In response to these trends, commercial shark and ray landings in US Atlantic waters rose from less than 900 MT (~20,000 lbs) in 1976 to more than 32,000 MT (70 million lb) by 1990 (see fig. 11.8B), an increase of 3,500%! This does not include recreational shark and ray landings, which were on par with commercial landings.

By 1989, 39% of US Atlantic landings were coastal and pelagic sharks caught in unmanaged fisheries to supply the fin trade, whereas 24% were Spiny Dogfish, and 36% were skates. Due to concerns over rapidly declining populations of large coastal shark species such as Dusky and Sandbar Sharks, among the most sought-after species in the shark fin soup market, a fishery management plan for the sharks, the first such effort, was implemented in 1993, with restrictive quotas. By 1996, the quota on large coastal sharks reduced landings to 25% of their peak.

However, the fishery for Spiny Dogfish to supply fish and chip markets in Europe continued to expand rapidly and, by 1996, Spiny Dogfish made up 60% of the shark and ray landings, skates made up 30%, and only 10% were coastal and pelagic sharks. The Spiny Dogfish stock was soon overfished, and a management plan was implemented in 1998, which began limiting their landings. By 2003, regulated Spiny Dogfish landings were less than 10% of the peak landings and contributed less than 10% to the overall US Atlantic shark and ray landings. Due to a steadily increasing fishery for bait as well as for "wings" for use as food, by 2003 skates constituted nearly 75% of shark and ray landings. Again, evidence of severe overfishing of at least three species of skates in this fishery led to skate management and a rebuilding plan, implemented in 2003.

Currently, all three groups of US Atlantic sharks (large coastal sharks, small coastal sharks, and pelagics) and rays are under management. By 1999, only 16 species of large coastal and pelagic sharks could be harvested, and 19 species were prohibited. By 2016, large coastal sharks were managed at a quota that was roughly 8% of their peak landings, and the most vulnerable species remained prohibited. Most fishers have exited this fishery as it became economically unsustainable and over the last decade, the quota for large coastal species has not been met in any year. By 2016, coastal and pelagic sharks contributed only 6% to the total US Atlantic shark and ray landings, whereas skates continued to be the dominant component at 53%. Due to the apparent recovery of the Spiny Dogfish stock, landings in this fishery increased, making up 39% of the 2016 commercial landings, but the most recent stock assessment (2022) suggests this stock is now fully exploited and is close to overfished thresholds. The astute mathematically-minded reader may note that these numbers add to 98%. The remaining 2% of 2016 shark and ray landings were Dusky Smoothhounds that were landed in a targeted fishery, which accelerated in the early 2000s, mostly off North Carolina. All the catch of this species went to Australia and the UK as *flake fish* (i.e., boneless fish fillets). Landings peaked for this single species at nearly 1800 tonnes (4 million lb) in 2010, surpassing landings of all large coastal shark species combined. The science-based advocacy group Shark Advocates International brought this fishery to the attention of NOAA and urged them to act to study and protect the stock. In a rare effort to manage shark fisheries before they are overfished, a 2014 stock assessment found that the population was healthy with no overfishing occurring, though it was close. As a small coastal species with about a 5-year doubling time (mature at three years, produce 15 pups annually), the species is more resilient. Landings in this fishery are now managed with a quota of 504.6 tonnes (1.11 million lb).

As you can see, the simple examination of landings does not present the whole story. In some cases, declines in landings may be signs of trouble for a stock, but in others they may reflect effective fisheries management.

It Is All in a Name

In our Authors' Note at the start of this book, we mentioned that common names were much less useful than scientific names. Here is a case where the use of common names was in fact *dangerous* for one population of sharks, namely Spiny Dogfish. Until about 2002, Spiny Dogfish and Dusky Smoothhound (also called the Smooth Dogfish) were both heavily fished along the US Atlantic Coast. Since both were small sharks often caught on the same gear, and both had *dogfish* as part of the common names, they were classified for fishery statistics in a single category as *dogfish*.

What harm, you might ask, could that possibly cause? Lots. The two species occupy two different superorders (Spiny Dogfish are squalomorphs, Dusky Smoothhounds are galeomorphs) separated by about 210 million years of evolution and, what is more, have drastically different life-history characteristics. Spiny Dogfish take as long as 20 years to mature and have only 4–6 pups every other year, after a 24-month gestation period. Dusky Smoothhounds mature in 3–4 years and have 15 pups annually after a gestation period of 8–9 months. Fortunately, after stocks of the former plummeted, their fisheries were classified and managed separately.

Save the Bay, Eat a Ray: A Cautionary Tale About Science and Elasmobranchs, with a Barb

We conclude this chapter with an introduction to a case study in which a paper[1] published in the journal *Science* in 2007 led to a wrongheaded movement to eradicate an elasmobranch with one of the lowest known reproductive potentials. The species is the Cownose Ray (*Rhinoptera bonasus*), and the paper, entitled *Cascading effects of the loss of apex predatory sharks from a coastal ocean*, claimed that large coastal sharks (Sandbar, Blacktip, Bull, Dusky, Scalloped Hammerhead, and Tiger Sharks [*Galeocerdo cuvier*]) along the east coast of the US had experienced huge population declines which led to what is called *predation release* and a *trophic cascade*.

Predation release here refers to removing predation pressure on smaller sharks and ray mesopredators (mid-level predators; e.g., Atlantic Sharpnose Sharks, Chain Catsharks, Little Skates, Butterfly Rays, and Cownose Rays) by removing their predators (i.e., the larger sharks). The ensuing trophic cascade refers to the reverberations further down the food chain when the population of the mesopredators *skyrocketed*, according to the paper, and the populations of *their* prey dropped.

According to the paper, the largest consequence of the predation release and ensuing trophic cascade was the population explosion of the Cownose Ray and the population crash of its chief prey, the Bay Scallop (*Argopecten irradians*), and the subsequent loss of the North Carolina fishery on which it was based. Another paper[2] by the same research group expanded the list of shellfish affected by the Cownose Ray to include oysters, as well as hard- and soft-shell clams.

A rebuttal to the *Science* paper was published in 2016 in the journal *Scientific Reports*.[3] Here, we summarize parts of that paper, along with additional information, by answering four major questions about the validity of the *Science* paper.

1 / Were the declines in large sharks as severe as reported?

In short, no. The main issue was reliance on a single data set, the University of (UNC) Carolina Shark Survey, for estimates of the declines in large sharks of between 87% and 99%. This data set was rigorously collected, but not appropriate for the breadth of the conclusions. Also, sample sizes of three of the seven large shark species were too small to allow the conclusions made by the *Science* paper. For example, the 99% decline of Smooth Hammerheads was based on a sample size of five sharks over a 32-year period. Only 23 Bull Sharks (also 99% decline) were caught by this survey over the 32-year period, and only 39 Tiger Sharks, for which the paper reported a 97% decline. Reliance on small sample sizes is dangerous as small changes in the catch one year can change the perception of both the direction and magnitude of the change of the population of these species.

The *Science* paper reported declines for Sandbar, Blacktip, and Dusky sharks of 87%, 93%, and 99%, respectively. The rebuttal paper published in *Scientific Reports* reanalyzed the UNC data and included the larger VIMS Shark Survey data set and concluded that there was a 50–60%, not 87%, decline in Sandbar Sharks over the time period. Blacktip Sharks showed the opposite trend. While declining in the UNC data set, Blacktip Sharks were increasing in the VIMS data set, likely because of warming coastal waters pushing them north. The important message is that a single survey is insufficient to draw such sweeping conclusions. Full stock assessments had been conducted on all three of these species, showing more modest population changes in using all available data sets, yet these assessments were ignored in the *Science* paper.

2 / Were the trophic links sufficient to elicit this cascade?

The first requirement for a trophic linkage between the large coastal sharks and the mesopredators is spatio-temporal overlap; that is, they must be in the same place at the same time.

Consider two of the mesopredators whose populations were supposed to have exploded after trophic (or predation) release, the Little Skate (*Leucoraja erinacea*) and Chain Catshark (*Scyliorhinus retifer*). Both of these species inhabit only water temperatures of 5–12°C (41–53.6°F). In the *Science* paper, the sharks considered predators of these two species are all found in warmer waters, with little to no overlap, and thus they could not be primary predators for the Little Skate and Chain Catshark because they rarely co-occur.

Let us look closely at the Cownose Rays. First, do the large coastal sharks eat them? Yes, but in small numbers. In fact, Cownose Rays are eaten more by Cobia than by sharks.

Are Cownose Rays significant natural predators of commercial bivalves? What is their natural diet? A variety of studies have shown that Cownose Rays, depending on the location, eat (in no particular order) a large range of prey, including bay scallops, soft-shell clams, other small bivalves, crustaceans, fish, and a number of other invertebrates, including cumaceans (small marine crustaceans), polychaetes, amphipods, and echinoderms. Commercial bivalves, with the exception of softshell clams, have been shown to be a minor part of the cownose ray diet.

3 / Were increases in smaller sharks and rays credible?

Not necessarily. Recall the case of the Little Skate and Chain Catshark mentioned earlier. Moreover, the Little Skate stock assessments showed either no trend or a declining trend. The Clearnose Skate (*Rostroraja eglanteria*), which is regularly eaten by large coastal sharks, did not show a clear trend in the paper's online supplemental material.

The *Science* paper also showed a clear increase in the Atlantic Sharpnose Shark population in the UNC survey, but the VIMS survey and stock assessment showed no change, which was in line with the federal stock assessment.

Multiple data sets were examined for Cownose Rays and most showed increases in the population. The problem with assessing Cownose Rays is that they travel in enormous schools of tens of thousands of individuals, and no fishing gear samples them effectively. The sample sizes for all except one of the surveys was small. The survey with the larger sample size showed a moderate increase in the population. Cownose Rays mature at 7–8 years, live to a maximum of 21 years, have a 12-month gestation period, and produce a single pup annually. The population doubling time for Cownose Rays is decades suggesting the population increases reported in the *Science* study are not biologically possible. What probably did occur was shifts in the population similar to what was suspected for Blacktip Sharks in the study.

4 / Was the stated increased Cownose Ray abundance responsible for declines in commercial bivalve populations?

Bay scallops and oysters have been subject to industrial scale fisheries for many decades. One of the more convincing lines of evidence refuting the claim that Cownose Rays were responsible for declines in bay scallops between 1970 and 2005 was that the bay scallops were in steep decline beginning in the 1980s; that is, *before* the purported increases in Cownose Ray populations in the 2000s occurred. One observer commented that the rays would have had to travel back in time to cause the crash. In addition, it was shown that oyster and bay scallop stocks had crashed coast-wide, mostly before 1990, including in regions outside of the range of Cownose Rays. The cause of the collapse of these stocks, including the century-long scallop fishery in North Carolina, was likely a combination of overfishing, loss of seagrass habitat, hurricanes, red tides, wasting disease, pollution, as well as a poorly regulated recreational fishery.

Save the Bay, Eat a Ray—The conflict between shellfish fisheries and Cownose Rays was nothing new. Since the 1970s when stocks of oysters and bay scallops began to collapse in North Carolina and Virginia, there have been calls every few years to implement fisheries for Cownose Rays to eliminate these perceived competitors for the valuable shellfish. These calls were usually dismissed over concerns for the vulnerability of the rays to overfishing, but the *Science* paper was taken by the industry as support for reducing the Cownose Ray population. A lot of effort was expended to develop commercial fisheries and associated markets for the meat and other products (e.g., leather from the skin). The meat was marketed domestically as well as in Europe and Korea. Bow-fishing tournaments were implemented to take place when pregnant female Cownose Rays entered the estuaries in order to kill both the mother and pup at once. These efforts were promoted under the guise of environmental sustainability through the *Save the Bay, Eat a Ray* campaign. Though no stock assessments had been conducted on Cownose Rays, it was known that they have among the most conservative life histories of any shark or ray, rendering their population extremely vulnerable to overfishing. The effects of these fisheries on the Cownose Ray population are unknown.

NOTES

1. Myers, R.A. et al. 2007. Science 315: 1846–1850.
2. Baum, J.K. and Worm, B. 2009. J. Anim. Ecol. 78: 699–714.
3. Grubbs, R.D. et al. 2016. Sci. Rep. 6: p.20970.

Concluding Comments

Shark and ray fisheries and bycatch remain complex issues of major concern, involving many species and stakeholders. Fishery managers have the knowledge required to assess the population status of shark and ray stocks and have used this knowledge to study and help rebuild some overfished stocks, both domestically and globally. However, these gains occur against the overwhelming backdrop of climate change and other human insults, which we will examine in the next chapter.

NOTES

1. Dent, F. and Clarke, S. 2015. State of the global market for shark products. FAO Fisheries and Aquaculture technical paper (590), Rome, FAO, 187 pp.
2. "International Shark Finning Bans and Policies" https://awionline.org/content/international-shark-finning-bans-and-policies. (Accessed 9/15/25).
3. Beverly, S. et al. 2003. Horizontal longline fishing methods and techniques: a manual for fishermen. Secretariat of the Pacific Community. 130 pp.
4. Finucci, B. et al. 2022. *Deepwater Chondrichthyans. In Biology of sharks and their relatives.* CRC Press. 603–634 pp.
5. Graham, K.J. et al. 2001. Mar. Freshwater Res. 52: 549–561.
6. Finucci, B. et al. 2024. Science, 383(6687): 1135–1141.
7. Talwar, B. et al. 2017. Mar. Ecol. Prog. Ser. 582: 147–161.
8. Finkelstein, M. et al. 2008. PloS One 3: p.e2480.
9. Waitling, L. and Norse, E.A. 1998. Conserv. Biol. 12: 1180–1197.
10. Ocean Studies Board and National Research Council. 2002. *Effects of trawling and dredging on seafloor habitat.* National Academy Press.
11. Talwar, B. et al. 2017. Mar. Ecol. Prog. Ser. 582: 147–161.
12. Gulak, S.J.B. et al. 2015. Afr. J. Mar. Sci. 37: 267–273.
13. Eddy, C. et al. 2016. Fish. Res. 174: 109–117.
14. Filmalter, J.D. et al. 2013. Front. Ecol. Environ. 11: 291–296.
15. Abel, D.C. and McConnell, R.L. 2012. J. Manage. Contr. 22: 481–485.
16. O'Boyle, R.N. et al. 1996. Atl. Fish. Res. Doc. 1996/024.
17. Stewart, E. 2008. World War II, Santa Barbara's Waterfront, and 2,000 Shark Livers a Day. https://www.independent.com/news/2008/aug/14/world-war-ii-santa-barbaras-waterfront-and-2000-sh/. (Accessed 9/15/25)

12 / Climate Change and Other Human Impacts

CHAPTER OUTLINE

Introduction

Overfishing may be the biggest current threat to sharks, but it is far from the only one. Sharks also face the same anthropogenic environmental insults as other marine organisms—climate change that causes waters to deoxygenate, warm, and acidify; nutrient pollution from agricultural runoff and undertreated human sewage; micro- and macro-plastic pollution; other types of pollution (other chemicals, noise, radioactivity, etc.); habitat destruction; and loss of biodiversity in the ecosystems they inhabit. What's more, the short- and long-term impacts of these are not all known. They may be negligible, or they can range from sublethal impacts that affect an organism's overall fitness (e.g., its physiological functions and behavior) to mortalities that cause declines in populations.

We lump these impacts together as *pollution*, plus physical habitat loss. We consider *climate change* below separately from pollution, although it is clearly a form of pollution.

As we survey these human impacts one-by-one, you may be frustrated by the unsatisfying lack of clarity about specific impacts on sharks. This lack of clarity is explained by several reasons, some of which are here:

- Many reports of impacts on sharks are anecdotal; for example, opportunistic encounters with one or more dead sharks, such as strandings. Conclusions based on anecdotal occurrences may be useful, but they typically involve low numbers of sharks, which may be at an advanced stage of decomposition, and which also may be temporally and physically distant from the source of the pollution.
- Controlled studies (i.e., exposing captive sharks to or administering doses of potentially harmful substances or biological agents) are logistically difficult and expensive and thus are too infrequently conducted.
- Even if controlled laboratory studies are undertaken, extrapolating from these to ecosystems or generalizing from these studies are both fraught with potential problems because of our limited knowledge of shark ecology.
- Studies on captive animals most often employ species or life stages capable of surviving in experimental systems, and these species do not represent the diversity of forms, physiology, and habitats of sharks as a group. Examples of more commonly studied sharks include juvenile Lemon Sharks (*Negaprion brevirostris*), Spiny Dogfish (*Squalus acanthias*), horn sharks (e.g., *Heterodontus francisci*), and Small-spotted Catshark (*Scyliorhinus canicula*).
- Correlations are easier to establish than cause-effect relationships. In other words, it may be fairly straightforward to relate the presence/absence of sharks or pathological conditions in individuals to levels of pollution, but it is much more difficult to demonstrate that some chemical or combination *caused* the absence or condition.
- Many sharks are migratory, and thus determining the level and duration of exposures to the suite of pollutants encountered during these movements is next to impossible.

A Few Principles of Toxicology

Let us begin with toxicology which, according to the American Society of Toxicologists, is the study of the adverse physicochemical effects of chemical, physical, or biological agents on living organisms and the ecosystem. The agent that causes the adverse effect is known as the *toxicant* or *pollutant.*

Chemicals, including radioactive ones (*radionuclides*), can affect organisms by being:

- *Toxins*—poisonous chemicals of biological origin (e.g., *brevetoxin* [see p. 440]), produced by the dinoflagellate *Karenia brevis*).
- *Carcinogens*—cancer-causing agents. An example is the class of chemicals known as polychlorinated biphenyls, more commonly called PCBs, which were widely used as electrical insulators. Radioactive chemicals like Cesium-137, which is released in nuclear weapon tests and reactor disasters like Chernobyl and Fukushima Daiichi, may also be carcinogenic.

- *Mutagens*—agents that cause changes in DNA (so named because they cause genetic mutations). Polyaromatic hydrocarbons (PAHs), another class of chemicals widespread in many marine and estuarine systems, and metals like arsenic, cadmium, chromium, and nickel, may also be mutagenic.
- *Teratogens*—agents that cause abnormalities or malformations during fetal development. In addition to being carcinogenic, PCBs are teratogenic, as is tributyl tin, a compound used as an antifouling agent on ships.

In all of these cases, toxicologists seek to determine the dose-response relationship; that is, how organisms respond over a range of concentrations or doses of toxicants. Toxicity may be *acute* (acting in the short term and usually fatal) or *chronic* (long term and typically sublethal). Even if the toxicant is sublethal, it could affect the overall *condition* (i.e., health) of an organism and thus reduce its *fitness* (i.e., its ability to grow and develop, forage, swim, sense its surroundings, avoid predation, mate, etc.). Moreover, sublethal toxicants may act on the offspring as mutagens or teratogens.

Toxicants also have different routes by which they enter organisms; for example, across gills, ingested as food, and absorbed across skin. Additionally, they may be systemic (i.e., acting on the entire organism) or they may be more localized, targeting and affecting different organs (e.g., liver, organs of the central nervous system) or tissues (e.g., muscle, nervous, connective).

Delayed mortality and sublethal effects may occur because organisms *bioaccumulate* some harmful substances. For example, mercury and fat-soluble organic compounds bioaccumulate in organisms by *bioconcentration* (i.e., being absorbed from water or prey and accumulating in tissue to levels greater than those found in surrounding water) and *biomagnification* (i.e., increasing in tissue concentrations as larger organisms consume smaller ones [those lower in the food web] and absorb and retain the harmful substances). In some cases, levels of these compounds decrease over time as a result of processes that detoxify, or depurate, the toxic compound.

In some cases, the toxicant may be detected by the animal and sidestepped, a phenomenon known as *behavioral avoidance*. While the short-term survival of the organism is enhanced, there may be other impacts on the organism, which may find itself outside of its accustomed surroundings.

As you might expect for a group as diverse as the sharks and batoids, thresholds at which the agents we describe here become toxic do not exist.

Categories of Human Impacts and Their Effects on Sharks

In general, water pollution refers to any hazardous or toxic material released by humans into the aquatic environment. We discuss the nine categories here, as well as an additional human impact not included as pollution—habitat destruction. We conclude with a section on climate change.

Disease-causing Agents

This category includes pathogenic bacteria, viruses, protozoans, and other parasites. Note that not all of these constitute pollution; most in fact are naturally occurring (i.e., not introduced by humans). The main anthropogenic source is untreated or improperly treated human waste, but animal waste from feedlots and meat-packing plants are also sources. Often these infections occur opportunistically after impacts of other stressors (e.g., thermal shock or other types of pollution).

The incidence of diseases in elasmobranch fishes apparently is low. Moreover, distinguishing normal shark pathogens from those considered pollutants is not always possible. Here are a few examples of pathogenic infestations in sharks.

- A 2018 study[1] of Leopard Sharks (*Triakis semifasciata*) that had died after stranding themselves on the shores of San Francisco Bay during the spring of 2017 found that they had severe meningoencephalitis—dangerous inflammation of the brain and parts of the spinal cord, likely but not definitively caused by the ciliated protozoan pathogen *Miamiensis avidus*. A study of 168 sharks and rays representing 18 species, mostly stranded animals but also a few aquarium mortalities, reported meningoencephalitis in 70%.[2]
- Nematode worms (also called roundworms) are common endo- and ecto-parasites of sharks (fig. 12.1). A 2002 paper[3] found what the authors characterized as *potentially debilitating lesions* on the pancreases of 10 of 20 Spiny Dogfish (*Squalus acanthias*) caused by the nematode *Pancreatonema americanum*. The authors could not establish a causal relationship between the pathology and pollution.
- Toxins from *harmful algal blooms* (HABs; see Plant Nutrients in this chapter), sometimes also called red or brown tides, have been shown to cause mortality in sharks (fig. 12.2). Increased anthropogenic nutrient input is leading to more frequent or larger scale HABs. In 2000, the cyanobacterium *Karenia brevis*, which produces a potent neurotoxin in the *brevetoxin* family, was responsible for a red

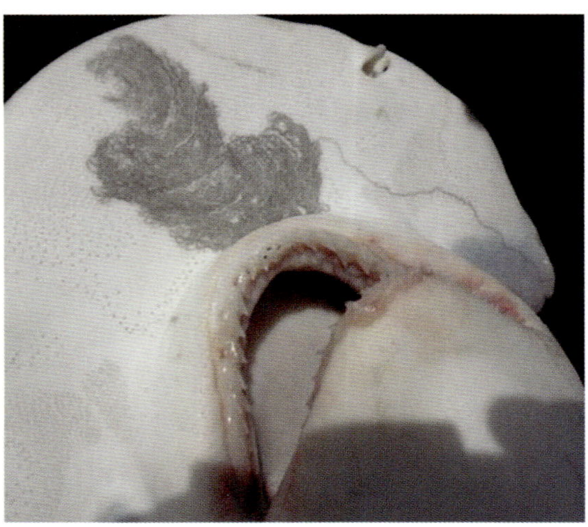

Figure 12.1. Skin of Atlantic Sharpnose Shark infected with nematode ectoparasites (Courtesy of Caroline Collatos).

Figure 12.2. 8 m (26 ft) Whale Shark presumably killed by the 2018 Florida red tide (Courtesy of the Florida Fish and Wildlife Conservation Commission).

tide in NW Florida that caused mass mortality in Blacktip Sharks (*Carcharhinus limbatus*) and Atlantic Sharpnose Sharks (*Rhizoprionodon terraenovae*).[4] Presumably, the toxin entered the sharks from their prey, but the sharks could have also died from the presence of anoxic water that often accompanies HABs.

In 2018, a dead 8 m (26 ft) male Whale Shark (fig. 12.2) washed ashore at Sanibel Island in Florida. Tissues tested positive for brevetoxin, but sufficient knowledge on toxic levels for this species do not exist and thus brevetoxin could not be definitively assigned as the cause of death.

- Sharks in marine aquaria often exhibit infectious and inflammatory diseases. We have observed Tiger Sharks in marine aquaria with prominent external bacterial or fungal infections likely caused by abrasions from contact with the walls of the aquaria. Infections such as these and others are described in the *Elasmobranch Husbandry Manual II*.[5]

Oxygen-demanding Wastes

One way to assess the health of aquatic ecosystems is to measure the dissolved oxygen (DO) in the water. Water with DO of > 3 to 8 parts per million (ppm) is said to be *normoxic* and can support virtually all aquatic life. Water with less than

2 to 3 ppm is said to be *hypoxic* and typically will support only anaerobic and/or relatively inactive fauna. Water devoid of DO is *anoxic* and cannot support vertebrate life except for brief periods.

Oxygen-demanding wastes are those that lead to decreases in the DO level. These include sewage (from both humans and animal factories, which are also considered in the chapter as plant nutrients), paper pulp (a byproduct of manufacturing paper products), and food processing wastes. These substances can either be decomposed by oxygen-consuming microbes (bacteria or fungi), or can be chemically oxidized, in both cases reducing DO.

Dissolved oxygen levels, particularly in nearshore environments, are known to affect the distribution of some elasmobranch fishes; for example, juvenile Bull (*Carcharhinus leucas*) and Sandbar Sharks (*C. plumbeus*) and adult Gray Smoothhounds (*Mustelus californicus*), but not Atlantic Sharpnose (*Rhizoprionodon terraenovae*) and Spinner Sharks (*C. brevipinna*).[6] The lack of an observed effect on the two latter species does not indicate that they can survive in anoxic conditions, but that they may be less sensitive to deviations in DO levels.

A recent, comprehensive review[7] of the responses of elasmobranchs to ocean deoxygenation, regardless of the cause, had the following conclusions:

- Many sharks and batoids avoid hypoxic waters.
- Many species, especially less active and/or benthic species, also tolerate mild to moderate hypoxia.
- Responses to hypoxia include "changes in activity, ventilatory responses, alterations to circulatory and hematological parameters, and morphological alterations to gill structures."
- More severe or prolonged hypoxia may exceed tolerance limits and may also cause shifts in the distribution of sharks and batoids inhabiting warmer waters.

Finally, it is worth emphasizing that environmental hypoxia may have ecosystem-wide impacts.

Water-soluble Inorganic Chemicals

This large category includes toxic metals, acids, bases, and salts. Examples include heavy metals like mercury, lead, tin, and cadmium, as well as selenium, arsenic, chlorine, and nitrates. These derive from the weathering of rocks accelerated by humans by mining, manufacturing processes, oil field drilling operations, and agricultural practices. A problem of increasing concern is ocean acidification due to increased CO_2 levels brought about by burning fossil fuels, deforestation, cement manufacturing, etc.

Let us focus on mercury here and in *Research, Conservation, and Education Focus* Box 12.1. The Food and Drug Administration (FDA) has cautioned pregnant women, women who may become pregnant, nursing mothers, and children on the hazard of consuming fish, including sharks, that may contain high levels of *methylmercury*, the positive ion that mercury is converted to by aquatic microbes.

Predatory fish like sharks accumulate the highest levels of methylmercury and therefore pose the greatest risk. Unlike *lipophilic* (fat-loving) substances like PCBs and dioxin (see pp. 447–449), which accumulate in fat, mercury accumulates in muscle tissue (the part of the fish we eat), so it cannot be avoided by removing fat and skin of fish.

Since industrialization, mercury in the environment has increased by 2–4 times. Sources include burning coal, burning medical and other waste that contain mercury, and gold and mercury mining.

Mercury causes changes in the function of cells, inhibits protein synthesis, and causes formation of reactive oxygen compounds or radicals, which can damage DNA and disrupt cell division. Mercury interferes with the development of the nervous system and affects brain function.

Methylmercury is rapidly taken up but only slowly eliminated from the body, so each step up in the food chain biomagnifies the concentration from the step below. Generally, sharks whose diets include larger predatory fish and mammals have higher levels of mercury and other metals, but this varies by species and metal. Sandbar Sharks, which are bottom associated in many ecosystems, show higher mercury levels because they are found near sediments containing methylmercury and other contaminants.[8]

Two studies[9, 10] examined methylmercury levels in 11 species of sharks in a Florida lagoon. As might be expected, there was significant variability in methylmercury levels among the species. In the Blacknose (*Carcharhinus acronotus*), Blacktip, Shortfin Mako (*Isurus oxyrinchus*), Blue (*Prionace glauca*), and Oceanic Whitetip (*C. longimanus*) Sharks, the methylmercury levels increased with body size and were at levels that could be harmful to these sharks. As meaningful as are the reports of high levels of mercury in sharks, not all studies of methylmercury levels in sharks at the time found high levels.[11] More recently, a review[12] of total mercury in sharks found a strong positive correlation between total mercury concentration and trophic level in elasmobranch muscle tissue, with the highest levels in lamniform and carcharhiniform sharks.

Finally, two studies of heavy metals in the Mediterranean Sea are noteworthy because they included deep-sea chondrichthyans. The first[13], in the Southeastern Mediterranean, found "exceptionally" high levels of total mercury over a four-decade period in the muscle tissues of five species of deep-sea sharks: Blackmouth Catshark (*Galeus melastomus*), Velvet Belly Lantern Shark (*Etmopterus spinax*), Little Sleeper Shark (*Somniosus rostratus*), Gulper Shark (*Centrophorus granulosus*), and Kitefin Shark (*Dalatias licha*). The area experiences a large mercury sink and has a long-term bottom accumulation of mercury.

The second study[14], in the Western Mediterranean Sea, of the heavy metals cadmium (Cd), lead (Pb), and copper (Cu) in muscle of eight species of sharks, eight batoids, and one chimaera, concluded that small-bodied, deep-sea species of low trophic level exhibited the highest levels. Among sharks, the Velvet Belly Lantern Shark for example, exhibited among the highest levels of all three metals,

BOX 12.1

Research, Conservation, and Education Focus

Are High Levels of Pollutant Exposure Affecting Sharks?

James Gelsleichter

Department of Biology
University of North Florida
Jacksonville, FL, US

As discussed in other sections of this chapter, sharks and rays tend to accumulate higher levels of many marine pollutants than most other fishes because of their relatively large size, long lifespans, and positions near or at the top of marine food chains. Still, whether exposure to such high levels of marine pollutants is affecting sharks and rays remains a highly debated question that is currently being researched in several laboratories throughout the world.

A general principle of toxicology, the scientific study of how harmful substances can affect living things, is that chemicals or other items suspected to be toxic can affect animals by interacting with important macromolecules (such as proteins, lipids, and nucleic acids) present in "target" cells. By doing so, toxic substances can interfere with the function and/or maintenance of these cells, causing harmful responses in tissues and organs. Therefore, to determine if marine pollutants are affecting sharks and rays, it is not enough to just study if these animals exhibit elevated levels of potentially toxic substances in tissues such as muscle or blood—we also need to determine whether these substances are "reaching" target cells and affecting the function of target tissues and organs.

A good example of this is research that has been conducted on the toxic metal, mercury, which is one of the most important marine pollutants because of its widespread distribution and tendency to *biomagnify*, or increase with trophic level, in marine habitats. Although it is a naturally occurring element that enters the marine environment through natural processes such as erosion and volcanic activity, mercury can become a pollutant in the environment because of its atmospheric release from several human activities, such as artisanal gold mining, coal combustion for power generation, and waste incineration. Following its release into the atmosphere, mercury vapor can become oxidized to divalent mercury, which can then be deposited in marine habitats and converted to methylmercury by bacteria. Methylmercury is highly persistent in marine ecosystems and readily absorbed into animal tissues, posing risks to marine wildlife as well as human seafood consumers.

Numerous studies have been conducted on mercury accumulation in sharks and have demonstrated that they often exhibit the highest mercury levels of all fishes occurring in a given habitat. However, although mercury levels in sharks are generally above the levels associated with health effects in other fishes, the potential effects of mercury on sharks remain unclear. To address this question more directly, Ehnert-Russo and Gelsleichter[1] studied whether the high levels observed in muscle of a small coastal shark, the Atlantic Sharpnose, also occur within a known mercury target organ—the brain—and whether mercury-related effects could be observed in brain tissue. To study this, they collected paired muscle and brain samples from Atlantic Sharpnose caught in fishery-independent surveys conducted in 12 locations within 9 US states, and measured tissue mercury levels along with known markers of mercury-induced brain cell damage. Like some earlier studies, they found that mercury levels in the brain were well below those observed in muscle of individual sharks, suggesting that there is limited transport to and/or active transport away from this important target organ. In addition, while some sharks did exhibit higher brain mercury levels than others, they also found limited evidence for mercury-induced effects on brain tissue. This was accomplished by measuring levels of brain antioxidants, which mercury can bind to and deplete in vertebrates, making brain cells more susceptible to cellular damage. The study also measured indicators of cellular damage, such as cell membrane destruction and the leakage of brain cell proteins.

Although the results of this study suggest that the shark brain may be less susceptible to mercury-related effects, it is important to remember that other tissues and organs can also represent targets for this harmful metal. This includes components of the immune system, which can be suppressed by mercury exposure in some vertebrates, making them more susceptible to other health maladies. This knowledge has led to other research on the effects of mercury in sharks, such as recent studies by Merly et al.[2] and Crawford et al.[3] that explored the relationship between mercury accumulation and general health of White Sharks—a species that has been shown to accumulate greater mercury levels than perhaps all other shark species. Interestingly, both studies found that despite having mercury levels that could be fatal in some other vertebrates, White Sharks did not exhibit apparent health-related complications, such as decreases in circulating immune cells or increases in blood markers of inflammation or organ damage. However, as illustrated by recent studies that have reported changes in gene expression in organs of bony fish exposed to elevated mercury levels,[4] it is possible that more sensitive tests are needed to uncover the effects of mercury and other pollutants on sharks.

Researcher Samantha Ehnert-Russo with an Atlantic Sharpnose Shark.

1. Ehnert-Russo, S. L. and Gelsleichter, J. 2020. Arch. Environ. Contam. Toxicol. 78(2): 267–283. doi:10.1007/s00244-019-00691-0.

2. Merly, L. et al. 2019. Mar. Poll. Bull. 142: 85–92. doi:10.1016/j.marpolbul.2019.03.018.

3. Crawford, L.M. et al. 2023. Mar. Poll. Bull 195: 115533. doi:10.1016/j.marpolbul.2023.115533.

4. Olsvik, P.A. et al. 2021. Chemosph. 279: 130588. doi:10.1016/j.chemosphere.2021.130588.

which the authors speculated could be due to the preponderance of cephalopods in their diet.

These studies are alarming not only because of the potential impacts on growth, reproduction, performance, etc. of the species studied, as well as others in their ecosystems, but also because of the potential for human health risks.

Plant Nutrients

Plant nutrients are water-soluble inorganic nitrate and phosphate salts. Sources include sewage, runoff from livestock feedlots, fertilized fields and lawns, and industrial wastes. The result of nutrient pollution ranges from plankton blooms to dead zones (e.g., large tracts of ocean with little or no DO).

While nutrients are essential for the growth of marine plants (especially microscopic forms), too many nutrients, a condition known as *eutrophication*, can lead to blooms, which can deoxygenate water when the phytoplankton die and decompose, using oxygen in the process. Hypoxic (low oxygen) or anoxic (no oxygen) conditions can result and any aerobic organism in this area must either leave or potentially die.

Over the last several decades, these hypoxic events have increased. The marked increase in human population in coastal regions worldwide has contributed to greater nutrient and organic matter inputs into coastal watersheds and thus these events.

The Gulf of Mexico Dead Zone covers an area within about 30 m (98 ft) of the bottom as extensive as 20,000 km^2 (9000 mi^2), about the size of New Jersey, seasonally, extending from the mouth of the Mississippi River. It is due to nutrient enrichment from agriculture in the Mississippi River watershed.

When they encounter areas of low DO, sharks tend to emigrate rather than adapt or succumb to the conditions, if the option exists. Anecdotal accounts of increases in sharks near beaches following nearby hypoxic events support their moving *en masse* from areas depleted of oxygen. However, as detailed earlier, some species, e.g., benthic elasmobranchs, exhibit increased tolerance of hypoxia compared to pelagic sharks.

On the other hand, dead sharks have been reported among fish kills associated with red tides. In some cases, toxins associated with the fish kills were responsible for the shark mortality, but in others whether the cause was a toxin or hypoxia/anoxia was not determined.

Sharks may also take advantage of kills of bony fishes caused by plankton blooms; for example, in the 1971 red tide in Tampa Bay, Florida, hundreds of sharks of several species were seen *gorging themselves on dead, decomposing, or dying fishes.*[15]

Organic Chemicals

This category includes a wide variety of both natural and synthetic chemicals such as oil, gas, solvents, pesticides, and even pharmaceuticals. Significant sources are improper disposal of industrial, household, and medical waste, and pesticide runoff from farms, forests, roadsides, and golf courses. We will focus on oil spills, the persistent organic pollutants (POPs) dichloro-diphenyl-trichloroethane (DDT) and polychlorinated biphenyls (PCBs), and some pharmaceuticals.

Oil

Oil receives perhaps the most attention in this category. Crude oil is a complex mixture of hydrocarbons. Crude oil is sticky, insoluble, and floats on water, although some components may sink or evaporate.

Polycyclic aromatic hydrocarbons (PAHs, a class of persistent organic pollutants) are considered the most toxic components of crude oil and are the common focus of studies on the ecotoxicology of marine oil spills. Sharks, like most fishes, can metabolize PAHs taken in with contaminated food and are capable of ridding themselves of these contaminants through the liver. However, episodic or prolonged PAH exposures that exceed the body's detoxification capabilities can result in cellular damage, mutagenic effects, reproductive impairment, and even death.

Oil exposure on sharks can be estimated by measuring concentrations of PAHs in the shark's tissues and by studying how hard the body is working to rid itself of these toxins. For example, exposure to PAHs in sharks has been studied by examining the activity levels of enzymes in the liver that detoxify it and measuring concentrations of PAH metabolites in the bile (i.e., the byproducts associated with metabolizing the toxins). In the study conducted by co-author Grubbs and colleagues on the 2010 Deepwater Horizon oil spill in the Gulf of Mexico, liver detoxification enzymes in the Little Gulper Shark continued to increase long after the spill was stopped, peaking three years later. Research into the long-term effects of this, the largest oil spill in US history, is ongoing.

Persistent Organic Pollutants (POPs)

Since the early twentieth century, humans have been synthesizing chemicals not naturally found on earth, highly stable chemicals that bacteria may be unable to render harmless. Concentrations of many of these chemicals have reached levels in the oceans that have begun to threaten the survival of many species of phytoplankton, invertebrates, large mammals, and fish. These chemicals are persistent because they tend to be difficult for organisms to decompose. Moreover, even if organisms can break down the chemicals, these decomposition products are often toxic as well. DDE (dichloro-diphenyl-dichloroethylene), a breakdown product from DDT, is one notorious example.

Persistent organic pollutants (POPs) include a bewildering array of synthetic organic compounds that persist in the marine environment, are easily distributed globally, and lodge and become concentrated in the fatty tissues of animals by bioaccumulation and biomagnification. Many, including DDT, are pesticides part of the "Green Revolution" that radically increased food production during the twentieth century. Dioxins and furans are industrial byproducts of combustion. Polychlorinated biphenyls (PCBs) were used for years in electrical insulation. The manufacture and use of all POPs (except for furans and dioxin, which are combustion by-products) have for years been banned in the United States.

POPs are of global concern because there is firm evidence of *global* transport of these substances, by air and water, to regions where they have never been used or produced, such as the North American Arctic.

They cause a range of harmful effects among humans and animals, including cancer, birth defects, damage to the nervous system, reproductive illnesses, immune system disruption, and in some cases, death. Individuals exposed to POPs, and their offspring too, can suffer damage to their reproductive and immune systems. Some POPs are *endocrine disruptors* (or endocrine-active chemicals). An endocrine disruptor is a chemical that interferes with the function of the endocrine system. It can mimic a hormone, block the effects of a hormone, or stimulate hormone production. Endocrine disruptors have been known to cause feminization of males, masculinization of females, and fertility problems in fishes.

Let us consider PCBs first. PCBs are a group of over 200 synthetic chemicals called *organochlorines*, which are complex compounds made of one or more chlorine (Cl) atoms attached to carbon-based structures such as chains and rings.

PCBs do not exist naturally on earth—they were synthesized during the late nineteenth century. By 1977, the manufacture of PCBs was banned in the US, the UK, and elsewhere. However, they are still being used in many developing countries. By 1992, scientists estimated that 1.2 million metric tons (264,500,000 lb) of PCB existed worldwide, while scientists estimate that 370,000 metric tons (810 million lb) have been dispersed globally, much of it into the oceans.

PCBs are lipophilic and thus tend to accumulate in the fatty tissues of animals. They enter food chains and webs mainly through the feeding of organisms called *sediment* or deposit feeders. PCBs move up the trophic tiers and become further concentrated.

PCBs have become widespread and are serious pollutants that have contaminated many marine trophic tiers. They are extremely resistant to breakdown (which, of course, was considered one of their virtues) and are known to be carcinogenic and probably mutagenic as well. In terms of their specific effect on life, PCBs have been shown to cause liver cancer and harmful genetic mutations in animals. PCBs have been linked to mass mortalities of striped dolphins in the Mediterranean, and to declines in Orca populations in Puget Sound and seal populations in the Baltic.

Although PCBs may threaten the entire ocean, the NW Atlantic is believed to be the largest PCB reservoir in the world because of the volume of PCBs produced in countries that border the region, particularly the US.

PCBs have been detected in a number of sharks and batoids at levels across the spectrum of those found in other marine organisms. PCBs have also been shown to transfer from embryo-to-embryo in some species. Virtually all of these studies report on the occurrence and concentration of PCBs in the liver and other tissues, but none have examined pathologies associated with PCBs in sharks and batoids.

Interpreting the meaning of the studies of PCBs in sharks and batoids is difficult, in part because there are numerous PCB compounds. Moreover, determining threshold levels for all of these compounds at which measurable damage is done is all but impossible because of interspecific differences in rates of accumulation and synergistic effects with one another and with other chemical compounds. In some species, males and females accumulate PCBs at different rates.

As stated earlier, thresholds for toxicity for agents toxic to sharks and their relatives are lacking. One study[16] applied toxicity thresholds for total PCB concentration in marine mammals to Shortfin Makos and Bull Sharks, both of which exceeded the lower toxicity threshold and nearly reached the upper threshold. Clearly, more work on the effects of PCBs on sharks and batoids would provide multiple conservation benefits.

DICHLORO-DIPHENYL-TRICHLOROETHANE (DDT)

DDT, whose production as one of the first synthetic pesticides dates back to the 1940s, is another persistent organochlorine and is similar in structure to PCBs. It was used effectively in insect control in agricultural settings as well as to kill insect vectors of malaria and other diseases, but at the same time had wide-ranging impacts on wildlife, especially birds. DDT is banned in the US and most other countries but its use to control mosquitoes that carry malarial parasites is still allowed in some countries because of its effectiveness and the severity of the problem.

Despite decreased production and use, some of the 1.5 Tg (teragrams; 3.3 billion lb) of DDT used or dumped since it was first produced still exists in the ocean, primarily in sediments and as a result of export to the deep sea. DDT persists longer in seawater than in soil or air.

Thus, it is not surprising that DDT still shows up in shark tissue. Let us consider two cases. In 2013, a 3.7 m (12.1 ft), 600 kg (1320 lb) female Shortfin Mako (*Isurus oxyrinchus*) was caught on hook-and-line (apparently for a TV show) near Huntington Beach, California and was donated to scientists.[17] The liver was analyzed for organic contaminants, and DDT was found to comprise 86% (an estimated 11.4 g or 0.5 oz) of the total contaminant load. The DDT concentration was nearly 100 times greater than the no-consumption level set for humans by the US Environmental Protection Agency (EPA). The researchers who measured the DDT as well as PCB levels noted that they were consistent with the shark feeding near a POP-contaminated area (i.e., the Palos Verdes Shelf Superfund site off the coast of Los Angeles County). Among the Mako's stomach contents was the remains of a juvenile California Sea Lion (*Zalophus californianus*) estimated to weigh about 67.6 kg (149 lb) before being consumed.

The second case involves DDT levels in the Greenland Shark (*Somniosus microcephalus*), a top predator found in cold waters of the North Atlantic and Arctic.[18] DDT concentrations in a sample of 15 Greenland Sharks were higher than those of other studies. The authors of the study speculated that the high DDT levels could be attributed to *continuous and chronic exposure* to DDT compounds, but other chemical indicators suggested an old contamination event. Although some of the results were conflicting, recent discoveries that this species may be capable of living as long as 400 years could easily mean that the DDT was acquired between the 1940s and 1970s, and that the lower metabolism of this species translated into slow depuration (declines due to metabolism or other loss).

High DDT levels have been found in a number of other sharks, including Whale Sharks, Basking Sharks, Scalloped Hammerheads (*Sphyrna lewini*), Pacific Angel Sharks (*Squatina californica*), Pacific Sharpnose Sharks (*Rhizoprionodon longurio*), Kitefin Sharks, and Blue Sharks, to name a few.

Pharmaceuticals

The last category of organic chemicals we discuss would have been considered a frivolous line of inquiry at one time but now constitutes a rapidly worsening problem. We refer to the exposure to and hazards of human, veterinary, and agricultural pharmaceuticals and personal care products that find their way into marine and coastal environments.

The list of pharmaceuticals in the marine environment is a veritable pharmacy checklist, and includes analgesics, antibiotics, antidepressants, antihistamines, antianxiety drugs, estrogens, contraceptives, bronchodilators, decongestants, nonsteroidal anti-inflammatory drugs (NSAIDs), and on and on.

Personal care and cleaning products include ingredients found in shampoos, sunscreens, fragrances, and cosmetics. These pharmaceutical and personal care product chemicals arrive in coastal and ocean waters and sediments through sewage and landfills, and, for pharmaceuticals only, from aquaculture, feedlots, and agriculture.

Once in marine waters or sediments, the ingredients in the pharmaceuticals and personal care products may enter the food chain, either as the parent product or after being degraded.

Initially, it was thought that the diluting effect of the large volumes of water into which the pharmaceutically active compounds and personal care products were discharged would result in negligibly low levels in the water column and sediments and would amount to little biological significance. Initially there was some truth to this, until chemicals from pharmaceuticals and personal care products began to be detected in marine organisms.

More important than the dilution factor are the rates of input, residence time, and toxicity. An important aspect of pharmaceuticals is that they specifically *target* biological functions, as well as possess a suite of so-called *side effects*, often at low concentrations. Also, long-term exposure to many of the chemicals is likely more harmful than acute impacts. Finally, lipophilic substances may bioconcentrate and biomagnify more so than water soluble ones.

Among marine organisms, the highest concentrations of pharmaceutically active compounds and those derived from personal care products have been found in molluscs, fishes, and mammals.[19] Studies on sharks have been more limited, and these have focused more so on detecting and quantifying a subset of these classes of chemical compounds in sharks and not on their impacts on their health and behavior.

Synthetic polycyclic musks, a group of compounds that are mainstays in the fragrance industry, have been detected in the livers of Scalloped Hammerheads (*Sphyrna lewini*).[20] Parabens, preservatives in pharmaceuticals, food, and personal care products, have been found in Atlantic Sharpnose, Bonnethead (*Sphyrna tiburo*), Shortfin Mako, Spinner, Bull, Blacktip, and Great Hammerhead Sharks (*S. mokarran*), as well as Atlantic Stingray (*Hypanus sabina*),[21] although the highest levels in the study were found in bony fishes.

Finally, researchers examined levels of 10 pharmaceutically active compounds (one synthetic estrogen contraceptive [e.g., Ortho Evra], impotence agent [Viagra], lipid-lowering drug [Lipitor], and eight antidepressants [including, but not limited to, Paxil, Effexor, Prozac, Zoloft, and Celexa]) in neonate Bull Sharks in pristine and wastewater-impacted systems in Florida.[22] They detected the estrogen and six of the antidepressants in sharks in the wastewater-impacted system, but only one antidepressant in a single shark in the pristine system. The authors concluded that the compounds detected in the Bull Sharks "were generally low in concentration . . . and not expected to pose acute health risks to these individuals." Whether these Bull Sharks were merely mellower is not known.

Addendum: A 2024 paper[23] reported the presence of cocaine and one of its metabolites (breakdown products) in Brazilian Sharpnose Sharks (*Rhizoprionodon lalandii*). The report's title included the words *cocaine shark*, thus ensuring disproportionate media attention, most of which framed the story as amusing and thus once again squandering an opportunity for meaningful enlightenment.

Sediments

Pollution by suspended sediment is a growing problem. Fine soil particles (silt and clay) along with organic and inorganic chemicals that originate as runoff from construction sites, logged forests, mining operations, and croplands block sunlight, foul gills of fishes and invertebrates, and can even smother bottom communities (e.g., seagrass beds, oyster reefs, and coral reefs).

Dredging channels to promote port operations enhances the transport of salty oceanic water into estuaries, in the process changing the salinity structure, circulation, flushing, and water residence times. Dredging also resuspends sediments into the water column and may contaminate the water column with toxic compounds from the substrate as well as add nutrients and organic matter. Such changes can have serious effects on the health of organisms as well as biological productivity and ecosystem function.

Most studies of the impacts of sediments on marine organisms have focused on channel dredging and beach augmentation (euphemistically better known as *renourishment*). Studies specifically focusing on sharks have been more limited.

Turbidity refers to the relative clarity of water and is one of the factors that influences the distribution of marine organisms. Relatively high turbidity (i.e., murkiness) is a normal persistent or episodic characteristic of ecosystems where elasmobranchs are found (e.g., along the coasts of Virginia, North Carolina, South Carolina, and Georgia in the United States). Additionally, naturally occurring high turbidity occurs seasonally or episodically at river mouths.

Dredging to deepen and widen ship channels or to mine sand for beach augmentation resuspends sediments and can result in large sediment plumes, the duration and extent of which are determined by a number of factors (e.g., type of dredge, sediment characteristics, depth, and current).

According to the US Army Corps of Engineers, more than 400 ports and 25,000 miles of navigation channels in the United States are periodically dredged. Many of these projects are controversial because of their environmental impact and cost. For example, the Port Miami Deep Dredge Project, completed in 2015, deepened the Government Cut seaport channel from 12.8 to 15.8 m (42 to 52 ft) to accommodate larger ships with deeper drafts. In doing so, sensitive coral reefs near the dredging site were damaged.

There are two major questions with respect to sediment pollution and sharks: Does sediment pollution associated with dredging and other projects affect sharks and batoids and, if so, in what ways?

Turbidity is known to influence the presence or absence of sharks. In one study of estuaries in the US state of Georgia,[24] subadult Bonnetheads were found in areas of lower turbidity than juvenile Sandbar Sharks, although other factors (e.g., salinity, depth, dissolved oxygen) also influenced their presence. In the same study, turbidity did not affect the presence of Blacktip and Atlantic Sharpnose Sharks. The study was not designed to assess human impacts, and the authors concluded that these results might represent habitat preferences for these species.

In another study, immature Scalloped Hammerheads were shown to prefer habitats with higher turbidity, perhaps as a mechanism to deter visually orienting predators.[25]

In a third study, turbidity was one of the factors, although not the most important, that predicted the presence or absence of three sharks (Bull, Bonnethead, and Blacktips) along the Gulf of Mexico coastline. Again, this study did not address anthropogenic sedimentation. Studying the direct effects of turbidity is difficult, as there are many covariates that also affect spatiotemporal ecology.

The most comprehensive studies of the impacts of dredging associated with development were conducted by scientists at the Bimini Biological Field Station (BBFS) and others beginning in 1997, when a large resort complex began construction on North Bimini (fig. 12.7). Initial plans called for a 930-room hotel, 3000 m^2 (32,300 ft^2) casino, 18-hole golf course, and two marinas on the small Bahamian island. Dredging began in 1999 and by 2006 about 750,000 m^3 (26,500,000 ft^3) of sediment had been excavated.

The BBFS was particularly poised to study the impacts of this development because they have been studying the ecology of the system since the early 1990s to understand the biology and ecology of Lemon Sharks, which utilize parts of the impacted area as a nursery ground. One study[26] found that the survival rate of juvenile Lemon Sharks in the system and the coverage of seagrass (habitat for their prey) both declined after dredging. Studies of impacts of the development on the biodiversity of the impacted area, as well as impacts on Lemon Sharks, are continuing. We further discuss Lemon Sharks in Bimini later in this chapter.

Another major case of sediment pollution is that which occurs at the Abbot Point coal loading port in Queensland, Australia, from which as much as 50 million tonnes (110 billion lb) of coal are exported annually. The terminal is 20 km (12 mi) from seagrass beds and is 40 km (24 mi) from the southern section of the Great Barrier Reef, an ecosystem already in poor health as a result of warming oceans, landscape changes causing runoff, predatory starfish, pesticides, and other stressors.

Sediment pollution associated with the port comes from a variety of sources, including dredging, dumping of sediment spoils, and fugitive coal dust (the dust that escapes as coal is moved to the terminal and loaded on to ships). Again, there have been no studies that focus on sharks, but numerous studies have documented the impacts of sediment on coral reef ecosystems, and it is reasonable to hypothesize that these might affect the health, behavior, distribution, and/or abundance of sharks at impacted areas.

Bull Sharks appear to thrive in areas altered by dredging, e.g., in marinas, and outfalls from power plants,[27] although the impacts of sediment pollution on their presence have not been examined.

Radioactive Substances

Radioactive substances are produced by the nuclear power industry, the military, electrical utilities, mining industries, and as a result of medical diagnostics and treatment. In the 1960s and 1970s, radioactive waste buried in shallow trenches leached into surrounding soil and groundwater. Modern disposal techniques have been engineered to minimize the potential for leakage. There is also natural background radiation in the ocean.

The meltdowns and use of seawater to cool reactor cores at the Fukushima Daiichi Nuclear Power Plant after the 2011 tsunami led to renewed public interest in radiation pollution. Shark species that foraged near the site of the Fukushima Daiichi plant included Mako, Blue, and Salmon (*Lamna ditropis*) Sharks, high trophic-level predators that might accumulate and transport radioactivity. Pacific Bluefin Tuna (*Thunnus orientalis*) were shown to transport radionuclides from Fukushima across the entire North Pacific Ocean.[28]

Although the amount of radiation, primarily cesium-134 and cesium-137, in the seawater was several orders of magnitude over safe levels after the event, the half-life of the radioactivity was short and the radiation dispersed quickly, minimizing concerns. Cesium is excreted relatively quickly, and biomagnification is minimal, so concerns about the effects of these radionuclides on marine organisms were minimal.

The United States detonated 23 nuclear devices between 1946 and 1958 on and near the Bikini Atoll in the Marshall Islands in the equatorial Pacific Ocean. Despite rumors of sharks missing their second dorsal fins, recent assessments of the biota of the atoll report a resilient, highly diverse, healthy ecosystem, and no validation of the shark rumor.

Heat

Heat as a form of water pollution is principally produced from the cooling of power plants and industrial processes, although anthropogenic climate change also heats bodies of water. Most threatened are tropical organisms (e.g., corals), which live closer to their upper thermal limit (CTmax), and thus the ecosystems in which the tropical marine organisms live (see chapter 8).

Studies of the effects of thermal effluent on sharks are lacking. Recall that often the first reaction to environmental stressors is behavioral avoidance, and this likely applies to sharks and batoids in these areas. Notable exceptions occur when there are advantages to warmer waters (e.g., abundant prey or refuge from temperature minima). An example of the latter is that Nurse Sharks (*Ginglymostoma cirratum*) were shown to aggregate at the thermal effluent from the Turkey Point power plant in the US state of Florida.[29] Sharks were also shown to gather at the thermal plume of the Kahe outfall on Oahu, Hawaii.[30]

Plastics

Figure 12.3 shows a Sandbar Shark that we caught on an experimental longline in 2016 in Winyah Bay, South Carolina with a plastic packaging strap fully encircling the shark near the gills. When we removed the plastic packaging ensnaring this shark around, we considered the incident anecdotal and rare, but such reports are now commonplace. This particular shark likely survived after we removed the strap, since it appeared otherwise healthy, was actively feeding (since it took the bait on our longline), and swam off strongly once released, but many sharks and other marine life do not survive similar entanglements. In one incident, for example, scientists discovered a Shortfin Mako so entangled in plastic debris that its growth was stunted and its spine deformed.

Plastic is so commonplace that we need not devote much space to describing its uses. We are in what has been called the *Age of Plastics*. Global production of plastic was more than 400 million metric tons (880 billion lb) in 2022, and is increasing exponentially.[31] By 2017, more than 8.3 billion tonnes (18.3 trillion lb) of virgin plastic had been produced, according to one scientific assessment,[32] which

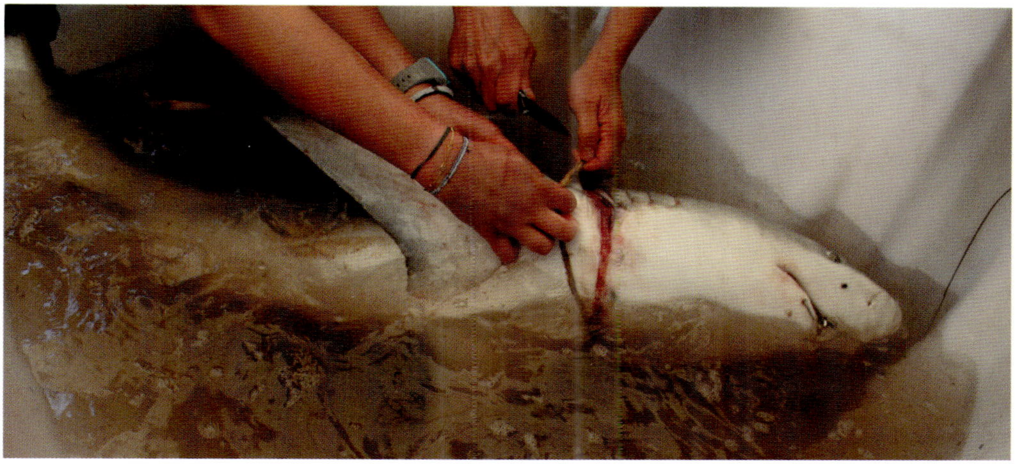

A

B

Figure 12.3. (A) Sandbar Shark with plastic strap encircling body, caught on an experimental longline. (B) Releasing the shark, which swam away strongly (Courtesy of Caroline Collatos).

also reported that since 2015, 9% of the plastic produced had been recycled, 12% incinerated, and 79% wound up in landfills or the environment.[33] Globally, between 60–80% of all litter is plastic.[34]

How much plastic is in the marine environment? Estimates of the amount of plastic that finds its way into the marine environment annually range from 1.8% to 10% of annual global plastic production. There are two broad categories of plastics in the marine environment, *macroplastics* and *microplastics* (fig. 12.4). The former category includes a wide variety of plastics ranging from beverage bottles and packaging to abandoned commercial fishing gear, some of which may fragment or degrade to smaller pieces. Microplastics are < 5 to 10 mm (0.2 to 0.4 in) in size and may be produced as degradation products of larger plastics in the marine environment or may be manufactured as microplastics (e.g., microbeads in cosmetic and personal care products).

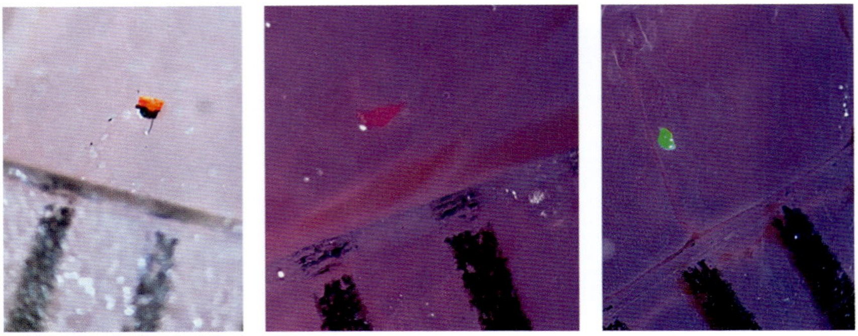

Figure 12.4. Microplastic fibers and fragments found in the gut of Atlantic Sharpnose Sharks. Scale represents 1 mm increments.

Once in the marine environment, plastics have been shown to cause a suite of problems at the organismal and ecosystem level, although much remains to be understood about these impacts. Let us focus on sharks. Broadly, problems are associated with either entanglement or ingestion.

Entanglement of sharks and batoids in plastics, not including bycatch in monitored nets, has been extensively reported anecdotally (like our earlier example), especially from abandoned drift nets.[35] However, the global extent and impact of entanglement on local populations or ecosystems is difficult to assess, although recent reports[36] have documented dozens of entanglements affecting hundreds of sharks and batoids, including some of the most threatened batoids, rays in the order Rhinopristiphormes.[37]

Ingestion of macro- and especially microplastics has been studied more rigorously than entanglement and is a major line of inquiry as we write this book. Direct ingestion of large plastic by sharks is rare. A 2018 review article[38] cites the personal comments of a shark biologist who found that less than 1% of shark stomachs contained ingested plastics, which included PVC fragments, and bottle and pen caps. As of late 2018, sharks found with ingested microplastics include Blackmouth Catshark (*Galeus melastomus*),[39] Blue Sharks,[40] Portuguese Dogfish (*Centroscymnus coelolepis*),[41] Velvet Belly Lantern Shark (*Etmopterus spinax*),[42] Spiny Dogfish (*Squalus acanthias*),[43] Longnose Spurdog (*Squalus blainville*),[44] Brazilian Sharpnose Shark (*Rhizoprionodon lalandii*),[45] and Atlantic Sharpnose Shark.[46] Additionally, Whale Shark (*Rhincodon typus*) skin biopsies showed chemical evidence of microplastic ingestion.[47] A 2024 review[48] of 48 publications on the subject expanded the list to 54 species, which included 41 sharks, 12 batoids, and a single chimaera.

Microplastic toxicity may be due either to adsorption of nonpolar chemical pollutants to the surface of the microplastics (e.g., PCBs, DDE, and DDT), or harmful chemicals in the plastics (e.g., flame retardants and plasticizers). However, studies on the effects of toxic microplastics on sharks and batoids are lacking.

Additionally, although the subject has not yet been thoroughly studied, evidence suggests that plastics are capable of moving up the food chain.

Given the pervasiveness of plastics in the marine environment, the continued growth in their production and use, the absence of meaningful recycling, and the continued enumeration of macro- and microplastics in sharks and other marine organisms, it is not farfetched to assert confidently that research in the area will expand considerably.

Noise

Anthropogenic noise in the marine environment is a significant problem to marine mammals, but does background noise from ships and boats, dock and other construction, air guns used in seismic exploration, windmills, and sonar affect sharks?

Unfortunately, studies on the impacts of noise on sharks are lacking. A 2019 study[49] on the behavioral responses of reef sharks to underwater sounds concluded that the behavior of these sharks was altered by relatively low sound levels, but that a *critical need* exists for more studies.

In the absence of such tests, the next best sources of information are comparing what is known about hearing in sharks with the range of marine anthropogenic noises to assess the likelihood of damage, and consulting studies on the impacts of noise on bony fishes.

Intense sounds have the ability to burst air-filled spaces, like swim bladders in fish, which can lead to hearing impairment or loss. Recall, however, that sharks lack swim bladders. Intense sounds have the potential to damage the hair cells of the lateral line and inner ear, usually causing temporary hearing threshold shifts (i.e., altering the intensity and frequency of sounds detected).[50]

Noise associated with underwater construction could lead to temporary decreases in hearing sensitivity, and high intensity noise when a pile is hit by the hammer could cause tissue damage in the inner ears and other organs of bony fishes and elasmobranchs.[51] Vibrations associated with pile driving and conveyed through the substrate might have greater impacts on benthic sharks and rays. Shipping noise is not considered dangerous to sharks, except perhaps by masking other sounds.[52]

Noises that do not cause short-term or permanent damage might affect sharks and batoids in other ways (e.g., by causing a startle or other avoidance response), affecting foraging or movement, or impacting prey or other organisms in the ecosystem.

One area of particular concern is noise from towed arrays of air guns used in seismic testing for oil and natural gas. These pulsed, high intensity, low frequency sounds are known to impact marine organisms from zooplankton to mammals. Among fishes, sharks and batoids are the most sensitive to low frequency sound and thus might be most susceptible to air gun blasts.

Physical Habitat Destruction

Sharks and batoids are found in a variety of marine ecosystems, including estuaries, salt marshes, intertidal zones (rocky, sandy, muddy), shallow subtidal benthic environments (kelp beds, seagrass communities), deep-sea, open ocean, and tropical ecosystems like coral reefs and mangrove swamps. None of these has escaped anthropogenic impact, and studies to more fully understand how these changes have specifically impacted sharks are lacking. We will examine one of the most biologically productive ecosystems on the planet, mangroves, which play important roles in the life histories of numerous sharks and batoids.

Mangroves

Mangroves are a broad group of tropical, salt-tolerant trees that grow at the water's edge between 32°N and 38°S latitude. Five countries—Indonesia, Australia, Brazil, Nigeria, and Mexico—are home to almost half of the entire global area of mangrove ecosystems.

Stilt roots of mangroves at the water's edge form complex networks, trapping sediment and preventing its transport further seaward. They also provide a myriad of hiding places for a great diversity of marine life, especially the juvenile and larval forms of fish, crustaceans, and mollusks, any of which constitute food for elasmobranchs.

Mangrove communities flourish mainly along the shores of estuaries, along coasts, and at the mouths of large rivers like the Irrawaddy in Myanmar. These locations mean that mangrove communities tend to be at risk from development, especially in countries with few if any land-use controls. Mangrove environments are readily cleared for aquaculture, resort developments, and housing. They are also threatened globally by rapid sea-level rise and locally by agricultural runoff, oil, deforestation for biomass fuel, and gas exploration and production, the latter for example, along the Persian Gulf and the coasts of Australia and Nigeria.

The southwest coast of Florida contains the most extensive mangrove forests in the continental United States, and yet even there, mangrove forests continue to decline because of development.

Aquaculture development along tropical coasts that converts mangroves to fish or shrimp farms directly threatens the survival of mangrove forests. Industrial shrimp farming along the NE coast of Brazil, for example, puts at risk much of that nation's one million surviving hectares (2.47 million ac) of mangroves, about 7% of the global total.

In addition to serving as habitat for sharks and batoids (e.g., the endangered Smalltooth Sawfish [*Pristis pectinata;* fig. 12.5], and juvenile Giant Shovelnose Ray [*Rhinobatos typus*]), mangroves and the waters they fringe serve a suite of critical ecosystem, social, and economic services. Mangrove forests, which are exclusively tropical, cover less than 14 million ha (34.6 million ac) globally, representing from 50–65% of historical levels. Additionally, an undetermined

Figure 12.5. Smalltooth Sawfish in shallow water adjacent to mangroves in the Bahamas.

percentage of remaining mangroves has been degraded by pollution. Losses have slowed in some areas as governments and residents recognize the value of mangroves to commercial, artisanal, and subsistence fishing, and as buffers against storms and tsunami.

A complete list of the sharks and batoids that reside in mangrove ecosystems as neonates, juveniles, and/or adults has not been published, although there are numerous publications on specific species, regional checklists of species, and specific ecosystems.[53]

One particularly well-researched species is the Lemon Shark of Bimini, Bahamas (fig. 12.6). Understanding the biology of the Lemon Shark was a major focus of noted shark biologist Dr. Samuel Gruber of the BBFS. Research conducted by Gruber and his coworkers over approximately two decades has shown that mangroves and the lagoon fringed by the mangroves are critical to the growth and survival of juvenile Lemon Sharks during their first several years of life, after which they move to more nearshore or coastal habitats. Young Lemon Sharks take advantage of the protection from predation offered by the mangroves (one of their main predators are larger Lemon Sharks), as well as the abundant food supply in the lagoon.

A 2008 study[54] concluded that the development of a resort in Bimini, Bahamas (fig. 12.7) that included removal of mangroves significantly damaged the nursery grounds of Lemon Sharks. By August 2010, more dredging had occurred

Figure 12.6. Juvenile Lemon Shark on flats adjacent to mangroves in Bimini, Bahamas (Courtesy of Annie Guttridge).

and about 67 ha (166 ac) representing 39% of the mangrove habitat surrounding the system had been removed[55] (fig. 12.7). This habitat was one of the most important Lemon Shark nurseries in the NW Bahamas, which also serves to recruit adult Lemon Sharks to SE US habitats.

After the development started, survival and growth rates of Lemon Sharks decreased, and sharks remaining in the area were less healthy than comparable sharks in undisturbed areas.

Finally, mangrove and coral reef communities (see Climate Change later in the chapter) are sometimes cited as examples of *linked habitats* because of their close association. Mangroves are often found landward near coral reefs. Although mangroves require quiet waters containing some sediment in which to proliferate, coral reefs require clear waters free of fine sediment that can smother the corals and kill the reef. The communities are also complimentary in that coral reefs protect mangroves from wave erosion and storm damage, whereas mangroves provide refuge for juvenile stages of numerous economically and/or ecologically important reef animals. The removal of mangroves often leads to the degradation or even disappearance of the reef. For a thorough review of elasmobranchs

Figure 12.7. (A) Photo from North Bimini, Bahamas taken before and after substantial development for the Bimini Bay Resort showing loss of mangroves (Courtesy of Kristine Stump). (B) Low altitude aerial view of the extent of the mangrove destruction (Courtesy of Kristine Stump).

in mangrove systems, we recommend the publication *Mangrove Use by Sharks and Rays: a Review.*[56]

Sharks and batoids face similar problems in salt marshes, which typically replace mangroves in protected intertidal areas along temperate coasts and play similar ecological roles as mangroves. Similarly, they are degraded by a suite of physical human activities (e.g., construction of docks, bulkheads, dikes, and so on), as well as long-term impacts associated with their presence (e.g., altering the hydrology of the system leading to changing erosion and deposition of sediments, relocation of channels, different salinity regimes, changing vegetation patterns, and dredging). Boat traffic can lead to erosion of banks. The pace of so-called *land reclamation*, more appropriately draining and filling, as well as removal for dock construction, etc., has slowed, but the process still occurs, albeit on a small scale in the United States. By the early 1990s, an estimated 50% of US salt marshes were lost principally due to draining and filling for agriculture, industry, and residential uses. Over 90% of California's coastal wetlands have been lost. Historical global losses are estimated to be between 25 and 50%. Estimates of remaining salt marshes globally range from 2.2 to 40 million ha (54 to 98 million ac), and a recent effort mapped about 5.5 million ha (13.6 million ac) in 43 countries.[57]

Climate Change and Sharks

There is complete and unambiguous agreement among scientists that human activity is causing global climate change. Scientific uncertainty remains only on issues that include the time frame and severity of climate change impacts, as well as the cost, impact, and effectiveness of mitigation and/or adaptation strategies.

What are some of the impacts of climate change on sharks? Since the first edition of this book was published in 2020, considerably more research has been conducted on climate change impacts on sharks and their relatives, as well as the ecosystems in which they live. By the time this second edition comes out in 2025, even more knowledge about impacts will accumulate. We urge you to do your own literature search. An excellent place to begin, and one of the resources that we have used in this book, is the chapter by Rummer et al.[58] in the comprehensive book *Biology of Sharks and Their Relatives* by Carrier et al.[59] Also, as this book went to press, a paper entitled *Global Risk Assessment of Sharks to Climate Change*[60] was published. The paper considered both short- and long-term risks to sharks under different climate change scenarios. We recommend this paper as well, especially to see how their assessments play out, and conservation efforts respond, as climate change relentlessly advances.

Let us start with acidification. The scale of impacts of ocean acidification has the potential to be enormous, causing changes in carbonate chemistry that have not been seen in 65 million years, and will affect the vitality and survival of virtually all taxonomic groups. A 2018 paper[61] discusses that the impacts of acidification on sharks may be direct (i.e., affecting the physiology [by acidifying blood and tissues] and behavior of sharks) or indirect (e.g., changing the community structure or prey). In their analysis of other studies, the paper notes that separating acidification from temperature increases may produce conflicting and confounding results.

As of 2018, physiological effects of relevant elevated CO_2 levels had been studied in only four species—Small-spotted Catshark (*Scyliorhinus canicula*), Port Jackson Shark (*Heterodontus portusjacksoni*), Bamboo Shark (*Cephaloscyllium punctatum*), and Epaulette Shark (*Hemiscyllium ocellatum*). Note that these are all benthic species and are in other ways unrepresentative of sharks as a group.

One area of interest is the impacts of acidification on embryonic development. On the one hand, exposure to elevated CO_2 levels in the Port Jackson, Bamboo, and Epaulette Shark did not affect their embryonic development (e.g., their survival, development time, and growth rates). On the other, in the Little Skate (*Leucoraja eglanteria*), high CO_2 levels intensified the effects of elevated temperature on embryonic development.

Among the physiological responses to elevated CO_2 levels in Small-spotted Catshark were changes in resting metabolic rate, scope for aerobic activity (an index of maximum aerobic metabolic performance; e.g., faster swimming), and increased bicarbonate and sodium levels. Growth rate was not affected. A suite of similar physiological responses was found in the Epaulette Shark, and included changes in resting oxygen consumption, acid-base balance, hypoxia tolerance, aerobic enzyme activity, and other hematology variables. Because Epaulette Sharks live in shallow reef habitats subject to swings in CO_2 levels, the authors of the study speculated that the species might be capable of tolerating the physiological changes associated with increased CO_2 concentration.

Both studies cited above exposed the sharks to the elevated CO_2 levels for periods of 30 to 60 days. Studies in which neonates were exposed to these conditions for > 200 days demonstrated more significant impacts (e.g., decreased condition, aerobic potential, peroxidative damage in the brain, and others).

Studies on the effects of elevated CO_2 levels on behavior of sharks have shown more direct impact, although results of elevated CO_2 and temperature together were somewhat contradictory. In one study, Port Jackson Sharks maintained in elevated CO_2 levels for > 65 days took about four times longer than those in controls to detect their prey. However, that time was reduced by one-third when the study was conducted in warmer than normal water. A follow-up study validated this finding but, at the same time, reported that elevated CO_2 levels inhibited chemical and visual sensory responses important for foraging. A study on the Dusky Smoothhounds (*Mustelus canis*) showed that high levels of CO_2 were associated with impaired ability to track prey using odors.

In contrast to the few studies of ocean acidification on sharks, there have been a large number of studies on the effects of elevated temperature on this group (e.g., as signals for movements, as well as impacts on metabolism, growth, reproduction, and foraging). However, fewer of these have focused on warming in the range associated with anthropogenic climate change.

To remind you, as we stated in chapter 8, temperature is one of the most important environmental factors for all living organisms. It affects not only their diversity, abundance, and distribution, but also their behavior, activity, growth, development, metabolism, heart rate, digestive rate, blood characteristics, and other aspects of their physiology at all levels from the molecular to organ systems.

Some additional climate change impacts that could affect sharks and rays include changes in precipitation patterns that alter the salinity structure of nearshore and oceanic systems, increased intensity and frequency of tropical storms, rising sea levels that coastal wetland communities may not be able to keep up with, and hypoxia leading to increased size and severity of dead zones. Some of these impacts may be on a large scale (e.g., changes in ocean circulation that could include a critical slowing of the Gulf Stream).

Moving to ecological impacts, many sharks and rays will likely migrate to higher latitudes or deeper water as a result of temperature increases. It is well-known that temperature (both warm and cold) is a major driver of shark movement.[62] For example, a 2018 study[63] showed the presence of juvenile Bull Sharks in North Carolina estuaries that had not been shown to be frequently used habitat and correlated their presence with the early arrival of summer temperatures. A 2024 report[64], also on Bull Sharks, found significant delays (up to 36 days) and reduced overwintering stays (up to 81 days) in 2021 compared to 1982 in the western Gulf of Mexico. This correlated with a 1.55°C (2.8°F) autumn temperature increase, and especially decreases in prey density attributed to climate change. The senior author of this study discusses these climate change impacts in the accompanying *Research, Conservation, and Education Focus* Box 12.2. Another study[65] projected poleward range extensions of 64 km (40 mi) per decade for two shark species near Australia for 2030 and 2070.

In moving to higher latitudes, sharks and rays may encounter systems biotically and/or abiotically foreign or novel to them. These may cause problems like changes in the demographic characteristics of shark populations, including abundance and size structure; changes in trophic structure; changes in behavior; and possibly mortalities, extirpations, and even extinctions for species for which migrations may be difficult or improbable or which may already be depleted or threatened by other stressors. These could include sharks and rays inhabiting coastal ecosystems (e.g., seagrass beds, mangroves, salt marshes, and coral reefs).

Climate change may lead to changes in fundamental large- and small-scale ecological processes (e.g., nutrient cycles), which can alter community structure, food availability, predator-prey dynamics, and so on.

In this book's first edition, we noted that only the New Caledonia Catshark (*Aulohalaelurus kanakorum*) was considered threatened directly by climate change.[66] Since then, numerous reports have expanded the list considerably. For example, a 2024 assessment[67] of 132 species of chondrichthyan fish in the Eastern Tropical Pacific (ETP) concluded that conservatively 30 species (23%), mostly batoids, were *Highly Vulnerable* and, most (76%) of the remainder, *Moderately Vulnerable*. The assessment concluded that the most vulnerable chondrichthyans in the ETP were coastal species, especially those that relied on inshore nurseries.

The list is growing as we write, and there are critical gaps at all levels, especially understudied species, a large group that includes inaccessible or rarely encountered sharks and batoids (e.g., deep-sea species), larger species, different life stages, and so on, as well as how different climate stressors interact.[68]

It is far beyond the scope of this book to attempt a comprehensive summary of additional ways in which climate change have already, and may in the future, impact sharks and rays, especially since they represent a rapidly moving target, with new developments and insights occurring with breakneck rapidity.

Finally, in 2024, the paper *Ecological Roles and Importance of Sharks in the Anthropocene Ocean*,[69] published in the journal *Science*, reviewed the existing data on the eponymous subject. The paper noted a critical gap in understanding the important roles that sharks play in deep-sea and polar ecosystems. Similar gaps exist, the paper noted, for "*smaller-bodied species, and for smaller age classes of large species in most ecosystems.*" The paper reaffirmed the outsized role of overfishing in damaging shark ecological roles, but noted other forces, including "*climate change, habitat loss, and the blue economy (such as energy, mining, shipping, and aquaculture)*" that will impact sharks and their prey.

Concluding Comments

Sharks have a 400+ million-year evolutionary history, but are they capable of surviving the current human-dominated era, the Anthropocene, and the current sixth great extinction we are currently causing? A 2014 report,[70] still relevant, by 23 shark specialists, led by Nicholas Dulvy, systematically assessed the threat to 1041 sharks, rays, and batoids. In the paper's abstract, they state:

> We estimate that one-quarter are threatened according to IUCN Red List criteria due to overfishing (targeted and incidental). Large-bodied, shallow-water species are at greatest risk and five out of the seven most threatened families are rays. Overall chondrichthyan extinction risk is substantially higher than for most other vertebrates, and only one-third of species are considered safe. Population depletion has occurred throughout the world's ice-free waters but is particularly prevalent in the Indo-Pacific Biodiversity Triangle and Mediterranean Sea. *Improved management of fisheries and trade is urgently needed to avoid extinctions and promote population recovery* (emphasis ours).

In 1992, 1700 scientists, including the majority of living Nobel laureates in the sciences, issued the *World Scientists' Warning to Humanity*:[71]

> Human beings and the natural world are on a collision course. Human activities inflict harsh and often irreversible damage on the environment and on critical resources. If not checked, many of our current practices put at serious risk the future that we wish for human society and the plant and animal kingdoms and may so alter the living world that it will be unable to sustain life in the manner that we know. Fundamental changes are urgent if we are to avoid the collision our present course will bring about.

A human population that is still growing, and which continually practices a resource-consumptive lifestyle, is the single biggest impediment to planetary sustainability. World trade has enormous potential to foster the objectives of development but can also be, and also has been, the source of massive environmental

BOX 12.2

Research, Conservation, and Education Focus

Climate Change is Increasing the Suitability of Bull Shark Nursery Habitats in the Gulf of Mexico

Philip Matich

Saving the Blue,
Davie, FL, US

Sharks face many threats in today's oceans. Overfishing, fisheries bycatch, and habitat degradation are the most immediately pressing.[1] However, less apparent threats, like climate change, are expected to exacerbate these human actions and challenge the management policies set in place to aid in the recovery of sharks, because of their uncontrollable nature.[2] Climate change is leading to the warming of seascapes around the world, which has implications for many species, including sharks.[3] Warming waters increase the energetic and oxygen needs of sharks because of their ectothermic nature.[4] If waters get too warm, habitats may become unsuitable for some species and can force species to seek new habitats because of their temperature tolerances. Controlled research studies show that this can even have implications for sharks before they are born.[5] But it's not clear that sharks will be uniformly affected by climate change.[6]

Scientists have been interested in the impacts of climate change on sharks for decades.[7] But only recently has there been adequate data to begin to understand these impacts. While laboratory studies show that sharks and their relatives can adjust their tolerances to warmer waters,[8] it is only in the last few years that studies have shown sharks are adjusting their behavior in the wild. Based on long-term monitoring and telemetry data, it is clear that at least some sharks are expanding or shifting their distributions poleward and into deeper waters[9] and changing when they make annual migrations.[10]

In the Gulf of Mexico, long-term monitoring data show that juvenile Bull Sharks have increased in their relative abundances within many of their nursery habitats, and climate change is at least in part responsible.[11] Warming waters have increased access to shallow waters of estuaries in the fall, allowing young sharks to remain in protected nurseries for longer periods of time, leading to increased first-year survival. These changes have been most prevalent in cooler, more northern Gulf estuaries like Galveston Bay, TX, and Mobile Bay, AL, which have experienced up to 600% increases in the relative abundance of juvenile Bull Sharks and can now be considered nurseries because of these changes. Consequently, increased abundances of juvenile sharks may lead to increased abundances of adults, contributing to recovery and conservation efforts in the region.

Yet it's unclear how climate change will impact these sharks moving forward. Warming waters have increased the suitability of habitats for juvenile Bull Sharks, but have also caused reductions in key prey populations due to reduced winter spawning success.[12] Bull Sharks feed on a variety of prey species[13] and exhibit flexibility in their diets.[14] Therefore, increasingly abundant juvenile Bull Sharks should be able to feed on other prey species, some of which have concurrently increased in abundance in the NW Gulf of Mexico in recent years (Gafftopsail Catfish; *Bagre marinus*[15]). But the long-term implications of these changes in shark abundance are unclear, including if food webs can support these growing nursery populations.

We know that sharks play important roles in ecosystems,[16] and Bull Sharks are the most abundant shark species in the estuaries of the NW Gulf of Mexico.[17] Uniquely, Bull Sharks move between freshwater and saltwater habitats,[18] and thus may provide an important service in connecting different habitats through their hunting behavior. Based on recent work using chemical tracers, we know that a variety of shark species,[19] including juvenile Bull Sharks,[20] can fill such a role, which helps stabilize ecosystems against changes in species composition. Will climate change affect this role Bull Sharks play? We don't know yet.

While the future is unclear, long-term monitoring coupled with directed research questions will offer more insight and guidance for policy makers concerning shark management and conservation in light of persistent and potentially accelerating climate change. Investigations of climate change impacts on sharks have been more prevalent in the developed world where shark management and conservation are prioritized (e.g., United States, Australia), but this pervasive threat will impact global shark populations.[21] Thus, more attention is warranted considering shifts in shark distributions and behaviors will likely necessitate changes in management approaches. Ultimately, we need to decide how to incorporate climate change predictions into current and new policies on the catch, consumption, and trade of sharks. To do this, we need more studies investigating how sharks respond to climate change.

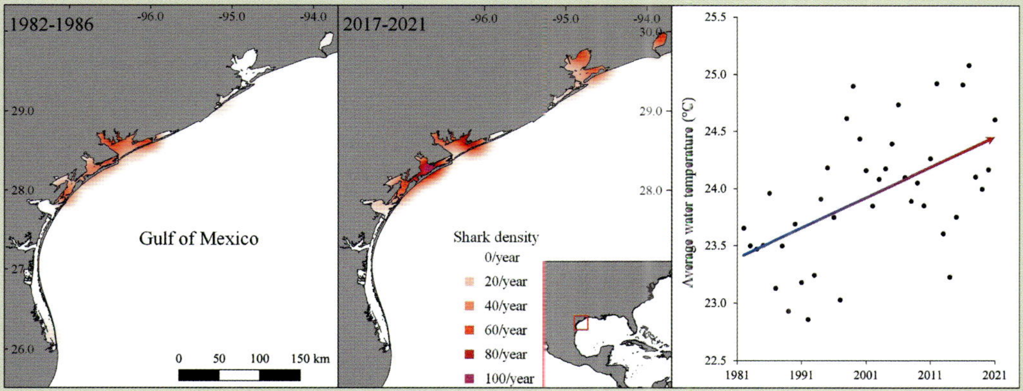

Changes in juvenile Bull Shark density in Gulf of Mexico.

1. Dulvy, N.K. et al. 202. Curr. Biol. 31: 4773–4787.

2. LeDee, O.E. et al. 2021. J. Wildl. Manag, 85: 7–16. and Cerutti-Pereyra F et al. 2024. Glob. Change Biol.: e17373.

3. Scheffers, B.R. et al, 2016. Science 354: 7671.

4. Pinsky, M.L. et al. 2020. Ann. Rev. Mar. Sci. 12: 153–179. and Rummer, J.L. et al. 2022. Chapter 25: Climate change and sharks. In: J.C. Carrier et al. (eds.), Biology of sharks and their relatives. (3rd ed.). CRC Press.

5. E.g., Rosa R., et al. 2014. Proc. Royal Soc. B: Biol. Sci. 281: 20141738; Musa, S.M. et al. 2020. J. Fish Biol, 97: 257–264; Wheeler, C.R. et al. 2020. Rev. Fish Biol. Fish. 30: 373–386; Wheeler, C.R. et al. 2021. Sci. Rep 11: 454; and Ripley, D.M. et al. 2021. Conserv. Physiol. 9: 1–7.

6. E.g., O'Brien S. et al. 2013. Conserv. Genet. 14: 125–144; Birkmanis, C.A. et al. 2020. Front. Mar. Sci, 7: 570; Diaz-Carballido, P.L. et al. 2022. Front. Mar. Sci. 8: 745501; and Cerutti-Pereyra F. et al. 2024. Glob. Change Biol.: e17373.

7. E.g., Cotton, P.A. et al. 2005. Fish. Oceanog. 14: 151–155 and Okey, T.A. et al. 2007. Fish Fisher. 8: 359–366.

8. E.g., Fangue, N.A. et al. 2003. Copeia 2003: 315–325.

9. E.g., Birkmanis, C.A. et al. 2020. Front. Mar. Sci, 7: 570; Braun, C.D. et al. 2023. Sci. Adv. 9: 2718; and Coulon N, et al. 2024. Glob. Change Biol. 30: e17157.

10. E.g., Hammerschlag, N. et al. 2022. Glob. Change Biol. 28: 1990–2005; Lubitz, N. et al. 2024. Nat Clim Change 14: 526–535 and Matich, P. et al. 2024. J. Anim. Ecol. doi:10.1111/1365-2656.14140.

11. Matich, P. et al. 2024. J. Anim. Ecol. doi:10.1111/1365-2656.14140 and Mullins, L. 2024. Sci. Rep. 14: 4100.

12. Erickson, K.A.2021. Glob. Change Biol. 27: 2520–2536.

13. Reviewed by Cottrant, E. et al. 2021. Mar. Ecol. Prog. Ser. 659: 127–141.

14. Matich, P. and Heithaus, M.R. 2014. J. of Anim. Ecol. 83: 199–213.

15. Matich, P. et al. 2021. J. Fish Biol. 98: 470–484.

16. Dedman, S. et al. 2024. Science 385 (6708), p.adl2362.

17. Plumlee, J.D. et al. 2018. Estuar. Coast. Shelf Sci. 204: 103–113.

18. Anderson, W.G. et al. 2006. Gen. Comp. Endocr. 147: 39–46.

19. Shipley, O.N.et al. 2023. Proc. Roy. Soc. B 290: 20230262.

20. Matich, P. and Heithaus, M.R. 2014. J. of Anim. Ecol. 83: 199–213.

21. Rummer, J.L. et al. 2022. Chapter 25: Climate change and sharks. In: J.C. Carrier et al. (eds.), Biology of sharks and their relatives. (3rd ed.). CRC Press.

degradation without multinational and international agreements. It remains to be seen if global agreements can stem the seemingly inexorable rise of greenhouse gas emissions. Major failures in the United States include top-down governmental intransigence or paralysis as well as an ignorant or indifferent citizenry. The system clearly is broken.

But there are also hopeful signs: the resilience of nature, the human spirit, effective management of fisheries, signs of recovery of some shark and even critically endangered sawfish populations, and you, whom we suspect are reading this book because of your interest and concern for the fate of sharks.

And do not overlook the value of environmental education. Every May for nearly 30 years we have taught a course in shark biology at the Biological Field Station in The Bahamas. The Shark Lab is a diminutive institution that punches way beyond its weight class in terms of its impact on the shark world. *Research, Conservation, and Education Focus* Box 12.3 offers insight into the Shark Lab.

While we all may put at least some of our faith in the ability of larger institutions to begin to turn the supertanker of human insults to the planet in general, and sharks and rays in particular, in the interim or even as an alternative, there are many actions that you can take.

First, you should get out into nature and, conversely, away from the addicting screens of your phone and laptop. Richard Louv, in his groundbreaking book, *Last Child in the Woods*, asserted that a new disorder, Nature Deficit Disorder (NDD), has arisen in an entire generation as a result of replacing outdoor experiences with staring at screens. NDD leads to a lack of interest and concern about environmental issues, and addiction to entertainment and electronics. Sound scholarship demonstrates that time spent on watching screens leads to shorter attention spans, diminished vocabularies, and impaired cognitive skills, and reinforces overconsumption, a key environmental problem. Children and young adults can name more corporate logos than sharks, as well as the trees and birds they see every day. They exhibit brand-loyalty instead of planet allegiance. To value nature, you must first experience it. Seeing a shark swimming in its environment can be a life-transforming experience, one that cannot be replicated on a screen.

Other actions are more intuitive. Actively participate in a movement to make the planet more sustainable.[72] If possible, reduce your ecological footprint by eating responsibly, being a more responsible consumer, and conserving energy. Reduce the amount of, eat ethically raised, or eliminate if possible, meat, dairy, and eggs in your diet, and buy local and, if possible, organic vegetables or grow your own. Avoid processed food in wasteful packaging. Educate yourselves about the social and environmental impacts of what you buy—and buy less. Use durable goods instead of disposables, skip the next computer and smartphone upgrades, and use mass transit.

Our last advice: make conservation biology and education a career, or at least a central part of your life. Remember that humans and sharks are both stakeholders in achieving and sustaining a livable planet.

BOX 12.3

Research, Conservation, and Education Focus

The Bimini Shark Lab and a Network of Shark Conservationists

Matthew Smukall

Bimini Biological Field Station
Bimini, The Bahamas

When Dr. Samuel "Doc" Gruber established the Bimini Biological Field Station (BBFS), more broadly known as the Shark Lab, in 1990, it was a small, rustic building on the sparse island of South Bimini, with a primary mission to provide a learning experience for early career marine scientists. As part of this mission, the Shark Lab has welcomed thousands of students to experience life at a remote field station, field research, and the unique ecosystems around Bimini. Shark Lab guests hail from countries around the world, many of whom have never previously seen sharks in the wild. Collectively, these people have been instrumental in contributing toward the missions of the Bimini Shark Lab and building one of the world's largest datasets and biological sample collections for elasmobranchs, especially Lemon Sharks and Tiger Sharks. These efforts have helped advance the overall understanding of life history, behavior, abundance trends, and movement of sharks and rays. As these people return home, they take with them lessons learned in Bimini and apply those to local projects or studies. A focus of the BBFS Foundation has been to help build a network of like-minded scientists, researchers, and educators. Today, there are hundreds of former "Shark Labbers" around the world focused on marine research and conservation.

A wide network of people is especially important given that the challenges facing marine conservation in the future decades will require a large collaborative approach, relying on the willingness of many entities to combine efforts for the greater good. The most effective conservation includes multiple stakeholders, including those from different backgrounds and perspectives. To best achieve these long-term conservation goals, it may require the compromise and collective efforts of people from diverse careers and expertise. While not immediately apparent to all, shark science and conservation should include people with experience in business, fishing industry, politics, social science, technologies, communications, etc. As many of these people may not be classically trained in sciences, it is our responsibility as scientists to help bridge this gap and explain the overall goals of research efforts.

Throughout daily life, people often remark "you have the coolest job." The truth is that many in the general public may have an oversimplified view of what "shark research" entails. To be sure, there are days on boats studying sharks in beautiful marine ecosystems. However, there are as many, if not more, days spent doing administrative tasks, writing grant proposals, seeking funding, preparing gear, analyzing data, drafting manuscripts, giving presentations, and finally trying to get the outcome of research to people that can influence policy or legislation. As scientists, most are not academically trained in business, accounting, mechanics, information technology, and politics. Therefore, it is essential that scientists and conservationists work with other professionals that have expertise in these disciplines.

During courses or public question and answer sessions, we are also asked "how people can get involved in shark research." My response to them is that because there are so many facets of "shark science," there are many ways for people to get involved. Not everyone may be fortunate enough to live near the ocean or be able to engage in the field aspects. However, there are other equally important ways to help work toward the collective goal of shark conservation. To support this end goal, we have been increasing our educational impact through the development of educator-focused shark biology courses hosted in Bimini. These educators from around the Bahamas, Canada, and United States have significant outreach capacity to reach multiple classes of students for several years to come. Importantly, many of these educators come from areas where marine sciences may be traditionally underserved. Not all students may want to become scientists, but it can be just as impactful if they develop an understanding and respect of sharks and take that with them into their career of choice. These students may become successful businesspeople, politicians, engineers, philanthropists, etc. who use their future positions to make decisions based upon a deep respect of the marine environment.

This book does a remarkable job of providing accessible and straightforward information that can be valuable to shark scientists, but also understood by people of non-scientific backgrounds, and we are always recommending it to students and educators alike. This approach does a tremendous job of bridging the gap between scientists and the more general public.

Dr. Matthew Smukall and Dr. Dean Grubbs take a group of educators to wade with Southern Stingrays near Bimini, The Bahamas.

NOTES

1. Retallack, H. et al. 2018. J. Wildlife Dis. 55: 375–386.

2. Chang, R.K. and Okihiro, M.S., 2024. Front. Veter. Sci, 11, p.1410332.

3. Borucinska, J.D. and Frasca Jr., S. 2002. J. Fish Dis. 25: 367–370.

4. Flewelling, L.J. et al. 2010. Mar. Biol. 157: 1937–953.

5. Smith, M. et al. 2017. *The Elasmobranch Husbandry Manual II.* Ohio Biological Survey, Inc. https://www.researchgate.net/profile/Lisa-Hoopes/publication/340675186_Chapter_13_Elasmobranch_mineral_and_vitamin_requirements/links/5e98637b92851c2f52a86161/Chapter-13-Elasmobranch-mineral-and-vitamin-requirements.pdf. (Accessed 10/2/24).

6. Drymon, J.M. et al. 2013. Fish. Bull. 111: 370–380.

7. Waller, M.J. et al. 2024.. J. Fish Biol. https://doi.org/10.1111/jfb.15830. (Accessed 8/18/24)

8. Clark, R.B. 2001. *Marine Pollution.* (5th ed.). Oxford University Press.

9. Matulik, A.G. et al. 2017. Mar. Pollut. Bull. 116: 357–364.

10. Kiszka, J.J. et al. 2015. Deep Sea Res. Pt. I: Oceanogr. Res. Pap. 96: 49–58.

11. See Fig. 4 in Gilbert, J.M. et al. 2015. Mar. Pollut. Bull. 92: 186–194.

12. Tiktak, G.P.et al. 2020. Mar. Poll. Bull., 160: 111701. doi: 10.1016/j.marpolbul.2020.111701. (Accessed 8/18/24).

13. Sisma-Ventura, G. et al. 2024. Environ. Intern. 187, p.108661.

14. Carrasco-Puig, P. et al. 2024. Mar. Poll. Bull. 199: 115942.

15. Steidinger, K.A. and Ingle, R.M. 1972. Env. Let. 3(4): 271–278.

16. Tiktak, G.P.et al. 2020. Mar. Poll. Bull., 160: 111701. doi: 10.1016/j.marpolbul.2020.111701 (Accessed 8/18/24).

17. Lyons, K. et al. 2015. J. Fish Biol. 87: 200–211.

18. Cotronei, S. et al. 2018. Bull. Env. Contam. Toxicol. 101: 7–13.

19. Cotronei, S. et al. 2018. Bull. Env. Contam. Toxicol. 101: 7–13.

20. Nakata, H. 2005. Env. Sci. Tech. 39: 3430–3434.

21. Xue, X. et al. 2017. Env. Sci. Tech. 51: 780–789.

22. Gelsleichter, J. and Szabo, N.J. 2013. Sci. Tot. Env. 456: 196–201.

23. de Farias Araujo, G. et al. 2024. Sci. Total Environ., 948: 174798.

24. Belcher, C.N. and Jennings, C.A. 2010. Env. Biol. Fish. 88: 349–359.

25. Yates, P.M. et al. 2015. PLoS One 10(4): p.e0121346.

26. Jennings, D.E. et al. 2008. Env. Biol. Fish. 83: 369–377.

27. Curtis, T.H. et al. 2013. Mar. Coast. Fish. 5: 28–38.

28. Madigan, D.J. et al. 2012. Proc. Natl. Acad. Sci. 109: 9483–9486.

29. Thorhaug, A. et al. 1979. Env. Conserv. 6: 127–137.

30. Coles, S.L. 2018. Mar. Res. Indonesia 19: 52–71.

31. Pilapitiya, P.N.T. and Ratnayake, A.S. 2024. Cleaner Materials, p.100220. https://www.sciencedirect.com/science/article/pii/S2772397624000042. (Accessed 10/2/24).

32. Geyer, R. et al. 2017. Sci. Adv. 3(7): p.e1700782.

33. Although we cite the specific sources of these data, we were informed by this excellent review: Bonanno, G. and Orlando-Bonaca, M. 2018. Env. Sci. Pol. 85: 146–154.

34. Derraik, J.G.B. 2002. Mar. Pollut. Bull. 44: 842–852.

35. Sazima, I. et al. 2002. Mar. Pollut. Bull. 44: 1149–1151; Wegner, N.C. and Cartamil, D.P. 2012. Mar. Pollut. Bull. 64: 391–394; and Colmenero, A.I. et al. 2017. Mar. Pollut. Bull. 115: 436–438.

36. Parton, K.J. et al. 2019.. Endang. Spec. Res. 39: 173–190.

37. Pytka, J.M. et al. R.W., 2024. Rev. Fish Bio. Fisher. 34: 131–160.

38. Thiel, M. et al. 2018. Front. Mar. Sci. 5: 238.

39. Alomar, C. and Deudero, S. 2017. Env. Pollut. 223: 223–229.
40. Bernardini, I. et al. 2018. Mar. Pollut. Bull. 135: 303–310.
41. Cartes, J.E. et al. 2016. Deep Sea Res. Pt. I: Oceanogr. Res. Pap. 109: 123–136.
42. Cartes, J.E. et al. 2016. Deep Sea Res. Pt. I: Oceanogr. Res. Pap. 109: 123–136.
43. Avio, C.G. et al. 2015. Mar. Env. Res. 111: 18–26.
44. Anastasopoulou, A. et al. 2013. Deep Sea Res. Pt. I: Oceanogr. Res. Pap. 74: 11–13.
45. Miranda, D.D.A. and de Carvalho-Souza, G.F. 2016. Mar. Pollut. Bull. 103: 109–114.
46. Elise Pullen, personal communication.
47. Fossi, M.C. et al. 2017. Comp. Biochem. Physiol. C: Toxicol. Pharmacol. 199: 48–58.
48. Gong, Y. et al. 2024. Rev. Fish Biol. Fisher.: 1–19.
49. Chapuis, L. et al. 2019. Sci. Rep.: 6924.
50. Cecilia Krahforst, personal communication.
51. Casper, B.M. et al. 2012. In: Popper, A.N. and Hawkins, A. (eds.). The effects of noise on aquatic life. Springer. 93–97 pp.
52. Casper, B.M. et al. 2012. In: Popper, A.N. and Hawkins, A. (eds.). The effects of noise on aquatic life. Springer. 93–97 pp.
53. Kanno, S. et al. 2023. Mar. Ecol. Prog. Ser. 724: 167–183.
54. Jennings, D.E. et al. 2008. Env. Biol. Fish. 83: 369–377.
55. Jennings, D.E. et al. 2012. J. Coast. Conserv. 16: 405–428.
56. Kanno, S. et al. 2023. Mar. Ecol. Prog. Ser. 724: 167–183.
57. Mcowen, C.J. et al. 2017. Biodiver. Data J.: 5.
58. Rummer, J.L et al., 2022. In: Carrier et al. (eds). Biology of sharks and their relatives). CRC Press. 767–793 pp.
59. Carrier, J.C. et al. (eds.). 2022. *Biology of sharks and their relatives.* CRC Press.
60. Santos, C.P. et al. 2024. Sci. Total Environ. 954: 76361.
61. Rosa, R. et al. 2017. Biol. Let. 13: 20160796
62. Schlaff, A.M. et al. 2014. Rev. Fish Biol. Fish. 24: 1089–1103.
63. Bangley, C.W. et al. 2018. Sci. Rep. 8: 6018.
64. Matich, P. et al. 2024. J. Anim. Ecol. doi: 10.1111/1365-2656.14140. (Accessed 8/25/24).
65. Robinson, L.M. et al. 2015. Deep Sea Res. II: Topic. Stud. Oceanogr. 113: 225–234.
66. Dulvy, N.K. et al. 2014. Elife 3: e00590.
67. Cerutti-Pereyra, F. et al. 2024. Glob. Change Biol. 30: e17373.
68. Rummer, J.L et al., 2022. In: Carrier et al. (eds). Biology of sharks and their relatives). CRC Press. 767–793 pp.
69. Dedman, S. at al. 2024. Science, 385(6708): adl2362. (Accessed 8/25/24).
70. Dulvy, N.K. et al. 2014. Elife 3: e00590.
71. Kendall, H., 1992. "1992 World Scientists' Warning to Humanity" https://www.ucsusa.org/resources/1992-world-scientists-warning-humanity. (Accessed 9/15/25).
72. Here are two recommendations: Support Shark Advocates International (for sharks), and Dogwood Alliance (for climate change and forest protection).

Appendix

Conservation: Efforts to Protect Sharks and Rays

The following 93 species of elasmobranchs are considered Critically Endangered by the IUCN.[1]

CRITICALLY ENDANGERED SHARKS AND BATOIDS*

*Stripenose Guitarfish, *Acroteriobatus variegatus*
*Duckbill Eagle Ray, *Aetomylaeus bovinus*
*Ornate Eagle Ray, *Aetomylaeus vespertilio*
*Narrow Sawfish, *Anoxypristis cuspidata*
*Spotback Skate, *Atlantoraja castelnaui*
Borneo Shark, *Carcharhinus borneensis*
Pacific Smalltale Shark, *Carcharhinus cerdale*
Pondicherry Shark, *Carcharhinus hemiodon*
Oceanic Whitetip, *Carcharhinus longimanus*
Lost Shark, *Carcharhinus obsoletus*
Smalltail Shark, *Carcharhinus porosus*
Sand Tiger, *Carcharias taurus*
Whitefin Swell Shark, *Cephaloscyllium albipinnum*
Reticulated Swell Shark, *Cephaloscyllium fasciatum*
Sarawak Pygmy Swell Shark, *Cephaloscyllium sarawakensis*
Indian Swell Shark, *Cephaloscyllium silasi*
Dwarf Gulper Shark, *Centrophorus atromarginatus*
*Longnose Skate, *Dentiraja confusa*
*Common Blue Skate, *Dipturus batis*
*Flapper Skate, *Dipturus intermedius*
*Menni's Skate, *Dipturus mennii*
*Colares Stingray, *Fontitrygon colarensis*
*Smooth Stingray, *Fontitrygon garouaensis*
*Wingfin Stingray, *Fontitrygon geijskesi*
*Thorny Whipray, *Fontitrygon ukpam*
Tope, *Galeorhinus galeus*
Ganges Shark, *Glyphis gangeticus*
*Sharpnose Guitarfish, *Glaucostegus granulatus*
*Halavi Guitarfish, *Glaucostegus halavi*
*Widenose Guitarfish, *Glaucostegus obtusus*
*Blackchin Guitarfish, *Glaucostegus cemiculus*
*Clubnose Guitarfish, *Glaucostegus thouin*

*Giant Guitarfish, *Glaucostegus typus*
*Bangladeshi Guitarfish, *Glaucostegus younholeei*
New Guinea River Shark, *Glyphis garricki*
Irawaddy River Shark, *Glyphis siamensis*
*Tentacled Butterfly Ray, *Gymnura tentaculata*
Natal Shyshark, *Haploblepharus kistnasamyi*
Whitefin Tope Shark, *Hemitriakis leucoperiptera*
*Korean Skate, *Hongeo koreana*
*Smalltooth Stingray, *Hypanus rudis*
Daggernose Shark, *Isogomphodon oxyrhynchus*
*Maltese Skate, *Leucoraja melitensis*
*Pakistan Whipray, *Maculabatis arabica*
*Shorttail Whipray, *Maculabatis bineeshi*
Striped Smoothhound, *Mustelus fasciatus*
Speckled Smoothhound, *Mustelus mento*
Narrownose Smoothhound, *Mustelus schmitti*
Humpback Smoothhound, *Mustelus whitneyi*
*Common Eagle Ray, *Myliobatis aquila*
*Shortnose Eagle Ray, *Myliobatis ridens*
*Starrynose Cowtail Ray, *Pastinachus stellurostris*
*Dwarf Sawfish, *Pristis clavata*
*Smalltooth Sawfish, *Pristis pectinata*
*Largetooth Sawfish, *Pristis pristis*
*Green Sawfish, *Pristis zijsron*
*Brazilian Guitarfish, *Pseudobatos horkelii*
Shorttail Nurse Shark, *Pseudoginglymostoma brevicaudatum*
*Bowmouth Guitarfish, *Rhina ancylostoma*
*Whitespotted Guitarfish, *Rhinobatos albomaculatus*
*Bengal Guitarfish, *Rhinobatos annandalei*
*Spineback Guitarfish, *Rhinobatos irvinei*
*Common Guitarfish, *Rhinobatos rhinobatos*
*Bottlenose Wedgefish, *Rhynchobatus australiae*
*Jimbaran Guitarfish, *Rhinobatos jimbaranensis*
*Smoothnose Wedgefish, *Rhynchobatus laevis*
*African Wedgefish, *Rhynchobatus luebberti*
*Smoothback Guitarfish, *Rhinobatos lionotus*
*Brown Guitarfish, *Rhinobatos schlegelii*
*Philippine Guitarfish, *Rhinobatos whitei*
*Lusitanian Cownose Ray, *Rhinoptera marginata*
*Clown Wedgefish, *Rhynchobatus cooki*
*Taiwanese Wedgefish, *Rhynchobatus immaculatus*
*Broadnose Wedgefish, *Rhynchobatus springeri*
*Whitespotted Wedgefish, *Rhynchobatus djiddensis*

*False Shark Ray, *Rhynchorhina mauritaniensis*
Scalloped Bonnethead, *Sphyrna corona*
Scalloped Hammerhead, *Sphyrna lewini*
Scoophead Shark, *Sphyrna media*
Great Hammerhead, *Sphyrna mokarran*
Smalleye Hammerhead, *Sphyrna tudes*
Sawback Angel Shark, *Squatina aculeata*
Argentine Angel Shark, *Squatina argentina*
Chilean Angel Shark, *Squatina armata*
Japanese Angel Shark, *Squatina japonica*
Indonesian Angel Shark, *Squatina legnota*
Hidden Angel Shark, *Squatina occulta*
Smoothback Angel Shark, *Squatina oculata*
Angel Shark, *Squatina squatina*
*Bignose Fanskate, *Sympterygia acuta*
*Argentine Torpedo, *Tetronarce puelcha*
*Red Sea Torpedo, *Torpedo suessii*
Spotted Houndshark, *Triakis maculate*
*Smalleye Round Ray, *Urotrygon microphthalmum*
*Reticulate Round Ray, *Urotrygon reticulata*

The list of elasmobranchs considered Endangered by the IUCN as of 2025 numbers 129 species, whereas an additional 178 species are listed as Vulnerable. Among the endangered species are the following iconic sharks:

Winghead Shark (*Eusphyra blochii*), Grey Reef Shark (*Carcharhinus amblyrhynchos*), Whale Shark (*Rhincodon typus*), Basking Shark (*Cetorhinus maximus*), Caribbean Reef Shark (*Carcharhinus perezi*), Dusky Shark (*Carcharhinus obscurus*), Blacknose Shark (*Carcharhinus acronotus*), Sandbar Shark (*Carcharhinus plumbeus*), Shortfin Mako (*Isurus oxyrinchus*), Pelagic Thresher (*Alopias pelagicus*), and other ecologically important but generally lesser-known species.

Protecting Sharks and Rays

In this section, we discuss some of the more prominent measures undertaken by governments and nongovernmental organizations, internationally and in the United States, in the name of shark and ray conservation and management.

International Efforts to Protect Sharks and Rays

Convention on International Trade in Endangered Species of Wild Fauna and Flora (CITES)

CITES is an agreement currently between 184 governments *to ensure that international trade in specimens of wild animals and plants does not threaten their survival.*

The following species of elasmobranchs are listed:[2] Basking Shark (*Cetorhinus maximus*); Whale Shark (*Rhincodon typus*); White Shark (*Carcharodon carcharias*); all species of sawfish; Oceanic Whitetip Shark (*Carcharhinus longimanus*); Porbeagle (*Lamna nasus*); Scalloped, Great, and Smooth Hammerheads (*Sphyrna lewini, S. mokarran,* and *S. zygaena,* respectively); manta and devil rays (Mobulidae); Silky Shark (*C. falciformis*); and threshers (Alopiidae). Sawfish (Pristidae) are listed in CITES Appendix I, *species threatened with extinction. Trade in specimens of these species is permitted only in exceptional circumstances.* All others are in CITES Appendix II, *those not necessarily threatened with extinction, but in which trade must be controlled in order to avoid utilization incompatible with their survival.*

In 2022, CITES added 60 species of requiem and hammerhead sharks and 37 species of guitarfish to Appendix II, thereby mandating that member governments document that trade is legal and sustainable. CITES measures do not directly affect fishing or finning restrictions, domestic use of sharks, or habitat conservation, and thus is a tool that should be complemented with others.

Convention on the Conservation of Migratory Species of Wild Animals (CMS)

Part of the United Nations Environment Programme, CMS works for *conservation and sustainable use of migratory animals and their habitats*... through... *internationally coordinated conservation measures.* CMS has 127 signatories, among which are notably absent the United States, the Russian Federation, and China.[3] In 2012, the *Memorandum on the Conservation of Migratory Sharks,* the first global document of this type for migratory species of sharks, included a conservation plan,[4] which was updated in 2023.[5] Thirty-eight species of sharks and rays are listed in Annex I of the Memorandum of Understanding. These include most of those listed for CITES earlier, as well as others.

Global Sharks and Rays Initiative (GSRI)

Begun in 2015 and sponsored by the UN, the GSRI is a collaborative global shark conservation strategy whose goal is to coordinate and catalyze the efforts of global, regional, and national nongovernmental organizations. The following organizations are collaborating in this effort: IUCN Species Survival Commission (SSC) Shark Specialist Group (SSG), Shark Advocates International (a project of The Ocean Foundation), Shark Trust, TRAFFIC, Wildlife Conservation Society, and World Wildlife Fund.

Their plan, documented in *Global Priorities for Conserving Sharks and Rays: a 2015–2025 Strategy,*[6] is that by 2025, *the conservation status of the world's sharks and rays has improved—declines have been halted, extinctions have been prevented, and commitments to their conservation have increased globally.* Among their "deliverables" for December 1, 2025 is: "Declines of the most endangered shark and ray species have been halted, and the conservation status of Data Deficient species is understood."[7] Other goals for this date include reducing overexploitation by fisheries and reducing demand for unsustainable shark and ray products.

International Union for the Conservation of Nature (IUCN) International Plan of Action for Conservation and Management of Sharks (IPOA-Sharks)

The IUCN, founded in 1948, is best known for its *Red List of Threatened Species*,[8] but it is also *the global authority on the status of the natural world and the measures needed to safeguard it*.[9] The IUCN Shark Specialist Group (SSG), one of whose members is co-author Grubbs, was established in 1991 by the IUCN Species Survival Commission to identify and prioritize species at risk, monitor threats, and evaluate conservation action.

In 1999, the SSG produced the Food and Agriculture Organization of the United Nations (FAO) International Plan of Action for the Conservation and Management of Sharks (IPOA-SHARKS).[10] The IPOA-Sharks recommended the following 10 principles:

1 / Ensure that shark catches from directed and nondirected fisheries are sustainable.

2 / Assess threats to shark populations, determine and protect critical habitats, and implement harvesting strategies consistent with the principles of biological sustainability and rational long-term economic use.

3 / Identify and provide special attention, in particular to vulnerable or threatened shark stocks.

4 / Improve and develop frameworks for establishing and coordinating effective consultation involving all stakeholders in research, management, and educational initiatives within and between States.

5 / Minimize the unutilized incidental catches of sharks.

6 / Contribute to the protection of biodiversity and ecosystem structure and function.

7 / Minimize waste and discards from shark catches.

8 / Encourage full use of dead sharks.

9 / Facilitate improved species-specific catch and landings data and monitoring of shark catches.

10 / Facilitate the identification and reporting of species-specific biological and trade data.

As of 2021 (see pp. 475–477), about one third of sharks, skates, rays, and chimaeras are listed by the IUCN as *Threatened*, a catchall category that includes *Critically Endangered*, *Endangered*, or *Vulnerable* (*Endangered* denotes a *very* high risk of extinction; *Critically Endangered* means a species is considered to be facing an *extremely* high risk of extinction). The risk of extinction is greatest for coastal species, especially tropical and subtropical sharks and rays, most of which are threatened.[11] Three species (Lost Shark, Red Sea Torpedo Ray, and Java

Stingaree) had not been reported for many decades and are now considered Possibly Extinct.[12] The most threatened chondrichthyan families include the sawfishes, wedgefishes, giant guitarfishes, devil rays, and pelagic eagle rays, all of which are batoids, as well as hammerheads and angel sharks.[13]

The IUCN Red List is not to be confused with species listed under the US Endangered Species Act (ESA). The Red List serves to highlight at-risk species on a global scale but carries no mandates for protections or policies. Different criteria are used to determine listing under the US Endangered Species Act.

Important Shark and Ray Areas (ISRA)[14]

A relatively new (2022) IUCN initiative, identifying Important Shark and Ray Areas (ISRA)[15] was begun to identify habitat "important for one or more species of shark, ray, or chimaera species, that has the potential to be delineated and managed for conservation."[16] An atlas of these areas exists on their website.[17]

International Shark Finning Bans and Policies

More than 30 countries, including the US (see US Laws below) have full or partial bans on shark finning (recall that *finning* refers to removing fins and discarding the carcass) and an additional 22 countries ban shark fishing.[18] Most international fishery bodies (e.g., the Inter-American Tropical Tuna Commission, or IATTC) have finning bans, but these rely on a complicated fin-to-body-weight ratio that is difficult to enforce. Except where shark fishing is banned, it remains legal to sell fins from sharks landed with intact fins attached, which is widely regarded as best practice for finning ban enforcement.

The Animal Welfare Institute maintains a website[19] listing the names of companies that ban or continue to serve shark fin soup, and airlines and shipping companies that will or will not transport shark fins.

US Laws

Magnuson-Stevens Fishery Conservation and Management Act

In federal waters of the United States (i.e., from 3 to 200 mi (4.8–322 km), the primary law governing fisheries management is the Magnuson-Stevens Fishery Conservation and Management Act. From the shore to three miles, management of marine resources is in the jurisdiction of the state, who typically work closely with federal managers.

According to the National Marine Fisheries Service (NMFS), the objectives of the Magnuson-Stevens Act are to prevent overfishing, rebuild overfished stocks, increase long-term economic and social benefits, and ensure a safe and sustainable supply of seafood.

The Magnuson-Stevens Act is administered by the NMFS, a part of the National Oceanic and Atmospheric Administration (NOAA). Fisheries in federal waters are managed by one of eight regional councils composed of stakeholders nominated by state governors and representing the public, scientists, commercial and recreational fishers, and the fishing industry.

The major means by which fisheries are regulated are *Fishery Management Plans* (FMPs), which conduct assessments of the health of the *stock* in question, and then establish scientifically sound regulatory measures.

Because many sharks, along with tunas, swordfish, and other billfish, migrate and thus cross numerous jurisdictional lines, they have been managed under the direct auspices of NOAA since 2006, specifically the Consolidated Atlantic Highly Migratory Species Fishery Management Plan.[20]

Three amendments to the Magnuson-Stevens Fisheries Act exclusively concerned sharks:

- The *Shark Finning Prohibition Act of 2000* which, according to NOAA, *prohibits any person under U.S. jurisdiction from engaging in the finning of sharks, possessing shark fins aboard a fishing vessel without the corresponding carcass, and landing shark fins without the corresponding carcass.*
- The *Shark Conservation Act of 2010* which, again according to NOAA, *requires that all sharks in the United States, with one exception, be brought to shore with their fins naturally attached*, closing a loophole in the finning prohibition act.
- *The Shark Fin Sales Elimination Act of 2021*, which passed the US Senate by a vote of 83–11, made it illegal to "possess, acquire, receive, transport, offer for sale, sell, or purchase a shark fin or a product containing a shark fin" in the United States except for certain spiny dogfish and smoothhound fins. Unlike the previous two shark-related acts, the Shark Fin Sales Elimination Act is very controversial. This legislation was popular with some conservationists but less so with many shark biologists, who think it might damage the already well-managed U.S shark fishery. This, in turn, may have the unintended consequence of encouraging increased trade of fins from unsustainable fisheries elsewhere to meet global demands.

US Endangered Species Act

Although seemingly perpetually in danger of being weakened by the federal government, the 1973 US Endangered Species Act still remains one of the most effective tools to protect species in the United States threatened with extinction.

The NMFS administers the program for marine and anadromous (those moving between fresh and salt water) organisms, while the US Fish and Wildlife Service is responsible for terrestrial and freshwater organisms.

As we write in 2025 three species of sharks and six batoids are listed. An organism listed as Endangered is in imminent danger of extinction whereas organisms listed as Threatened are considered likely to become endangered. The first sharks listed as Endangered, in 2014, were two distinct populations of the Scalloped Hammerhead, those in the Eastern Pacific and Eastern Atlantic Oceans. Populations of Scalloped Hammerheads in the Central and Southwest Atlantic and Indo-West Pacific were listed as Threatened. This listing was followed by the addition of the Angel Shark as Endangered and, in early 2018, the Oceanic Whitetip Shark as Threatened. Among batoids, five species of sawfish are listed as Endangered, and these were the first elasmobranchs listed on the US Endangered Species List. In 2018, the Giant Manta Ray was listed as Threatened.

In addition to protecting species, the idea of ecological restoration is also built into the Endangered Species Act, which requires that *the ecosystems upon which [threatened and endangered species] depend* be protected.

If you ever encounter an endangered elasmobranch (e.g., a sawfish) in the wild, you should not approach it or attract it, but you should report it to the NMFS or, specifically for sawfish, to the International Sawfish Encounter Database (ISED).[21]

NOTES

1. https://www.iucnredlist.org/search?query=sharks&searchType=species. (Accessed 9-17-25).
2. Sharks and rays. https://www.cites.org/prog/shark. (Accessed 9-17-25).
3. Convention on the Conservation of Migratory Species of Wild Animals. Parties and Range States. https://www.cms.int/en/parties-range-states. (Accessed 9-17-25).
4. CMS Memorandum on the Conservation of Migratory Sharks. Annex 3 to the MoU: Conservation Plan. https://www.cms.int/sites/default/files/document/Sharks_Conservation_Plan_E_0.pdf. (Accessed 9/17/25).
5. CMS Memorandum of Understanding on the Conservation of Migratory Sharks. https://www.cms.int/sharks/en/page/sharks-mou-text. (Accessed 9/17/25).
6. Bräutigam, A. et al. 2016. Global Priorities for Conserving Sharks and Rays: A 2015–2025 Strategy. https://portals.iucn.org/library/node/45894. (Accessed 9-17-25).
7. UN Global Priorities for Conserving Sharks and Rays. A 2015 2025 Strategy. https://sdgs.un.org/partnerships/global-priorities-conserving-sharks-and-rays-2015-2025-strategy#deliverables. (Accessed 9/17/25).
8. The IUCN Red List of Threatened Species. https://www.iucnredlist.org. (Accessed 9/17/25).
9. About IUCN. https://www.iucn.org/about. (9/17/25).
10. UN FAO International Plan of Action for Conservation and Management of Sharks. https://www.fao.org/ipoa-sharks/en/. (Accessed 9/17/25).
11. Dulvy, N.K. et al. 2021. Cur. Biol. 31(21): 4773–4787.
12. Dulvy, N.K. et al. 2021. Cur. Biol. 31(21): 4773–4787.
13. Dulvy, N.K. et al. 2021. Cur. Biol. 31(21): 4773–4787.
14. Important Shark and Ray Areas (ISRA). https://sharkrayareas.org/. (Accessed 9/17/25).
15. Important Shark and Ray Areas (ISRA). (Accessed 9/17/25).
16. Frequently Asked Questions About ISRAs. https://sharkrayareas.org/isra/faq/#toggle-id-1. (Accessed 9/17/25).
17. The Web-GIS ISRA Atlas. https://sharkrayareas.org/e-atlas/. (Accessed 9/17/25).

18. International Shark Finning Bans and Policies. https://awionline.org/content/international-shark-finning-bans-and-policies. (Accessed 9/17/25).

19. International Shark Finning Bans and Policies. https://awionline.org/content/international-shark-finning-bans-and-policies. (Accessed 9/17/25).

20. Consolidated Atlantic Highly Migratory Species Management Plan. https://www.fisheries.noaa.gov/management-plan/consolidated-atlantic-highly-migratory-species-management-plan. (Accessed 9/17/25).

21. Report a Sawfish Encounter. https://www.floridamuseum.ufl.edu/sawfish/report-encounter/. (Accessed 9/17/25).

INDEX

Page numbers in *italics* signify figures.

CAMPAIGN 431

FLORENCE 1944

The Allied Advance in Central Italy

PIER PAOLO BATTISTELLI ILLUSTRATED BY JOHNNY SHUMATE

OSPREY PUBLISHING
Bloomsbury Publishing Plc
Kemp House, Chawley Park, Cumnor Hill, Oxford OX2 9PH, UK
Bloomsbury Publishing Ireland Limited,
29 Earlsfort Terrace, Dublin 2, D02 AY28, Ireland
Bloomsbury Publishing Inc.
1359 Broadway, 12th Floor, New York, NY 10018, USA
E-mail: info@ospreypublishing.com
www.ospreypublishing.com

OSPREY is a trademark of Osprey Publishing Ltd

First published in Great Britain in 2026

A catalogue record for this book is available from the British Library.

ISBN: PB 9781472870506; eBook 9781472870513; ePDF 9781472870483; XML 9781472870490

26 27 28 29 30 10 9 8 7 6 5 4 3 2 1

Maps by Bounford.com
3D BEVs by Paul Kime
Index by Fionbar Lyons
Typeset by Lumina Datamatics Ltd
Printed by Repro India Ltd

MIX
Paper
FSC FSC® C047271

Osprey Publishing supports the Woodland Trust, the UK's leading woodland conservation charity.

To find out more about our authors and books visit **www.ospreypublishing.com**. Here you will find extracts, author interviews, details of forthcoming events and the option to sign up for our newsletter.

For product safety related questions contact productsafety@bloomsbury.com

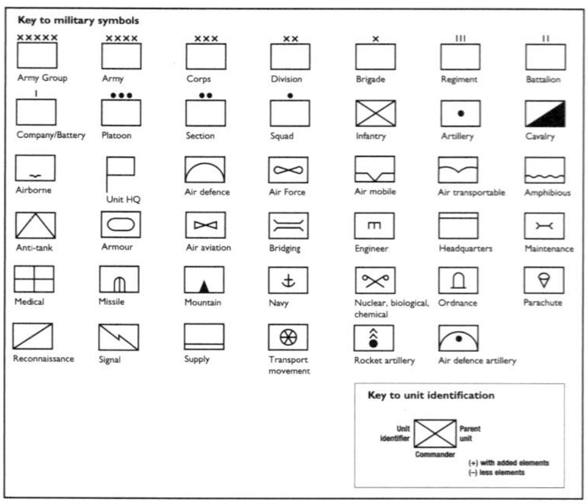

Acknowledgements

The author gratefully acknowledges the help and support provided by Marc Romanych of the Digital History Archive, Dottor Piero Crociani (Rome), Dr Christopher Pugsley (Waikanae, New Zealand), Dr Klaus Schmider (Sandhurst) and the invaluable help provided by Liz Evans with her research at The National Archives. Last but not least a big thank you goes to the series editor, Brianne Bellio.

Abbreviations

A, Armd	Armoured
AAI	Allied Armies in Italy
AOK	Armee Oberkommando (Army command)
CCA/CCB	Combat Command A/B
CDN	Canadian
CIL	Corpo Italiano di Liberazione (Italian Liberation Corps)
DIA	Division d'Infanterie Algérienne (Algerian infantry division)
DIM	Division d'Infanterie Motorisée (motorized infantry division)
DMI	Division d'Infanterie Marocaine (Moroccan infantry division)
FB	Festungs Bataillon (fortress battalion)
FEC	French Expeditionary Corps
FJ	Fallschirmjäger (paratrooper)
FJR	Fallschirmjäger Regiment (paratrooper regiment)
Flak	Flieger Abwehr Kanone (anti-aircraft gun)
FR	France
Gds	Guards
Geb	Gebirgs (mountain)
GR	Grenadier Regiment (German infantry regiment)
HG	Hermann Göring (division)
HGr	Heeresgruppe (army group)
HQ	Headquarters
ID	Infanterie Division
IND	Indian
IR	Infanterie Regiment (infantry regiment)
KG	Kampfgruppe (battle group)
LSB	Landesschützen Battalion (territorial unit)
LW	Luftwaffe
NZ	New Zealand
Pak	Panzer Abwehr Kanone (anti-tank gun)
PG	Panzergrenadier
PGR	Panzergrenadier Regiment (mechanized infantry regiment)
POL	Poland
Pz	Panzer (armoured)
PzKpfw	Panzerkampfwagen (tank)
RCT	Regimental Combat Team
RTS	Regiment Tiralleurs Senegalais
SA	South Africa
SP	Self-Propelled
SS	Waffen-SS unit
TF	Task Force
UK	United Kingdom
US	United States

Unit designations

For simplicity, Allied units are usually denoted by their nationality (e.g. BR, US, CDN, NZ). German units retain their denomination such as Armee Oberkommando (army) and Korps (corps), or Panzer and Fallschirmjäger (armoured and parachute), mostly related to the various types of divisions: Infanterie (infantry, whose units were re-designated Grenadier), Panzer (armour), Panzergrenadier (mechanized infantry) and Fallschirmjäger (paratrooper). German units at battalion level within regiments are usually written like this: I./FJR 1, meaning the first battalion of Fallschirmjäger Regiment 1.

Front cover main illustration: The attack at Viterbo, 8–9 June 1944. (Johnny Shumate)
Title page photograph: A column of Sherman tanks belonging to the South African Prince Alfred's Guard and Light Horse/Kimberley Regiment have reached the main square of Impruneta on 3 July. (NARA via Digital History Archive)

CONTENTS

ORIGINS OF THE CAMPAIGN

This part of the Italian campaign, which ended with the battle for Florence, the last step of a long march that had started in May 1944 to the south of Rome, clearly shows how the 1944 battles for Cassino and Anzio simply produced a stalemate – basically delaying both Allied and German strategy. Shortly after the Salerno battle in September 1943, when the Allied forces started their advance along the Italian peninsula, the prevailing idea on both sides was that the decisive battle would be fought on the northern Apennines, on what would be known as the 'Gothic Line'. From the German side, the basic strategic concept was to delay the enemy advance so as to be able to prepare the defences of the Gothic Line, while keeping an eye to any possible development in the Balkans (meaning a possible Allied landing). From the side of the Allied commanders, the

Ponte Vecchio in Florence after the battle. Note the buildings destroyed at the northern end to prevent its use; the same happened at the southern end. (Captain Tanner/Imperial War Museums via Getty Images)

concept was basically the same: the Germans would withdraw north of Rome, and make a stand on the northern Apennines. This required preparing an attack plan in order to break through into the Po River valley. Both strategic concepts were shelved between December 1943 and May 1944, when the attention focused on breaking the stalemate and fighting for Anzio and Cassino.

Following the Allies' entry into Rome on 4 June 1944, the Italian campaign developed again into a war of movement that saw, on the one side, the Germans withdrawing north and, on the other side, the Allies pursuing them. Once again, uncertainty prevailed on the Allied side. While emphasizing the need to destroy as many German forces as possible, in particular those facing the US 5th Army, the Allied plan still focused on reaching the area needed to prepare the final assault to the Gothic Line and preventing the Germans from creating another defence line in central Italy. The aim of gaining ground, based on both time and space, also influenced the German strategy. As in 1943, the Germans needed to gain time to prepare the defences of the Gothic Line and therefore used space to achieve their goal. Eventually, the 1944 campaign – culminating in the battle for Florence – could be seen as nothing more than a simple pursuit: one side pushing hard against the withdrawing enemy forces to prevent their reorganization and the development of a strong defence line; the other side intent on a delaying withdrawal all along the area between Rome and Florence.

In reality, on a closer look, the 1944 Florence campaign can be considered one of missed opportunities. Following the breakthrough at Cassino and Anzio, and particularly after the battle for Rome, the German forces were at their lowest ebb, weaker than ever. The Allies could now rely on the most effective elements of their superiority, speed (guaranteed by their high degree of motorization and mechanization) and interdiction of enemy movements, the latter being attained thanks to their air superiority. Both factors could have been used to manoeuvre the available forces with the aim of actually destroying the German ones, if not in their entirety at least in such a proportion as to make an actual defence of the Gothic Line impossible. Had the Allied forces pursued the aim of destroying the German forces rather than gaining ground as fast as possible, the outcome of the entire campaign might have been completely different. Eventually the attack on the Gothic Line would have faced much less opposition, and a breakthrough into the Po valley might have been achieved already in 1944. What might sound like a 'what if' history exercise is in fact an analysis of the campaign fought between June and August 1944 between Rome and Florence. Had this been fought by the Allies relying on speed and manoeuvre rather than with the aim of gaining ground, the Germans might have been inflicted severe losses and made unable to defend the Gothic Line for a protracted period of time. There are several reasons that this never happened, not least the decision to land in southern France, which greatly weakened the Allied forces in Italy. Most importantly, this makes the summer 1944 campaign in Italy decisive in a way, as it may have definitively influenced the campaign that followed. Once again, the Germans dominated the battlefield with their withdrawals, somehow imposing their own strategy on the Allies in spite of their marked inferiority. Which makes this campaign yet another example of how to conduct a withdrawal, and not how to conduct a pursuit.

CHRONOLOGY

1944

11 May	Start of Operation *Diadem*, the fourth Cassino battle, which ends with the Polish seizure of the town and abbey on 18 May
22–23 May	US VI Corps breaks out from the Anzio bridgehead
27 May	Decision to switch from the advance towards Valmontone in favour of the advance towards Rome, the battle for Rome begins
4 June	The battle of Rome ends with the German withdrawal and the entry into the city of American troops
5 June	Start of the German retreat north
6 June	Allied landing in Normandy
8–9 June	The US 5th Army reaches Viterbo
10 June	The newly formed Corpo di Liberazione Italiano reaches Pescara, the US VI Corps is withdrawn from the front
13 June	The British 8th Army troops reach Terni and Narni
15 June	The US 5th Army troops reach Grosseto
16–17 June	French forces land in the islands of Pianosa and Elba
20 June	German forces deploy on the Trasimene Line
25 June	The US 5th Army reaches Piombino
19–28 June	Battle of Lake Trasimene
1 July	German withdrawal to the Florence intermediate defence line, the city is declared open
2–3 July	The Germans withdraw from Siena allowing French troops to enter the city
8 July	US 5th Army units enter Volterra, the Polish II Corps starts its offensive towards Ancona
16 July	British 8th Army troops reach Arezzo
17–18 July	Battle for Ancona
19 July	US 5th Army units enter Leghorn
20 July	The British 8th Army starts its offensive towards Florence
23 July	Allied forces reach the Arno River, US 5th Army troops enter Pisa
23 July–8 August	The battle of Florence is fought by the British 8th Army
11 August	Allied forces cross the Arno River at Florence
25 August	Start of Operation *Olive*, the battle for the Gothic Line

OPPOSING COMMANDERS

ALLIED

Of the many leaders who fought in Italy in the summer of 1944, one stands out for his decision-making on the battlefield: **Mark Wayne Clark (1896–1984)**. Born on 1 May 1896 in New York State, Mark Wayne Clark graduated at West Point in 1917, two years after his friend Dwight David Eisenhower. Commissioned as a second lieutenant in the infantry, Clark was promoted to captain in August 1917, a few months before arriving on the Western Front in Europe. Wounded shortly after his units were deployed to the front, Clark was deemed unfit for frontline duties and was transferred to the Supply Section of the newly formed First Army. This enabled Clark to serve after the war in several staff duties at high level, including aide to the Assistant Secretary of War and, from 4 August 1941, as assistant chief of staff at the General Headquarters of the US Army. This also saw Clark rise in rank, as he was promoted to temporary brigadier general. In January 1942, Clark was appointed deputy chief of staff of the Army Ground Forces, becoming chief of staff in May. Promoted to major general in April 1942, Clark joined Eisenhower at the command of US II Corps first, then of the Army Forces in the European Theatre of Operations. In October, Clark followed Eisenhower again, now as his deputy, to the Mediterranean Theatre of Operations, with the latter being now Supreme Allied Commander in the area. Promoted to lieutenant general in November 1942, in January 1943 Clark was given command of the 5th Army deployed in French North-West Africa. Clark commanded 5th Army during the September 1943 Salerno landings, and held the command until December 1944 when he took over command of the Allied Armies in Italy Headquarters (AAI HQ), as well as 15th Army Group, from Field Marshal Harold Alexander, a position he held until the end of the war. After commanding US occupation forces in Austria after the war, in 1952–53 Clark commanded the United Nations forces in Korea. He retired on 31 October 1953 after serving as Commander of the Far East Command. Between 1954 and 1965, Clark commanded the South Carolina Military College (the Citadel) before his final retirement in 1965. He died on 17 April 1984. Much criticized for the January 1944 Rapido (actually the Gari) River crossing, Clark also made a fatal decision in May 1944 as he switched the 5th Army's axis of advance from Valmontone to the north, towards Rome. This is an instance of what was surely a lack of determination and ingenuity in Clark's command during the summer of 1944, even though one has to take

Group photo in Rome at the Campidoglio on 5 June 1944. From left: Major General Geoffrey Keyes, Lieutenant General Mark Clark, Major General Lucian Truscott, General Alphonse Juin, Major General Alfred M. Gruenther, chief of staff expeditionary army. (US National Archives and Records Administration, hereafter NARA)

into account the fact that, at the same time, his army was being depleted of its best forces.

Much of 5th Army's successes on the battlefield were due to the high quality of its corps commanders. First of all was **Lucian K. Truscott (1895–1965)**. Born on 9 January 1895 in Texas, he was commissioned in 1917 as a cavalry officer serving on the Mexican border during World War I. A graduate of the US Army Command and General Staff College, in August 1940 Truscott was promoted to lieutenant colonel and served as a staff officer with the 9th Army Command, which put him in touch with Dwight Eisenhower and enabled him to develop the foundation of the US Army Rangers, modelled after the British Commandos. Temporarily promoted to colonel in December 1941, in May 1942 Truscott was assigned as a staff officer to the Allied Combined Arms Staff and was eventually promoted to major general and given command of the elements of 9th Infantry Division, which landed in French North-West Africa in November 1942. As commander of 3rd Infantry Division in March 1943, he took part in the invasion of Sicily and subsequent operations in Italy, until landing at Anzio in January 1944. On 17 February 1944, Truscott was given command of VI Corps following the removal of General Lucas, and led it in the Italian campaign until 15 August 1944, landing in southern France. Promoted to lieutenant general in September 1944, Truscott took over command of 15th Army until replacing Clark as commander of 5th Army in Italy on 3 December 1944. After the war, Truscott held several army commands until his retirement in September 1947. He died on 12 September 1965. Albeit a major mistake, the switching of the advance axis at Valmontone on 27 May was also a spectacular achievement, VI Corps literally turning on its heels to move north. With this, Truscott proved his capabilities and surely might have done better under different circumstances.

As commander of II Corps, **Geoffrey T. Keyes (1888–1967)** played a main role during the summer 1944 campaign in Italy. Born on 30 October 1888 in New Mexico, he was commissioned in 1913 as a cavalry officer, seeing action in the raid against Pancho Villa. A language instructor at West Point during World War I, Keyes served as a staff officer with the Panama Canal Division before attending the Army War College and becoming chief of staff of the 2nd Armored Division under General Patton. Promoted to brigadier general in December 1941, Keyes took over command of Combat Command B (CCB) of 3rd Armored Division, being eventually promoted to major general and given command of 9th Armored Division. In November 1942, Keyes was deputy commander of I Armored Corps, later becoming the 7th Army during the July 1943 invasion of Sicily. On this occasion, Keyes led the advance to Palermo in charge of a Provisional Corps, until taking over command of II Corps in September 1943. He led the corps until 1945, and in due course was promoted to lieutenant general in April 1945. After commanding

7th and 3rd Armies, Keyes retired from active duty in 1954 and died on 17 September 1967.

A newcomer to the Italian campaign, **Willis D. Crittenberger (1890–1980)** was born on 2 December 1890 in Maryland. After graduating from West Point in 1913 as a cavalry officer, Crittenberger served as a training officer in the United States during World War I, eventually attending the Army Command and Staff College, which helped him develop an interest in armoured warfare. Promoted to temporary colonel in June 1941, and a brigadier general in July, Crittenberger commanded the 2nd Armored Brigade, part of 2nd Armored Division, before being promoted to major general in February 1942 and taking over command of the division from Patton. From August 1942, Crittenberger oversaw the creation of III Armored Corps, which in October 1943 became XIX Corps and was deployed to Britain in January 1944. In March, Crittenberger assumed command of IV Corps in Italy, being deployed at the front in June–July to replace VI Corps. Crittenberger, who held the command until the end of the war, was promoted to lieutenant general in June 1945 and, after the war, commanded the Caribbean Defense Command before retiring in December 1952. He died on 4 August 1980. His reputation as a corps commander was defended by Truscott, who ultimately considered him a better commander than Keyes. Still, the lack of determination and initiative in IV Corps during the summer 1944 campaign in Italy was noticeable.

Major General Willis D. Crittenberger, second from left, photographed in April 1945 as he accepts the surrender of the Army of Liguria. The German general close to him is the chief of staff, Generalmajor Max Pemsel. (NARA)

Born on 16 December 1888 in Bône, Algeria, **Alphonse Juin (1888–1967)** graduated at the Saint-Cyr Military Academy in 1912, taking part in the 'pacification' of Morocco before leading Moroccan troops during World War I. After completing his studies at the École Supérieure de Guerre in 1921, Juin served in Tunisia and then again in Morocco, taking part in the Rif campaign. After teaching at the École Supérieure de Guerre in 1933, Juin was then appointed second-in-command of the 3e Régiment de Zouaves, which he commanded from 1935, in due course being promoted to colonel in 1937. A brigadier general from 1938, following the outbreak of World War II, Juin asked to serve in the homeland and was given command of the 15e Division d'Infanterie Motorisée in December 1939. Taken prisoner in May 1940, Juin was released by request of the Vichy government, which made him an aide to the commander-in-chief of the troops in Morocco. Promoted to full generalship in July 1941, Juin replaced General Weygand in command of the French forces in North Africa in November 1941. Facing the Allied landing in North-West Africa in November 1942, Juin advocated for an immediate ceasefire and was then made commander of the Détachement d'Armée Française (DAF), being also promoted to colonel general in December 1942. As such, Juin organized the Corps Expéditionnaire Français (FEC, French Expeditionary Corps), which he led in Italy from November 1943. In August 1944, Juin succeeded Charles de Gaulle as Chief of the Armed Forces Staff, dedicating himself to the reorganization of the French Army. In 1946, Juin was sent to Indochina (then Vietnam), before being posted again to French North Africa (Rabat, Morocco) until 1950. After serving as Inspector General of the French Armed Forces from 1951, Juin was promoted to Maréchal de France in May 1952. Retired from active duty in 1957, he died on 27 January 1967.

Even though the Polish II Corps only played a minor role in the summer 1944 campaign, a mention has to be made of **Władysław Anders (1892–1970)**, born on 11 August 1892 in Warsaw, then part of Russia. Anders served with the Russian Army until 1918, when he joined the Polish Army (or Polish Land Forces), eventually commanding the 15th Ulan Cavalry Regiment in the 1919–21 war between Poland and Russia. After attending the French Military Academy at Saint-Cyr, in 1920 Anders was part of the staff of General Rozwadowski, the Chief of the General Staff, and, after opposing the 1926 attempted coup, he was finally given command of a cavalry brigade in 1931. Promoted to full generalship in 1934, Anders was in command of a cavalry brigade in September 1939 and, after fighting the Germans, was taken prisoner by the Soviets on 29 September while trying to reach the Hungarian border. Released following the German attack against the Soviet Union, Anders led the newly built Polish Army, which was evacuated in March 1942 to Palestine via the Persian Corridor. Here, the army was transformed into the Polish II Corps, which Anders led during the Italian campaign until 1945. Deprived of rank and citizenship by the new Polish government after the war, Anders remained in the United Kingdom acting as Inspector General of the Polish Armed Forces in exile. He died on 12 May 1970.

The British 8th Army played a lesser role during the summer 1944 campaign, which was partly because of its commander, **Oliver Leese (1894–1978)**. Born on 27 October 1894 in London, Leese attended Eton and joined the Officers' Training Corps, thereafter joining the British Army in September 1914 as a reserve officer. Serving with a battalion of the Coldstream Guards in France during the first stage of World War I, Leese was wounded but eventually returned to the front, also serving with Guards units. Awarded the Distinguished Service Order (DSO) and Mentioned in Despatches, after the war Leese remained with the Army and was promoted to captain in 1921, before attending the Staff College Camberley. After serving as a staff officer, in 1938 Leese was made a brevet colonel and was posted to India, where he served as an instructor at the Quetta Staff College. In 1937, he inherited the baronetcy from his father and, in March 1940, he returned to Britain, taking over command of 20th Independent Infantry Brigade, which was sent to Norway. Posted to the General Headquarters of the British Expeditionary Force in France as a staff officer, Leese was promoted to acting major general on 11 May 1940 (he had been promoted to brigadier nine days earlier), taking part in the evacuation from Dunkirk. After leading a brigade group, on 30 December 1940, Leese was

Lieutenant General Władysław Anders (sitting on a Sherman gun barrel) in April 1945 when the Polish II Corps faced its last battle in Italy. (Narodowe Archiwum Cyfrowe, hereafter NAC)

given command of the 15th Infantry Division, then of the Guards Armoured Division. In December 1941, he was promoted to major general and, in September 1942, was dispatched to Egypt to take over command of XXX Corps. After fighting at El Alamein and Tunisia, Leese commanded the corps during the July 1943 invasion of Sicily, being promoted to lieutenant general in September 1943. In December 1943 Leese replaced Montgomery in command of 8th Army, a position he held until September 1944, when he was made commander-in-chief of 11th Army Group in India. Knighted in the field in July 1944 by King George VI, Leese was removed from command following a clash with General Slim, and retired from active duty in 1947. He died on 22 January 1978.

Lieutenant General Oliver Leese being knighted in the field by King George VI in Italy, near Orvieto. (Keystone/Hulton Archive/Getty Images)

Amongst the most prominent corps fighting under Leese's command during the campaign was the XIII Army Corps, led by **Sidney Kirkman (1895–1982)**. Born on 29 July 1895 in Bedfordshire, Kirkman attended the Royal Military Academy Woolwich, and was commissioned in February 1915. An artillery officer, he served with distinction during World War I, which he ended with a promotion to acting major. After obtaining a regular Army commission, between the wars Kirkman served as an officer in various areas of the British Empire until 1932, when he attended the Staff College Camberley, being then made General Staff Officer at the War Office in 1934. After a staff posting with the Royal Air Force between 1938 and 1940, Kirkman commanded a Territorial Army artillery regiment until 1940, when he was sent to France as part of the British Expeditionary Force. Following the evacuation from Dunkirk, Kirkman was awarded an OBE and promoted to acting brigadier in March 1941, being given command of the Royal Artillery, then of the 56th Infantry Division, of I Corps, and of various corps commands in Britain until he was sent to Egypt in September 1942. After serving as 8th Army's artillery commander, in April 1943 Kirkman was promoted to acting major general and given command of the 50th Infantry Division, which he led in Tunisia and in Sicily. Promoted to acting lieutenant general in January 1944, Kirkman was given command of XIII Corps, which he led until the end of the war. After the war, Kirkman held various staff positions until retiring from active duty in 1950. He died on 29 October 1982.

A portrait of Lieutenant General Sidney Kirkman, XIII Corps' commander. (© National Portrait Gallery, London)

Another prominent role was played by X Corps, led by **Richard McCreery (1898–1967)**. Born on 1 February 1898 in Leicestershire, he was commissioned in 1915 as a cavalry officer, and was deployed to the Western Front in 1916. Serving as infantry during World War I, McCreery was severely wounded and, after the war, became adjutant to the 12th Lancers Regiment. After attending the Staff College Camberley in 1928, in 1930 McCreery was made brigade major and given command of his regiment. In 1940, McCreery fought in France commanding the 2nd Armoured Brigade and, after Dunkirk, in December 1940 was promoted to acting major general and given command of 8th Armoured Division. Now an expert in mechanized warfare,

in March 1942 McCreery became an advisor in armoured fighting vehicles (AFVs) for the 8th Army commander, Sir Claude Auchinleck, before becoming part of Harold Alexander's staff at the Middle East Command. He was chief of staff of 18th Army Group in 1943, during the campaign in Tunisia, still under Alexander's command. In July 1943, McCreery was given command of VIII Corps, then in August he took over command of X Corps, which took part in the campaign in Italy. On 1 October 1944, McCreery replaced Oliver Leese as commander of 8th Army, a position he held until the end of the war. Commander of the British forces in Austria until 1946, McCreery then replaced Montgomery as commander of the British Army of the Rhine until 1948 before retiring in 1949, after promotion to full generalship. He died on 18 October 1967.

Commander of the British X Corps, the future commander of 8th Army, Lieutenant General Richard McCreery. (Public domain via Wikimedia Commons)

GERMAN

As the commander-in-chief in Italy, Feldmarschall Albert Kesselring imposed a basic strategy of staged withdrawal in order to gain time. From the German side, the summer 1944 campaign became more a matter of small unit tactics, which saw the army and corps commanders dealing with the situation first-hand. Facing the more critical situation was the new Armee Oberkommando (AOK) 14's commander, **Joachim Lemelsen (1888–1954)**. Born on 26 September 1888 in Berlin, Lemelsen got a commission as an artillery officer in 1907, taking part in World War I and subsequently joining the Reichswehr. Promoted to Oberst in 1934, he commanded various artillery units until 1935, when he took over command of the Artillery School. Promoted Generalmajor in 1937, in 1938 he was given command of the 29. Infanterie Division, which he led in the campaigns against Poland and France. At the end of May 1940, he was given command of 5. Panzer Division, which he led during the second half of the 1940 campaign in the West. Promoted to General der Artillerie on 1 August 1940 (in June 1941 the designation was altered to General der Panzertruppe), in October 1940 Lemelsen was given command of the XXXXVII. Armee Korps, which he led in the war against the Soviet Union. Awarded the Ritterkreuz (Knight's Cross) in July 1941, in July 1943 Lemelsen took part in the battle of Kursk before being given command of the LXXXVII. Armee Korps in Italy in September 1943. In November and December 1943, he was temporary commander of AOK 10 in the absence of von Vietinghoff, being then transferred to the Führerreserve (temporary leave) and being appointed commander of AOK 1 in May 1944. Following General von Mackensen's removal from command of AOK 14, on 7 June 1944 Lemelsen took over leading the army until 24 October 1944, when he again took over command of AOK 10. Lemelsen became commander of AOK 14 once more in February 1945, ending the war as a British prisoner of war. He died on 30 March 1954.

General der Panzertruppe Joachim Lemelsen, who took over command of AOK 14 from Eberhard von Mackensen on 7 June 1944. (NAC)

One of the leading corps under AOK 14's command was the LXXVI. Panzer Korps, led by **Traugott Herr (1890–1976)**. Born on

16 September 1890 in Weferlingen, he was commissioned in 1911 as an infantry officer. After distinguishing himself during World War I, when he was wounded more than once and received several awards, he joined the Reichswehr, commanding several infantry units. Promoted to Oberst in August 1939, he was given command of the Infanterie Ersatz Regiment 13, then of the Infanterie Regiment (motorisiert) 66, which he led in the war against Poland, then in the 1940 campaign in the West. Commander of 13. Schützen Brigade from October 1940, he took part in the 1941 campaign in the East. He was given command of 13. Panzer Division in December 1941. Promoted to Generalmajor in April 1942, he led the division in the advance into the Caucasus, and was wounded in October 1942. Promoted to Generalleutnant in December 1942, while still in hospital, in June 1943 he was given command of LXXVI. Panzer Korps in France, then in Italy. Promoted to General der Panzertruppe in September 1943, Herr distinguished himself at Anzio and, in November 1944, temporarily replaced Lemelsen in command of AOK 14. In February 1945, he was given command of AOK 10 replacing von Vietinghoff, and ended the war in British captivity. He died on 13 April 1976.

General der Panzertruppe Traugott Herr, the skilled commander of LXXVI. Panzer Korps. (Heinrich Hoffmann/ullstein bild via Getty Images)

As commander of the I. Fallschirm Korps, **Alfred Schlemm (1894–1986)** faced serious situations more than once during the summer of 1944. Born on 18 December 1894 in Rudolstadt, he was commissioned in 1913 as an artillery officer. After distinguishing himself during World War I, Schlemm joined the Reichswehr, serving first as an artillery commander, then as a staff officer in various positions. In 1937, he was transferred to the Luftwaffe, and promoted to Oberst in February 1938. After serving again as a staff officer, in December 1940 (after being promoted to Generalmajor in March) Schlemm became chief of staff of the XI. Fliegerkorps, the airborne command, being eventually given command of the Luftwaffen Gefechtsverband Schlemm in February 1942. This was an ad hoc unit employed on the Eastern Front, which Schlemm led until June 1942 when he was promoted to Generalleutnant and was given command of 1. Flieger Division. In October 1942, Schlemm returned to the Eastern Front as commander of II. Luftwaffen-Feld Korps, being promoted to General der Flieger on 1 January 1943. His command was then transferred to Italy in September–October 1943, being re-designated I. Fallschirm Korps in January 1944. Schlemm distinguished himself during the battle of Anzio, and on 11 June 1944 was awarded the Ritterkreuz. In October 1944, he was removed by Kesselring, disappointed by the performance of the corps on the battlefield, and was to be given command of III. Fallschirm Korps. Since this was never formed, on 4 November 1944 Schlemm was promoted to General der Fallschirmtruppe and was given command of the 1. Fallschirm Armee. Wounded on 21 March 1945 during an air attack, he ended the war in the Führerreserve. He died on 24 January 1986.

General der Fallschirmtruppe Alfred Schlemm (right), commander of I. Fallschirm Korps, seen here while visiting troops at Anzio along with General Eberhard von Mackensen. (© Wolfgang STOCKER/Domaine public/ECPAD/Défense)

Generaloberst Heinrich von Vietinghoff-Scheel, here with Feldmarschall Albert Kesselring and, in the background, the chief of staff of HGr C, Siegfried Westphal. (NAC)

Along with Kesselring, **Heinrich von Vietinghoff-Scheel (1887–1952)** was one of the German commanders who endured for a long time during the Italian campaign. Born on 6 December 1887 in Mainz, von Vietinghoff-Scheel was commissioned in 1907 as an infantry officer. After serving with distinction during World War I and joining the Reichswehr, he held various appointments as commander of infantry units until promoted to Generalmajor in April 1936, following which he relinquished command of the 1. Schützen Brigade, becoming the Inspekteur der Panzertruppen und Heeresmotorisierung (Inspector of Armoured Troops and Army Motorization, specifically in Inspektion 6). Promoted in March 1938 to Generalleutnant, in November he was given command of 5. Panzer Division, which he led during the Polish campaign before being given command of XIII. Armee Korps in October 1939. After the campaign in the West, on 1 July 1940 von Vietinghoff was promoted to General der Panzertruppe and awarded the Ritterkreuz on 24 June. Commander of XXXXVI. Armee Korps from November 1940, von Vietinghoff fought in the Balkans and then against the Soviet Union until replacing General Model in command of AOK 9 in June 1942. In December 1942, he was given command of AOK 15 in France, being put in the Führerreserve in August due to illness. On 15 August 1943, von Vietinghoff was given command of AOK 10 in Italy, and on 1 September 1943 he was promoted to Generaloberst. He held the command of AOK 10 (with some interruptions) until 15 January 1945, when he returned to the Eastern Front now in command of Heeresgruppe (HGr) Kurland. This continued until he replaced Kesselring as Oberbefehlshaber Süd and commander of Heeresgruppe C (HGr C) on 15 March 1945. After he was released from captivity in September 1947, von Vietinghoff took a leading role in the rebuilding of the post-war German Army. He died on 23 February 1952.

Von Vietinghoff's leading corps commander was **Frido (Fridolin) von Senger und Etterlin (1891–1963)**. Born on 4 September 1891 in Waldshut, he was commissioned in 1910 as an artillery officer. He joined the Reichswehr after serving with distinction in World War I, and in March 1939 von Senger was promoted to Oberst and, as commander of Kavallerie Regiment 3, took part in the campaign against Poland. He next had command of Reiter Regiment 22, then of 2. Reiter Brigade, from February 1940. After the campaign in the West, von Senger served as chief of the German delegation to the commission for the implementation of the German–French armistice, being then promoted to Generalmajor on 1 September 1941. Commander of 17. Panzer Division from October 1942, he fought on the Eastern Front and was awarded the Ritterkreuz on 8 February 1943, being then promoted to Generalleutnant in May 1943. In June 1943, he was liaison officer with the Italian 6th Army in Sicily, then commander of the German troops in Sardinia before being given command of XIV. Panzer Korps on 8 October 1943. He held the command for the duration of the war, being promoted to General der Panzertruppe on 1 January 1944 and being the temporary commander of AOK 14 from 11 March 1945. He died on 9 January 1963.

A portrait of Generalleutnant Frido von Senger und Etterlin, who was best known for being the defender of Monte Cassino. (Public domain via Wikimedia Commons)

OPPOSING FORCES

ALLIED

By early June 1944, the balance of power in Italy had definitely shifted to the Allied side. Of the various formations that had started Operation *Diadem* in May, only the Polish II Corps had suffered severe losses which, above all, could hardly be replenished since the manpower pool was not available. On the other hand, the Allies enjoyed almost complete superiority on any other field: armour, motorization, artillery and supplies, especially following the capture of the harbours of Leghorn and Ancona. Air superiority guaranteed both direct fire support on the battlefield and an almost constant interdiction of the German lines of communication and supply. From 6 June 1944, the 8th Army was supported by the Desert Air Force (DAF), and the 5th Army by the US XII Tactical Air Force (TAF), whose operations struck immediately at the Germans while they withdrew and attempted to move XIV. Panzer Korps to the western coast. At the same time, the combined efforts of the DAF and the XII TAF hit the area between Pisa–Florence–Fano with the aim of interrupting the flow of supplies for the German forces, who could only rely on Flak (anti-aircraft artillery) to defend them. The situation enabled the Allied commanders to rotate troops, allowing rest and replenishment for some units while employing other units at the front, but it was not without consequences. As the advance progressed, the 8th Army – and to a lesser extent the 5th Army as well – faced a terrain made difficult first by the persistent Allied bombing, then by the several destructions implemented by the withdrawing German forces, which would use minor demolitions to cover for isolated tanks or self-propelled guns firing at the advancing enemy. As a result, it took time to clear the area and allow the sappers to make the way safe for the advancing troops. It is worth noting that even the British Official History remarks how 8th Army possessed one arm that would remain unused during the campaign, at a time when it could have been decisive: the mountain warfare-trained Indian divisions (three available: 4th, 8th and 10th),

'Rover Joe' in action: Corporal H. Langham, from Yorkshire, reporting over the phone information from Major Francis Basil O'Brien, who coordinates the reports from the jeep radio stations to the air controllers. (NARA via Digital History Archive)

A column of troops from the 92nd Infantry Division marching past a burned-out Tiger I tank near the Arno. (NARA via Digital History Archive)

Engineers from the Brazilian Expeditionary Force building a Bailey bridge. The arrival of reinforcements in the 5th Army area belatedly followed the departure of units used for the landing in southern France. (NARA via Digital History Archive)

all experienced and adept in this kind of warfare, which might have enabled a drive across the Apennines.

From the point of view of the opposing armies, from the Allied side, the most important issue was the depletion of the US 5th Army to the advantage of the planned landing in southern France, Operation *Anvil*. Italy was the only theatre of war that could supply the troops needed for the operation, given the lack of reserves, and all those needed were taken from 5th Army, which saw its strength drop considerably. Despite the fact that, during the summer campaign, 5th Army suffered considerably fewer losses than its opponents, it is noticeable that by 1 August its strength had dropped to less than half of the available strength on 4 June. Its strength on 1 June was 248,989, dropping to 205,992 on 1 July and then to 153,323 on 1 August. In June and July, nine infantry divisions were withdrawn, along with the equivalent of a tenth, with the 3rd Infantry Division being the first to leave on 24 June (it had temporarily served as a Rome garrison). While the British 1st Infantry Division was transferred to the British V Corps on 15 June, the British 5th Division was, at the same time, transferred to the Middle East and, five days later, the US 36th Infantry Division was transferred as well.

It was then the turn of the 1st Special Service Force and of the 509th Parachute Battalion, the French units following in succession. On 24 June, the 1ère Division de Marche d'Infanterie was transferred, followed a few days later by the 3e Division d'Infanterie Algérienne. The French Expeditionary Corps was then relieved in early July, with the following divisions leaving the Italian theatre: 4e Division de Montagne Marocaine, 2e Division d'Infanterie Marocaine, along with corps troops and three groups of Tabors. Commands were also withdrawn, the HQ of US VI Corps being assigned to 7th Army on 15 June, followed on 22 July by the FEC HQ. Corps troops reassigned included one reconnaissance squadron, three tank and three tank destroyer battalions, 12 artillery battalions and two combat engineers regiments, amongst others. Replacements were limited, both because of lack of experience and, in some cases, of inadequate training.

On 21 June, 5th Army received the 91st Infantry Division; its 361st Regimental Combat Team (RCT), having preceded the main body of the unit, had been deployed to the frontline on 11 June attached to the 36th Division. This was followed on 5 July by the 363rd RCT, the 91st Division being part of the US IV Corps on 10 July. Only the 370th RCT of the US 92nd Infantry Division was deployed in time to take part in the operations along the Arno River in early August 1944, the 517th Parachute Infantry RCT being temporarily assigned from 26 May to 2 July, while for the first time 5th Army deployed the 442nd RCT (specifically made up of Americans of Japanese descent). The Brazilian Expeditionary Force, including the Brazilian Infantry Division, started to arrive in Italy in July, but there was clearly no hurry to employ them at the front and by 15 August only one RCT was available, which was not sent to the frontline. In order to make good for the shortage of infantry, two Antiaircraft Artillery Groups (91st and 107th) were reorganized as provisional infantry regiments, forming Task Force 45, which also incorporated the 91st Cavalry Reconnaissance Squadron and the 2nd Armored Group, thus creating a provisional division named after the 45th Antiaircraft Artillery Brigade.

American tanks resting amongst the ruins of Coreno before resuming their advance towards Rome. Note the censored unit markings. (NARA)

GERMAN

Given the chaos and confusion dominating in both AOK 10 and, to a much greater extent, in AOK 14, it is hard to provide an accurate analysis of the German forces in Italy during the summer 1944 campaign. Some data is available, nonetheless, enabling some sort of picture to be formed. On 8 June, the HGr C provided an overall view of the situation which was as follows: 14 divisions were still operational (six with AOK 14, eight with AOK 10), five could be replenished in a short period of time (two and three, respectively), three others required a major reorganization (one with AOK 10, two with AOK 14). The reserves, deployed in the north under the Armee Abteilung von Zangen, amounted to three weak infantry divisions and one mountain division, along with the troops deployed on the Elba Island and the Festungs Brigade 135. On 10 June, HGr C deployed a grand total of 25 divisions, one (26. Panzer) under direct command, nine with AOK 14 (five infantry, one paratrooper, three Panzergrenadier), 11 with AOK 10 (eight infantry or equivalent, one paratrooper, one Panzergrenadier, one Panzer), and four in the

GIs entering a village, which suffered heavily in the war. The Italian houses, mostly built of stone, provided excellent cover for the German defenders. (NARA)

Two Polish soldiers using a 2in mortar under cover of a destroyed Italian *Semovente* (self-propelled gun) in German use. (NARA via Digital History Archive)

rear areas (including one Panzergrenadier). By 23 June, the number had risen to 27 divisions, broken down as follows: two infantry, directly subordinated, being reorganized, 11 with AOK 14 (five infantry, one of which reorganizing, one paratrooper, one Panzer and four Panzergrenadier), 11 with AOK 10 (eight infantry or equivalent, one of which reorganizing, one paratrooper, one Panzer and one Panzergrenadier), and three infantry or equivalent divisions in the rear area. The impression of a growing strength is merely illusory, as the data below reveals.

On 14 June, AOK 14 reported that its divisions were basically a Kampfgruppe (KG) with a fighting strength between 150 and 300 men, as evidenced by the matching list of units on the same date: 162. Infanterie Division (Turkistan) was composed of three regimental Kampfgruppen; 90. Panzergrenadier Division included two regimental Kampfgruppen and the Sperrverband Baade (blocking detachment); and 16. SS-Panzergrenadier Division and the 42. Jäger Division forming the Sperrverband Hoffmann. Two battalions from 4. Fallschirmjäger Division reported a strength of 60 and 75, respectively, and on 25 June the situation of AOK 14 was the following: its division deployed a total of 13 strong battalions, which was 400 or more men, mostly with the 16. SS-Panzergrenadier Division, the 19. Luftwaffen-Feld Division and the 356. Infanterie Division. The rebuilt 356. Infanterie Division had three medium-strength battalions, and the total of 12 average-strength battalions was split between the latter, the 29. and 90. Panzergrenadier Divisions and the two Luftwaffen-Feld Divisions. The bulk of the divisions had 24 weak battalions (100 men or less), plus 15 worn out ones, which is more than half of the total of 67 infantry battalions. The AOK 14 had at its disposal 20 medium and 95 heavy Pak, or anti-tank

Anti-aircraft gun emplacement for an Italian 90mm gun. The emplacement has been positioned in such a way as to cover the road as well, in case of use against ground targets. (NARA via Digital History Archive)

guns, plus 181 light and 60 heavy artillery guns. On 1 July, the army divisions within AOK 14 reported a total actual strength of 97,653 all ranks, the divisions having an average of 12,000 men each with some remarkable exceptions (19. Luftwaffen-Feld Division some 7,200 strong, 20. Luftwaffen-Feld Division some 4,500 strong). On 1 August, the overall strength of the army units was 98,989, the actual combat strength being 40,561 (SS and Fallschirmjäger units not included).

Even though AOK 10 had suffered much less than AOK 14, its units were only just in slightly better condition. Between 9 and 10 June, the following combat strengths were reported: Fallschirm Panzer Division 'Hermann Göring' some 800, 15. Panzergrenadier Division some 2,500, 1. Fallschirmjäger Division some 900, 29. Panzergrenadier Division some 2,200, 26. Panzer Division some 1,400. At the end of June, AOK 10 reported the loss of some 24,900 men, against the arrival of roughly 13,600 replacements. The only division with a relatively high number of strong battalions was, at the end of June, 278. Infanterie Division, followed by 5. Gebirgs and 114. Jäger Division. All others had an average of weak battalions, in particular the 1. Fallschirmjäger Division, which could rely only on two weak and two depleted battalions. On 22 July, the situation only saw limited improvements, with the 715., 305., 71. and 278. Infanterie Divisions having some strong battalions, along with 5. Gebirgs Division. All other divisions still had medium or low-strength battalions. The tank situation was even worse. On 20 June, HGr C reported that, since 1 May, a total of 215 Panzer and Sturmgeschütz, plus 26 self-propelled (SP) anti-tank guns, had been received, including 33 Tiger I, 38 PzKpfw VI Panther, 75 PzKpfw IV and 53 Sturmgeschütz III. Tank and SP

Soldiers from the 16. SS-Panzergrenadier Division 'Reichsführer-SS' are examining the remnants of two destroyed Sherman tanks. This was the only Waffen-SS unit to fight in Italy. (NAC)

Mussolini inspecting Italian troops training in Italy. The lack of reliability of the Italian formations, proved at Elba Island, led to their disbanding in June 1944. (NAC)

Men of the Organization Todt (OT) parading in front of two Italian L-3 tankettes, which have an OT symbol on their front. The OT was the organization responsible for the construction of the Gothic Line. (NAC)

gun losses in May included 13 PzKpfw IV, another 23 being lost in June along with three PzKpfw III and seven Panther. On 23 July, AOK 14 reported the following tank situation with its divisions: 26. Panzer Division with 23 Panzers, schwere Panzer Abteilung 504 (Tiger) with 13 Panzers, 3. Panzergrenadier Division with 19 Sturmgeschütze, 90. Panzergrenadier Division with three Sturmgeschütze, 16. SS-Panzergrenadier Division with six Sturmgeschütze. For the rest, most of the divisions relied on the Italian SP guns, which, albeit in some cases effective, were definitely outdated.

ORDERS OF BATTLE

Notes: The first order of battle can be used as a reference. Details such as unit composition and commanders may be omitted on subsequent orders of battle (except for newly arrived units and commands).

ORDERS OF BATTLE, 27 MAY 1944

Allied Armies in Italy (AAI),[1] General Sir Harold Alexander

US 5th Army, Lt Gen Mark W. Clark

US VI Corps, Maj Gen Lucian K. Truscott
45th US Infantry Division, Maj Gen William W. Eagles
(157th US Infantry Rgt, 179th US Infantry Rgt, 180th US Infantry Rgt)
34th US Infantry Division, Maj Gen Charles W. Ryder
(133rd US Infantry Rgt, 135th US Infantry Rgt, 168th US Infantry Rgt)
36th US Infantry Division, Maj Gen Fred L. Walker
(141st US Infantry Rgt, 142nd US Infantry Rgt, 143rd US Infantry Rgt)
1st US Armored Division, Maj Gen Ernest N Harmond, from 17 July Maj Gen Vernon E. Prichard
(1st US Armored Rgt, 13th US Armored Rgt, 6th Armored Infantry Rgt)
3rd US Infantry Division, Maj Gen John W. O'Daniel
(7th US Infantry Rgt, 15th US Infantry Rgt, 30th US Infantry Rgt)
1st British Infantry Division, Maj Gen Ronald Penney, from 24 July Maj Gen Charles Loewen
(2nd Infantry Bde, 3rd Infantry Bde, 24th Infantry Bde Guards)
5th British Infantry Division, Maj Gen P. G. S. Gregson-Ellis
(13th Infantry Bde, 15th Infantry Bde, 17th Infantry Bde)

US II Corps, Maj Gen Geoffrey T. Keyes
85th US Infantry Division, Maj Gen John B. Coulter
(337rd US Infantry Rgt, 338th US Infantry Rgt, 339th US Infantry Rgt)
88th US Infantry Division, Maj Gen John F. Sloan
(349rd US Infantry Rgt, 350th US Infantry Rgt, 351st US Infantry Rgt)

French Expeditionary Corps, Gen Alphonse Juin
1st French Infantry Motorized Division, Maj Gen D. Brosset
(1st Infantry Bde, 2nd Infantry Bde, 4th Infantry Bde)

1 The 15th Army Group was formed on 11 July 1943 to command the US 7th Army and British 8th Army, the former replaced by the 5th Army in September. On 11 January 1944 the 15th Army Group was re-designated Allied Armies in Italy (AAI), also known as Allied Forces Headquarters (AFHQ), but the name was changed a week later to Allied Central Mediterranean Force only to revert to the designation of AAI/AFHQ on 9 March 1944. As Gen Clark took over command on 16 December 1944, the command reverted to the designation of 15th Army Group.

2nd Moroccan Infantry Division, Maj General A. W. Dody
(4th Moroccan Rifle Rgt, 5th Moroccan Rifle Rgt, 8th Moroccan Rifle Rgt)
3rd Algerian Infantry Division, Maj Gen de Goislard de Monsabert
(3rd Algerian Rifle Rgt, 4th Tunisian Rifle Rgt, 7th Algerian Rifle Rgt)
4th Moroccan Infantry Division, Maj Gen F. Sevez
(1st Moroccan Rifle Rgt, 2nd Moroccan Rifle Rgt, 6th Moroccan Rifle Rgt)

British 8th Army, Lt Gen Sir Oliver Leese

Directly subordinated:
6th South African Armoured Division, Maj Gen W. H. Evered Poole
(11th South African Armoured Bde, 12th South African Infantry Bde)

Canadian I Corps, Lt Gen Eedson L. M. Burns
5th Canadian Armoured Division, Maj Gen Bert M. Hoffmeister
(5th Canadian Armoured Bde, 11th Canadian Infantry Bde)
1st Canadian Infantry Division, Maj Gen Christopher Vokes
(1st Canadian Infantry Bde, 2nd Canadian Infantry Bde, 3rd Canadian Indian Infantry Bde)

British XIII Corps, Lt Gen S. C. Kirkman
6th British Armoured Division, Maj Gen Vyvyan Evelegh
(26th Armoured Bde, 1st Infantry Bde Guards)
8th Indian Infantry Division, Maj Gen Dudley Russell
(17th Indian Infantry Bde, 19th Indian Infantry Bde, 21th Indian Infantry Bde)
78th British Infantry Division, Maj Gen C. F. Keightley
(11th Infantry Bde, 36th Infantry Bde, 38th Irish Infantry Bde)
4th British Infantry Division, Maj Gen Alfred Dudley Ward
(10th Infantry Bde, 12th Infantry Bde, 28th Infantry Bde)

Polish II Corps, Lieutenant General Władysław Anders
3rd Carpathian Division, Maj Gen Bronislaw Duch
(1st Carpathian Rifle Bde, 2nd Carpathian Rifle Bde)
5th Kresowa Division, Maj Gen Nikodem Sulik-Sarnowski
(5th Wilenski Infantry Bde, 6th Lwowska Infantry Bde)

British X Corps, Lt Gen Richard McCreery
2nd New Zealand Division, Lt Gen Sir Bernard Freyberg
(4th NZ Armoured Bde, 5th NZ Infantry Bde, 6th NZ Infantry Bde)

British V Corps, Lt Gen Charles W. Allfrey
4th Indian Infantry Division, Maj Gen Arthur W. W. Holworthy
(5th Indian Infantry Bde, 7th Indian Infantry Bde, 11th Indian Infantry Bde)
10th Indian Infantry Division, Maj Gen Denis Reid
(10th Indian Infantry Bde, 20th Indian Infantry Bde, 25th Indian Infantry Bde)

Heeresgruppe C/Oberbefehlshaber Südwest, Feldmarschall Albert Kesselring

AOK 14, GenOb Eberhard von Mackensen, from 7 June Gen d.PzTr Joachim Lemelsen

Directly subordinated:
92. Infanterie Division, GenLt Werner Goeritz[2]
(Grenadier Rgt 1059, 1060)

I. Fallschirm Korps, Gen d.FschTr Alfred Schlemm
4. Fallschirmjäger Division, GenMaj Heinrich Trettner
(Fallschirmjäger Rgt 10, 11, 12)
65. Infanterie Division, GenMaj Hellmuth Pfeifer
(Grenadier Rgt 145, 146, 147)
3. Panzergrenadier Division, GenMaj Hans Hecker
(Panzergrenadier Rgt. 8, 29)

2 Disbanded 20 June, remnants into 362. Infanterie Division

362. Infanterie Division, GenLt Heinrich Greiner
(Grenadier Rgt 954, 955, 956)

LXXVI. Panzer Korps, Gen d.PzTr Traugott Herr
Fallschirm Panzer Division 'Hermann Göring', GenMaj Wilhelm Schmalz
(Fallschirm Panzer Rgt 'HG', Fallschirm Panzergrenadier Rgt 'HG' 1–2)
715. Infanterie Division, GenMaj Hanns von Rohr
(Grenadier Rgt 725, 735)

AOK 10, GenOb Heinrich von Vietinghoff-Scheel

XIV. Panzer Korps, GenLt Frido von Senger und Etterlin
29. Panzergrenadier Division, GenMaj Walter Fries
(Panzergrenadier Rgt 15, 71)
94. Infanterie Division, GenLt Bernhard Steinmetz
(Grenadier Rgt 267, 274, 276)
15. Panzergrenadier Division, GenMaj Eberhard Rodt
(Panzergrenadier Rgt 104, 115)
305. Infanterie Division, GenMaj Friedrich-Wilhelm Hauck
(Grenadier Rgt 576, 577, 578)
26. Panzer Division, GenMaj Smile von Lüttwitz
(Panzer Rgt 26, Panzergrenadier Rgt 9, 67)
90. Panzergrenadier Division, GenMaj Ernst-Günther Baade
(Panzergrenadier Rgt 200, 361)
1. Fallschirmjäger Division, GenLt Richard Heidrich
(Fallschirmjäger Rgt 1, 3, 4)
5. Gebirgs Division, GenLt Max Schrank
(Gebirgsjäger Rgt 85, 100)
114. Jäger Division, GenMaj Alexander Bourquin
(Jäger Rgt 721, 741)
44. Infanterie Division, GenLt Bruno Ortner
(Grenadier Rgt 131, 132, 134)

LI. Gebirgs Korps, Gen d.GebTr Valentin Feuerstein
334. Infanterie Division, GenLt Walter Scheller
(Grenadier Rgt 754, 755, 756)
278. Infanterie Division, GenLt Harry Hoppe
(Grenadier Rgt 992, 993, 994)

ORDERS OF BATTLE, 14 JUNE 1944

Allied Armies in Italy (AAI), General Sir Harold Alexander

Directly subordinated:
Polish II Corps, Lt Gen Władysław Anders
3rd Carpathian Division, Maj Gen Bronislaw Duch
(1st Carpathian Rifle Bde, 2nd Carpathian Rifle Bde)
5th Kresowa Division, Major General Nikodem Sulik-Sarnowski
(5th Wilenski Infantry Bde, 6th Lwowska Infantry Bde)

US 5th Army, Lt Gen Mark W. Clark

Directly subordinated (Rome garrison):
3rd US Infantry Division, Maj Gen John W. O'Daniel
(7th US Infantry Rgt, 15th US Infantry Rgt, 30th US Infantry Rgt)

US IV Corps, Maj Gen Willis D. Crittenberger
34th US Infantry Division, Maj Gen Charles W. Ryder
(133rd US Infantry Rgt, 135th US Infantry Rgt, 168th US Infantry Rgt)
36th US Infantry Division, Maj Gen Fred L. Walker
(141st US Infantry Rgt, 142nd US Infantry Rgt, 143rd US Infantry Rgt)

French Expeditionary Corps, Gen Alphonse Juin
1st French Infantry Motorized Division, Maj Gen D. Brosset
(1st Infantry Bde, 2nd Infantry Bde, 4th Infantry Bde)
2nd Moroccan Infantry Division, Maj Gen A. W. Dody
(4th Moroccan Rifle Rgt, 5th Moroccan Rifle Rgt, 8th Moroccan Rifle Rgt)
3rd Algerian Infantry Division, Maj Gen de Goislard de Monsabert

(3rd Algerian Rifle Rgt, 4th Tunisian Rifle Rgt, 7th Algerian Rifle Rgt)
4th Moroccan Infantry Division, Maj Gen F. Sevez
(1st Moroccan Rifle Rgt, 2nd Moroccan Rifle Rgt, 6th Moroccan Rifle Rgt)

Army reserves:
1st US Armored Division, Maj Gen Ernest N. Harmond, from 17 July Maj Gen Vernon E. Prichard
(1st US Armored Rgt, 13th US Armored Rgt, 6th Armored Infantry Rgt)
1st British Infantry Division, Maj Gen Ronald Penney, from 24 July Maj Gen Charles Loewen
(2nd Infantry Bde, 3rd Infantry Bde, 24th Infantry Bde Guards)
5th British Infantry Division, Maj Gen P. G. S. Gregson-Ellis
(13th Infantry Bde, 15th Infantry Bde, 17th Infantry Bde)

US II Corps, Maj Gen Geoffrey T. Keyes
85th US Infantry Division, Maj Gen John B. Coulter
(337rd US Infantry Rgt, 338th US Infantry Rgt, 339th US Infantry Rgt)
88th US Infantry Division, Maj Gen John F. Sloan
(349rd US Infantry Rgt, 350th US Infantry Rgt, 351st US Infantry Rgt)

US VI Corps, Maj Gen Lucian K. Truscott
45th US Infantry Division, Maj Gen William W. Eagles
(157th US Infantry Rgt, 179th US Infantry Rgt, 180th US Infantry Rgt)

British 8th Army, Lt Gen Sir Oliver Leese

Directly subordinated:
2nd New Zealand Division, Lt Gen Sir Bernard Freyberg
(4th NZ Armoured Bde, 5th NZ Infantry Bde, 6th NZ Infantry Bde)
10th Indian Infantry Division, Maj Gen Denis Reid
(10th Indian Infantry Bde, 20th Indian Infantry Bde, 25th Indian Infantry Bde)

British XIII Corps, Lt Gen S. C. Kirkman
6th South African Armoured Division, Maj Gen W. H. Evered Poole
(11th South African Armoured Bde, 12th South African Infantry Bde)
78th British Infantry Division, Maj Gen C. F. Keightley
(11th Infantry Bde, 36th Infantry Bde, 38th Irish Infantry Bde)
4th British Infantry Division, Maj Gen Alfred Dudley Ward
(10th Infantry Bde, 12th Infantry Bde, 28th Infantry Bde)

British X Corps, Lt Gen Richard McCreery
6th British Armoured Division, Maj Gen Vyvyan Evelegh
(26th Armoured Bde, 1st Infantry Bde Guards)
8th Indian Infantry Division, Maj Gen Dudley Russell
(17th Indian Infantry Bde, 19th Indian Infantry Bde, 21th Indian Infantry Bde)

British V Corps, Lt Gen Charles W. Allfrey
4th British Infantry Division, Maj Gen Alfred Dudley Ward
(10th Infantry Bde, 12th Infantry Bde, 28th Infantry Bde)
Corpo Italiano di Liberazione, Gen B. Utili

Heeresgruppe C/Oberbefehlshaber Südwest, Feldmarschall Albert Kesselring

Directly subordinated:
715. Infanterie Division, GenMaj Hanns von Rohr
(Grenadier Rgt 725, 735)
16. SS-Panzergrenadier Division 'Reichsführer-SS', SS-Gruppenführer Max Simon
(SS-Panzergrenadier Rgt 35, 36)

AOK 14, Gen d.PzTr Joachim Lemelsen

Directly subordinated:
19. Luftwaffen-Feld Division, GenLt Erich Baessler
(Jäger Rgt 37, 38, 45)
65. Infanterie Division, GenMaj Hellmuth Pfeifer
(Grenadier Rgt 145, 146, 147)
362. Infanterie Division, GenLt Heinrich Greiner
(Grenadier Rgt 954, 955, 956)
92. Infanterie Division, GenLt Werner Goeritz[3]
(Grenadier Rgt 1059, 1060)

XIV. Panzer Korps, GenLt Frido von Senger und Etterlin
162. (Turk) Infanterie Division, GenMaj Ralph von Heygendorff
(Infanterie Rgt 303, 314, 329)
90. Panzergrenadier Division, GenMaj Ernst-Günther Baade
(Panzergrenadier Rgt 200, 361)
29. Panzergrenadier Division, GenMaj Walter Fries
(Panzergrenadier Rgt 15, 71)
20. Luftwaffen-Feld Division, GenMaj Wilhelm Crisolli
(Jäger Rgt 39, 40)

I. Fallschirm Korps, Gen d.FschTr Alfred Schlemm
3. Panzergrenadier Division, GenMaj Hans Hecker
(Panzergrenadier Rgt 8, 29)
26. Panzer Division, GenMaj Smile von Lüttwitz
(Panzer Rgt. 26, Panzergrenadier Rgt 9, 67)
4. Fallschirmjäger Division, GenMaj Heinrich Trettner
(Fallschirmjäger Rgt 10, 11, 12)
356. Infanterie Division, GenLt Karl Faulenbach
(Grenadier Rgt 869, 870, 871)

AOK 10, GenOb Heinrich von Vietinghoff-Scheel

Directly subordinated:
114. Jäger Division, GenMaj Alexander Bourquin
(Jäger Rgt 721, 741)

LXXVI. Panzer Korps, Gen d.PzTr Traugott Herr
1. Fallschirmjäger Division, GenLt Richard Heidrich
(Fallschirmjäger Rgt 1, 3, 4)
334. Infanterie Division, GenLt Walter Scheller
(Grenadier Rgt 754, 755, 756)
Fallschirm Panzer Division 'Hermann Göring', GenMaj Wilhelm Schmalz
(Fallschirm Panzer Rgt 'HG', Fallschirm Panzergrenadier Rgt 'HG' 1–2)
15. Panzergrenadier Division, GenMaj Eberhard Rodt
(Panzergrenadier Rgt 104, 115)
94. Infanterie Division, GenLt Bernhard Steinmetz
(Grenadier Rgt 267, 274, 276)

LI. Gebirgs Korps, Gen d.GebTr Valentin Feuerstein
305. Infanterie Division, GenMaj Friedrich-Wilhelm Hauck
(Grenadier Rgt 576, 577, 578)
44. Infanterie Division, GenLt Bruno Ortner
(Grenadier Rgt 131, 132, 134)
5. Gebirgs Division, GenLt Max Schrank
(Gebirgsjäger Rgt 85, 100)
71. Infanterie Division, GenMaj Wilhelm Raapke
(Grenadier Rgt 191, 194, 211)
278. Infanterie Division, GenLt Harry Hoppe
(Grenadier Rgt 992, 993, 994)

ORDERS OF BATTLE, 21 JUNE 1944

Allied Armies in Italy (AAI), General Sir Harold Alexander

US 5th Army, Lt Gen Mark W. Clark

Directly subordinated (Rome garrison):
3rd US Infantry Division, Maj Gen John W. O'Daniel
(7th US Infantry Rgt, 15th US Infantry Rgt, 30th US Infantry Rgt)

US IV Corps, Maj Gen Willis D. Crittenberger
36th US Infantry Division, Maj Gen Fred L. Walker
(141st US Infantry Rgt, 142nd US Infantry Rgt, 143rd US Infantry Rgt)
34th US Infantry Division, Maj Gen Charles W. Ryder

3 Disbanded 20 June, remnants into 362. Infanterie Division

(133rd US Infantry Rgt, 135th US Infantry Rgt, 168th US Infantry Rgt)
1st US Armored Division, Maj Gen Ernest N. Harmond, from 17 July Maj Gen Vernon E. Prichard
(1st US Armored Rgt, 13th US Armored Rgt, 6th Armored Infantry Rgt)

French Expeditionary Corps, Gen Alphonse Juin
1st French Infantry Motorized Division, Maj General D. Brosset
(1st Infantry Bde, 2nd Infantry Bde, 4th Infantry Bde)
2nd Moroccan Infantry Division, Maj Gen A. W. Dody
(4th Moroccan Rifle Rgt, 5th Moroccan Rifle Rgt, 8th Moroccan Rifle Rgt)
3rd Algerian Infantry Division, Maj Gen Joseph de Goislard de Monsabert
(3rd Algerian Rifle Rgt, 4th Tunisian Rifle Rgt, 7th Algerian Rifle Rgt)
4th Moroccan Infantry Division, Maj Gen F. Sevez
(1st Moroccan Rifle Rgt, 2nd Moroccan Rifle Rgt, 6th Moroccan Rifle Rgt)

Army reserves:
US II Corps, Maj Gen Geoffrey T. Keyes
85th US Infantry Division, Maj Gen John B. Coulter
(337rd US Infantry Rgt, 338th US Infantry Rgt, 339th US Infantry Rgt)
88th US Infantry Division, Maj Gen John F. Sloan
(349rd US Infantry Rgt, 350th US Infantry Rgt, 351st US Infantry Rgt)

US VI Corps, Maj Gen Lucian K. Truscott
45th US Infantry Division, Maj Gen William W. Eagles
(157th US Infantry Rgt, 179th US Infantry Rgt, 180th US Infantry Rgt)

British 8th Army, Lt Gen Sir Oliver Leese

British XIII Corps, Lt Gen S. C. Kirkman
6th South African Armoured Division, Maj Gen W. H. Evered Poole
(11th South African Armoured Bde, 12th South African Infantry Bde)
78th British Infantry Division, Maj Gen C. F. Keightley
(11th Infantry Bde, 36th Infantry Bde, 38th Irish Infantry Bde)
4th British Infantry Division, Maj Gen Alfred Dudley Ward
(10th Infantry Bde, 12th Infantry Bde, 28th Infantry Bde)

British X Corps, Lt Gen Richard McCreery
6th British Armoured Division, Maj Gen Vyvyan Evelegh
(26th Armoured Bde, 1st Infantry Bde Guards)
8th Indian Infantry Division, Maj Gen Dudley Russell
(17th Indian Infantry Bde, 19th Indian Infantry Bde, 21th Indian Infantry Bde)

Army reserves:
Canadian I Corps, Lt Gen E. L. M. Burns
5th Canadian Armoured Division, Maj Gen Bert M. Hoffmeister
(5th Canadian Armoured Bde, 11th Canadian Infantry Bde)
1st Canadian Infantry Division, Maj Gen Christopher Vokes
(1st Canadian Infantry Bde, 2nd Canadian Infantry Bde, 3rd Canadian Indian Infantry Bde)

British V Corps, Lt Gen Charles W. Allfrey
10th Indian Infantry Division, Maj Gen Denis Reid
(10th Indian Infantry Bde, 20th Indian Infantry Bde, 25th Indian Infantry Bde)
4th Indian Infantry Division, Maj Gen Arthur W. W. Holworthy
(5th Indian Infantry Bde, 7th Indian Infantry Bde, 11th Indian Infantry Bde)
1st British Infantry Division, Maj Gen Ronald Penney, from 24 July Maj Gen Charles Loewen
(2nd Infantry Bde, 3rd Infantry Bde, 24th Infantry Bde Guards)
1st British Armoured Division, Maj Gen A. Galloway, Maj Gen R. A. Hull
(2nd Armoured Bde, 18th Infantry Bde, 43rd Gurkha Infantry Bde)

Directly subordinated to Allied Armies in Italy HQ:
Polish II Corps, Lt Gen Władysław Anders
3rd Carpathian Division, Maj Gen Bronislaw Duch
(1st Carpathian Rifle Bde, 2nd Carpathian Rifle Bde)
5th Kresowa Division, Maj Gen Nikodem Sulik-Sarnowski
(5th Wilenski Infantry Bde, 6th Lwowska Infantry Bde)
Corpo Italiano di Liberazione, Gen B. Utili

Heeresgruppe C/Oberbefehlshaber Südwest, Feldmarschall Albert Kesselring

Directly subordinated:
715. Infanterie Division, GenMaj Hanns von Rohr
(Grenadier Rgt 725, 735)
65. Infanterie Division, GenMaj Hellmuth Pfeifer
(Grenadier Rgt 145, 146, 147)
16. SS-Panzergrenadier Division 'Reichsführer-SS', SS-Gruppenführer Max Simon
(SS-Panzergrenadier Rgt 35, 36)
19. Luftwaffen-Feld Division, GenLt Erich Baessler
(Jäger Rgt 37, 38, 45)

AOK 14, Gen d.PzTr Joachim Lemelsen

Directly subordinated:
362. Infanterie Division, GenLt Heinrich Greiner
(Grenadier Rgt 954, 955, 956)

XIV. Panzer Korps, GenLt Frido von Senger und Etterlin
162. (Turk) Infanterie Division, GenMaj Ralph von Heygendorff
(Infanterie Rgt 303, 314, 329)
90. Panzergrenadier Division, GenMaj Ernst-Günther Baade
(Panzergrenadier Rgt 200, 361)
20. Luftwaffen-Feld Division, GenMaj Wilhelm Crisolli
(Jäger Rgt 39, 40)
3. Panzergrenadier Division, GenMaj Hans Hecker
(Panzergrenadier Rgt 8, 29)

I. Fallschirm Korps, Gen d.FschTr Alfred Schlemm
29. Panzergrenadier Division, GenMaj Walter Fries
(Panzergrenadier Rgt 15, 71)
26. Panzer Division, GenMaj Smile von Lüttwitz
(Panzer Rgt 26, Panzergrenadier Rgt 9, 67)
4. Fallschirmjäger Division, GenMaj Heinrich Trettner
(Fallschirmjäger Rgt 10, 11, 12)
356. Infanterie Division, GenLt Karl Faulenbach
(Grenadier Rgt 869, 870, 871)

AOK 10, GenOb Heinrich von Vietinghoff-Scheel

Directly subordinated:
94. Infanterie Division, GenLt Bernhard Steinmetz
(Grenadier Rgt 267, 274, 276)

LXXVI. Panzer Korps, Gen d.PzTr Traugott Herr
305. Infanterie Division, GenMaj Friedrich-Wilhelm Hauck
(Grenadier Rgt 576, 577, 578)
Fallschirm Panzer Division 'Hermann Göring', GenMaj Wilhelm Schmalz
(Fallschirm Panzer Rgt 'HG', Fallschirm Panzergrenadier Rgt 'HG' 1–2)
1. Fallschirmjäger Division, GenLt Richard Heidrich
(Fallschirmjäger Rgt 1, 3, 4)
334. Infanterie Division, GenLt Walter Scheller
(Grenadier Rgt 754, 755, 756)
15. Panzergrenadier Division, GenMaj Eberhard Rodt
(Panzergrenadier Rgt 104, 115)

LI. Gebirgs Korps, Gen d.GebTr Valentin Feuerstein
71. Infanterie Division, GenMaj Wilhelm Raapke
(Grenadier Rgt 191, 194, 211)
278. Infanterie Division, GenLt Harry Hoppe
(Grenadier Rgt 992, 993, 994)
5. Gebirgs Division, GenLt Max Schrank
(Gebirgsjäger Rgt 85, 100)
114. Jäger Division, GenMaj Alexander Bourquin
(Jäger Rgt 721, 741)
44. Infanterie Division, GenLt Bruno Ortner
(Grenadier Rgt 131, 132, 134)

ORDERS OF BATTLE, 19 JULY 1944

Allied Armies in Italy (AAI), General Sir Harold Alexander

Directly subordinated:
1st British Infantry Division, Maj Gen Ronald Penney, from 24 July Maj Gen Charles Loewen
(2nd Infantry Bde, 3rd Infantry Bde, 24th Infantry Bde Guards)
1st British Armoured Division, Maj Gen A. Galloway, Maj Gen R. A. Hull
(2nd Armoured Bde, 18th Infantry Bde, 43rd Gurkha Infantry Bde)

US 5th Army, Lt Gen Mark W. Clark

US IV Corps, Maj Gen Willis D. Crittenberger
34th US Infantry Division, Maj Gen Charles W. Ryder
(133rd US Infantry Rgt, 135th US Infantry Rgt, 168th US Infantry Rgt)
91st US Infantry Division, Maj Gen William G. Livesay
(361st US Infantry Rgt, 362nd Infantry Rgt, 363rd US Infantry Rgt)
88th US Infantry Division, Maj Gen John F. Sloan
(349rd US Infantry Rgt, 350th US Infantry Rgt, 351st US Infantry Rgt)

French Expeditionary Corps, Gen Alphonse Juin
2nd Moroccan Infantry Division, Maj Gen A. W. Dody
(4th Moroccan Rifle Rgt, 5th Moroccan Rifle Rgt, 8th Moroccan Rifle Rgt)
4th Moroccan Infantry Division, Maj Gen F. Sevez
(1st Moroccan Rifle Rgt, 2nd Moroccan Rifle Rgt, 6th Moroccan Rifle Rgt)

Army reserves:
1st US Armored Division, Maj Gen Ernest N Harmond, from 17 July Maj Gen Vernon E. Prichard
(1st US Armored Rgt, 13th US Armored Rgt, 6th Armored Infantry Rgt)

US II Corps, Maj Gen Geoffrey T. Keyes
85th US Infantry Division, Maj Gen John B. Coulter
(337rd US Infantry Rgt, 338th US Infantry Rgt, 339th US Infantry Rgt)

British 8th Army, Lt Gen Sir Oliver Leese

British XIII Corps, Lt Gen S. C. Kirkman
6th South African Armoured Division, Maj Gen W. H. Evered Poole
(11th South African Armoured Bde, 12th South African Infantry Bde)
4th British Infantry Division, Maj Gen Alfred Dudley Ward
(10th Infantry Bde, 12th Infantry Bde, 28th Infantry Bde)
6th British Armoured Division, Maj Gen Vyvyan Evelegh
(26th Armoured Bde, 1st Infantry Bde Guards)
2nd New Zealand Division, Lt Gen Sir Bernard Freyberg
(4th NZ Armoured Bde, 5th NZ Infantry Bde, 6th NZ Infantry Bde)

British X Corps, Lt Gen Richard McCreery
4th Indian Infantry Division, Maj Gen Arthur W. W. Holworthy
(5th Indian Infantry Bde, 7th Indian Infantry Bde, 11th Indian Infantry Bde)
10th Indian Infantry Division, Maj Gen Denis Reid
(10th Indian Infantry Bde, 20th Indian Infantry Bde, 25th Indian Infantry Bde)
8th Indian Infantry Division, Maj Gen Dudley Russell
(17th Indian Infantry Bde, 19th Indian Infantry Bde, 21th Indian Infantry Bde)

Army reserves:
Canadian I Corps, Lt Gen E. L. M. Burns
5th Canadian Armoured Division, Maj Gen Bert M. Hoffmeister
(5th Canadian Armoured Bde, 11th Canadian Infantry Bde)
1st Canadian Infantry Division, Maj Gen Christopher Vokes
(1st Canadian Infantry Bde, 2nd Canadian Infantry Bde, 3rd Canadian Indian Infantry Bde)

Directly subordinated to Allied Armies in Italy HQ:
Polish II Corps, Lieutenant General Władysław Anders
3rd Carpathian Division, Maj Gen Bronislaw Duch
(1st Carpathian Rifle Bde, 2nd Carpathian Rifle Bde)
5th Kresowa Division, Maj Gen Nikodem Sulik-Sarnowski
(5th Wilenski Infantry Bde, 6th Lwowska Infantry Bde)
Corpo Italiano di Liberazione, Gen B. Utili

Heeresgruppe C/Oberbefehlshaber Südwest, Feldmarschall Albert Kesselring

Directly subordinated:
94. Infanterie Division, GenLt Bernhard Steinmetz
(Grenadier Rgt 267, 274, 276)
162. (Turk) Infanterie Division, GenMaj Ralph von Heygendorff
(Infanterie Rgt 303, 314, 329)

AOK 14, Gen d.PzTr Joachim Lemelsen

LXXV. Armee Korps, Gend.GebTr Hans Schlemmer
19. Luftwaffen-Feld Division, GenLt Erich Baessler
(Jäger Rgt 37, 38, 45)
20. Luftwaffen-Feld Division, GenMaj Wilhelm Crisolli
(Jäger Rgt 39, 40)
65. Infanterie Division, GenMaj Hellmuth Pfeifer
(Grenadier Rgt 145, 146, 147)

XIV. Panzer Korps, GenLt Frido von Senger und Etterlin
16. SS-Panzergrenadier Division 'Reichsführer-SS', SS-Gruppenführer Max Simon
(SS-Panzergrenadier Rgt 35, 36)
26. Panzer Division, GenMaj Smile von Lüttwitz
(Panzer Rgt 26, Panzergrenadier Rgt 9, 67)
362. Infanterie Division, GenLt Heinrich Greiner
(Grenadier Rgt 954, 955, 956)
90. Panzergrenadier Division, GenMaj Ernst-Günther Baade
(Panzergrenadier Rgt 200, 361)
Kampfgruppe Crisolly, composed of remnants of:
3. Panzergrenadier Division, GenMaj Hans Hecker
(Panzergrenadier Rgt 8, 29)

I. Fallschirm Korps, Gen d.FschTr Alfred Schlemm
29. Panzergrenadier Division, GenMaj Walter Fries
(Panzergrenadier Rgt 15, 71)
4. Fallschirmjäger Division, GenMaj Heinrich Trettner
(Fallschirmjäger Rgt 10, 11, 12)
356. Infanterie Division, GenLt Karl Faulenbach
(Grenadier Rgt 869, 870, 871)

AOK 10, GenOb Heinrich von Vietinghoff-Scheel

Directly subordinated:
715. Infanterie Division, GenMaj Hanns von Rohr
(Grenadier Rgt 725, 735)

LXXVI. Panzer Korps, Gen d.PzTr Traugott Herr
Fallschirm Panzer Division 'Hermann Göring', GenMaj Wilhelm Schmalz
(Fallschirm Panzer Rgt 'HG', Fallschirm Panzergrenadier Rgt 'HG' 1–2)
1. Fallschirmjäger Division, GenLt Richard Heidrich
(Fallschirmjäger Rgt 1, 3, 4)
334. Infanterie Division, GenLt Walter Scheller
(Grenadier Rgt 754, 755, 756)
15. Panzergrenadier Division, GenMaj Eberhard Rodt
(Panzergrenadier Rgt 104, 115)

LI. Gebirgs Korps, Gen d.GebTr Valentin Feuerstein
305. Infanterie Division, GenMaj Friedrich-Wilhelm Hauck
(Grenadier Rgt 576, 577, 578)
44. Infanterie Division, GenLt Bruno Ortner
(Grenadier Rgt 131, 132, 134)
114. Jäger Division, GenMaj Alexander Bourquin
(Jäger Rgt 721, 741)
5. Gebirgs Division, GenLt Max Schrank
(Gebirgsjäger Rgt 85, 100)
71. Infanterie Division, GenMaj Wilhelm Raapke
(Grenadier Rgt 191, 194, 211)
278. Infanterie Division, GenLt Harry Hoppe
(Grenadier Rgt 992, 993, 994)

OPPOSING PLANS

ALLIED

The main issue from the Allied point of view was the lack of a plan for the advance north of Rome. Only on 7 June did Field Marshal Alexander issue a directive emphasizing the need to advance swiftly in order to reach the northern Apennines, where the decisive assault was to be launched. Even though Alexander aimed at inflicting as many losses to the Germans as possible, there was no 'annihilation' concept in his orders. The Germans were to suffer their losses mostly thanks to the air forces and the swift advance of the ground troops; the idea of attempting to surround and destroy at least part of the withdrawing German units not being considered. Taking geography into account, Alexander ordered an 'oblique' advance with the AAI's left wing, mostly composed of the US 5th Army, moving fast to the Pisa–Lucca–Pistoia line while, on the right wing, the British 8th Army was to advance to the Florence–Bibbiena–Arezzo line. The basic idea was to tie up the still relatively strong AOK 10 while striking hard at the weaker AOK 14, possibly with the aim of destroying it and creating the basis for an envelopment of AOK 10 west of the Tiber River with an outflanking movement.

A group of American paratroopers ready for action in Italy. In spite of the availability of the 517th Parachute Infantry Regiment in June 1944, no airborne operation was ever considered by the Allied commands. (NARA)

In order to achieve these aims, the Allied commanders were given almost complete freedom of action, with the advance of both 5th and 8th Army to proceed independently and to take advantage of every opportunity which might arise to inflict losses on the enemy. 5th Army's first objectives were the Civitavecchia harbour and the Viterbo airfields, both of strategic value for supplies and for the air offensive against Germany. The possibility of an amphibious landing was taken into account, albeit only with limited aims (the idea at first was to execute it in the Civitavecchia area), while the seizure of Elba Island was also considered. A second

A formation of American B-25 Mitchell medium bombers flying over Italy in the summer of 1944. Since spring the Allied air forces had been hammering the main railway centres in Italy as part of Operation *Strangle*, with the aim of cutting the German lines of supply. (NARA)

order, issued on 14 July, aimed at outlining the long-range planning for the Italian campaign. By then, more than a month after the first order, Alexander had to acknowledge the fact that the Germans were resisting all along the front while attempting to preserve their forces, which required a series of operations in three different stages: the first aimed at driving the German forces back to the Gothic Line while inflicting as many losses as possible; the second aimed at breaking through the Gothic Line in order to, as envisaged by the third stage, exploit the success and reach the Po River. Clark's orders, issued on 6 June, matched the spirit of Alexander's ones. No further directives would be issued until the Arno River Line was reached; the operations by the units on the field were reacting to the existing situation and dealt with by the commanders, who were urged to confer amongst themselves. The aim was to advance to the north-west with armoured units supported by mobile forces, compelling the Germans to withdraw by threatening to envelop their positions. A pace of advance was outlined, soon to reveal that it was too

The remnants of a German position in central Italy, summer 1944. The odd mixture of the German Sturmgeschütz and the Italian 105/32 field gun was rather common as the German troops used all the weapons and equipment they had taken from the Italians in September 1943. (NARA via Digital History Archive)

slow compared to the actual successes on the battlefield. This eventually led Clark to issue the order to not consider the 'phase lines' as 'stop lines' but to maintain the pace of advance with a steady effort.

GERMAN

Facing a critical situation, on 9 June, Kesselring issued a simple order that stated the need to preserve control of as much Italian territory as possible before withdrawing. A final withdrawal to the Gothic Line was unavoidable, but both AOK 10 and AOK 14 had to buy time in order to enable its completion and deny at the same time the gains to the advancing Allied forces. The basic idea was, as in the winter of 1943, to enforce a staged withdrawal along a series of pre-arranged defence lines before establishing on the main defence line along the northern Apennines. The first delaying, or provisional, defence line was the Dora, running all across the Italian peninsula some 100km north of Rome from Orbetello to the southern part of Lake Bolsena, and from there to the Apennines and Narni, then to the south of Rieti and from there to L'Aquila and the Gran Sasso massif on the eastern coast. A second provisional line, the Frieda (subsequently developed into the Albert Line, better known as the Trasimene Line), was to run just to the north of the Dora one, from Piombino to Radicondoli, Lake Trasimene, Perugia, to the north-east of Foligno on the Apennines, and from there to Porto Civitanova on the eastern coast, south of Ancona. In order to slow down the Allied advance and to assist the German defences, a large-scale series of demolitions was to be carried out while retreating so as to impede enemy movements. In any case, Kesselring considered it essential to prevent the defeat of his forces before they could redeploy on the Gothic Line, something which did not really agree with Hitler's orders. After a conference held on 13 June, Hitler issued his own orders to Kesselring to hold the line south of the Gothic Line as long as possible in order to enable its completion, a task which would require several months and that was considered the only possibility to prevent an Allied breakthrough into the Po River valley. In order to erase the concept that the Gothic Line was some sort of impregnable defence like Cassino, Hitler ordered its name to be changed to Green Line. Hitler's order set the pace for the rest of the summer 1944 campaign, Kesselring ordering both AOK 14 and AOK 10 to hold onto the Frieda – or Albert – Line as long as possible, a necessity considering that the Dora Line had already been breached in several points. This provided the basis for the Trasimene Line battle, and for the slowed pace of the German withdrawal to the Arno River Line.

A group of captured Fallschirmjäger pose defiantly for the photographer. In spite of the heavy losses suffered, the two Fallschirmjäger divisions in Italy remained formidable opponents. (NARA)

THE CAMPAIGN

THE BATTLE FOR ROME, 27 MAY–4 JUNE

After months of stalemate, the Allied breakthroughs at Cassino and Anzio marked the peak of their leadership during the Italian campaign. The final assault, Operation *Diadem*, started on 11 May leading to the fall of Cassino between 15 and 18 May, when 5th Army had broken through the Gustav Line and advanced along the coast approaching the second German line of defence, the Hitler Line. Success was not without issues, as both Clark and Juin remarked upon the slow pace of 8th Army's advance, the former being also irritated by Alexander's order to start the breakthrough attack from the Anzio beachhead already on 22 May. Every other consideration apart, Clark remarked that an early attempt to break through from the beachhead before all the available German reserves were committed to the south could end with a disaster, but he still complied with the order. All that Clark was to achieve was a one-day postponement of the breakthrough, now scheduled for the 23rd. By then the Hitler Line had been breached, 5th Army advancing along the Tyrrhenian coast with its II Corps while inland the FEC swarmed into the Liri River valley west of Cassino, with I Canadian Corps barely supporting its advance on the right wing.

The breakthrough from the Anzio beachhead aimed not, as the Germans might have imagined, to the north, towards Rome, but to the east, to the south of Velletri and Highway 7 (the Via Appia), leading from Rome to Cisterna and the coast. The attack, led by four American divisions, started the battle for Cisterna, which ended two days later with the VI Corps advancing towards Valmontone, thus threatening Highway 6, which led from Rome to Cassino. At this stage, with the bulk of AOK 10 still deployed on the same positions it had occupied at the beginning of Operation *Diadem*, an Allied advance to the east would have yielded positive results. Had the 5th Army advanced across the mountains from Valmontone to the Liri River north of Cassino, AOK 10's lines of communication would have been cut and the army itself compelled to withdraw east, across

GIs run to take cover when some isolated German soldiers suddenly start firing. Note the 100mm German field gun in the background. (NARA)

The battle for Rome, 27 May–4 June

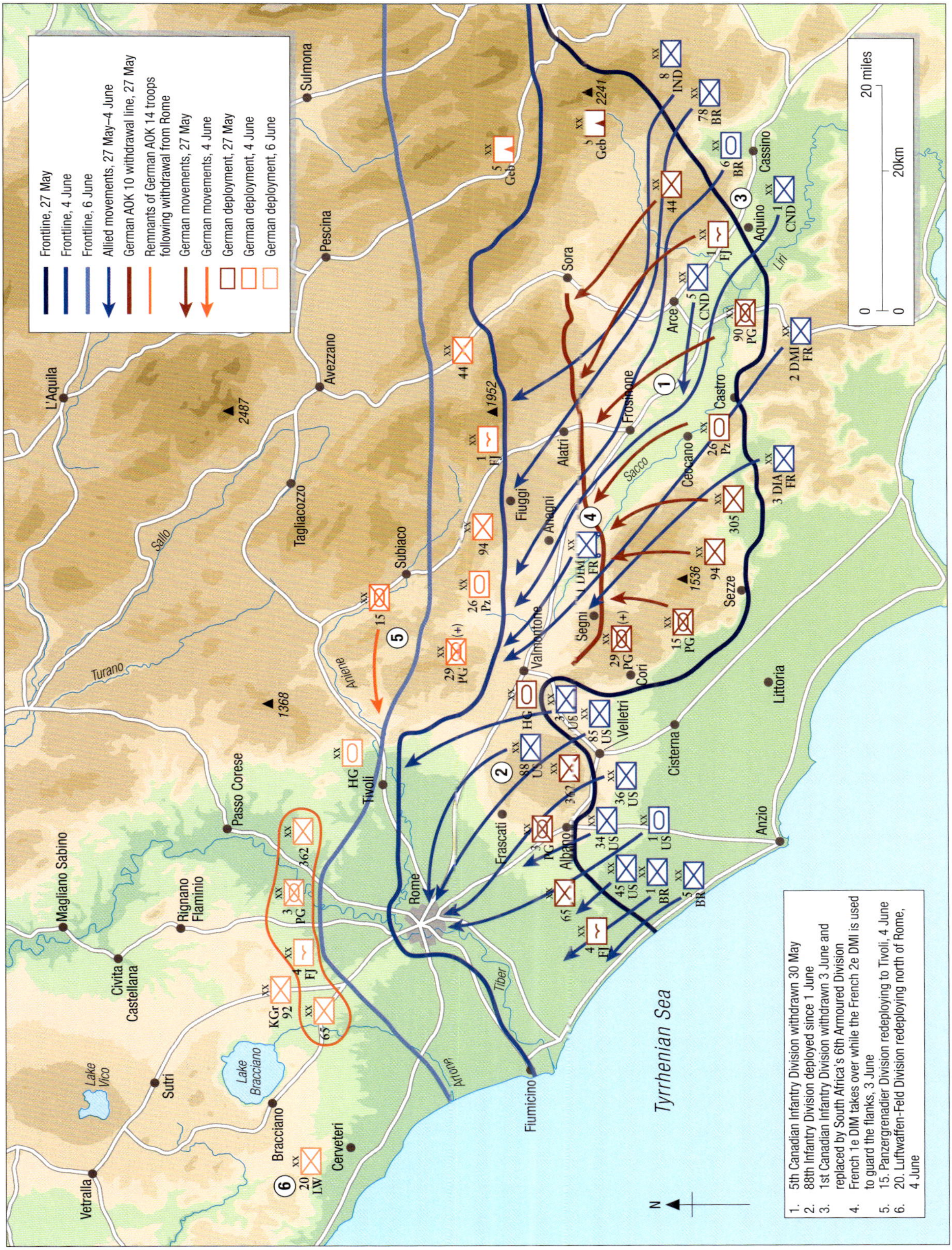

the Apennines, towards the Adriatic coast. It should be considered that at this stage the Allied forces could move faster than the German ones, even on mountainous terrain. The destruction of the German forces in Italy was, at this stage, Alexander's aim and was fully shared by General Truscott, both convinced that if the German forces were destroyed, the seizure of Rome would follow as a natural consequence. Von Vietinghoff, acting on Hitler's orders, did play into their hands as he ordered to hold the positions as long as possible, conducting an early 'staged withdrawal' from the portions of the Gustav Line still in German hands. The only reaction to the VI Corps' advance at Valmontone was the redeployment of the 'Hermann Göring' Division, the line south of the city still intact as General von Mackensen believed that the final assault would be on Rome.

Quite suddenly, on 25 May, General Clark ordered Truscott to shift the axis of advance of VI Corps to the north-west along the axis north and south of the Colli Albani (Albano hills) and Lake Albano, attacking with the 34th Division, which was to shift from Valmontone (the target being left to the 3rd Division along with the 1st Special Service Force and Task Force Howze from the 1st Special Service Force), attacking with 45th Division and the corps' reserve, 36th Division. This started the final stage of the Allied breakthrough, and the battle for Rome, in spite of the protests from Clark's subordinates. VI Corps' advance faced the well-prepared Caesar Line, with mines and barbed-wire-protected positions concealed in isolated farms and hamlets, the kind of terrain the American troops would face again north of Rome. The natural consequence was that VI Corps' attack soon became bogged down, merely advancing by 31 May as far as the north of Velletri and a few kilometres west of Aprilia. At this point, the situation was changed by Kesselring's decision to withdraw from the positions still held, which he made on 26 May. AOK 10 was to pull back, albeit slowly, in order to link up with AOK 14 and eventually deploy along the Caesar Line south of Rome. The slow pace of the German withdrawal was matched by the disappointingly slow pace of 8th Army's advance, in spite of their overwhelming air superiority. Eventually, VI Corps' attack was renewed on 31 May, also thanks to the deployment of two new divisions (85th and 88th Infantry), which moved along Highway 6 towards the south-eastern gate of Rome. Velletri soon fell, and on 2 June, the Caesar Line was broken through in the area between the Colli Albani and Highway 6 west of Valmontone. On that same day, Kesselring, authorized by Hitler who did not want to turn Rome into a battlefield, ordered the withdrawal. What did follow was almost complete chaos. South of Lake Albano, the 45th Division advanced straight towards Rome along Highway 7, with the 88th Division (now spearheaded by the 1st Special Service Force) advancing along Highway 6 (via Casilina) and eventually winning the prize of being the first Allied unit to enter Rome on 4 June. The remnants of I. Fallschirm and LXXVI. Panzer Korps

An 88mm Pak 43-41 still on its trailer is being inspected by an American soldier near Viterbo. (NARA via Digital History Archive)

mostly withdrew across the only bridges still intact, the ones in Rome, the city itself falling deaf to Alexander's repeated appeals to rise and prevent the German withdrawal with every possible means. At the same time, XIV. Panzer Korps started withdrawing north, in an attempt to establish a link with AOK 14 north of Rome. As the bulk of AOK 10 was still mostly deployed along the Gustav Line, the result was a diagonal frontline running from the north of Rome to the Apennines and to the south of Pescara. With the German forces in disarray and facing severe losses, disorganization and panic, there were plenty of opportunities that could be exploited, which led to the first phase of the summer 1944 campaign in Italy.

An abandoned SdKfz 7/2 anti-aircraft half-track with a 37mm Flak 38 gun north of Rome, clearly unable to protect the withdrawing troops from the Allied *Jabos*. (NARA)

THE 5TH ARMY AT THE DORA LINE, 4–21 JUNE

Seriously depleted by losses and with a terrain unfavourable for defence, AOK 14 faced a most critical situation. The army, along with AOK 10, was therefore to conduct a staged withdrawal, disengaging the bulk of its divisions while leaving behind strong rearguards, deploying at first on an intermediate line but aiming at deploying soon to the Dora Line because of the terrain. The intermediate line, running from Maccarese to the Isola Farnese (north of Rome) to Tivoli, could in fact not be held for long, especially considering that the arrival of reinforcements would take time. These included the 356. Infanterie Division from Genua and the 20. Luftwaffen-Feld Division, to be deployed on the army's right wing along the coast. AOK 10 was to redeploy the battered 26. Panzer Division, sending it to Orvieto as a reserve for AOK 14. The problem was that, given its current situation, AOK 14 was not much concerned about Kesselring's orders. As he assumed command of the army on 7 June, replacing General Eberhard von Mackensen, Joachim Lemelsen found a situation much worse than he had expected. No clear picture of the situation could be obtained due to the breakdown of communications; units were withdrawing independently and the concerning signals of chaos were spreading, with rear units withdrawing as far as Florence. This was in spite of the poor road conditions due to Allied air activity, which delayed the withdrawal of the German units. For this reason, AOK 14 was ordered by 5 June to redeploy as soon as possible on the Dora Line, ignoring any attempt to

A destroyed PzKpfw IV tank is being inspected on the road north of Rome. Most German tanks were either destroyed by air attacks or abandoned due to breakdowns or lack of fuel. (NARA)

US

THE ATTACK AT VITERBO, 8–9 JUNE 1944 (PP.32–33)

The 5th Army advance north of Rome was chaotic to a certain degree, the American units moving forward to face either the withdrawing Germans or occasional stubborn resistance. Attached to Combat Command A (CCA) of the 1st Armored Division on 2 June, the 701st Tank Destroyer Battalion was part of the leading units that entered Rome on the evening of 4 June 1944. The next day, the battalion crossed the Tiber, though the advance was hampered by a lack of maps and by unknown terrain. The 701st Battalion then merged with the 3rd Battalion, 1st Armored Regiment, to create Task Force Cole, intended to reconnoitre the terrain north of Rome. Facing little or no German resistance, the group was fast in its movement, only being slowed down by demolition and terrain. The advance along the Aurelia highway saw the group reach Bracciano by 7 June, when a new Doran Force was formed under command of the CCA. Its task for 8–9 June was the seizure of Viterbo, defended by elements of the 3. Panzergrenadier Division. A first attempt to advance into the town was made on the afternoon of 8 June, but the 701st faced strong German resistance and requested infantry, having the 1st Battalion, 135th Infantry Regiment, 34th Division, attached. The second attack, on the morning of 9 June, was successful and, after a brief but fierce struggle, Viterbo was seized. A good many Italian towns were in ruins due to the Allied bombings, which made things complicated for the open-topped M10 tank destroyer **(1)**, which required infantry support in close combat. In the summer of 1944, many GIs would fight in shirtsleeve order **(2)**, the German soldiers wearing their denim summer uniforms or even the Afrika Korps ones. The use of Sturmgeschütze, hiding inside destroyed buildings **(3)**, was rather common and, for the first time, the German troops in Italy started making use of the Panzerfaust anti-tank rocket.

hold intermediate delaying positions, in order to reorganize its remnants while receiving reinforcements, which now also included the 162. Infanterie Division (Turkestan), to join 356. Infanterie Division on the Dora Line east and west of the Lake Bolsena, while the 20. Luftwaffen-Feld Division would be deployed between the Tyrrhenian coast and Lake Bracciano.

This was a necessity given the situation, as General Schlemm presented in a detailed report describing the conditions of the German forces in the area. Both 3. Panzergrenadier and 65. Infanterie Divisions had managed to cross the Tiber south of Rome, their withdrawal being covered by 4. Fallschirmjäger and 362. Infanterie Divisions, which faced the American troops into Rome until they too withdrew across the Tiber. American armoured troops had already started their pursuit on 5 June, moving along the Flaminia and Cassia highways north of Rome and soon penetrating the boundary between the 3. Panzergrenadier and the 4. Fallschirmjäger Divisions at Torre Vecchia and in other places, compelling the German units to deploy into the intermediate Line II north of the capital. There were no doubts that this line would not hold for long, especially because of the supply situation. Movement along the roads was now almost impossible due to Allied air activity, several burned-out vehicles congesting the roads themselves along with the columns of retreating troops. Schlemm estimated that the combat strength of the divisions under his command was at its lowest, with some 600 men in 4. Fallschirmjäger, 500 in 3. Panzergrenadier, 400 in 65. Infanterie and just 100 in 362. Infanterie Division. Without reinforcements, it would have been impossible to check the enemy advance and a withdrawal to the intermediate line (simply named E) would have been mandatory.

There was a distinct advantage for the Germans: the chaos and the situation created by the Allied advance. The 5th Army units spent the rest of 4 and most of 5 June consolidating their positions in Rome, being unable themselves to cross the Tiber other than in the city. The 'bottleneck' created by the systematic destruction of bridges and roads slowed down their advance, and only on 6 June did the full pursuit north of Rome start with Clark alerting the FEC to prepare to relieve II Corps, while IV Corps was to prepare itself to take over from VI Corps. First to move was the 1st Armored Division, with its CCA (Combat Command A, 1st Armored Regiment, 135th Infantry Regiment from 34th Infantry Division) followed by the 36th Division advancing inland while, along the coast, the CCB (13th Armored Regiment, 6th Armored Infantry Regiment) was to advance with 34th Division, the 45th Division being held in reserve. As these formations assembled, the 1st Reconnaissance Regiment pushed north of Rome as far as some 10km, encountering little or no resistance at all. Late on 5 June, VI Corps was instructed to seize Civitavecchia as soon as possible, and for this

A Sherman from the 1st Armored Division struggles to move into the narrow streets of Paganico, 21 June. (NARA via Digital History Archive)

US 5th Army advance, 5–21 June

purpose was to cooperate closely with the CCB, whose tanks were to advance during daytime, the infantry taking over and advancing at night. Although slowed down by several obstacles, the American units made no contact with the German troops and moved some 40km north of Rome by 6 June. The morning after, Civitavecchia was easily seized, the Germans offering only a limited firefight north of the city. At this point, CCB handed over the area to the 34th Division, shifting itself into the 36th Division area to join the CCA. Also on 7 June Tarquinia was seized, US troops advancing almost 10km north of the town without encountering opposition. This eventually materialized the next day, mostly from elements of the 65. Infanterie and 20. Luftwaffen-Feld Divisions (forming the Gruppe Göritz) which had reached the area. At the same time the 45th Division was moved east of Rome, before being sent to Naples to join 7th Army. The only event of some relevance was the attachment of 361st RCT (the first element of 91st Division to arrive in Italy) to 34th Division on 9 June, shortly before 36th Division took over in the area and 34th Division assembled near Civitavecchia.

A 152/45 Italian gun pressed into German service and used to defend the Punta dei Ripalti area on Elba Island. (NARA via Digital History Archive)

Inland, the 36th Division and CCA also made good progress moving past Bracciano on 7 June, before seizing Manziana the next day after a short battle. The American units advanced then along the Manziana–Tolfa–Civitavecchia road facing no resistance, the 36th Division simply preparing to take over from the 34th in the area. The first serious clash took place on 8 June as CCA met a strong German rearguard at Vetralla, which was eventually seized. Other German troops were encountered some 3km south of Viterbo, which was seized without fighting by Task Force C early on 9 June. That same night, the CCB assembled close to the CCA, before moving to Vetralla and Tuscania. Meanwhile, the FEC started to deploy to relieve II Corps, CCB having a small clash with rearguards close to Tuscania, which was seized on 10 June. This was the last engagement for 1st Armored Division, which had gone into army reserve on 11 June, as IV Corps took over from VI Corps which was then assigned to 7th Army. As Tarquinia was seized, a minor German counter-attack on the night of 10/11 June clearly showed that the days of swift advance were now over. Meanwhile, II Corps moved north of Rome with the 85th and 88th Divisions, the 3rd Division acting as Rome

French troops marching across Portoferraio on Elba Island. The battle for the island was brief, but still bitter and hard fought. (NARA via Digital History Archive)

French troops in the ruins of Portoferraio. After the seizure of the island, the Moroccan troops terrorized the civilian population, robbing, attacking and assaulting, and those actions are still remembered there today. (NARA via Digital History Archive)

garrison before joining 7th Army. The advance along the Orte–Soriano–Viterbo road created some issues with the VI Corps boundary, while the 85th and 88th Divisions advanced astride Highway 2. Only on 7 June did minor clashes occur with the German rearguard, the 85th Division being relieved on 9 June by FEC units just as it reached a point some 10km from Bracciano. The advance of 88th Division had been halted on 7 June, as the new boundaries between 5th and 8th Army enforced on that day cut across its line of advance. Since 9 June, Task Force Ellis had screened the area between Orte and Soriano while the British units deployed, and 5th Army's advance was only resumed on 11 June with IV Corps (deploying only the 36th Division) along the coast and the FEC inland with the 1ère Division d'Infanterie Motorisée and 3e Division d'Infanterie Algérienne.

Still facing no determined German resistance, 5th Army units continued their advance on 10–11 June with the 36th Division moving towards Grosseto. The first serious German resistance was encountered close to Orbetello, with artillery fire followed by a counter-attack, but the advance was resumed on 12 June, facing no opposition while Capalbio was seized. The next German defences were met at Magliano, seized on the 14th, with American reconnaissance units entering Grosseto the next day without opposition. IV Corps' advance continued over the next days while the Grosseto area was mopped up, the boundary between the corps and the FEC becoming effective on the 16th. By 20 June, IV Corps had reached the area north of Grosseto, to the intersection between Highways 1 and 73 and at Paganico, now well beyond the Dora Line. A period of rest followed, units preparing to start a new phase of battle. The FEC, operational since 10 June, advanced rapidly, facing no serious opposition until reaching Montefiascone on 11 June. On the next day, a German counter-attack and a destroyed bridge slowed down the advance, the 1ère Division d'Infanterie Motorisée reaching the northern end of Lake Bolsena by 14 June. On the western side of the lake, the 3e Division d'Infanterie Algérienne also faced a German counter-attack on 11–12 June, before completing their seizure of the western shores of the lake. The advance, now facing more serious German opposition as it moved past the Dora Line, continued

French troops landing at Marina di Campo on Elba. Despite their efforts, the landing almost resulted in failure as they faced German opposition. (NARA via Digital History Archive)

with the French troops approaching the Orcia River by 20 June.

In the early hours of 17 June, a French amphibious force mainly composed of units from the 9e Division d'Infanterie Coloniale and with a large array of special forces, for a total strength of some 9,700, landed on Elba Island. What seemed to have been conceived as some sort of mopping-up operation soon clashed with the strong resistance of the 2,000 German defenders, whose determination was undermined by the defection of two Italian battalions, 500 strong combined. By 18 June, the French troops were able to advance into the island, the Germans opting for evacuation on the next day, which was completed early on the 20th. The operation, under direct command of the AAI HQ, had little if any influence for the 5th Army and was clearly not coordinated with Clark. Its main aim seems to have been boosting French morale, or just to please the French commander. The Germans, with an eye to more sensible strategic matters, became concerned about a possible Allied landing along the coast which might have had serious repercussions on the defence of the main lines. It never happened. In fact, Lemelsen faced a serious situation until 10 June when the 90. Panzergrenadier Division was able to intervene in support of the weakened 162. Infanterie Division and 20. Luftwaffen-Feld Division, the situation west of Lake Bolsena being serious as well given the state of exhaustion of both the 3. Panzergrenadier Division and 4. Fallschirmjäger Division. Thanks to the intervention of the 356. Infanterie Division, the Germans were able to deal with the FEC's assault from Viterbo, which on 12 June compelled Kesselring to put XIV. Armee Korps under the command of AOK 14. Facing an enemy which, as he estimated, was 3:1 superior in manpower, 10–15 times stronger in armour and 200:1 in terms of airpower, Lemelsen asked for a tactical withdrawal only to be informed by Kesselring on 14 June that AOK 14 was now required to make a long-lasting stand on the Albert (formerly Frieda) Line. The situation worsened between 15 and 18 June, as 8th Army made its advance deep into AOK 10's lines and the battle for Elba Island started. The latter had one major consequence: afraid of an Allied landing, Kesselring ordered 16. SS-Panzergrenadier Division to deploy in coastal defence between Viareggio and Leghorn, the 20. Luftwaffen-Feld Division being used to defend the harbour of Piombino.

The landing at Marina di Campo on 17 June. The task force carrying the French troops was American and composed of both American and British craft. (NARA via Digital History Archive)

A German 50mm Pak 38 gun overlooking the Marina di Campo bay from its position in Strongpoint 20, which caused a good many problems for the Allies during the landing. (NARA via Digital History Archive)

THE LANDING AT ELBA ISLAND, 17–20 JUNE 1944

On 17 June, French troops under General Joseph Magnan (commander of 9e Division d'Infanterie Coloniale) landed on Elba Island, starting Operation *Brassard* ('Armband'). They faced German defences under command of General Franz Gall, which fiercely opposed

GERMAN*
Kommandant Elba (HQ at Villa Napoleone)
Festungs Bataillon 902
Strongpoints 34, 15 – 16 – 17, 10 – 13 – 14, 3 – 4 – 5 – 6 – 7 – 8
Festungs Bataillon 908
Strongpoints 25, 22 – 22a – 23 – 24 – 29, 18 – 19 – 20 – 21a
Italian 6th Coastal Battalion
Strongpoints 14, 29, 35 – 36
Marine Artillerie Abteilung 616
Strongpoints (batteries) 1 – 2, 8 – 9, 11 – 12, 21
Heeres Artillerie Abteilung z.b.V. Sterz
Strongpoints (batteries) 28, 30, 33
Flak Abteilung 192
Strongpoints (batteries) 31 – 32
**Note: not all strongpoints are labelled on BEV*

EVENTS

17 June

1. 0100hrs: The 7th Detachment of the Bataillon de Choc lands at Capo d'Enfola to attack the 152mm gun battery at Strongpoint 1. The French commandos are repulsed and are compelled to re-embark after suffering heavy losses: 23 out of 48 men.

2. The 6th Detachment lands at Punta Acquaviva with the aim of reaching the German HQ at Villa Napoleone. After losing three hours in attempting to climb Mount Poppe, it also re-embarks.

3. 0350hrs: The 3rd and 4th Detachments land at Capo Fonza and Capo Stella respectively, only to discover the gun batteries they were supposed to attack are non-existent.

4. The 5th Detachment is landed in the wrong area west of Capolivieri, which requires a long journey to reach the battery in Strongpoint 11. It is attacked, without success, later that same morning. The detachment withdraws to the north.

5. 0350hrs: The 1st and 2nd Detachments, under direct command of Major Gambiez, land west and east of Strongpoint 21. The two detachments attack the strongpoint at 0530hrs, which is subsequently seized by the main landing force.

6. 0400hrs: The first wave of the 13e RTS lands at the Golfo di Campo. The German resistance is strong and, at 0430hrs, the German guns hit landing craft *303* and *274*, one of which catches fire and explodes. The third assault wave withdraws.

7. The landing of the Groupe de Commandos d'Afrique at Capo di Fonza faces a weaker German resistance. At 0500hrs, the third wave, followed by reinforcements, is landed there to consolidate the beachhead.

8. 0610hrs: The 13e RTS seizes Strongpoint 19 and, despite not knowing about the progress of the landing, starts advancing north. The seizure of Strongpoint 18 at 0800hrs consolidates the beachhead and enables further advances inland.

9. 1000hrs: Strongpoint 26 is attacked by 4e RTS. The Italians manning the defences surrender at once.

10. 1000hrs: The 13e RTS attacks Strongpoint 25. Its defenders, following the fall of Strongpoint 26, are ordered to withdraw.

11. 1700hrs: The 13e RTS attacks and seizes Procchio, breaking through the attempted German line of defence. The defenders of Strongpoints 23 and 24 withdraw to Portoferraio.

12. 1800hrs: The 4e RTS, supported by the Commandos d'Afrique, attacks Strongpoint 28 at Mount San Martino. It is seized by 2000hrs.

18 June

13. 1200hrs: After attacking and seizing the outer line of strongpoints, the 13e RTS attacks Portoferraio, which is seized in half an hour.

14. After a swift advance to Strongpoint 14, seized without effort, the 4e RTS and the Tabors advance north and attack Strongpoints 33 and 34. The northern prong, attacking Strongpoint 31, is halted by a strong resistance. Strongpoints 33 and 34 fall in the early afternoon, opening the way to Porto Longone.

the chaotic landing. In spite of some serious setbacks, mostly due to the lack of knowledge of the enemy defences, the French troops managed to gain control of most of the island after three days of intense fighting, even though they did not succeed in preventing the German evacuation. The landing, its actual value being debatable, was the last large-scale Allied amphibious operation in the Italian campaign.

FRENCH
Bataillon de Choc
- **A.** Groupe Gambiez, dét. Carbonnier
- **B.** Groupe Gambiez, dét. Lefort
- **C.** Groupe Lamy
- **D.** Groupe Lamy, dét. Durieux
- **E.** Groupe Lamy, dét. Jacobsen
- **F.** Groupe de Commandos d'Afrique
- **G.** 13e Regiment Tirailleurs Senegalais
- **H.** 4e Regiment Tirailleurs Senegalais
- **I.** 4e Regiment Tirailleurs Senegalais, 2e Groupe de Tabors Marocain, Groupe de Commandos d'Afrique, elements Bataillon de Choc

Note: gridlines are shown at intervals of 5km (3.11 miles).

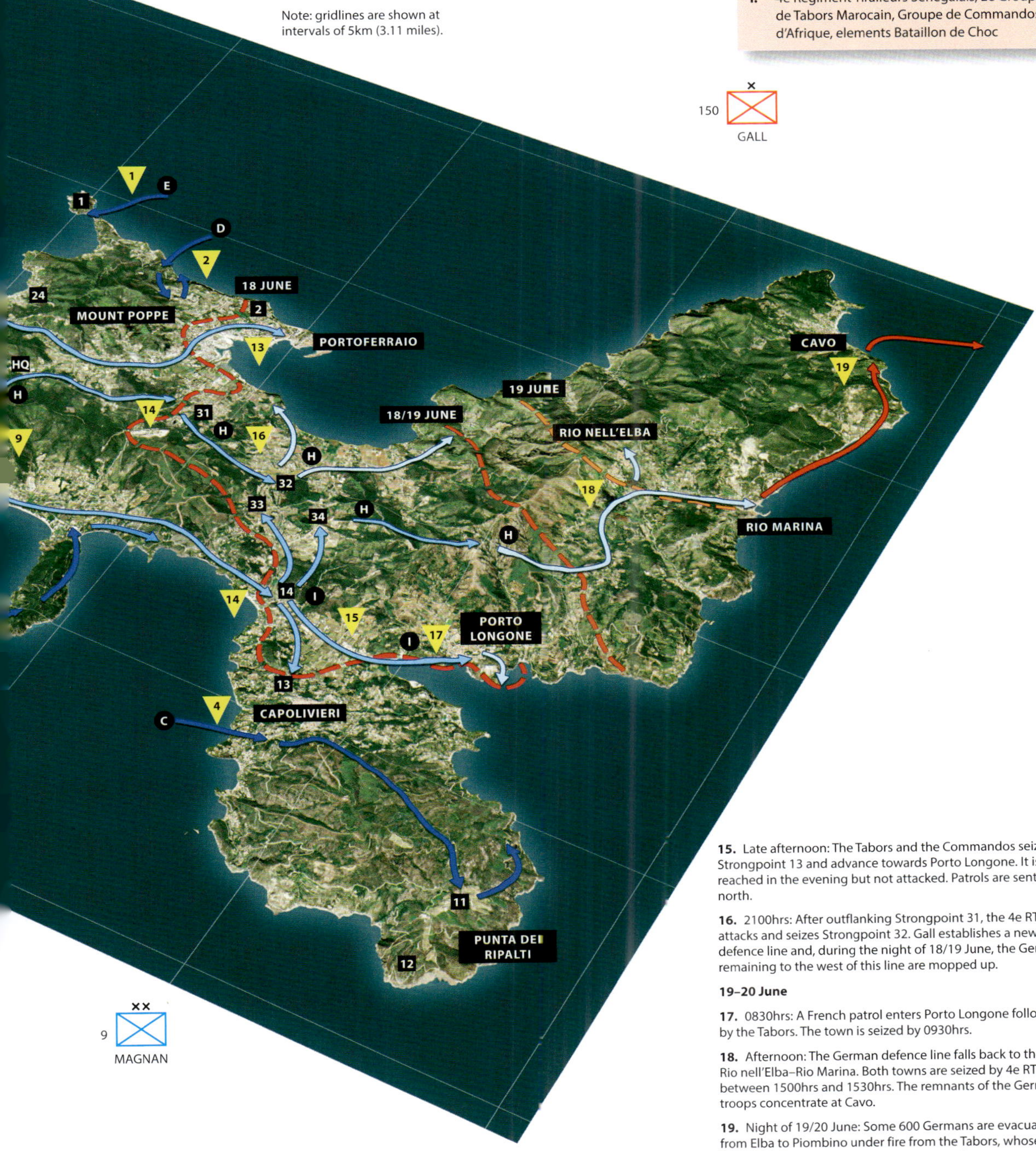

15. Late afternoon: The Tabors and the Commandos seize Strongpoint 13 and advance towards Porto Longone. It is reached in the evening but not attacked. Patrols are sent north.

16. 2100hrs: After outflanking Strongpoint 31, the 4e RTS attacks and seizes Strongpoint 32. Gall establishes a new defence line and, during the night of 18/19 June, the Germans remaining to the west of this line are mopped up.

19–20 June

17. 0830hrs: A French patrol enters Porto Longone followed by the Tabors. The town is seized by 0930hrs.

18. Afternoon: The German defence line falls back to the line Rio nell'Elba–Rio Marina. Both towns are seized by 4e RTS between 1500hrs and 1530hrs. The remnants of the German troops concentrate at Cavo.

19. Night of 19/20 June: Some 600 Germans are evacuated from Elba to Piombino under fire from the Tabors, whose patrols had reached Cavo. Overall German losses amounted to some 2,000 men, while French losses are estimated at some 800–900.

THE 8TH ARMY ADVANCE, 4–21 JUNE

According to Field Marshal Alexander's directives, issued before the start of *Diadem*, the task of 8th Army following the seizure of Rome was to advance along the Terni–Perugia axis, which is along a stretch of relatively flat terrain east of the Tiber and of the mountainous area south of Lake Trasimene. This effectively split the Allied advance in two, and left an entire sector without the presence of Allied troops: the stretch of mountainous terrain further to the east, the main Apennines, and the Adriatic coastal area. As in the case of 5th Army, the advance was slowed down by the amount of disruption to communications caused both by the Allied bombing and by German destruction. Adding to that the fact that Leese seemed completely unaware, and unconcerned, of the potentialities offered by the Indian divisions for mountain warfare, the natural consequence was the loss of many good opportunities. I Canadian Corps would be held in reserve, while XIII and X Army Corps would advance abreast up the Tiber valley, XIII Corps operating both on the west and on the east of the river (6th South African and British 6th Armoured Division), with the 4th Infantry Division advancing along minor routes on the eastern flank. The task assigned to X Corps, with the 8th Indian and 2nd New Zealand Divisions, was minor, which was to advance through the central Apennines more or less following the German pace of withdrawal. On the Adriatic coast V Corps, under direct command of the AAI HQ, was to watch for any sign of a German withdrawal and react accordingly, while the Polish Corps reorganized after the battle of Cassino and the Italian Corpo Italiano di Liberazione (CIL, Italian Liberation Corps), an impromptu unit, was being organized to replace V Corps' 10th Indian Division. Most importantly, the fact that the bulk of AOK 10's units were still deployed well south of AOK 14's defence line seems to have been ignored, since no serious attempt was made to engage them in order to prevent their withdrawal.

Advance in XIII Corps' area was slowed down by congestion and German resistance. The 6th South African Armoured Division moved from Rome on 6 June to Orte only to find all the bridges on the Tiber destroyed, which compelled a swing towards Orvieto west of the Tiber. To the east, the 6th Armoured Division clashed with the rearguards from the 'Hermann Göring' Division and 15. Panzergrenadier Division protecting the Rieti–Terni road in order to allow the withdrawal of LI. Gebirgs Korps at Avezzano. A major clash took place at Monterotondo, which paved the way for the advance to Passo Corese, reached on 8 June. At the same time 4th Infantry Division was also advancing at a slow pace dealing with the rearguards from 15. Panzergrenadier Division and 1. Fallschirmjäger Division, while the 8th Indian Division, leading X Corps' pursuit, advanced steadily but without much energy along with the 2nd New Zealand Division to its right. By

German Pioniere preparing the destruction of a stretch of road during the summer. The many obstacles created by the Germans, combined with the destruction from air attacks, slowed down the Allied advance. (NAC)

9 June the former had reached Highway 5 (Rome–Pescara), the latter Avezzano on the same route, which was found clear of Germans. The fact that the German withdrawal set the pace for 8th Army's advance became noticeable along the Adriatic coast, where only on 7–8 June did V Corps notice that the opposing forces were moving back (which they had started doing since the 6th). Pescara and Chieti were seized without opposition on 10 June, but the AAI HQ ordered not to press on in order to save bridging equipment. This greatly helped the withdrawal of LI. Gebirgs Korps, dramatically short of motor vehicles. In fact, Allied commanders were more concerned about the reorganization of forces for the southern France landing and for the planned offensive in the northern Apennines than for the actual situation at the front, with the result – remarked by the official history – that a good many chances were lost. To what extent this kind of foresight influenced the operations at the front is clearly shown by 8th Army's order to wait and see if the Germans withdrew from Ancona. Only if they did not was the Polish Corps to be brought forward and used to seize the city.

A Sturmgeschütz III hidden inside a house is ready to fire upon the enemy. The German self-propelled guns proved to be more effective than the tanks. (NAC)

From the German point of view, the AOK 10 situation was critical, albeit not as much as that of AOK 14. The terrain and the extensive demolitions slowed down the enemy advance, and on 10 June all the units west of the Tiber were put under AOK 14's command, the 26. Panzer Division being sent to the Civita Castellana–Orte area in order to fill a dangerous gap between the two armies. This weakened AOK 10 even further, which could only rely on the 278. Infanterie Division at almost full strength, while the 94., 362. and 92. Infanterie Divisions had to be pulled out of the frontline as they needed a complete reorganization. Thanks to the lack of initiative on 8th Army's side, the withdrawal proceeded almost unhindered, apart from the constant air threat. While LXXVI. Panzer Korps, in part still deployed on the Caesar Line, withdrew, XIV. Panzer Korps deployed the 26. Panzer Division to halt the enemy advance along the Casilina highway and, on 7 June, assembled this and 29. Panzergrenadier Division in the Attigliano–Orte area while the 334. Infanterie Division was to protect the Orte–Terni route. On 8 June, AOK 10 ordered the assembly in the Orte–Orvieto–Todi area of XIV. Panzer Korps (soon to be put under AOK 14 command) in order to screen the right flank of the army, while the LXXVI and LI. Gebirgs Korps were to proceed with the withdrawal, relying on demolitions to slow down the enemy pace of advance. Here too, the possibility of Allied tactical landings on the Adriatic coast was a concern, and a mobile force was set up to deal with the possibility. In any case, the army proposed to resist on the Dora Line starting from 12 June, while planning a further line of resistance along the Lake Bolsena–Terni–Teramo line.

British 8th Army advance, 4–21 June

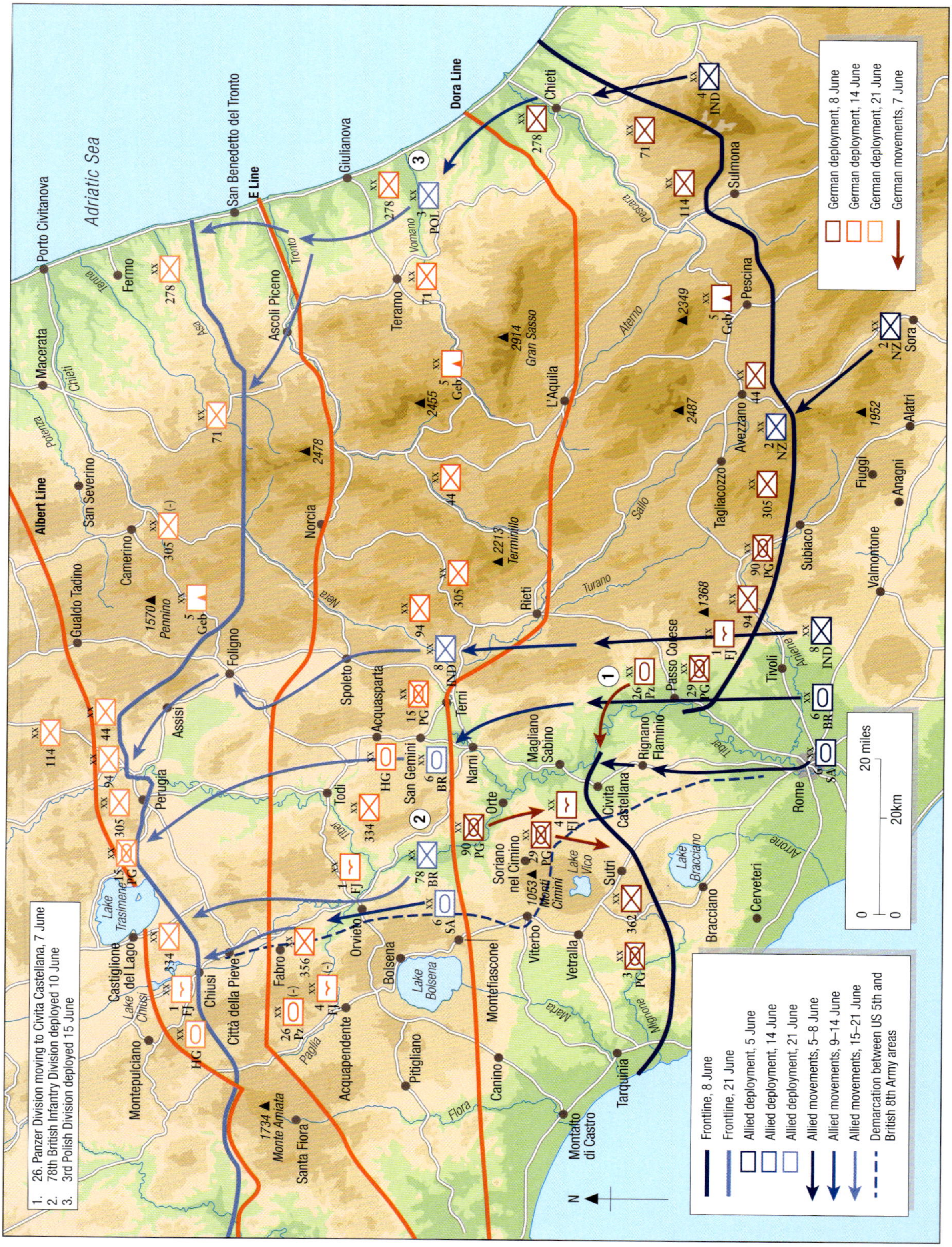

The 8th Army's advance became a matter of slow, methodical clearing up of the obstructions and of the main roads, with only XIII Corps being actually engaged. On 10 June, the 6th South African Division moved towards Orvieto flanked on the right by the 78th Infantry Division, supported by 9th Armoured Brigade, to the west of the Tiber. At first resistance from the rearguards of 356. Infanterie Division was dealt with without problems but, on the 11th, the South Africans faced more determined resistance at Bagnoregio and, after some probes on the 12th, decided to mount a full-scale divisional attack the next day. The attack actually took place on 13 June, but by then I. Fallschirm Korps had already abandoned Orvieto facing the growing pressure of both XIII Corps and of the FEC. On 14 June, the South Africans entered Orvieto unopposed, continuing their advance northwards in the days that followed until, on 16 June, Leese ordered XIII Corps to advance west of Lake Trasimene and X Corps to advance to the east, both corps to converge at Arezzo. Still, the pace of 8th Army's advance slowed down again and on the night of 16/17 June the 78th Division failed to dislodge the German defenders from Città della Pieve, attempting then an outflanking manoeuvre. Only on the 18th did the rearguards of 1. Fallschirmjäger Division abandon the city, still defending the area to the north of it and, most importantly, the approaches to the stretch of terrain between Lake Chiusi and Lake Trasimene. Likewise, the South African advance was checked south of Chiusi. It was on 21 June that Leese's command realized that XIII Corps was now dealing with the Albert Line which, it was agreed, required a full-scale corps' operation. The German withdrawal was in fact successful, in spite of the almost complete lack of resistance on the Dora Line. On 11 June, Lemelsen ordered 29. Panzergrenadier Division and 26. Panzer Division to move into a new defence position and, while holding as long as possible south-east of Orvieto, the divisions in need of reorganization (92., 65. and 362. Infanterie Divisions) were to be moved back to the Gothic Line in order to be reorganized. Basically, AOK 14 accepted the idea of a faster withdrawal while AOK 10, deployed south of its neighbouring command and withdrawing at a slower pace, was anything but pleased. As a consequence, on 12 June von Vietinghoff started to plan a protracted defence along Lake Trasimene–north of the Perugia–Porto Civitanova line, a variation of the Frieda (or Albert) Line. On 14 June, the enemy advance north of Orvieto became a reason of concern, especially considering the simultaneous threat against Todi, which was interpreted as an attempt to break through to Perugia. To deal with the situation, the 'Hermann Göring' Division was to join the 1. Fallschirmjäger Division near

German soldiers in an impromptu dugout ready to face the enemy. Note the one in the foreground who not only has a Panzerfaust but also prepared a couple of stick grenades. (NAC)

British infantrymen advance into an Italian village under cover of a Bren gun on the right. Note how all of them are in shirtsleeve order, almost a necessity in the summer Italian climate. (NARA via Digital History Archive)

British troops using an 81mm mortar to blast the German strongpoint obstructing the advance of their unit. In June 1944, mortars were used more than the artillery, given the fast pace of the advance. (NARA via Digital History Archive)

Città della Pieve and, while rearguards were to defend the present positions, a new defence line was to be formed along Città della Pieve–Deruta–Assisi.

A sudden crisis emerged when, after Narni was seized without opposition on 13 June by the 6th Armoured Division, Terni was seized as well only to find all the bridges on the Nera destroyed. The 8th Indian Division, advancing along a different route, entered Terni the next day and decided to continue its advance along secondary routes. Led by the armoured cars of the 12th Lancers, the division moved along the mountain roads while the 6th Armoured Division waited for the Bailey bridge at Terni to be completed, which did happen on the 17th. With a rather extraordinarily swift advance, the Lancers and the Indian troops reached Spoleto and advanced across a secondary road to enter Foligno on 16 June, the Lancers seizing a bridge still intact. The move thwarted AOK 10's planned defences, compelling it to hastily withdraw north, temporarily holding the positions south of Assisi but, since it was now clear that Perugia was threatened by both the 6th Armoured and the 8th Indian Divisions, the city was eventually abandoned on the night of 19/20 June while LXXVI. Panzer Korps deployed on the Albert Line. The Lancers and Indian troops attempted a push across the Apennines east of Foligno in the direction of Macerata and the coast, but the German opposition prevented any exploitation. The advance along the Adriatic coast continued at the slow pace imposed by the LI. Gebirgs Korps.

THE BATTLE FOR THE ALBERT LINE, 22 JUNE–3 JULY

The German lines of defence were mostly nothing more than lines drawn across a map, lacking any prepared defence or even actual positions. The Dora Line was just a withdrawal line where the units were supposed to reorganize while checking the enemy advance, and it mostly failed its purpose. The AOK 14 area on the Tyrrhenian coast, deemed the most endangered and potentially suitable for an enemy breakthrough, was to see a series of defence lines in a row: first the Frieda Line, north of the Dora, stretched along the Piombino–north-west of Grosseto–Lake Trasimene–Perugia–Porto Civitanova line. This roughly matched the Albert Line, whose name was subsequently adopted, which ran effectively from Castiglione della Pescaia

An improvised way to use self-propelled artillery – 75mm field guns mounted on M3 halftracks. This was a temporary solution, soon replaced by other vehicles. (NARA via Digital History Archives)

on the Tyrrhenian coast (west of Grosseto) to the north of Grosseto to Magione, east of Lake Trasimene, then across the central Apennines to Mount Conero, south of Ancona on the Adriatic coast. Preparing for the eventuality of an enemy breakthrough, both AOK 14 and AOK 10 reconnoitred further defence lines, AOK 14, for example, outlining a line – named Anton – running from Follonico (south of Piombino) to the River Orcia, where it merged with the Albert Line. Most importantly, Kesselring attempted to turn these lines drawn on the maps into actual defence lines by having the divisional reconnaissance select the most suitable (tank proof) defence positions and by having some sort of prepared positions built, mostly thanks to Italian labourers and even relying on compulsory labour. By 16 June, both AOK 14 and AOK 10 were instructed to pull back to the Albert Line which, as ordered by Hitler, was to become a line of protracted resistance.

A 40mm Bofors anti-aircraft gun ready to deal with a German aircraft. In fact, the almost complete disappearance of the Luftwaffe from the sky soon had these units turned into infantry. (NARA via Digital History Archive)

On 11 June, the 5th Army advance became more organized, now with the US IV Corps along the coast and, inland, the FEC to the west of the Tiber. Already on that day General Crittenberger had planned a further advance along the coast with 36th Division reinforced by 361st RCT plus one tank and one tank destroyer battalion, the 34th Division to remain in reserve. 36th Division was to advance along the coastal road supported on its right wing by the Task Force Ramey (built around the 91st Reconnaissance Squadron), whose main task was to screen the area and maintain contact with the FEC. Here, Juin created a pursuit corps led by General de Larminat and included the 1ère Division d'Infanterie Motorisée and 3e Division d'Infanterie Algérienne, which were to advance north of Lake Bolsena. On its own, 36th Division's task was to reach Grosseto, seizing it on the 15th, and from there move on to Piombino, Cecina and Leghorn. By 21 June, 5th Army was still some 50km south of its intermediate goal, the frontline extending from the south of Piombino and Roccastrada more or less parallel to the Orcia River, to the west of Mount Amiata. East of the Amiata, the FEC was still south of the intersection between Highway 1 and 73, and Montepulciano. Thus, 5th Army's immediate goal was a line running from Cecina on the coast to Volterra, the road junction south of Poggibonsi and, to the east, the Chianti mountain range just to the south of Florence. This area, dominated by the famous Tuscan hills, would provide the springboard for the further advance north to the Arno River, from Pisa to Florence. Crittenberger, realizing the nature of the terrain, alerted the 34th Division to prepare to take over from the 36th thus envisaging an advance no longer

A Sherman tank advancing at the crossroad before Orvieto (note the road sign), while a group of British soldiers guard some German prisoners of war. (NARA via Digital History Archive)

Near Pontedera, a jeep drives past a destroyed PzKpfw IV, certainly belonging to the 26. Panzer Division. This was the Panzer unit that fought in Italy from 1943 to 1945. (NARA via Digital History Archive)

facilitated by tank superiority. In fact, the 36th Infantry Division, now without the attached 517th Parachute Regiment (PIR), continued its advance astride the coastal highway seizing Piombino without difficulty, but was slowed down by persistent rain. On 23 June, the 517th PIR deployed along the coast, on the left flank of the 36th Division, while the 1st Armored Division (in line since 22 June) relieved Task Force Ramey. The rearguard in the area, elements from the 19. Luftwaffen-Feld Division, withdrew unhindered, and on 25 June, the 36th Division was pulled out of the line to be attached to 7th Army for the invasion of southern France. The next step was attacking the Tuscan 'Iron Hills', just to the east, which saw the 1st Armored Division prepare a full assault that started on 23 June and lasted for four days, the division advancing for an average of 4km a day. The pace of advance slowed down for the 34th Division as well since, after deploying on the line on 26 June with the attached 442nd Infantry Regiment (replacing 517th PIR), following an initial swift advance compelling XIV. Panzer Korps to withdraw to the Cecina River, an entire day was needed to cover the distance of less than 10km to the river itself. Also, the presence in the area of the 16. SS-Panzergrenadier Division was noted, revealing a strengthening of the German defences. This led the 34th Division commander, General Ryder, to opt for an attack to the east of Cecina, without realizing that 26. Panzer Division was defending the area. As a result, the first attempt to establish a bridgehead on the Cecina on 30 June failed, the single infantry company across the river being isolated until 2 July, when it was joined by reinforcements thanks to the massive intervention of artillery and air support. At this point, Ryder saw no other alternative than a direct attack against Cecina, in spite of the fact that any attempt to advance along the coastal road had faced strong enemy opposition. The renewed attempt to dislodge the Germans from their positions south of the Cecina failed, and only the massive intervention of artillery support saved the situation.

Deployed along the Albert Line from 21 June, Lemelsen's attention focused mostly inland where attacks were expected between Montepulciano and Montalcino, along the main route to Siena, and at Roccastrada, to the east of Tuscany's 'Iron Hills'. The possibility of a landing to the north of Piombino was expected, the US divisions regrouping for the southern France landing having been mistaken for an assault force directed against Italy. Most concerning of all was the increased rate of desertion amongst the ranks of the 162. Infanterie Division (Turkestan), which compelled the redeployment of 26. Panzer Division in order to protect the coastal route. Unable to halt the American attacks at Roccastrada and along the coast, the Germans had no alternative but to resort to 16. SS-Panzergrenadier Division in spite of the potential threat of a landing. A first withdrawal took place on 23 June, another one being allowed late on the next day. Still, on the 25th, the massive attack started by the FEC both west and east of Montalcino caused a crisis,

US 5th Army advance, 21 June–23 July

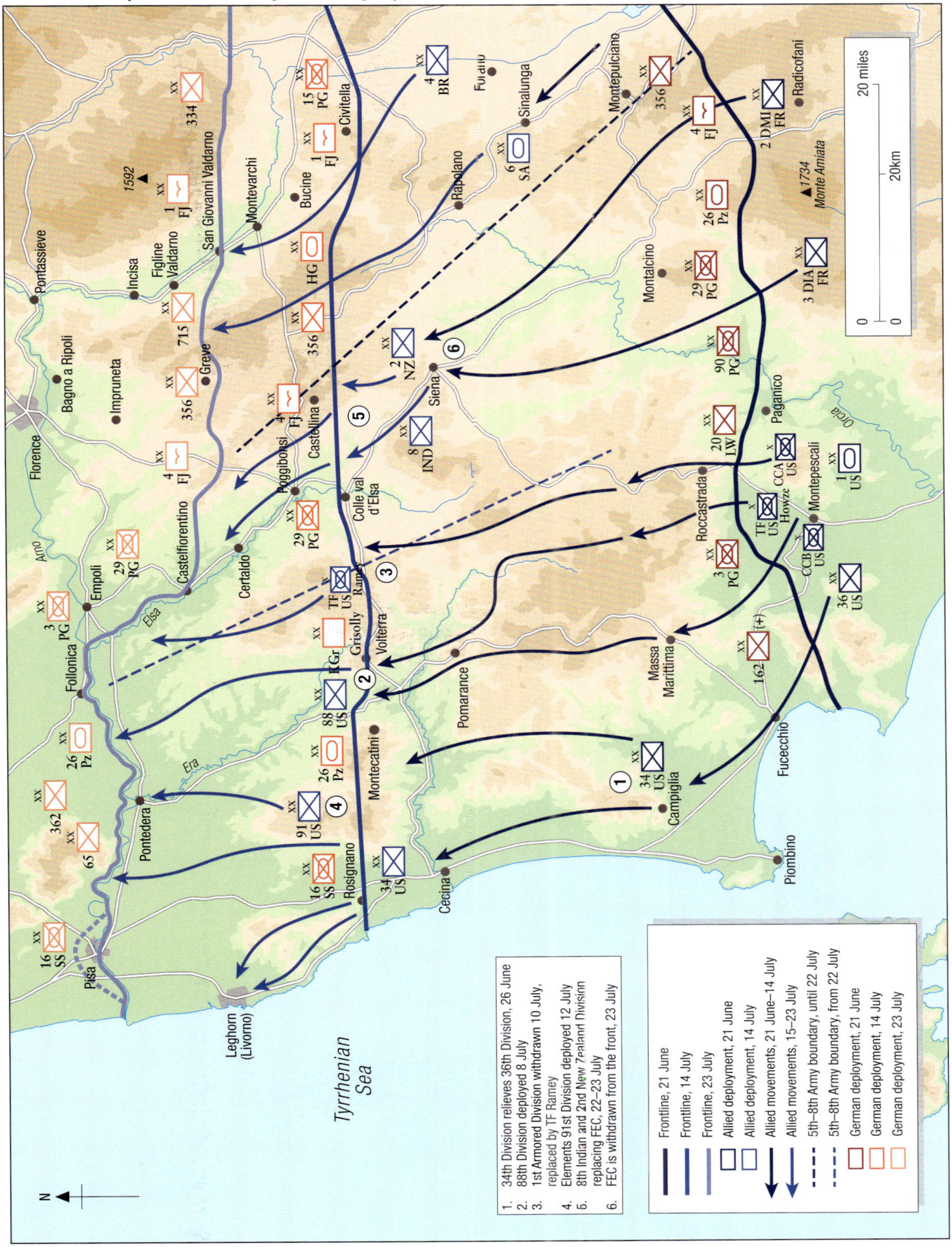

which led Lemelsen to ask again for permission to withdraw. Even the intervention of the 14 Tiger I tanks, which according to estimates were facing some 200 enemy tanks, failed to provide relief given the loss of ten of the heavy tanks, and thus leaving only nine operational plus another 15 in repair. Still, on 25 June, the situation eased as 29. Panzergrenadier Division successfully repulsed an attack south of Montalcino and 4. Fallschirmjäger Division launched a successful counter-attack which did cost heavily, its Fallschirmjäger Regiment 12 losing some 60 per cent of its strength. Nevertheless, on 26 June, AOK 14 was authorized to withdraw to the Anton Line, now about to face the 34th Division's attack. In order to deal with the situation, Kesselring considered reinforcing AOK 14 with more units taken from AOK 10, but von Vietinghoff replied that, instead, he would rather take over part of the area now under control of the former. Thus, on 28 June, Kesselring extended to the west the area under command of AOK 10 (mostly by putting 356. Infanterie Division at its disposal), while authorizing AOK 14 to prepare further withdrawals north to a series of intermediate lines up to a point to the north-west of Montepulciano.

A light M5 tank of the 1st Armored Division entering Montaione in July, just after the unit had taken part in the seizure of the town. (NARA via Digital History Archive)

A destroyed Sherman being inspected near Viterbo. This tank was hit by German shells, which hit the ammunition stored inside and set the tank ablaze. (NARA via Digital History Archive)

As 26. Panzer Division took over the area of the 162. Infanterie Division, which von Senger now deemed unsuitable for the front and only suitable for anti-partisan warfare, on 1 July the leading elements of 34th Division attacked Cecina again, this time facing the determined resistance of the German rearguards as they entered the city, AOK 14 having in fact started withdrawing. The costly seizure of Cecina (almost 400 men lost by a single regiment) was matched by the advance of 1st Armored Division's CCB that, on 30 June, moved to the north-east of Volterra while Task Force Howze was halted by artillery fire. The intervention of the CCA on the 3rd solved the situation, even though it suffered heavily when attacking Casole d'Elsa (south-east of Volterra), seized after three days of fighting. This marked the end of the 1st Armored Division's activity, the 88th Infantry Division replacing it in early July, and starting its new task by seizing Volterra on 8 July. Even the FEC's attack, started on 21 June, faced strong and determined resistance thanks to I. Fallschirm Korps, which checked its advance until 26 June, when the

advance to the west of 1st Armored Division compelled the Germans to withdraw. The next objective presented an unusual, albeit not less important, difficulty: General de Gaulle had to reassure the Pope that the city of Siena, historically of great importance, would be spared by the war and therefore Juin planned an enveloping manoeuvre, to the east and the west of Siena, which was to be bypassed. This was an eventuality the Germans had taken into account and General Schlemm, witnessing the battle approaching Siena, decided not to defend the city with Kesselring's blessing. Only the nearby bridges could be blown up and, in order to prevent a surprise French thrust, the approaches to Siena could be mined and blocked. In fact, also thanks to the low infantry strength of I. Fallschirm Korps, Siena had been evacuated already on 2 July, French troops entering it in the early hours of the next day without firing a single shot. At the same time, Juin was instructed to prepare for the redeployment of the FEC, which was to take part in the landing in southern France. The 2e Division d'Infanterie Marocaine and the 4e Division de Montagne Marocaine were now deployed at the front, facing the strong German strongpoint at Colle val d'Elsa and Poggibonsi, north of Siena, blocking the road to Empoli also thanks to the Tuscan mountain range to the east. Colle val d'Elsa fell on 7 July, after a brief clash, but as the French advanced towards Poggibonsi, the entire strategic situation on the Italian front changed.

THE 8TH ARMY AT THE TRASIMENE LINE, 21 JUNE–2 JULY

The 8th Army's lack of pressure on the withdrawing AOK 10 eventually enabled Kesselring to move units back and forth and redeploy them according to his needs. In June three divisions were pulled out of the line in the AOK 10 area to be sent to AOK 14: the 26. Panzer, 29. and 90. Panzergrenadier Divisions. While withdrawing north, AOK 10 also had the chance to reorganize its own frontline, moving them from the less threatened areas to the more sensitive ones. Early in June, AOK 10 deployed six divisions between the Apennines and the Adriatic coast, which is east of Cassino, but at the end of the month the same area was defended by less than four: 278. Infanterie Division along the coast, 71. Infanterie Division inland, and 5. Gebirgs Division further inland, with elements from 305. Infanterie Division. The 114. Jäger Division had been pulled out of the line to reorganize and act as a reserve, while the area between Perugia and Lake Chiusi was now defended by seven divisions. In early June, the area near Cassino roughly matching the same stretch of the front was defended by just four or five divisions. The relative freedom of movement enjoyed by AOK 10 somehow made good for the lack of strength of its units; in June the

A column of American field artillery (here a 105mm field howitzer is being towed) advances under the protection of the artificial fog created by the M2 smoke generator. (NARA via Digital History Archive)

army had lost a total of 15,565 men, mostly with Fallschirm Panzer Division 'Hermann Göring' (2,175 losses), 94. Infanterie Division (2,716 losses), 15. Panzergrenadier and 1. Fallschirmjäger Divisions (1,627 and 1,271, respectively), and with 44., 305. and 334. Infanterie Divisions (which lost 1,584, 1,296 and 1,452 men, respectively). At the end of June, AOK 10 reckoned it had lost a total of 28,908 men since 11 May, getting back at the same time only 19,034 replacements for an imbalance of 9,874 men. Clearly, AOK 10 could not withstand a prolonged pressure, as it was not able to reinforce AOK 14, take over portions of its front and continue to fight all at the same time.

Mostly, the crisis in the AOK 10 area had been caused by the sudden advance of X Corps in the central area of 8th Army's front to Foligno, and from there to Perugia. This accelerated the withdrawal along the coast of LI. Gebirgs Korps, LXXVI. Panzer Korps being at the same time busy with the XIII Corps advance north of Orvieto.

To make things worse, German intelligence reported that I Canadian Corps was being held in reserve. In the view of the German commanders, such a crack formation could only be spared if being prepared for a major offensive, and the question was where and when. It was thanks to the capture of a dispatch rider that von Vietinghoff was able to form a clearer, and more precise, view of the situation. Since 16 June, the idea was that 8th Army would concentrate on the most logical approach, which was to continue with its attacks to the north using the favourable terrain with the aim of driving for Arezzo. This was even seen as a prelude to a further strategic move aimed at the Po valley, which would have as its first move an attack in the area between Montepulciano and Lake Trasimene. The reaction was to accelerate the redeployment of 114. Jäger Division while preparing a series of defences along the Albert Line, mostly aimed at making it as 'tank proof' as possible. The adjustment of the boundary with AOK 14 also served the purpose of bringing the entire area under AOK 10's command, even though von Vietinghoff had to warn that, should AOK 10 face an all-out assault, there was no guarantee that this could be halted on the defence line as both Hitler and Kesselring had requested.

GIs entering Volterra just after it had been vacated by the Germans. In most cases, there was no fighting inside the towns, which would have destroyed a good many artistic and historical buildings and objects. (NARA)

It is worth noting that, after the war, British and American historians acknowledged the German ability in moving and shifting their understrength, immobile units across the front in spite of a daily average of some 1,000 air sorties, which greatly impacted on communications. Being able to preserve the manoeuvrability of their units was essential for the Germans, but it was also an issue for the 8th Army, which relied heavily on supplies and on motor transport. By 18 June, General Leese had to reckon with the fact that the Chiusi–Sinalunga road west of Lake Trasimene was not capable of supporting the weight of the entire XIII Corps with its heavy motor traffic. This compelled him to revise his orders and order X Corps to advance up the Tiber valley to the north of Perugia, aiming first at Sansepolcro and, from there, striking at Arezzo from the east. Highway 71 was assigned for exclusive use by XIII Corps, which was to advance towards Arezzo from the south along the Chiana Valley west of the Trasimene, which had effectively separated the axes of advance of the two corps since the very beginning of the offensive. Since the mountain watershed to the north also separated geographically the XIII and X Corps' axes of advance, the two were bound to get in contact only when approaching Arezzo, which resulted in the practical impossibility of mutual support. In other words, 8th Army had sacrificed the manoeuvrability so dear to the Germans. Still, on paper the balance of power was fully in favour of 8th Army, which could deploy the following units against the weakened German divisions: XIII Corps with the 6th South African Armoured Division and the 78th Infantry Division, plus the 4th Division held in reserve and bound to replace the 78th Division, which was intended to be withdrawn to the Middle East for a prolonged period of rest and refit. Armoured support was provided by the 9th Armoured Brigade (less the 3rd Hussars), and by the Canadian 11th Armoured Regiment, with the Canadian 1st Armoured Brigade held in reserve. This was clearly the spearhead of 8th Army's offensive, X Corps being considerably weaker with the 6th Armoured Division and the 8th Indian Infantry Division lacking reserves and having only the 3rd Hussars as armoured support. To face them von Vietinghoff deployed to the east of Lake Trasimene, facing X Corps, 15. Panzergrenadier and 305. Infanterie Division, the latter reinforced by having absorbed the remnants of the disbanded 94. Infanterie Division.

A French M5 light tank entering Siena on 3 July 1944, cheered by the customary crowd of Italians who, it should be noted, are all dressed up for the occasion. (NARA via Digital History Archive)

Terrain also had its role in the battle. To the west, facing the XIII Corps, were deployed the 'Hermann Göring', the 1. Fallschirmjäger and the 334. Infanterie Divisions. The central sector of AOK 10 was left to the single 5. Gebirgs Division, while the 71 and 278. Infanterie Division defended the left wing on the Adriatic coast.

Terrain and the season had their role to play. The Trasimene Line, as the British called this portion of the Albert Line, was now highly cultivated, with crops at their peak

French troops, helped by Italian partisans, advance north of Siena in early July. This was the last engagement of the FEC in Italy. (NARA via Digital History Archive)

British 8th Army advance to Florence, 16 June–8 August

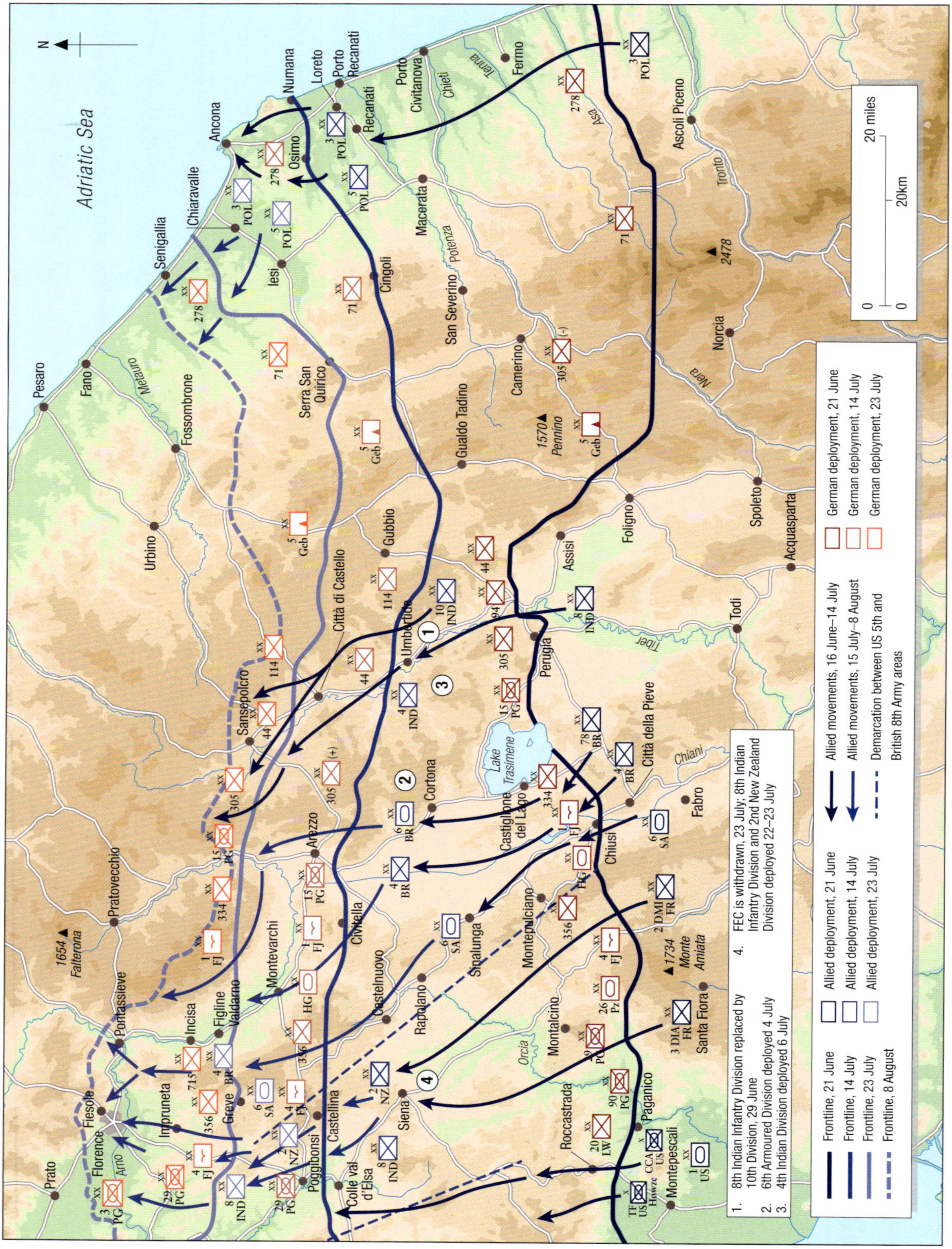

along with extensive vineyards groomed by the numerous, dispersed farms scattered all across the countryside. The series of small villages and isolated farms provided the Germans with excellent observation points and with a long number of strongpoints, further strengthened by the poor conditions of the available roads, which were mostly farm tracks not suitable for the tanks. They were also impeded in their movements by obstacles such as numerous streams and irrigation canals, but these were not as much of an issue for the German infantry, which could easily ford them. Even tanks experienced lack of visibility due to the fully grown crops, the entire area being described as ideal for a defence in depth. This was part of the German plan, AOK 10 having been given priority in the assignment of tanks and anti-tank guns, along with mortars, artillery and Nebelwerfer rocket launchers. Most important of all, neither side had a clear idea of what the battle would mean, which eventually turned out to be more of a disadvantage for 8th Army than for AOK 10. Only Alexander seemed to have sensed the fact that the battle would be decisive in a way, and that the Germans were determined to defend their positions south of the Gothic Line. On 15 June, having realized that the Germans would not abandon Ancona without fighting, he ordered General Anders to bring his Polish II Corps to the front, the Polish units moving on the next day to replace the British V Corps, which left in place the Italian CIL. Between 16 and 21 June, the Polish II Corps advanced for more than 70km crossing the Chienti River, and getting as far as less than 50km away from Ancona, at which point the German resistance stiffened, which required reinforcements for the Polish Corps along with a two-week pause to reorganize. This too made it clear that the Allied pursuit of the withdrawing German forces had come to an end.

It did not take long before Kirkman realized that the battle for Trasimene Lake would not be an easy business. On 19 June, the 78th Division advanced to the south-western end of the lake only to face determined resistance by 334. Infanterie Division, which checked its advance on the next day. At this point, Kirkman decided to bring 4th Division into the battle from the reserve, with the aim of deploying it on the left flank of 78th Division, but also taking into account the necessity of taking over the right flank as well if the battle had not been won by the time the division was intended to be withdrawn to be sent to the Middle East. With 4th Division deploying on 21–24 June facing the bulk of 1. Fallschirmjäger Division, the XIII Corps was now pressing along three axes: the 6th South African Division along the Chiusi–Sinalunga road, the 4th Division east of Lake Chiusi and the 78th Division right to the west of Lake Trasimene. This time the Germans were determined to resist, as the South Africans experienced at Chiusi. A first attempt to seize the city was repulsed by the 'Hermann Göring' Division which, facing an infiltration during the night of 21/22 June by the 1st City of Capetown Highlanders moving

An Italian *Semovente* 75/18 in German service, ready for action. These vehicles, less effective than the Sturmgeschütz, were employed by the infantry divisions in the anti-tank units (Panzerjäger). (NAC)

FIGHTING AT THE TRASIMENE LINE, 24 JUNE 1944 (PP.56–57)

During the Allied advance in central Italy, the Germans realized the progress made by the Allied armoured formations. Although terrain still posed several obstacles, the Allied tanks managed to move along hills and broken earth, all too often taking the German defences by surprise. The lack of anti-tank guns, as well as of tanks, on the German side was decisive on the battlefield. The Allies on the other hand developed a very effective coordination with their artillery hitting the German positions, which would then be attacked by tanks with infantry support. This is what happened during the fight for the Trasimene (or Albert) Line, when the British 78th Division was supported by the Canadian 1st and British 9th Armoured Brigades. After breaking through the first line of defence along the Sanfatucchio–Pucciarelli line, the 78th Division advanced north-east, avoiding the German strongpoint positioned on the high ground to the west. This brought the German artillery in to hit the advancing columns, as the 38th (Irish) Brigade moved towards Pescia and Ranciano. The terrain, dominated by isolated farms, was made difficult by poor visibility due to the non-harvested wheat fields **(1)**, and the lack of close cooperation between tanks and infantry on the field only made things more muddled. On 24 June, as the 5th Northamptonshire Regiment advanced **(2)**, supported by Shermans of A Squadron, Ontario Regiment (11th) **(3)**, Panthers of I./Panzer Regiment 4 **(4)** launched a counter-attack, supported by the infantry of 15. Panzergrenadier Division. The events proved that, with the right handling, the Sherman could stand up to the Panther, as the tanks of the Ontario Regiment managed to destroy one Panther and damage the other, which was forced to withdraw. Ultimately, the German counter-attack met with failure, and in the morning, Pescia was seized without a fight.

from the railway station, launched a strong counter-attack that eventually compelled the South Africans out from Chiusi. The South Africans practically abandoned their attempts to get into Chiusi, which they attempted to surround by closing the western road towards Chianciano. In the meantime, the 24th Guards Brigade made slow progress on the XIII Corps' left wing, mostly due to the poor roads and the heavy rains. Only by 23 June did it come into sight of Sarteano, to the south-west of Chiusi, at the extreme wing of 8th Army.

Between 21 and 22 June, the 78th Division fought against the stubborn resistance of the 334. Infanterie Division, which also launched a series of local counter-attacks. Eventually the villages of Sanfatucchio and Pucciarelli were seized, breaking the first defence line and paving the way for a further advance. This was not easy, as the Germans held the line with both the 1. Fallschirmjäger Division and the 334. Infanterie Division, while bringing in reinforcements from 15. Panzergrenadier Division and deploying the Panther tanks of I./Panzer Abteilung 4, which launched isolated counter-attacks and were badly mauled by the Shermans of the 1st Canadian Armoured Brigade. It is worth noting that, as both the British and the Americans remarked, the Germans practically wasted their tanks by using them in penny packet formations, either to try and halt the enemy advance or to launch isolated counter-attacks. This was certainly the wrong way of using tanks, but one has to reckon with the fact that there was little if any alternative. A large-scale, divisional-level counter-attack was practically impossible and the only suitable use for the few available tanks was to deploy them where the situation required, all too often with poor results when compared to the Sturmgeschütz – the assault guns (many of which were Italian made) – which supported most of the infantry divisions. After a pause between 22 and 23 June, on the 24th the 78th Division attacked again, driving into the villages of Pescia and Ranciano, now close to the eastern end of Lake Trasimene and to Castiglione del Lago. However, once again, the tanks were unable to follow the pace of advance of the infantry due to the terrain and the Pescia River, the 78th Division having to wait for the 4th Division's advance on the left wing. The latter was fighting fiercely against the hard-pressed 1. Fallschirmjäger Division, which, in any case, was able to reinforce the badly battered 334. Infanterie Division along with 15. Panzergrenadier Division. On 25 June, the Germans withdrew to a last line of resistance on the ridge to the north of the lake, which led to a fierce struggle on 26 June. At the same time the 4th Reconnaissance Regiment was approaching the northern end of Lake Chiusi, so the 'Hermann Göring' Division was left with no alternative than to withdraw north to avoid encirclement and, on the night of 25/26 June, evacuate Chiusi. The final assault on the Frattavecchia ridge took place on 28 June, when the Trasimene Line was broken through and the Germans were compelled to withdraw north to the intermediary line of defence. The battle for the Trasimene Line had been decisive, but it also showed that the Allied advance had entered a new phase. The 6th Inniskilling Regiment of the 36th Brigade had to be pulled out of the line and disbanded, its remnants being absorbed by the 2nd Inniskilling.

Nevertheless, the breakthrough at Lake Trasimene paved the way for the further advance of XIII Corps, the 6th South African Armoured Division seizing Chianciano first, then Montepulciano by 29 June. The battle was certainly important also because of the lack of success by X Corps, whose

THE BATTLE OF LAKE TRASIMENE, 19–28 JUNE 1944

Two weeks after the withdrawal from Rome, the German forces attempted a last stand on the Albert Line, known by the Allies as the Trasimene Line. This was a necessary measure needed to allow the completion of the fortifications on the Gothic Line to the north. The Albert Line had little in terms of fortifications or prepared positions,

EVENTS

1. 19 June: The 78th Division's 11th Brigade reaches Macchie with the 5th Northamptonshire and 1st East Surrey, supported by the Warwickshire's tanks. Units consolidate there under heavy artillery fire.

2. 20 June: The 11th Brigade attacks Sanfatucchio with the 1st East Surrey to the left, the 2nd Lancashire directly to the town and the 5th Northamptonshire to the right, all supported by the Warwickshire. The infantry is pinned down and the attack is called off with troops withdrawing at 1930hrs. Sanfatucchio is then heavily hit by the 78th Division's artillery.

3. 20 June: Leading elements of the 36th Brigade advance towards Strada facing heavy fire from the rearguards of FJR 4. A prong to the outskirts of Vaiano is repulsed by a strong counter-attack.

4. 21 June: The 2nd Lancashire Fusiliers (11th Brigade) manages to establish a foothold in Sanfatucchio but is pinned down by the German defences. The town is then attacked and cleared by 1130hrs by the 2nd London Irish Rifles (attached to 38th Brigade), the advance continuing north towards San Felice. It is checked by a counter-attack by IR 754.

5. 21 June: Following the seizure of Sanfatucchio, the 36th Brigade's 6th Royal Inniskilling, supported by the Ontario Armoured Regiment, moves across route 71 towards Pucciarelli, facing light machine gun fire. The town is seized, but German opposition prevents any further advance.

6. 21 June: 78th Division's 36th Brigade is relieved by the 28th Brigade, 4th Division. The advance to Vaiano is stalled.

7. Night of 21/22 June: The South African 1st City of Capetown Highlanders, supported by 11th Armoured Brigade, attacks Chiusi. It manages to enter the town, but is repulsed by a counter-attack from the 'HG' Division. Any further progress in the area is stalled.

8. 24 June: The 28th Brigade (supported by the tanks of the Ontario Regiment) attacks Vaiano with the 2nd Somerset in the lead. Vaiano is seized by 1600hrs, when the tanks push forward as far as the crossroad (Podere del Papa), only to withdraw lacking infantry support.

9. 24 June: LXXVI. Pz Korps orders a withdrawal to the defence line Pescia–Badia–north of Vaiano, while preparing to counter-attack.

10. 24 June: The 78th Division attacks the German defences with the 1st Royal Irish (with a company from 6th Royal Inniskilling) moving on to Ranciano (just west of Pescia), and the 5th Northamptonshire astride the railway supported by the tanks of Ontario Regiment. At 0900hrs the Germans launch a failed counter-attack, supported by ten Panthers from I./Pz Regiment 4. Both Pescia and Ranciano are seized.

relying mostly on a series of strongpoints built in small villages. It was soon clear that the Germans intended to defend their positions, as the 6th South African Division was halted at Chiusi, to the west of the Trasimene Line. Supported by the tanks of the 1st Canadian and 9th British Armoured Brigades, the British 78th Division started the attack in the area between Lakes Trasimene and Chiusi on 19 June. Two days later, facing the stubborn German opposition, the British 4th Division joined in, eventually breaking through the line after one week of battle.

11. Night of 24/25 June: Following the enemy advance in the Pescia–Ranciano area, the 1. FJ Division withdraws from Vaiano, while 334. ID deploys north of the Pescia River. The remnants of IR 754 form the KGr Ortlieb, while I./IR 755 with IR 756 form KGr Foss. The 334. ID is reinforced by II./FJR 3, the Aufklärung Abteilung 'HG' and III./PGR 104 from 15. PG Division.

12. Night of 24/25 June: The 6th Royal West Kent, supported by Wiltshire Yeomanry, crosses the Pescia, laying a Bailey bridge. The 6th Royal West Kent advances seizing La Bandita, while the 5th Buffs advances towards the Madonna del Vitellino. An attempt by the 8th Argyll and Sutherland Highlanders to advance along the road to Castiglione del Lago is halted.

13. 26 June: The 4th Division's 28th Brigade, after mopping up the Vaiano area, advances north towards La Villa. The Hampshires, supported by the Three Rivers tanks, seize La Villa, while 2nd Somerset seizes Badia that same night. The Germans withdraw, establishing a new line 1km north of La Bandita–Vitellino–Casamaggiore–Gioiella.

14. Night of 25–26 June: Facing the withdrawal of 1. FJ Division, the 'HG' Division evacuates Chiusi, pursued by the 6th South African Division.

15. 27–28 June: The 10th Brigade (supported by the Three Rivers) passes through the 28th Brigade, pressing on along the Lopi–Gioiella road. On 28 June, the 1st/6th East Surrey seizes Casamaggiore, while the 2nd Duke of Cornwall and 2nd Bedfordshire advance north, reaching the Pozzuolo–Castiglione road.

16. 27–28 June: The 36th Brigade breaks through the German defences advancing to the Pozzuolo–Castiglione road. The LXXVI. Pz Korps orders withdrawal to the Frieda Line.

A Lince (Lynx) scout car in German use is bringing Indian prisoners of war away from the front. This vehicle was produced by the Italians, copying the Daimler Scout Car captured in North Africa. (NAC)

drive along the Tiber River valley north of Perugia was frustrated by the German resistance. Here too the planned reorganization of frontline troops was of foremost importance, since the 6th Armoured Division was scheduled to replace the 78th Infantry Division from the end of June while the battle-worn 8th Indian Division was to be replaced by the fresh 10th Indian Infantry Division. The latter was the first to attack on 26 June to the east of the Tiber and, having caught the defences of 114. Jäger Division unawares, succeeded in inflicting severe losses but failed to achieve a breakthrough. The attack to the west of the Tiber by 6th Armoured Division was much less successful, the division not having the necessary infantry strength to fight in the mountain terrain in the area. The division succeeded in advancing to Magione, on the eastern side of Lake Trasimene, but failed to seize it before being withdrawn from the line. Between 28 June and 1 July, both the 6th Armoured and the 8th Indian Divisions were pulled out of the line, being replaced by the 10th Indian Division, which deployed at the end of the Trasimene battle (needless to say, this helped the Germans in bringing in reinforcements given the lack of commitment east of the lake). The 10th Indian Division started to advance on 30 June, as the Germans started to redeploy on the intermediary defence line north of the Trasimene, which accounts for the 12km gain by 2 July. The division halted south of Umbertide, which was seized without fighting on 5 July. On the next day, the Indian advance was checked by strong resistance to the north, along the Nestore River, about halfway between Umbertide and Città di Castello. By 5 July, AOK 10 was holding a line running from the south of Arezzo, east of the Chianti Mountain range, to the north of Umbertide, and from there across the Apennines to Gualdo Tadino–Fabriano–Numana and Osimo, on the Adriatic coast just south of Ancona. Following Kesselring's orders, von Vietinghoff – who remarked his sadness about having to 'move back' in spite of the good performance of his troops – started reconnoitring and preparing a series of intermediate defence lines running parallel to the Green or Gothic Line, to the south. The situation was such that neither Kesselring nor von Vietinghoff were able to look forward more than a day or two, the overall situation being eased by the constant flow of supplies, which saw the fuel situation being now less tense, while the supply of ammunition even exceeded the capability of the fighting troops to absorb them. The real concern, as Kesselring pointed out to Hitler on 1 July, was the shortage of

Fallschirmjäger in position to face the advancing enemy, a MG 42 on a tripod (right foreground) prepared for use. Many are on alert, but one (left foreground) is taking the chance for a nap. (NAC)

manpower and the lack of weapons and equipment, given the heavy losses suffered in terms of anti-tank and field guns, and armoured vehicles. Still, early in July, one could come to the conclusion that the German forces in Italy had taken a severe battering since the beginning of Operation *Diadem* on 11 May but also that, after some two months of protracted fighting withdrawals, they were in a better condition, as the events that followed would prove.

THE 5TH ARMY'S ADVANCE TO THE ARNO, 3–23 JULY

In his post-war history of the Italian campaign, British historian W. G. F. Jackson relied on the 'division count' to try and make some sense of the phase of the campaign that followed the seizure of Rome. In his analysis, he reckoned that from May 1944 the Germans had in Italy more divisions than the Allies and that, therefore, the latter had succeeded in tying down a sizeable enemy force with relatively few troops, so as to prevent the Germans from reinforcing the vital Normandy front. The analysis is clearly flawed, not only because the 'division count' cannot be made between units with such different strengths (the German ones almost unavoidably understrength, the Allied ones not quite so), let alone the amount of support troops in the rear areas. It also does not take into account the fact that the Allies, and Alexander in particular, had a strategy for the Italian front which, in the summer of 1944, clashed with the broader Allied strategy in Europe. The idea of diverting German troops from northern France was in fact a bait for the other Allied commanders, the one Alexander offered to the Allied planners focused on the landing in southern France. On this occasion, the Allied strategy for Italy was summarized as follows: the primary objective of the campaign was to complete the destruction of the German armed forces in Italy so as to force the enemy to draw as many reserves as possible from other fronts to commit them to this particular theatre. The basic objection one can move to this strategy, as outlined, is that on the battlefield the AAI behaved as if their primary objective was to seize ground and places of international relevance, rather than destroy the enemy forces. This is particularly true when we look at Valmontone, but it is also true if we consider the many faults in the advance north of Rome.

The point is that most of these considerations are based on inaccurate data. While the AAI HQ believed that, following the Anzio landings, the Germans had committed six divisions to Italy, they in fact had only committed four. Likewise, Allied intelligence estimated that the Germans had six divisions or equivalent in northern

A truck loaded with Japanese-American soldiers from the 442nd Regiment entering Leghorn, now vacated by the Germans. The usual cheering crowd welcomes them. (NARA)

GIs cautiously patrol the area near Pisa in July, following the approach of the 34th Infantry Division to the Arno River. (NARA via Digital History Archive)

Italy, only two of which were close to full combat effectiveness, but in reality the Germans committed four of them, while withdrawing five divisions from the front for rest and refit (a sixth was disbanded). Still, in Alexander's view, one could reckon that by the time Rome had been surpassed, Kesselring would have been left with no more than ten fully efficient divisions which he could use to defend the Gothic Line, while according to estimates he would need a dozen at least. This meant that, facing a successful Allied offensive in central Italy, the Germans would have been compelled to bring to the theatre eight or even ten new divisions taken from other fronts, mostly the Western Front, with the result of weakening them. This was the background to Alexander's argument which stressed the need for a continuation of the Allied offensive in Italy, which one could consider to be of greater help to the drive in northern France than a landing in the south of the country. This is without taking into account the strategic aims Alexander had never set apart, which was the drive into the Ljubljana Gap and the advance right into Austria – or southern Germany. In his estimate, the seizure of Leghorn and Ancona would have enabled the sustainment of an Allied force of 20 divisions, which were supposed to face 24 German divisions of lesser strength. Therefore, the Allied armies had to fight to seize the ground needed to mount a full-scale attack against the Gothic Line, which meant they had to advance north from the approximate Grosseto–Perugia line while inflicting as many losses as possible on the enemy. These arguments and calculations faced the dire reality of Allied planning, mostly dominated by the Americans who were determined to land in southern France. The situation was made clear to Alexander from 12 June, and the argument continued until the end of the month while units were being withdrawn from the front in what can only be described as an uncoordinated way. This seems to be the key to the entire affair, which culminated on 5 July when Alexander was instructed by the Allied Command in the Mediterranean that an overriding priority was to be given to Operation *Dragoon*, the landing in southern France, against the operations in the Italian theatre, the former requiring the creation of an assault force of at least ten divisions, which could only be drawn from Italy. The directive was not unexpected, rather the contrary, but it still caused a shock as it proved that Italy had lost the sort of priority it had been given thus far. Nobody realized that the balance of power in Italy was to tip against the Allies only temporarily and to a minimal extent, most of the reorganization having been started since the fall of Rome, while the Germans had to surrender four divisions for employment in other fronts. In a way, a swift success in Italy, the kind that gained ground while inflicting losses to the enemy, was now needed more than ever in order to prove that Italy still had a strategic value and that a wide-ranging success could still be achieved in the theatre.

This was precisely what Kesselring and his army commanders were to prevent, while pursuing what still was the basic German strategy in the area: to keep the 5th and 8th Armies pinned down south of the Po River valley, while at the same time trying to work out whether the Allied forces being regrouped were intended for a large-scale landing either in southern France or in Liguria, in northern Italy. The potential threat did not mean that any reinforcement was planned for the Italian theatre of war, rather the contrary. In July, the Fallschirm Panzer Division 'Hermann Göring' was withdrawn from the front to be sent to the east, Kesselring being promised two divisions (the 98. and 237. Infanterie Divisions) from the Balkans in return. Most importantly, Kesselring assumed that now the Allied aim would be to seize Florence, which put the emphasis on 8th Army while it could be reckoned that 5th Army would have been weakened and therefore its advance slowed down (a point confirmed by intelligence information on the withdrawal of the FEC from the front). Attention focused on the AOK 10 sector, in particular on Ancona, which Hitler deemed of strategic importance as he believed its seizure by the Allies could represent the prelude for a landing in the Balkans, a point of view neither Kesselring nor von Vietinghoff shared, however, as both were convinced that Ancona was quite without value and that its defence was not worth an entire German division. In practical terms, the conduct of the operations was to match what had been done until then; fight a protracted withdrawal offering as much resistance as possible in order to buy time, while trying to spare forces and complete the construction of the Green or Gothic Line so as to prevent any Allied breakthrough into the Po valley.

An M10 tank destroyer enters Leghorn following the German evacuation – the Italian crowd more stunned than cheerful this time. The large arrow on the wall shows the way to the nearest air raid shelter. (NARA via Digital History Archive)

The 5th Army's next objective, the important city of Leghorn required yet another effort by Crittenberger's IV Corps, this time without incurring the same mistakes that had been made when fighting for Cecina, when a series of frontal assaults had dissipated the combat strength of the corps. This time the 34th Infantry Division was strengthened by having the 442nd RCT attached, along with the 804th Tank Destroyer Battalion and the 363rd RCT from 91st Infantry Division just arrived in Italy, the second component of the division to be deployed in the area. The manoeuvre attack envisaged by Crittenberger was rather simple, which

The crew of an M10 tank destroyer and medical personnel are entertaining themselves with Italian partisans on the outskirts of Leghorn. (NARA via Digital History Archives)

was to attack Leghorn – the most important harbour north of Naples – by outflanking it from the east, threatening the Germans from the flank, thus forcing them to withdraw from the city. Here too, terrain was to play a decisive role. The front of IV Corps ran between Rosignano on the coast to Volterra, some 10km north of the Cecina River, and was flanked to the east by the Era River running north to south from the Arno, at the height of Pontedera. The rugged terrain was not much suitable for tanks, and required infantry to do most of the fighting, supported mainly by the artillery. Three different routes were available for the attack: one was along Highway 1 running along the coast to the south of Leghorn, where it broke down by connecting a series of small coastal towns. The second approach was a secondary route leading through a valley flanked, to the left, by the coast and, to the right, by a high ridge. The route ended at Colle Salvetti, to the east of Leghorn, and from there to Pisa. The third route was a minor road running parallel to the former, further inland until it linked with route 206 to the south of Colle Salvetti, at the height of Leghorn. Crittenberger decided to send the bulk of the 34th Division along the two inland routes – the 135th and 442nd Infantry Regiments moving abreast along with the 168th Infantry Regiment to the right, while the coastal road would be checked by the 804th Tank Destroyer Battalion and 34th Cavalry Reconnaissance Troop. The 133rd Infantry Regiment, badly mauled at Cecina, was held in reserve. Further to the east, 1st Armored Division carried on with its probes until relieved on 8 July by the first elements of 88th Infantry Division (itself to be subsequently relieved by the 91st Infantry Division), whose objective was to cross the Era from the west and proceed north along its eastern bank, to the Arno River between Pontedera and Empoli. While these operations were being carried out, the FEC was to continue its drive north along the right wing of the 5th Army north of Siena, advancing along the Elsa River valley from Poggibonsi to Certaldo and to Empoli while waiting to be relieved by the II Corps, which had been held in reserve since the seizure of Rome. The 5th Army's aim was not just the seizure of Leghorn, essential for supplies, but also to advance north to the Arno River, even though no plans were made to cross it by establishing a bridgehead.

The German defences, albeit not particularly strong, included the reinforced 19. Luftwaffen-Feld Division with elements from 16. SS-Panzergrenadier Division (which, it must be remembered, was still forming). Inland from Leghorn, along the main route from Volterra to Pontedera, was the 26. Panzer Division, reinforced by elements from 90. Panzergrenadier Division. The area between Volterra and Poggibonsi was defended by an array of units, including the remnants of 20. Luftwaffen-Feld Division (soon to be disbanded), known as Kampfgruppe Crisolly, the remnants of 3. Panzergrenadier Division along with elements from both the 19. and 20. Luftwaffen-Feld Divisions, and a fresh regiment from

GIs digging in at Leghorn under the shade. From left to right (front): Private First Class Frank Navarro from San Antonio, Texas; Private First Class Jack Allen, Urbana, Oregon (tending the mortar); Corporal Bill Cannon, Raynecliff, New York; and Sergeant Thad Zdzieblowski, Milwaukee, Wisconsin. In the rear: Pfc Milo Dayles, Minneapolis, Minnesota; Private Walter H. Sherman, Minneapolis; and Private Joe Schianone, Kingston, New York. (NARA via Digital History Archive)

362. Infanterie Division. Facing the FEC on the road to Poggibonsi was the 29. Panzergrenadier Division, with on its left Kampfgruppe 4. Fallschirmjäger and 356. Infanterie Divisions, at the demarcation between AOK 14 and AOK 10. The only reserve available was the rest of 90. Panzergrenadier Division, forming a Kampfgruppe deployed at Castelfiorentino to the south of Empoli. The German defences were echeloned in depth, as the 34th Division discovered while approaching the starting point on 2 July. On the next day, 135th Infantry made initially good progress while advancing inland but, as it approached the village of Le Presselle, the German defences reacted violently, soon halting the American advance. The same fate befell the 442nd Infantry Regiment advancing on the right flank, while 3rd Battalion of the 135th Infantry Regiment, supporting the advance along the coast of 804th Tank Destroyer Battalion, got entangled in a vicious fight for the town of Rosignano, defended by a unit from 16. SS-Panzergrenadier Division. The typical Italian town, built on a hilltop and made of houses built of stone, Rosignano offered the most suitable stronghold and the Germans exploited it until driven out from the southern end on 4 July, which prompted a small counter-attack supported by four PzKpfw IV tanks. Three of the tanks were destroyed and, after the 3rd Battalion of the 442nd Infantry had taken over, American pressure on the town continued until about half of it was in American hands by the 5th. That the advance to Leghorn would not be a cake walk was proved by the stubborn German defence of Rosignano, which had to be cleared almost house by house with the 3rd Battalion reaching its northern end by late on 7 July, only to discover that the Germans had taken position in a series of houses to the north. Only by 9 July were both 135th and 442nd Infantry able to take position some 7km to the north of Rosignano, still mopping up the area from isolated pockets of resistance. Only the speedy advance along the ridge on the east eased the situation, as it eliminated flanking fire. This was achieved by 168th Regiment that, after a good start on 3 July which brought it about 1.5km from Castellina, was joined by 363rd Regiment to the right, the two forming the spearhead of the 34th Division advance. Castellina, at about the same height as Rosignano, proved an equal obstacle thanks to the German defence even though the combined action of both the 168th and 363rd Regiments managed to dislodge the defenders and check the isolated counter-attacks, led by small parties of 50 to 70 men. After seizing Castellina on the 6th, the two American regiments faced the counter-attack from 9. Panzergrenadier Regiment of the 26. Panzer Division, which ensued in a bitter struggle that halted the American advance. Only by 9 July was the main thrust of the 168th and 363rd Regiments able to make progress, seizing the high ground to the south-east of Leghorn after yet another fierce struggle with the troops from 26. Panzer Division. On 10 July, as the area was being mopped up by 168th Regiment, the 133rd Regiment moved forward bypassing the frontline positions and continued with its advance along with

Japanese-American soldiers from the 100th Infantry Battalion of the 34th Infantry Division rest after entering Leghorn. (NARA via Digital History Archive)

A PzKpfw V Panther in one of the customary positions used in the campaign. The tank was hidden in the courtyard of a farm, hull down, waiting for approaching enemy tanks. All too often German tanks like this one were destroyed by air attacks. (Archiwum Polskije Prowincji Dominikanów)

the 135th Regiment close to the coast, now free from the German flanking fire. A German strongpoint, the town of Pastina, was bypassed and left to the 442nd Regiment to clear up while both the 135th and the 133rd Regiments continued with their advance to the north, the 363rd Regiment making progress at a slower pace to the right. This was eventually relieved on 12 July by elements from the 91st Infantry Division, just committed, while Task Force Ramey was pulled out of the line and put in reserve.

The fight for the northern end of the ridge lasted until 12–13 July, when the Germans suddenly withdrew, allowing a speedy American advance which brought the 133rd Regiment close to Lorenzana, a town inland at about the same elevation as Leghorn, the 168th Regiment being brought back to action on 15 July deploying between the 133rd, on the right flank of the 34th Division, and the 363rd about to approach Leghorn from the east. The seizure of Lorenzana, on 16 July, was not as difficult as with the previous towns, and the 168th Infantry was able to continue with its advance to the north-east of Leghorn along with the 133rd, which by then was some 5km to the south of the Arno. The last German defences before the Arno lasted through 17 July, the 26. Panzer Division counter-attacking with the support of seven Tiger I tanks, which were repulsed by the massive intervention of the artillery. At this point, the two regiments were close to the road linking Leghorn's northern end to the Arno River east of Pisa, which gave the signal for the final assault on the harbour city. This had been approached from the south by the 135th Infantry and the 442nd Infantry to the right, both making slow progress due to stubborn German resistance until 12 July, when the Germans deployed their available Sturmgeschütze self-propelled guns and PzKpfw IV tanks. By then the American advance was still stuck at Castiglioncello, just to the north of Rosignano. On 13 July, progress on the left flank of the German defences prompted a sudden German withdrawal, which allowed both 135th and 442nd Infantry to advance almost unhindered, the two regiments making an 8km gain on the 14th. The next day, as the Americans approached Leghorn, the Germans counter-attacked, giving the 135th Regiment a hard time, mostly because the American tanks had not been able to keep the same pace of advance as the infantry. The German attack was eventually repulsed by bringing forward all the available artillery, including 36 guns from the British 66th Anti-Aircraft Brigade. On 16 July, the 135th Infantry seized Hill 449, or Monte Maggiore to the east of Leghorn, thus almost completing the envelopment of the city. Major difficulties were encountered along the coastal road by the tank destroyers and the armoured cars, which were slowed down by poor roads and unsuitable terrain. On 18 July, when the 363rd Infantry Regiment returned to the 34th Division's area as Task Force Williamson, the final plan for the seizure of Leghorn was drafted. The 135th Infantry Regiment was to continue with its advance towards the south-eastern part of the city while, from the east, the 363rd Infantry Regiment was to deliver the main thrust.

Only one battalion of the 135th Regiment faced some determined resistance by elements from 16. SS-Panzergrenadier Division who, for the most, were simply evacuating the city and withdrawing north. As the American advance was delayed mostly by the terrain, the Germans were able to withdraw almost unhindered, allowing the 135th Infantry Regiment to enter Leghorn – almost without fighting – in the early hours of 19 July, the 363rd following just two hours later.

The main task of seizing Leghorn accomplished, even though the Germans had carried out a systematic destruction of all the harbour installations, left IV Corps no other task than to continue its advance northwards towards the Arno River. Already on 19 July the 363rd Infantry Regiment had sent its patrols to the north, encountering no more than minor German outposts acting as a rearguard while the bulk of the German forces had clearly withdrawn beyond the Arno. In fact, the German withdrawal had been started on 15–16 July with the aim of deploying to a defence line to the south of the Arno, which soon proved untenable. Thanks to some sort of renewed freedom of movement, the German high command no longer asking for protracted resistance south of the Gothic/Green Line, Lemelsen was able to consider a definitive withdrawal to the north of the Arno, which was started on 19 July by I. Fallschirm Korps. As the elements of 16. SS-Panzergrenadier Division withdrew from Leghorn, the German front was already in movement and by 20 July LXXV. Armee Korps had deployed to the north of the Arno as well. The last German troops south of the river were the rearguard of 16. SS Division, chased by 34th Infantry Division which, on 21 July, acquired a new commander with Major General Charles L. Bolte. On the next day, IV Corps ordered all its units to advance to the Arno River, which saw 34th Division attacking while giving space to 91st Division to deploy along the front. On 23 July, the leading elements of 363rd Infantry Regiment entered Marina di Pisa at the mouth of the Arno, the 2nd Battalion entering the southern end of Pisa by midday. The Americans found the southern part of the 'leaning tower' city to be heavily mined and booby-trapped, and also under heavy German artillery fire. The advance to the Arno basically marked the end of 5th Army's campaign in central Italy in the summer of 1944. Almost at once, a major reorganization was started, with the anti-aircraft units being transformed into infantry units, while 91st Division completed its deployment at the front. Until 13 July, only the 361st and 363rd RCTs of 91st Division had been in action, the division being deployed in its entirety thanks to the arrival of the divisional HQ, the unit being deployed between 34th Division to the left and 88th Division to the right. Both 91st and 88th Divisions proceeded with their advance to the Arno River, moving from the east of Volterra along the Era River valley, the former attacking since 13 July only to face the customary stubborn resistance

Carefully camouflaged with nets and tree branches, this 75mm Pak 40 lies in wait, a deadly enemy given its low profile and the narrow, tree-lined road. Note the crew, all in shirtsleeves. (NAC)

of the German rearguard, minefields and demolition, and the usual unsuitable terrain. The element which slowed down the American pace of advance was the German artillery, now used to support the withdrawal to the north of the Arno. By 16 July, the leading elements of the 91st Division approached the Arno and Pontedera facing a last German counter-attack supported by tanks. On 18 July, the Americans entered Pontedera, taking some time to consolidate their gains and secure the city, which was found to be full of snipers. Like the 91st, the 88th Division had attacked from 8 July with the aim of reaching the Arno. The division had been resting as a reserve until that day, when it relieved the 1st Armored, which had put itself in reserve. After a limited advance on 9 July, the division found an obstacle in the 'fortress town' of Laiatico, to the north of Volterra close to the Era. A battle ensued, lasting for two days on 12 and 13 July, the 88th Division continuing with a steady advance for the next three days until reaching Palaia, just to the south of the Arno, by the 17th. This practically marked the end of the advance, apart from patrols sent forward to the Arno. Task Force Ramey advanced on the right flank of the 88th Division, at the very border between IV Corps and the FEC. The unit, composed of elements from the 1st Armored Division (Company F of 1st Armored Regiment, 81st Reconnaissance Battalion, 27th Armored Field Artillery Battalion, Company B of 16th Armored Engineers, Company A of 776th Tank Destroyer Battalion), lacked infantry and was to act mostly as a flanking unit, which advanced from 9 July making slow progress until reaching a position some 12km to the east of Palaia by 20 July. Task Force Ramey was relieved on the next day, just as 8th Army took over from the FEC (now with 2e Division d'Infanterie Marocaine and 4e Division de Montagne Marocaine).

The FEC maintained a low activity level following the seizure of Siena, mostly focusing on its imminent withdrawal from the front to prepare for the landing in southern France. Plans for its relief had already been made by 7 July, as the FEC units entered Colle val d'Elsa, activity being mostly aimed at giving the Germans the impression that the advance was being slowed down by their resistance rather than by the new plans. On 8 July, as the 88th Division attacked, the Moroccans of the 4e Division started their advance along the Elsa River valley facing initially light resistance, their action being supported on the right flank by the units of the 2e Division d'Infanterie Marocaine. A renewed push was made on 11 July in coordination with the American advance, and the French were able to move past the renowned town of San Gimignano moving towards Santo Stefano and San Miniato. With the Germans starting their withdrawal on 13 July, the French advanced to the Tavarnelle–Castelfiorentino line while mopping up the area to the south. With plans for the relief being completed on 16 July, the FEC went mostly on the defensive but, facing yet another German withdrawal on the 18th, pursuit was started again down to

Two Canadian soldiers have taken position in a shell hole during the advance to the Gothic Line, August 1944. The seizure of Florence, prelude to the Gothic Line campaign, led to the deployment of the Canadian Corps, which had been previously held in reserve. (NARA via Digital History Archive)

Certaldo, which was seized on 19 July along with San Donato. On the next day the advance continued, mostly to keep up with the German withdrawal, Castelfiorentino being seized on the 20th. The relief of the leading elements of 2e Division d'Infanterie Marocaine started on the night of 21/22 July, their positions being taken over by the 8th Indian Division and 2nd New Zealand Division. On 22 July, command in the area passed to the British 8th Army, the French participation in the Italian campaign having come to its end.

THE 8TH ARMY ADVANCE: THE BATTLE OF ANCONA AND THE APPROACH TO FLORENCE, 5–22 JULY

On 3 July, Kesselring flew to Hitler's headquarters for a face-to-face analysis of the situation in Italy, which resulted in Hitler allowing greater freedom of manoeuvre than in June, when he had insisted on the need not to lose ground to the south of the Gothic/Green Line. The progressive weakening of 5th Army (confirmed by the eventual withdrawal of the FEC) strongly suggested that the Allies were to focus on 8th Army, with the aim of seizing Ancona and Florence. As AOK 10 redeployed to the line running from the north of Montepulciano to the north of Lake Trasimene, and from there to Gualdo Tadino, German tactics had been consolidated: there would be a series of intermediate lines south of the Arno River, and from there to the Gothic/Green Line intended to slow down the enemy advance until the preparation of the latter had been completed, which was expected to take place in a period of time between four and six weeks (to mid-August). Rearguard actions had to be fought, but no longer with the determination shown in June and, above all, with a focus on sparing forces for future actions. The enemy rate of advance had to be slowed down, but combat efficiency had to be spared and commanders could rely on units being withdrawn from the front to be reorganized and replenished. Accordingly, von Vietinghoff prepared a series of switch lines to the south of the Gothic/Green one, a 'Vorfeld' (advanced area) likely to adapt to the development of the situation. The first of these was the Georg Line, running from the south of Montevarchi and of Arezzo to Umbertide–Gubbio–Gualdo Tadino and Fabriano. Von Vietinghoff estimated that some resistance could be put up in this area, thus slowing down 8th Army for some time. This was partly undermined by the acknowledgement of the fact that the area south of Arezzo was favourable for the enemy to successfully deploy tanks and that, for this reason, it was hardly suitable for a protracted defence. For this reason, von Vietinghoff authorized General Herr to withdraw

A jeep driven by Polish soldiers moves through the city of Ancona, in ruins after the heavy bombardment and the battle. The railway station, visible in the background, has been razed to the ground. (NAC)

The crew of a Sherman tank from the Canadian 1st Armoured Brigade at Volturara is using the 75mm gun of the tank to dry its washing while taking a rest and having tea. (NARA via Digital History Archive)

north of the city, where he presumably would be able to enjoy some rest. The most imminent danger was a possible offensive by 5th Army to the north-east of Siena, towards the southern bend of the Arno, which required the creation of reserves. LXXVI. Panzer Korps then deployed a Kampfgruppe near San Giovanni Valdarno, while 114. Jäger Division was ordered to deploy north of Umbertide. Still, the Wehrmacht High Command did not fully agree with these concerns as its 5 July directives requested that the most likely objectives of the Allied advance, Leghorn and Ancona, had to be defended as long as possible. The latter was considered more important, as Hitler believed that the Allies could use its harbour to support a landing in the Balkans, an idea dating back to 1943, which apparently never fully vanished.

In a way, the situation might have been gloomier for the Allies than for the Germans at the beginning of July. With the planned landing in southern France having been confirmed, and the resulting withdrawal of units from Italy soon to be completed, the idea of destroying the German forces in the theatre apparently disappeared to make room for a simpler attack plan. This aim was to bring the final approach to the Gothic Line, and from there start a major offensive aimed at breaking through the line and reaching for the Po valley without pausing. Suddenly, in spite of the fact that the southern France landing had been taken into account since early June, Alexander found himself facing a strategic problem instead of the opportunity to destroy the enemy forces. According to the early July estimates, the Germans had 12 to 14 divisions or equivalent in Italy, with a further six to seven that could be sent from outside the theatre as a reinforcement. To oppose the 18–21 German divisions the Allies could muster 18 divisions, four of which were armoured, plus seven independent armoured brigades deemed capable only of supporting infantry. In order to achieve their aims, the Allies had to deploy all their divisions and prepare for a three-stage offensive. The first stage was to penetrate along the Rimini–Pisa line with the aim of exploiting the breakthrough reaching the Po, and secure bridgeheads across it. The second phase aimed at an advance to the north-east, towards the Adige River, while the third phase required the forcing of the German defences in the area and an advance to the Ljubljana Gap. In order to keep the pace of the offensive rolling, one-third of the Allied forces had to be rotated to rest and refit. In order to make good for the lack of troops, as a first step the 78th Division was to be kept in Italy instead of the Middle East, which would be reinforced with troops from the United Kingdom. Most importantly, this required a swift pace of advance, which had to be followed at each stage by a period of rest in order to secure supplies and the administrative situation. Alexander's plea for reinforcements fell on deaf ears, however, even though 78th Division was allowed to remain in Italy 'on call' while the Americans promised the availability of 92nd Infantry

Division along with the arrival and completion of the Brazilian Expeditionary Force. Otherwise, the reinforcements available would be limited: the 43rd Gurkha Lorried Brigade, the 3rd Greek Brigade and the new 66th Brigade from Gibraltar. Meanwhile, the Polish Corps was to be reinforced using Polish prisoners of war taken in France and a series of Italian formations had to be created. From this point of view, the approach to the Gothic Line – otherwise ignored in the series of directives outlined above – had become more important than it had been before.

A Polish Daimler Dingo scout car moving past the town of Recanati in July. (NAC)

Facing a stiffening German resistance, helped by a greater freedom of manoeuvre, the 8th Army was to advance rapidly so as to allow the early start of the offensive against the Gothic Line, spare forces and secure its lines of communications by preventing their destruction both by the Germans and the Allied air forces. This made the Florence–Arezzo area of vital importance, and much better than the coastal approaches on both sides. From Florence a further advance could take place towards Faenza, north-west of Rimini, and along the route Pistoia–Bologna which would enable a swift approach to the Po. This made Florence strategically important and, quite suddenly, part of the process of facing up to the centre of the Gothic Line. This required the seizure of Arezzo first, which should have been fought as part of a greater battle also including Florence while avoiding the Pratomagno mountains to the east of the Arno River bend, south-east of Florence. This left only one axis of advance, the main route from Perugia to Arezzo and then to Florence in a type of terrain not suitable for massed attacks, given the restrictions imposed by terrain and poor communications. Considering the logistic problems faced by the Allied forces in Italy, the struggle to repair road and railway networks required bringing in supplies from Civitavecchia (the closest available harbour), and both Leghorn and Ancona were also essential. Above all, a clear picture of the situation and a definitive attack plan to reach Florence were needed. XIII Corps went for a straight head-on attack on Arezzo abreast the route from Cortona with the 6th South African Armoured Division on the left, 4th Infantry Division at the centre and 6th Armoured Division on the right, supported by the 1st Canadian Armoured Brigade. The hope of a German withdrawal disappeared as early as 5 July as the leading brigades clashed with the

An Italian 75mm self-propelled gun in use by the 278. Infanterie Division has been abandoned in Ancona and is being inspected by Polish troops. Note the Sherman in the background, with the tank tracks on the front hull. (NAC)

BUZENKA

POLISH TANKS ADVANCING TO ANCONA, 17 JULY 1944 (PP.74–75)

The battle for Ancona offered a rare opportunity for a tactical move, aimed at surrounding an entire German division, which could be destroyed by having its way of retreat cut off. This was the plan of General Anders, commander of the Polish II Corps, as he aimed at using the infantry of the 3rd Carpathian and 5th Kresowa Divisions to engage the 278. Infanterie Division along the frontline, while the Polish 2nd Armoured Brigade was to strike from the west, cross the Musone River and strike north to the Adriatic coast, reaching Falconara. The massed use of tanks, hundreds grouped together in one of the strongest tank attacks of the entire Italian campaign, was favoured as there was a lack of German armoured units in the area, and a widespread shortage of anti-tank guns in the 278. Infanterie Division. Terrain, once again, provided the advantage the Germans needed to slow down the enemy advance while withdrawing unhindered. On 17 July, Shermans of the 2nd Brigade **(1)** successfully crossed the Musone, advancing towards Polverigi, only to be halted (albeit temporarily) by German resistance on the way to Agugliano. The mountainous roads, often tree-lined **(2)**, offered unique opportunities for the few German anti-tank guns to block the Allied advance since the tanks could hardly manoeuvre in such tight terrain, and also often lacked close infantry support. The 75mm Pak 40 **(3)** was the standard German anti-tank gun in 1944. Considered a heavy gun, it proved its effectiveness along with the steady nerves of its crews, who were capable of dealing with enemy tanks. Using mostly the Panzergranate 39, the standard armour-piercing round, they successfully dealt with the Polish Shermans **(4)**, which reached Falconara only after the bulk of 278. Infanterie Division had managed to withdraw north.

German positions. Holding a crest to the south of Arezzo, the Germans put up a determined resistance that lasted for two days, until the advancing forces were compelled to consolidate their positions on 7–8 July. Once again, 8th Army faced a lack of infantry as its only suitable formation, 4th Division, was being employed east of Arezzo thus making progress in the other areas difficult. This compelled Leese to bring the 2nd New Zealand Division out of reserve; it had been intended for use in the attack to the Gothic Line, deploying it on the western flank on 7 July. This led to a stalemate lasting until 13 July, when the intensified air activity on the Allied side marked the prelude to the offensive. On 15 July, both the 6th Armoured and the 2nd New Zealand Divisions attacked towards Arezzo, this time only to find a less-determined resistance counterbalanced by strong counter-attacks and heavy artillery fire. This was correctly recognized as a signal for an imminent withdrawal, which took place during the night. Early on the morning of 16 July, the leading elements of 6th Armoured Division entered Arezzo and reconnoitred the ground around there, seizing a bridge on the Arno intact. The advance pushed as far as Montevarchi, and was in fact helped by the planned German withdrawal from the area. This was also the consequence of the order, issued on 11 July, to transfer the Fallschirm Panzer Division 'Hermann Göring' to the Eastern Front, intended to be replaced by the weak 715. Infanterie Division used until then for coastal defence. The German withdrawal and the British advance actually coincided, AOK 10 now deploying on a line between San Giovanni Valdarno–Montevarchi–north of Arezzo–Città di Castello.

A 50mm Pak 38 anti-tank gun ready for action. These guns, taking advantage of the terrain and the poor visibility due to vegetation, did slow down the massive Polish tank attack to the north of Ancona, enabling the German withdrawal. (NAC)

In the X Corps sector, the 4th Indian Division deployed west of the Tiber, allowing the 8th Indian Division to switch to the eastern side of the river. The defenders of Città di Castello, the 44. Infanterie Division supported by the 114. Jäger Division, were facilitated by the rough terrain, empty of suitable roads and dominated by mountains. The initial advance of 4th Indian Division was successful on 10 July; the division was able to seize Mount Favalto on the 13th to the left of Arezzo, and the 8th Army HQ ordered X Corps to support XIII Corps' advance to help with the capture of the city by pushing north. This led 4th Division to advance, threatening Città di Castello from the west, 8th Indian Division advancing and threatening the city from the east from 17 July. The situation was eased by the fall of

Sherman tanks from the Polish 2nd Armoured Brigade near Ancona in July. Following the example set by the New Zealanders, the corps created an armoured division to make up for the lack of infantry. (NAC)

THE BATTLE FOR ANCONA, 1–19 JULY 1944

Following the Allied advance into Rome, the left wing of the German deployment in Italy carried out a slow and methodical withdrawal along the Adriatic coast, pursued since the end of June by the Polish 3rd Carpathian Infantry Division. At the end of the month, as the Germans deployed along their defence

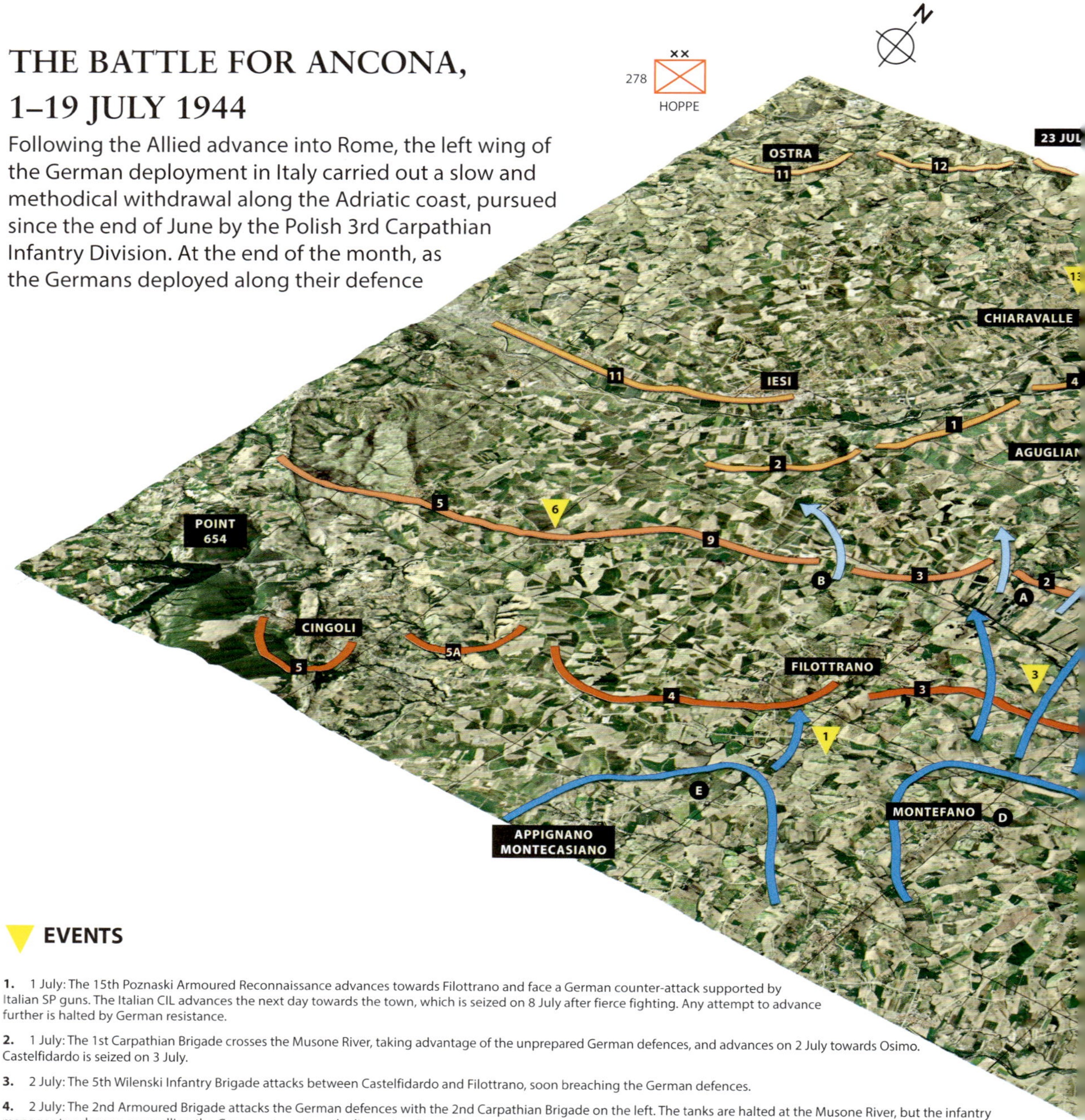

EVENTS

1. 1 July: The 15th Poznaski Armoured Reconnaissance advances towards Filottrano and face a German counter-attack supported by Italian SP guns. The Italian CIL advances the next day towards the town, which is seized on 8 July after fierce fighting. Any attempt to advance further is halted by German resistance.

2. 1 July: The 1st Carpathian Brigade crosses the Musone River, taking advantage of the unprepared German defences, and advances on 2 July towards Osimo. Castelfidardo is seized on 3 July.

3. 2 July: The 5th Wilenski Infantry Brigade attacks between Castelfidardo and Filottrano, soon breaching the German defences.

4. 2 July: The 2nd Armoured Brigade attacks the German defences with the 2nd Carpathian Brigade on the left. The tanks are halted at the Musone River, but the infantry manages to advance compelling the Germans to restore the line using Gruppe Peter (LSB 676, FB 903).

5. 4–5 July: After a fierce struggle, the 1st Carpathian Brigade, supported on the right by the 2nd Carpathian Brigade, seizes Osimo and compels GR 992 to withdraw.

6. 7 July: After having prevented an enemy breakthrough, the 278. ID withdraws to a new defence line between Monte Conero and Monte della Crescia. The claim for more than 40 enemy tanks destroyed is counterbalanced by heavy losses, which require pulling out of the front to reorganise three battalions (II./GR 994, I–II./GR 993).

7. 17 July: Preceded by a strong artillery barrage, the 5th Infantry Brigade, supported by the 15th Poznaski Armoured Reconnaissance and the 7th Hussars Regiment, attacks Monte della Crescia, which is seized by midday. Facing a breakthrough, the 278. ID commits its I./GR 992 only to temporarily fill the gap.

8. 17 July: The 2nd Armoured Brigade attacks Polverigi, which is seized in the afternoon securing a complete breakthrough of the German front. The attempt to halt the enemy advance by bringing in the last reserve, the I./GR 993, is unsuccessful. The fall of Agugliano paves the way to a swift Polish advance to the north.

9. 18 July: The entire front is in movement. On the right side of II Corps, the 1st Carpathian Brigade and the 12th Reconnaissance Regiment attack, breaking through the Gruppe Peter defences.

10. 18 July: Some 200 tanks of the 2nd Armoured Brigade are halted by demolitions and a strongpoint north of Agugliano, thus enabling the German withdrawal.

11. 18 July: Facing the loss of Ancona, the commander of 278. ID orders the abandonment of the city and a withdrawal west. After a swift advance along the coast, while mopping up the remnants of the Gruppe Peter, the 12th Reconnaissance Regiment enters Ancona at 1430hrs.

12. 18 July, evening: The pincer closes at Ancona without German troops in the pocket. Losses had been heavy for the Germans, with about 800 killed, 2,400 wounded and 800 prisoners, but the 278. ID had not been destroyed. Polish losses are also high with 496 killed, 1,789 wounded and 138 missing.

13. 19 July: The crossing of the Esino by the 2nd Armoured Brigade puts an end to the battle with the establishment of a bridgehead near Chiaravalle. The commander of the 278. ID is granted authorisation to withdraw. A new defence line is established further north, between Ostra and Filetto, to enable the division to recover.

line south of Ancona, a major attack was planned with the aim of seizing the town and its harbour and destroying the German 278. Infanterie Division in the area. The first phase of the battle took place between 1 and 9 July, the Polish II Corps approaching its starting line for the final battle. This started on 17 July, with the Polish 3rd Division carrying out a feint attack along the coast while, further inland, the 5th Kresowa Infantry Division carried out the main attack with the support of the 2nd Armoured Brigade. Even though the II Corps managed to seize Ancona, the aim of destroying the 278. Infanterie Division was missed as it managed to escape the trap by withdrawing north.

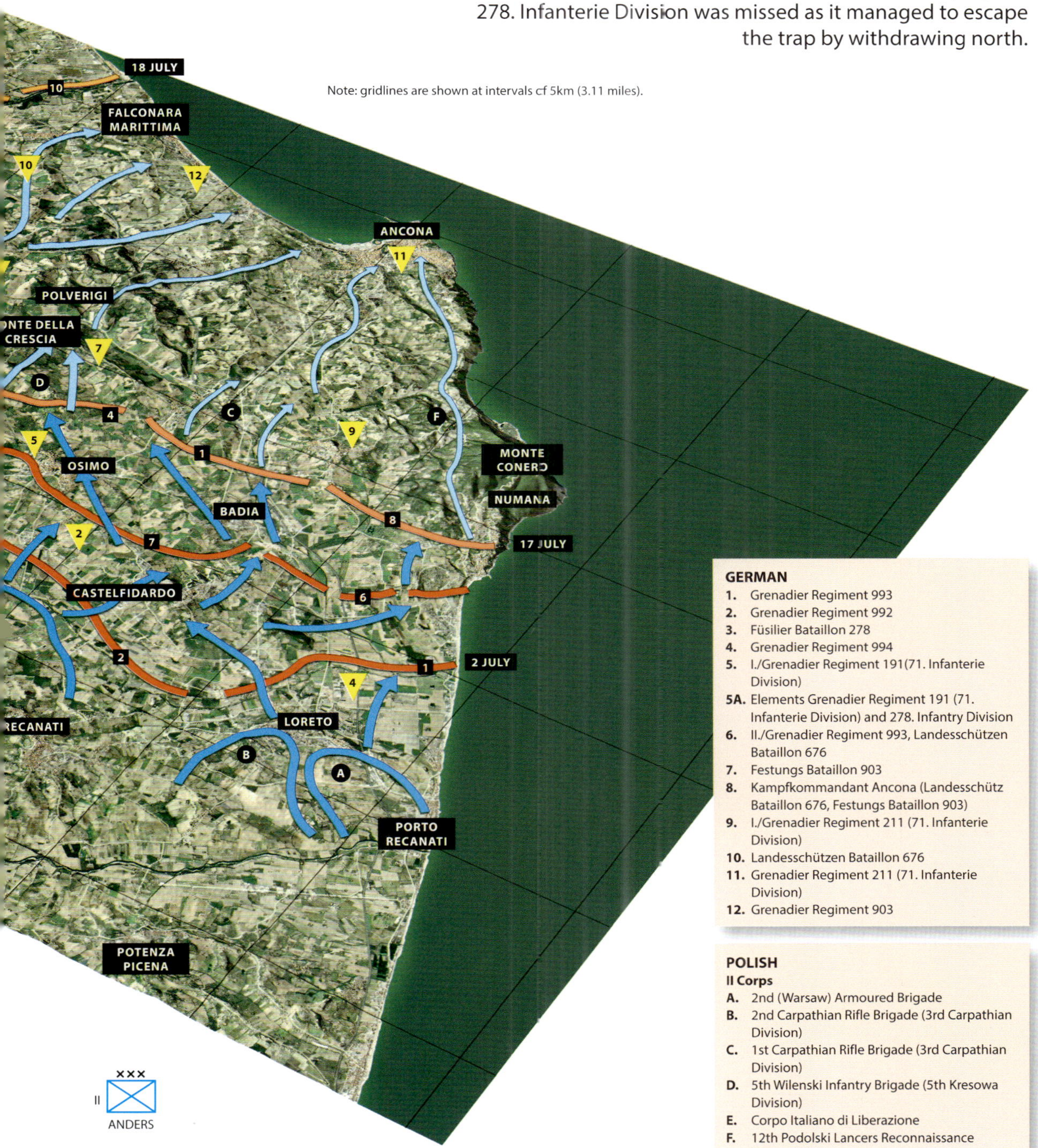

GERMAN

1. Grenadier Regiment 993
2. Grenadier Regiment 992
3. Füsilier Bataillon 278
4. Grenadier Regiment 994
5. I./Grenadier Regiment 191(71. Infanterie Division)

5A. Elements Grenadier Regiment 191 (71. Infanterie Division) and 278. Infantry Division

6. II./Grenadier Regiment 993, Landesschützen Bataillon 676
7. Festungs Bataillon 903
8. Kampfkommandant Ancona (Landesschütz Bataillon 676, Festungs Bataillon 903)
9. I./Grenadier Regiment 211 (71. Infanterie Division)
10. Landesschützen Bataillon 676
11. Grenadier Regiment 211 (71. Infanterie Division)
12. Grenadier Regiment 903

POLISH

II Corps

A. 2nd (Warsaw) Armoured Brigade
B. 2nd Carpathian Rifle Brigade (3rd Carpathian Division)
C. 1st Carpathian Rifle Brigade (3rd Carpathian Division)
D. 5th Wilenski Infantry Brigade (5th Kresowa Division)
E. Corpo Italiano di Liberazione
F. 12th Podolski Lancers Reconnaissance Regiment

Ancona, which prompted von Vietinghoff to redeploy 1. Fallschirmjäger Division to the Adriatic coast on 19 July, while Lemelsen had to cede 90. Panzergrenadier Division to form a strategic reserve. This was also the consequence of the expected Allied landing, which the Germans thought could take place in the area of Genua. In the meantime, the Adriatic sector was reinforced by the remnants of 162. and 94. Infanterie Divisions, plus 98. Infanterie Division transferred from the Balkans. The moment of lull also saw reorganization in the 8th Army front, with the tired 8th Indian Division being redeployed to relieve the FEC along with 2nd New Zealand Division by 22 July. The battle for Florence was about to begin.

In line since 15 June, facing the Chienti River, Polish II Corps advanced steadily following the German withdrawal along the Adriatic coast, the Italian CIL on its left flank. By 5 July, the line running from Cingoli, on the Apennines, to Filottrano–Osimo–Numana had been reached, paving the way for the final attack against Ancona. The attack, planned for 10 July, had to be postponed due to the need to wait for the CIL to deploy, and on 17 July the Polish attack started with the clear aim of destroying 278. Infanterie Division while seizing Ancona, thanks to a wide-ranging left hook formed by the available tank units. The task was given to the 5th Kresowa Division, supported by the Polish 2nd Armoured Brigade, while the 3rd Carpathian Division was to advance along the coast engaging the enemy and preventing it from escaping the trap. The CIL had the subsidiary role of advancing in the mountain area, towards Filottrano, in this case facing the 71. Infanterie Division which had taken over the right wing of 278. Division west of Ancona. With the 5th Wilenski Brigade in the lead, 5th Kresowa Division was able to secure its objective and the crossing of the Musone River on the 17th. This paved the way for the advance of the 2nd Armoured Brigade, now facing a broken enemy defence line and being able to advance steadily to the north-west of Ancona with a massed strength of some 200 tanks supported by 96 light and medium bombers and 215 fighter-bombers attacking

A Polish 2nd Armoured Sherman on the move across the narrow streets of Civitanova Marche, July 1944. (NAC)

the German positions. It is revealing of the Allied inability to conduct manoeuvre warfare and of German skills in defence that the massive hook failed in its aim, as 278. Infanterie Division started its withdrawal in the morning of 18 July thereby preventing its complete destruction. The Polish troops entered Ancona from the south, in the afternoon of the same day, as the remnants of the German forces deployed at first along the Iesi–Chiaravalle line before withdrawing at the end of the month along the Misa River at Senigallia. The fighting strength of the three regiments of the 278. Infanterie Division was now down to 150, 190 and 230, but the Polish forces did not press on hard and the line was held until the offensive on the Gothic Line started.

THE BATTLE FOR FLORENCE, 23 JULY–8 AUGUST

As the last week of July approached, Kesselring's command had reasons to feel reassured about the situation. On his left wing, the situation appeared under control, the Polish advance having been checked and reinforcements deployed on the coast. On the right wing, AOK 14 was not fully deployed along the Arno River Line up to Fucecchio, west of Empoli, the area facing Florence being defended by Schlemm's I. Fallschirm Korps with (from west to east) 3. and 29. Panzergrenadier Divisions, 4. Fallschirmjäger Division and 356. Infanterie Division. East of the boundary marked by the Chianti Mountains, AOK 10 held a line running south of the Apennine range, which made an enemy breakthrough quite unlikely. Foreseeing a major assault towards Florence, Schlemm prepared a series of 'map lines' intended to enforce the tested tactics of staged withdrawal in spite of Hitler's request to hold a line at least 10–12km south of the city. Florence itself would, like Rome, not

British troops entering Montespertoli following the German withdrawal, caused by the Canadian tanks of the 1st Armoured Brigade, which heavily blasted the town. (NARA via Digital History Archive)

An M10 tank destroyer of the 1st Canadian Armoured Brigade on the move north of Montevarchi. Due to the lack of tanks on the German side, these tank destroyers would eventually be used as artillery. (NARA via Digital History Archive)

be turned into a battlefield. Already declared an open city on 23 June, the cultural and artistic jewel of Tuscany was not spared by air attacks, but both Kesselring and Lemelsen were determined to avoid any fight in the city, the only measure to be taken being the destruction of the bridges on the Arno, with the exception of Ponte Vecchio (the approaches to the bridge would be destroyed instead, which blocked access). Also, once the battle had reached the area, the Germans planned a further withdrawal north to the pre-Gothic/Green Line, the Arno Line being considered not suitable for a protracted defence.

The task of advancing to Florence fell on Kirkman's XIII Corps, the plan being a two-pronged armoured thrust with the 6th South African Armoured Division advancing on the western side of the Chianti Mountains on to Greve, the 6th Armoured Division advancing on the right flank along the eastern bank of the middle (or lower) Arno, the two divisions being linked at the centre by the 4th Infantry Division advancing along the mountain range while the 8th Indian Division was to cover the left flank. The 2nd New Zealand Division, initially held in corps' reserve, was to be brought in to reinforce the 6th Armoured Division's advance in case of a strong German resistance. The plan revealed its weaknesses by having assigned to the 6th Armoured Division a terrain unsuitable for tank operations, and as such its progress was not likely to provide direct support to the 6th South African Division's advance. On their own, the Germans were prepared for the customary staged withdrawals with the first defence line (Irmgard) running some 35km to the south-east of Florence blocking the middle Arno valley, followed by the Karin and Lydia Lines, running respectively some 18km and 25km further back from the Irmgard Line. The Olga Line, roughly 15km to the south of Florence, blocked the approach to the city along the main routes, with the Paula Line being the last line of defence running 5km south of Florence and of the Arno.

Following the approach phase between 17 and 20 July, Kirkman's XIII Corps faced initial success with the 6th Armoured Division approaching Castiglion Fibocchi on the 18th, while the 4th Division approached Montevarchi. However, both seizure attempts failed and the advance stalled in spite of the efforts to outflank the German defences. To the west, the 6th South African Division made a good advance seizing Radda, west of the Chianti Mountains, on 17 July, with the 24th Guards Brigade able to seize Mount Maione by surprise on the night of 18/19 July, the advance being forestalled in the nearby Mount San Michele, the last peak dominating the Chianti Mountains. The seizure of the latter on 20 July compelled 715. Infanterie Division to withdraw to the next position, now covering the approach to San Giovanni in Valdarno. This made Kirkman focus on the 6th South African Armoured Division, whose advance towards Florence was to be supported by the 2nd New Zealand Division brought out of reserve, with the aim of a direct thrust along the main Highway 2 along the San Casciano–Signa line (the Arno River crossing some 8km west of

The battle for Florence, 23 July–8 August

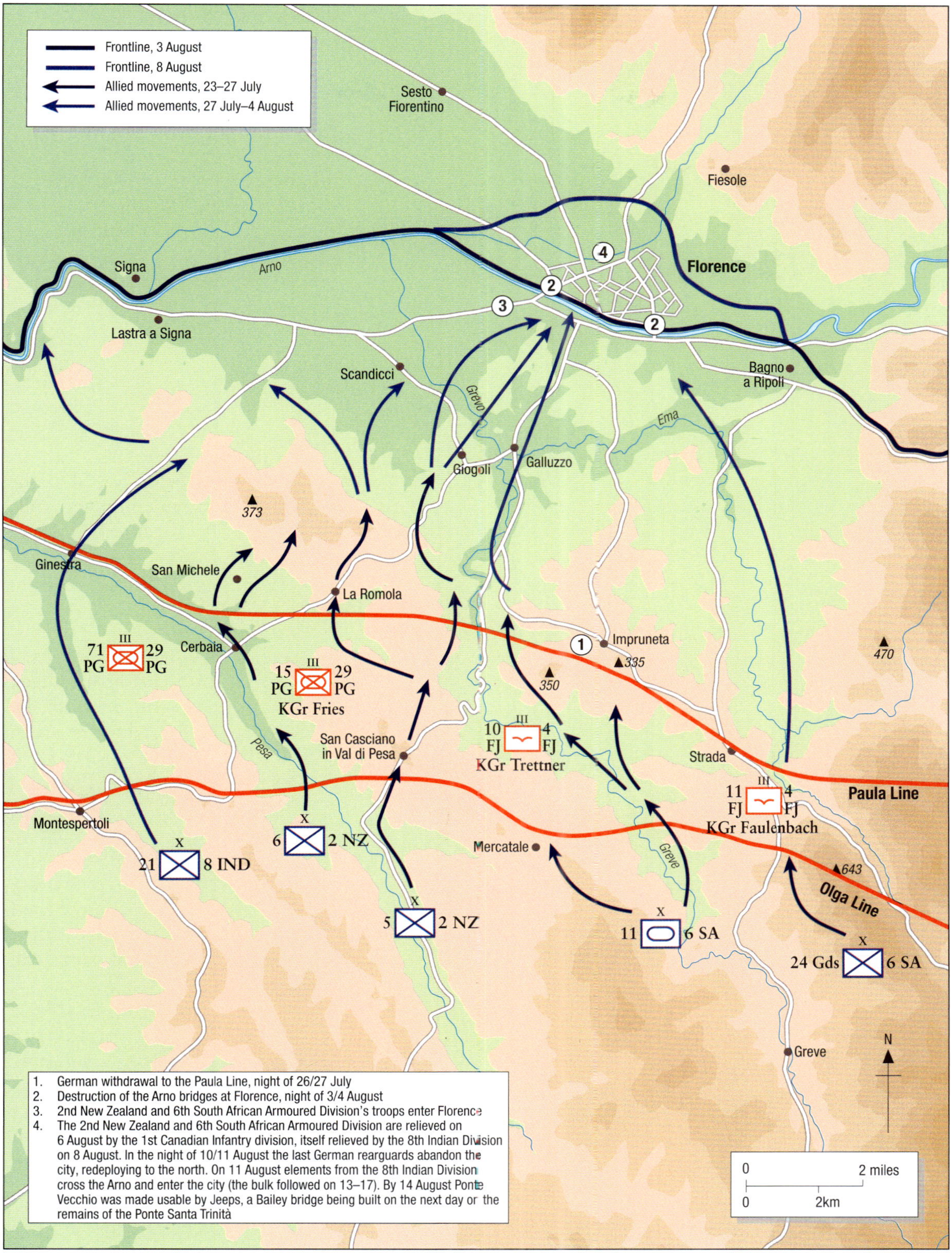

A column of Sherman tanks from the 1st Armoured Brigade in Montespertoli, to the west of Florence, ready to move towards the Arno supporting British troops. (NARA via Digital History Archive)

Florence), and the Greve–Impruneta line with the 6th South African Division, whose right flank was to be covered by the 4th Division and, further east, the 6th Armoured Division in the middle Arno valley, the 8th Indian Division approaching the upper Arno from the west in the Castelfiorentino–Empoli area. As the 8th Indian and the 2nd New Zealand Divisions relieved the FEC, the 6th South African Division continued with its advance north of Mount San Michele approaching Greve, which was found defended by deep minefields costing the loss of several tanks. A full attack was launched on 23 July, with I. Fallschirm Korps counter-attacking before withdrawing north during the night of 23/24 July. Mercatale was seized on 26 July in spite of the German defences, the 356. Infanterie Division supported by Tiger tanks, the advance being now checked by the defences of 4. Fallschirmjäger Division along the main route and, to the west, by 29. Panzergrenadier Division.

Right after replacing the French troops, on 22 July the leading brigade (5th) of the 2nd New Zealand Division advanced astride Highway 2 seizing Tavarnelle early on the 23rd, but soon faced determined German resistance to the north when the tanks of 18th Armoured Regiment dealt with the German strongpoint (part of the Olga Line) at the Villa Strada hamlet. The German defences, backed by armour, proved too strong to be easily dismissed and the advance came to a halt while, to the west, B Squadron of 18th Armoured Regiment dealt with some Tiger I tanks, which were eventually compelled to withdraw. The situation did not change for the 6th South African Division, which faced resolute German resistance at Greve, and found a solution in an outflanking manoeuvre along the nearby Monte Domini, which was seized on 23 July, the last peak being seized by 25 July, thus bringing the battle to the north of Greve. At the same time, other units were flanking to the west and the town was finally seized on the 23rd. The advance proceeded with the South Africans overrunning a German position some 3km south of Mercatale on the next day, only to face a counter-attack supported by a Tiger I tank, which checked any further advance. The South Africans had been halted at the Olga Line, while the New Zealanders continued their advance taking advantage of a limited withdrawal by the 4. Fallschirmjäger Division on the night of 24/25 July, the 5th Brigade approaching San Casciano, thus levelling with the South Africans. Facing the German positions on the Olga Line, the New Zealanders' advance was checked too at Montespertoli and Poppiano on the 25th. While the morale of the Germans was raised by local successes, the losses suffered by the already depleted units could not be ignored. Therefore, Kesselring authorized Schlemm to withdraw from the Olga to the Paula Line on the night of 26/27 July, practically sanctioning the future surrender of Florence. This enabled the

New Zealanders to enter San Casciano, now emptied of German troops, on 26 July, the only obstacles being snipers and the customary series of mines and booby-traps. The South Africans kept up with the German withdrawal as well, the 11th Armoured Brigade, after entering Mercatale late on 26 July, being checked by the rearguard of 4. Fallschirmjäger Division at the Greve River, on the southern outposts of the Paula Line. The most important success came at Poppiano, where the New Zealanders managed to catch the Fallschirmjäger while they were withdrawing and, exploiting the opportunity, brought forward 6th Brigade which bypassed 5th Brigade on 27 July starting an attack across the Pesa River. Thanks to the remnants of the bridge, and of the low level of the water, the New Zealanders were able to cross the river and establish a bridgehead, now facing the town of Cerbaia just as the South Africans approached Strada. The final battle would be coming next.

By then, Kesselring's main concern was for the Allied landings, the 'bridgehead' south of Florence being given pride of place but at the same time having little if any relevance, both AOK 14 and AOK 10 being instructed to secure the coasts while providing for reserves in case of need. This did not influence the resolution of the men of 4. Fallschirmjäger and 29. Panzergrenadier Divisions, determined to prevent the Allied advance towards Florence. As they moved north of Strada, the South Africans faced the destruction of three Sherman tanks by mines, while the Germans opened fire with their Tiger I and Sturmgeschütz from the nearby hill. Eventually, the Germans halted the enemy advance at Strada, inflicting heavy casualties in the process. Here too, the South Africans brought forward fresh units, with the 1st Scots Guards relieving the Coldstream while German fire hit the Sherman tanks of the Pretoria Regiment. Ultimately the Scots Guards were able to clear the ridge south of Strada with a bayonet charge, but failed to completely dislodge the Germans and seize the town. The unavoidable conclusion

French troops in Poggibonsi fire at the enemy relying on the cover provided by the house and the demolished sign pointing towards Florence. (NARA via Digital History Archive)

A tripod-mounted Bren gun checking enemy movement from an abandoned house. The presence of 8th Army forces in Florence did not last long, the area soon falling under the US 5th Army. (NARA via Digital History Archive)

was that any further advance along the 6th South African Armoured Division's axis towards Florence was impossible at the moment. The Paula Line, the last line of defence before Florence, ran from a bend on the Arno to the west at Montelupo more or less parallel to the north of the Pesa River with its western strongpoint at the towns of San Michele and La Romola, both dominated by the high ground to the north-west of the Cerbaia–Giogoli road. Faltignano was the main strongpoint between this and the Greve River to the east, flanked by Highway 2 marking the boundary between the 2nd New Zealand and the 6th South African Armoured Divisions. The latter faced the strongpoint of Point 350 (Poggio alle Carraie) and of Strada in Chianti on the road leading to Florence, the town of Impruneta to the north. Facing the 2nd New Zealand Infantry Division were the men of 29. Panzergrenadier Division, the 4. Fallschirmjäger Division opposing the advance of the 6th South African Armoured Division.

Thanks to the German withdrawal, the 2nd New Zealand Division started the last phase of its advance with a two-brigade front, the 6th to the left and the 4th Armoured to the right, the infantry moving faster than the tanks with the result that the 26th Battalion found itself isolated, and facing a German counter-attack on 28 July. The infantry eventually withdrew, the tanks approaching Cerbaia, only to be halted by mines and demolitions. The same fate befell the 3rd Company which, escorted by eight Sherman tanks from B Squadron, 20th Armoured Regiment, managed to advance halfway to La Romola only to be driven back by the usual German counter-attack. It was thanks to the news of the 26th Battalion advance that 4th Armoured Brigade, while having breakfast, decided to continue with its advance in order to level with the neighbouring 6th Brigade. The tanks, advancing over the ridge of Pisignano to the east of the divisional flank, fell into some sort of a trap and three Shermans were destroyed in rapid succession. Trying to break the stalemate, 6th Brigade was then ordered to attack San Michele, and 4th Armoured Brigade to attack La Romola if possible. The attack took place on the morning of 29 July, 6th Brigade succeeding in seizing San Michele, where the New Zealanders dug in. As the sun rose, the Germans launched a massive counter-attack supported by artillery and mortar fire, penetrating into the New Zealanders'

Manning a PIAT (Projector Infantry Anti-Tank) gun. This cumbersome and hard-to-use gun proved effective on several occasions, particularly when used by the New Zealand troops advancing across the Paula Line south of Florence. (NARA via Digital History Archive)

positions but without succeeding in dislodging them. Supporting tanks also faced fire from German Sturmgeschütze, themselves supporting their infantry. By midday, the German reaction mostly came in the form of heavy artillery barrage, the town of San Michele being reduced to a pile of rubble without compelling the New Zealanders to withdraw. After a new counter-attack at dusk, which resulted in a two-hour battle, the Germans gave up and ceased their attempts to dislodge the New Zealanders from San Michele, which was reinforced early on 30 July by the infantry and tanks from 6th and 4th Armoured Brigades. The South African advance was checked as well with the 11th South African Armoured Brigade's attempt to seize the village of Santa Lucia in Greve turning into a failure thanks to the stubborn German defence and to the fact that the infantry soon lost contact with the tanks. What did follow was an intense concentration of artillery fire from both sides, Santa Lucia being eventually seized but at the cost of heavy casualties, the South African advance having stalled.

A Sherman tank from the 6th South African Division entering Allerona, paving the way for an entire column of vehicles, all cheered on by the Italians, who were clearly ignoring the level of destruction brought to their village. (NARA via Digital History Archive)

At this point, Kirkman decided to concentrate XIII Corps' efforts with a massive attack astride Highway 2 (via Cassia), to be supported on the right flank by 4th Division and, on the left flank, by the 8th Indian Division. Both the 2nd New Zealand and the 6th South African Divisions were to deliver the decisive blow on the night of 30/31 July, this time on a narrow front and avoiding the heights to the west and east of the road. Once the New Zealanders had broken through the Paula Line, the South Africans were to advance east of the road and seize the high ground near Impruneta. Already by 29 July, the 5th New Zealand Brigade had been brought forward to secure the gap between the division and the 6th South African and, while the latter were holding their ground, the former advanced seizing a village against strong opposition. This preluded the main attack on the night of 30/31 July, 2nd New Zealand Division deploying from the west with 6th Brigade, 4th Armoured and 5th Brigade which managed to advance with difficulty. La Romola was attacked on 31 July, not without difficulties. The situation was stabilized thanks to the arrival of the tanks, the second phase of the attack being postponed due to supply problems. I. Fallschirm Korps faced similar problems, and with Schlemm being informed of heavy casualties, the decision was made to give up the fight due to lack of the necessary reinforcements. The order to hold the Paula Line was not altered, but following Lemelsen's advice on 30 July, Schlemm was authorized to reconnoitre a fall-back position closer to Florence, which effectively meant abandoning the bridgehead south of the Arno. The definitive decision came with Hitler's order, issued on the 31st, to destroy all the bridges on the Arno with the exception of Ponte Vecchio, whose approaches had to be destroyed instead, a task that fell to Oberst Adolf Fuchs, commander of Fallschirmjäger Regiment 10 and Stadt Kommandant of Florence.

The German artillery fire coming to a halt on 31 July resulted in the 6th South African Division attacking with the 24th Guards Brigade, relieved

A column of Sherman tanks belonging to the South African Prince Alfred's Guard and Light Horse/Kimberley Regiment have reached the main square of Impruneta on 3 August and are preparing to advance towards Florence. (NARA via Digital History Archive)

by the 4th Division at Strada, heading towards Impruneta. It was seized on 2 August without opposition due to the German withdrawal, even though the South African probes north of the town were soon halted by the German rearguard. The 2nd New Zealand Division renewed its attack at the Paula Line positions on the night of 1/2 August, again on a three-brigade front, 6th Brigade seizing almost at once the Castellare–Poggio Cigoli road, thus paving the way to the crossing of the River Ema and to the town of Galluzzo, south of Florence. Both the 4th Armoured and 5th Brigades' advance was halted by the unexpected resistance at Point 350, the dominating ground in the area, but the high casualties suffered at that point induced Schlemm on 2 August to order a withdrawal to the River Ema Line, less than 5km south of Florence, while leaving a strong rearguard. The last German strongpoint at La Poggiona was seized on 3 August, making the way to Florence practically open. On that same day the order to withdraw north of the Arno brought the battle for Florence to an end; first to enter the southern part of the city was the 4th Light Horse/Kimberley Regiment, which reached Ponte Vecchio at about 0430hrs on 3 August, also discovering the destruction of all other bridges. Later that same morning, they were joined by C Company, 23rd New Zealand Battalion. The Allied troops halted south of the Arno, while in Florence a vicious firefight ensued with snipers firing from multiple locations. On 4 August, the Germans started their evacuation of the city centre, and the Allied troops (elements from 8th Indian Division) entered the city on 11 August following the withdrawal of the last German rearguard.

A Polish patrol on the move north of Ancona. As reported by the original caption, this patrol captured 16 German soldiers while patrolling the area between the Metauro and Foglia Rivers. (NARA via Digital History Archive)

AFTERMATH

A look at the German losses during the summer 1944 campaign in Italy will disclose some interesting issues. The overall losses suffered by HGr C between May and July 1944 amounted to 76,605, of which 61,513 were lost in June and July alone, 15,092 being lost in May. Killed in action amounted to a total of 9,271, of which 2,262 were in May, 1,220 in June and 5,789 in July. The wounded in May were 8,070, shrinking to 4,006 in June and rising to 19,986 in July (total 32,062). Missing in May were 4,760, 4,579 in June and 25,933 in July, for a total of 35,272. The AOK 14 bore the brunt of the battle with a grand total of 42,739 losses in June and July, 4,536 of which were killed, 15,883 wounded and 22,320 missing. AOK 10 lost a total of 17,578 men, of which 2,182 were killed, 7,338 wounded and 8,058 were missing. The first remark is about the relatively low level of losses suffered in June as opposed to May and, in particular, in July, even though one can reckon that the latter figures are probably inflated by delayed reports and may include losses actually suffered in June. Nevertheless, this suggests that the German losses increased when fighting, but dropped in June when both AOK 14 and AOK 10 were withdrawing. This means the Allied pursuit failed in inflicting casualties right at the time when the enemy was at its weakest, and in the most vulnerable position.

Another important remark is about the losses suffered by the various commands. Both AOK 10 and AOK 14 accounted for most of the losses suffered in the theatre (60,317 in June and July as opposed to overall HGr C losses of 61,513), the latter suffering twice the number of killed and wounded, and as much as almost three times the number of missing in action. One could argue that AOK 14 came close to breaking point, but was eventually able to recover as, in fact, the entire German army in Italy did. The only available figures, those for US 5th Army, make some comparison possible. Between 5 and 30 June, the army lost 612 killed, 3,065 wounded and 164 missing. In July, the losses amounted to 1,276 killed, 5,460 wounded and 335 missing, figures confirming that June was a month of low combat level as opposed to July, 5th Army losses dropping at the end of the month matching the end of

German soldiers loading a Kfz 15 staff car with Panzerschrek, the 88mm anti-tank rocket launcher, which, along with other individual anti-tank weapons like the Panzerfaust, accounted for at least 37 per cent of the enemy tanks destroyed or immobilized. (NAC)

A British Sherman moving through an Italian village on the Adriatic coast under the watchful eye of an Italian partisan in August 1944. In spite of the many post-war claims, the Italian partisans did not play a major role during the summer 1944 offensive and mostly liberated the Italian cities following the German withdrawal. (NARA via Digital History Archive)

operations. Comparing the overall losses of 5th Army in June and July (a grand total of 10,912) with those suffered by AOK 14 (42,739), it seems clear that the former emerged victorious from the confrontation, but at the same time surely failed to inflict the number of losses it might have, thus crippling its combat effectiveness.

As these figures suggest, the summer 1944 campaign in Italy was one of lost opportunities for the Allies, who might have actually destroyed

An American P-40 Warhawk fighter-bomber flying over the Mount Vesuvius volcano in the summer of 1944. The overwhelming Allied air supremacy in Italy enabled even these aircraft to be used, in spite of the fact that they were largely outdated and outclassed by the German ones. (NARA)

the German capability to defend, thus securing the attainability of their aim: the breakthrough into the Po valley. The May decision at Valmontone is only the most blatant example, which should be followed by the indecisive and uncoordinated pursuit in June, also lacking a decisive 'focal point' aimed not just at following the retreating enemy, but rather at destroying it thanks to the decisive superiority in air forces, armour, motorization and mobility. Without transcending into a 'what if' sort of history, it is clear that the decision to land at Elba Island did not produce the same effects a landing along the coast might have, especially if coordinated with a decisive assault by 5th Army effectively supported by 8th Army's actions aimed at pinning down the German forces. This did not happen and, already in July, the Allied armies found themselves fighting the same kind of war they had been fighting in the winter of 1943/44, and also the same kind of war they would be fighting in the ensuing autumn and winter. The Germans, on the other hand, managed to handle defeat in quite a good way. First, in spite of the relatively high losses suffered and the subsequently low combat effectiveness of their units, they managed to impose their pace of movement on the Allies, forcing them to slow down or even to halt. This was achieved thanks to a skilful handling of the available forces, which were moved and reorganized according to need, the Allies clearly proving unable to prevent the Germans from having relative freedom of movement. This is particularly true of the situation in the AOK 10–8th Army area, where the lack of a decisive advance and engagement with the withdrawing enemy forces permitted the Germans to pull out units from the line, and deploy them in the more threatened AOK 14 area.

German losses were surely high, but Alexander's aim to destroy the enemy forces clearly was not accomplished. The territorial gains were also not of great value. Until mid-June, attention focused on the western coast before switching to the inland area at Trasimene. In July, the focus was again on the western and eastern coasts, until it switched again to the inland area at Florence. The decision to focus on the latter should have been the prelude to Alexander's planned attack against the Gothic Line across the Apennines. Instead, as early as 4 August, the attack plan was changed and the focus switched to the Adriatic coast. Needless to say, had the jump-off area been secured earlier, thanks to more determined and foresightful planning, the Gothic Line offensive might have started earlier and potentially avoided the weather difficulties it encountered, with the result of its eventual forestalling. The overall impression of the entire summer 1944 campaign in central Italy is, once again, a show of great German skill and professionalism as opposed to a lack of determination and the necessary planning and foresight on the Allied side. This is evidenced by the plans for the landing in southern France which, having been discussed before the breakthrough at Anzio, were blatantly ignored until the last moment, thus preventing an effective use of the available forces. In conclusion, one could say that the battles fought in Italy in June and July 1944 might have been decisive, but that they were just another wasted occasion on the Allied side.

THE BATTLEFIELD TODAY

Local memory of the Allied advance in Italy in 1944 is dominated by the idea of the *Liberazione* ('liberation') from the German occupation, usually with an eye to the role the Italian partisans played and to the post-war creation of the Italian Republic. No monuments were built, but there is an abundance of memorial plaques like this one on the northern edge of Palazzo Vecchio in Florence, remembering the words of Italian politician Piero Calamandrei. It reads: 'Since 11 August 1944 freedom, not gifted but conquered at the price of ruins, torture and blood became, thanks to the insurrection of the population and to the victory of the Allied armies, the only master of justice'. (Francesco Bini, CC BY-SA 4.0 https://creativecommons.org/licenses/by-sa/4.0, via Wikimedia Commons)

Describing the area where the battles in this book took place is not easy for a variety of reasons. First of all, most of the places are still the same as they were some 80 years ago, at least in their city centres. The main highways have received new pavements but, all too often, still resemble the versions of 1944; in most cases the traffic is diverted to post-war freeways (which in Italy require a toll). Entire areas, such as the Aurunci Mountains between Rome and Cassino and the Sibillini Mountains between Ancona and Terni, have been declared national heritage parks, thus the entire areas have been preserved. Apart from the development of new industrial and commercial areas, most of the places in the area between Rome and Florence have preserved the same view as of 80 years ago, the once famous 'Chiantishire' where people used to acquire their summer residences for a spot of the sun. The war left some deep scars and some places had to be rebuilt entirely. This is the case with Terni, which was almost completely razed to ground by air bombing, and Ancona, where the city centre was rebuilt with an eye of recreating its former status.

The fact that most of the area has seen little changes, or at least no major ones, doesn't mean that the events that took place in 1944 are remembered as they could, and should, be. The swift pace of the Allied advance and the lack of a major battle, such as Cassino, Anzio or the Gothic Line, has not helped memory focus on a determined event. The day of the 'liberation', which is when the city or town in question was reached by the Allied troops, is usually remembered at a local level, but little more. Italy's memory of the events following the surrender and the German occupation is still divided, the country falling (once again) into some sort of civil war no one wants to remember. The areas under political administrations tied to the left (once Communist) meant that even the word 'war' was banned, and the memory altered to serve political purposes. It is not uncommon to find places celebrating their 'liberation' by claiming it did not take place at the hand of the Allies but thanks to local partisans which, in most cases, is pure fantasy.

Some memories have been forgotten to avoid irksome issues. The island of Elba remembers the French landing in June 1944, but mostly because of the widespread and violent rapes and robberies that followed, making it anything but a 'liberation'.

Anyone seriously interested in tracing back the events of the summer of 1944 must take into account this sort of quagmire, which is typically Italian. There are museums in the area, but they are mostly dedicated to other, more famous battles or in general to the Italian war of liberation in 1944–45. Aside from the museums at Anzio and Cassino, the Museo Piana delle Orme near Rome has an interesting show of vehicles and uniforms mostly related to the above-mentioned battles. This is a private museum charging a fee, which is part of a large entertainment area. Most of the museums, funded by the local governments, are either small or not too rich. Some can be misleading: for example, the Museo della Liberazione in Rome (Liberation Museum) is anything but; it mostly remembers the German occupation being based in the former Gestapo HQ in town. Most of the museums in the area between Rome and Florence are dedicated to the more famous Gothic Line battles, in some cases with traces of the battles fought in the summer of 1944. Since the Corpo Italiano di Liberazione fought on the Adriatic (eastern) side of the front, this area is relatively rich. Orsogna, near Ortona with its museum of the famous battle, hosts the War Museum Gustav Line, which is a rather interesting museum that includes interesting relics and items. The Museo delle Due Guerre Mondiali (Two World Wars Museum) is in the nearby Loro Piceno; it is a generic museum like the Museo della Guerra di San Severino. The battle of Filottrano is remembered by the aptly named museum near Ancona, which has its own liberation museum (Museo della Liberazione) at Offagna. Falconara Marittima, just to the north, also has a Museo della Resistenza (Resistance Museum). On the Tuscany side of the Apennines, almost all the museums are dedicated to the Gothic Line battles: the Museo Gotica Toscana (Tuscany's Gothic Line Museum) in Scarperia offers both a museum and re-enactments, but only focuses on the events of September 1944. The same goes for the Museo Linea Gotica at Pianosinatico and the Museo Storico della Linea Gotica at Montemurlo.

A plaque on the wall of the Florence army garrison remembering the Italian volunteers who joined the Allied forces following the liberation of Florence. (Manuelarosi, CC BY-SA 4.0 https://creativecommons.org/licenses/by-sa/4.0, via Wikimedia Commons)

Even memorial plaques are scarce, especially in Rome which, one ought to remember, doesn't celebrate 4 June as liberation day. There is a memorial plaque dedicated to the 1st Special Service Force at Porta San Paolo, with another plaque at the embassy on Via Veneto, but otherwise the liberation day is almost completely forgotten. Florence has the same attitude. Most of its plaques and memorials are dedicated to the partisans and the resistance, other than to the destruction of the bridges. Only in August 2016 was a plaque remembering the Sikh soldiers of the 8th Indian Division installed near the Leopolda Station, the memory of the events of the summer 1944 still creating a rift amongst the Florentines.

FURTHER READING

Anders, Władysław, *An Army in Exile: The Story of the Second Polish Corps*, London: Macmillan (1949)
Campana, Giuseppe, *1943–1947: Il secondo corpo d'armata polacco in Italia*, Ancona (2009)
Ellis, John, *Cassino: The Hollow Victory*, New York: McGraw Hill (1984)
Fisher, Ernest F., Jr., *Cassino to the Alps*, Washington, DC: Center of Military History US Army (1993)
Gouthard, Adolphe, *Le Corps Expéditionnaire Français dans la campagne d'Italie (1943–1944)*, Paris-Limoges-Nancy: Charles-Lavauzelle & Cie (1947)
Hinsley, F. H., *British Intelligence in the Second World War*, Volume Three – Part II, London: HMSO (1988)
Jackson, Sir William, *The Mediterranean and the Middle East*, Volume VI – Part II: *Victory in the Mediterranean – June to October 1944*, London: HMSO (1987)
Laurie, Clayton D., *Rome – Arno: 22 January–9 September 1944*, Washington, DC: Center of Military History US Army (2019)
Pallud, Jean Paul, 'The Invasions of Elba Island', in: *After the Battle* issue 173, Old Harlow: Battle of Britain Press (2016)
Plowman, Jeffrey, 'The Battle for Florence', in: *After the Battle* issue 129, Old Harlow: Battle of Britain Press (2005)
Pugsley, Christopher, *A Bloody Road Home: World War Two and New Zealand's Heroic Second Division*, Auckland: Penguin (2014)
Sparapani, Sergio, 'The Battle of Ancona', in: *After the Battle* issue 169, Old Harlow: Battle of Britain Press (2015)
Starr, Chester G., (editor), *Fifth Army History*, Part VI: *Pursuit to the Arno*, Milan, n.d.

INDEX

Figures in **bold** refer to illustrations.